Alberto Isidori

Nonlinear Control Systems

An Introduction

2nd Edition

With 43 Figures

Springer-Verlag
Berlin Heidelberg NewYork
London Paris Tokyo Hong Kong

Prof. ALBERTO ISIDORI
Università degli Studi di Roma
«La Sapienza»
Dipartimento di Informatica
e Sistemistica
Via Eudossiana 18
I-00184 Roma

1st Edition 1985 under same title in the series
Lecture Notes in Control and Information Sciences Vol. 72

ISBN 3-540-50601-2 2. Aufl. Springer-Verlag Berlin Heidelberg NewYork
ISBN 0-387-50601-2 2nd ed. Springer-Verlag NewYork Berlin Heidelberg

ISBN 3-540-15595-3 1. Aufl. Springer-Verlag Berlin Heidelberg NewYork
ISBN 0-387-15595-3 1st ed. Springer-Verlag NewYork Berlin Heidelberg

Library of Congress Cataloging–in–Publication Data

Isidori, Alberto.
Nonlinear control systems : an introduction / Alberto Isidori.
(Communications and control engineering series)
Includes bibliographical references.
ISBN 0-387-50601-2
1. Feedback control systems. 2. Geometry, Differential.
QA402.3.I74   1989
629.8'3–dc20   89-19665

Printing: Mercedes-Druck, Berlin; Binding: Lüderitz & Bauer, Berlin
2161/3020 543210

For Maria Adelaide

Preface

The purpose of this book is to present a self-contained description of the fundamentals of the theory of nonlinear control systems, with special emphasis on the differential geometric approach. The book is intended as a graduate text as well as a reference to scientists and engineers involved in the analysis and design of feedback systems.

The first version of this book was written in 1983, while I was teaching at the Department of Systems Science and Mathematics at Washington University in St. Louis. This new edition integrates my subsequent teaching experience gained at the University of Illinois in Urbana-Champaign in 1987, at the Carl-Cranz Gesellschaft in Oberpfaffenhofen in 1987, at the University of California in Berkeley in 1988. In addition to a major rearrangement of the last two Chapters of the first version, this new edition incorporates two additional Chapters at a more elementary level and an exposition of some relevant research findings which have occurred since 1985.

In the past few years differential geometry has proved to be an effective means of analysis and design of nonlinear control systems as it was in the past for the Laplace transform, complex variable theory and linear algebra in relation to linear systems. Synthesis problems of longstanding interest like disturbance decoupling, noninteracting control, output regulation, and the shaping of the input-output response, can be dealt with relative ease, on the basis of mathematical concepts that can be easily acquired by a control scientist. The objective of this text is to render the reader familiar with major methods and results, and enable him to follow the new significant developments in the constantly expanding literature.

The book is organized as follows. Chapter 1 introduces invariant distributions, a fundamental tool in the analysis of the internal structure of nonlinear systems. With the aid of this concept, it is shown that a nonlinear system locally exhibits decompositions into "reachable/unreachable" parts and/or "observable/unobservable" parts, similar to those introduced by Kalman for linear systems. Chapter 2 explains to what extent global decompositions may exist, corresponding to a partition of the whole state space into "lower dimensional" reachability and/or indistinguishability subsets. Chapter 3 describes various "formats" in which the input-output map of a nonlinear system may be represented, and provides a short description of the fundamentals of realization theory. Chapter 4 illustrates how a series of relevant design problems can be solved for a single-input single-output nonlinear system. It explains how a system can be transformed into a linear and controllable one by means of feedback and coordinates transformation, discusses how the nonlinear analogue of the notion of "zero" plays an important role in the problem of achieving local asymptotic stability, describes the problems of asymptotic tracking, model matching and disturbance decoupling. The approach is somehow "elementary", in that requires only standard mathematical tools. Chapter 5 covers similar subjects for a special class of multivariable nonlinear systems, namely those systems which can be rendered noninteractive by means of static state feedback. For this class of systems, the analysis is a rather straigthforward extension of the one illustrated in Chapter 4. Finally, the last two Chapters are devoted to the solution of the problems of output regulation, disturbance decoupling, noninteracting control with stability via static state feedback, and noninteracting control via dynamic feedback, for a broader class of multivariable nonlinear systems. The analysis in these Chapters is mostly based on a number of key differential geometric concepts that, for convenience, are treated separately in Chapter 6.

It has not been possible to include all the most recent developments in this area. Significant omissions are, for instance: the theory of global linearization and global controlled invariance, the notions of left- and right-invertibility and their control theoretic consequences. The bibliography, which is by no means complete, includes those publications which were actually used and several works of interest for additional investigation.

The reader should be familiar with the basic concepts of linear system theory. Although the emphasis of the book is on the application of differential geometric concepts to control theory, most of Chapters 1, 4 and 5 do not require a specific background in this field. The reader who is not familiar with the fundamentals of differential geometry may skip Chapters 2 and 3 in a first approach to the book, and then come back to these after having acquired the necessary skill. In order to make the volume as self-contained as possible, the most important concepts of differential geometry used throughout the book are described—without proof— in Appendix A. In the exposition of each design problem, the issue of local asymptotic stability is also discussed. This also presupposes a basic knowledge of stability theory, for which the reader is referred to well-known standard reference books. Some specific results which are not frequently found in these references are included in Appendix B.

I wish to express my sincerest gratitude to Professor A. Ruberti, for his constant encouragement, to Professors J. Zaborszky, P. Kokotovic, J. Ackermann, C.A. Desoer who offered me the opportunity to teach the subject of this book in their academic institutions, and to Professor M. Thoma for his continuing interest in the preparation of this book. I am indebted to Professor A.J. Krener from whom—in the course of a joint research venture—I learnt many of the methodologies which have been applied in the book. I wish to thank Professor C.I. Byrnes, with whom I recently shared intensive research activity and Professors T.J. Tarn, J.W. Grizzle and S.S. Sastry with whom I had the opportunity to cooperate on a number of relevant research issues. I also would like to thank Professors M. Fliess, S. Monaco and M.D. Di Benedetto for their valuable advice.

Rome, March 1989   Alberto Isidori

Contents

1 Local Decompositions of Control Systems 1
  1.1 Introduction 1
  1.2 Notations 5
  1.3 Distributions 14
  1.4 Frobenius Theorem 23
  1.5 The Differential Geometric Point of View 36
  1.6 Invariant Distributions 44
  1.7 Local Decompositions of Control Systems 53
  1.8 Local Reachability 56
  1.9 Local Observability 74

2 Global Decompositions of Control Systems 82
  2.1 Sussmann's Theorem and Global Decompositions 82
  2.2 The Control Lie Algebra 88
  2.3 The Observation Space 92
  2.4 Linear Systems and Bilinear Systems 97
  2.5 Examples 105

3 Input-Output Maps and Realization Theory 112
  3.1 Fliess Functional Expansions 112
  3.2 Volterra Series Expansions 120
  3.3 Output Invariance 124
  3.4 Realization Theory 129
  3.5 Uniqueness of Minimal Realizations 141

4 Elementary Theory of Nonlinear Feedback for Single-Input Single-Output Systems **145**
  4.1 Local Coordinates Transformations 145
  4.2 Exact Linearization via Feedback 156
  4.3 The Zero Dynamics 172
  4.4 Local Asymptotic Stabilization 183
  4.5 Asymptotic Output Tracking 190
  4.6 Disturbance Decoupling 196
  4.7 High Gain Feedback 202
  4.8 Additional Results on Exact Linearization 208
  4.9 Observers With Linear Error Dynamics 217
  4.10 Examples 226

5 Elementary Theory of Nonlinear Feedback for Multi-Input Multi-Output Systems **234**
   5.1 Local Coordinates Transformations 234
   5.2 Exact Linearization via Feedback 243
   5.3 Noninteracting Control 259
   5.4 Exact Linearization of the Input-Output Response 266
   5.5 Exercises and Examples 281

6 Geometric Theory of State Feedback: Tools **289**
   6.1 The Zero Dynamics 289
   6.2 Controlled Invariant Distributions 310
   6.3 The Maximal Controlled Invariant Distribution in $\ker(dh)$ 320
   6.4 Controllability Distributions 338

7 Geometric Theory of State Feedback: Applications **344**
   7.1 Asymptotic Stabilization via State Feedback 344
   7.2 Tracking and Regulation 347
   7.3 Disturbance Decoupling 363
   7.4 Noninteracting Control with Stability via Static Feedback . . . 365
   7.5 Achieving Relative Degree via Dynamic Extension 382
   7.6 Examples 391

Appendix A **403**
   A.1 Some Facts from Advanced Calculus 403
   A.2 Some Elementary Notions of Topology 405
   A.3 Smooth Manifolds 406
   A.4 Submanifolds 411
   A.5 Tangent Vectors 415
   A.6 Vector fields 425

Appendix B **434**
   B.1 Center Manifold Theory 434
   B.2 Some Useful Lemmas 442
   B.3 Local Geometric Theory of Singular Perturbations 447

Bibliographical Notes **459**

References **464**

Subject Index **475**

1   Local Decompositions of Control Systems

1.1 Introduction

The subject of this Chapter is the analysis of a *nonlinear* control system, from the point of view of the interaction between input and state and—respectively—between state and output, with the aim of establishing a number of interesting analogies with some fundamental features of *linear* control systems. For convenience, and in order to set up an appropriate basis for the discussion of these analogies, we begin by reviewing—perhaps in a slightly unusual perspective—a few basic facts about the theory of linear systems.

Recall that a linear multivariable control system with m inputs and p outputs is usually described, in state space form, by means of a set of first order linear differential equations

$$\dot{x} = Ax + Bu \qquad\qquad (1.1a)$$

$$y = Cx \qquad\qquad (1.1b)$$

in which x denotes the *state* vector (an element of $\mathbf{R}^n$), u the *input* vector (an element of $\mathbf{R}^m$) and y the *output* vector (an element of $\mathbf{R}^p$). The matrices A, B, C are matrices of real numbers, of proper dimensions.

The analysis of the interaction between input and state, on one hand, and between state and output, on the other hand, has proved of fundamental importance in understanding the possibility of solving a large number of relevant control problems, including eigenvalues assignment via feedback, minimization of quadratic cost criteria, disturbance rejection, asymptotic output regulation, etc. Key tools for the analysis of such interactions—introduced by Kalman around the 1960—are the notions of *reachability* and *observability* and the corresponding *decompositions* of the control system into "reachable/unreachable" and, respectively, "observable/unobservable" parts. We review in this Section some relevant aspects of these decompositions.

Consider the linear system (1.1), and suppose that there exists a d-dimensional subspace V of $\mathbf{R}^n$ having the following property:

(i) V is *invariant* under A, i. e. is such that $Ax \in V$ for all $x \in V$.

Without loss of generality, after possibly a change of coordinates, we can assume that the subspace V is the set of vectors having the form $v = \mathrm{col}(v_1, \ldots, v_d, 0, \ldots, 0)$, i.e. of all vectors whose last $n - d$ components are zero. If this is the case, then, because of the invariance of V under A, this matrix assumes necessarily a block triangular structure

$$A = \begin{bmatrix} A_{11} & A_{12} \\ 0 & A_{22} \end{bmatrix}$$

with zero entries on the lower-left block of $n - d$ rows and d columns.

Moreover, if the subspace V is such that:

(ii) V contains the *image* (i.e. the range-space) of the matrix B, i.e. is such that $Bu \in V$ for all $u \in \mathbf{R}^m$,

then, after the same change of coordinates, the matrix B assumes the form

$$B = \begin{bmatrix} B_1 \\ 0 \end{bmatrix}$$

i.e. has zero entries on the last $n - d$ rows.

Thus, if there exists a subspace V which satisfies (i) and (ii), after a change of coordinates in the state space, the control system (1.1a) exhibits a decomposition of the form

$$\dot{x}_1 = A_{11}x_1 + A_{12}x_2 + B_1 u$$
$$\dot{x}_2 = A_{22}x_2.$$

By x_1 and x_2 we denote here the vectors formed by taking the first d and, respectively, the last $n - d$ new coordinates of a point x.

The representation thus obtained is particularly interesting when studying the behavior of the system under the action of the control u. At any time T, the coordinates of $x(T)$ are

$$x_1(T) = \exp(A_{11}T)x_1(0) + \int_0^T \exp(A_{11}(T - \tau))A_{12}\exp(A_{22}\tau)\,d\tau\, x_2(0) +$$
$$+ \int_0^T \exp(A_{11}(T - \tau))B_1 u(\tau)\,d\tau$$
$$x_2(T) = \exp(A_{22}T)x_2(0).$$

From this, we see that the set of coordinates denoted by x_2 does not depend on the input u but only on the time T. In particular, if we denote by $x^\circ(T)$ the point of $\mathbf{R}^n$ reached at time $t = T$ when $u(t) = 0$ for all $t \in [0, T]$, i.e. the point

$$x^\circ(T) = \exp(AT)x(0)$$

we observe that any state which can be reached at time T, starting from $x(0)$ at time $t = 0$, has necessarily the form $x^\circ(T) + v$, where v is an element of V.

This argument identifies only a necessary condition for a state x to be reachable at time T, i.e. that of being of the form $x = x^\circ(T) + v$, with $v \in V$.

However, under the additional assumption that:

(iii) V is the *smallest* subspace which satisfies (i) and (ii) (i.e. is contained in any other subspace of $\mathbf{R}^n$ which satisfies both (i) and (ii)),

then this condition is also sufficient. As a matter of fact, it is known from the theory of linear systems that (iii) occurs if and only if

$$V = \mathrm{Im}(B \ AB \ \ldots \ A^{n-1}B)$$

(where $\mathrm{Im}(\ldots)$ denotes the image of a matrix) and, moreover, that under this assumption the pair (A_{11}, B_1) is a *reachable* pair, i.e. satisfies the condition

$$\mathrm{rank}(B_1 \ A_{11}B_1 \ \ldots \ A_{11}^{d-1}B_1) = d$$

or, what is the same, has the property that for each $x_1 \in \mathbf{R}^d$ there exists an input u, defined on $[0, T]$, satisfying

$$x_1 = \int_0^T \exp(A_{11}(T - \tau))B_1 u(\tau)\, d\tau.$$

Then, if V is such that the condition (iii) is also satisfied, starting from $x(0)$ it is possible to reach at time T every state of the form $x^\circ(T) + v$, with $v \in V$.

This analysis suggests the following considerations. Given the linear control system (1.1), let V be the smallest subspace of $\mathbf{R}^n$ satisfying (i) and (ii). Associated with V there is a *partition* of $\mathbf{R}^n$ into subsets of the form

$$S_p = \{x \in \mathbf{R}^n : x = p + v, v \in V\}$$

characterized by the following property: the set of points reachable at time T starting from $x(0)$ coincides exactly with the element—of the partition—which contains the point $\exp(AT)x(0)$, i.e. with the subset $S_{\exp(AT)x(0)}$. Note also that these sets, i.e. the elements of this partition, are d-dimensional planes parallel to V (see Fig. 1.1).

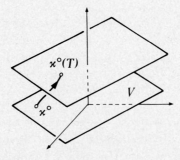

Fig. 1.1

An analysis similar to the one developed so far can be carried out by examining the interaction between state and output. In this case one considers instead a subspace W of $\mathbf{R}^n$ characterized by the following properties:

(i) W is invariant under A

(ii) W is contained in the *kernel* (the null-space) of the matrix C (i.e. is such that $Cx = 0$ for all $x \in W$)

(iii) W is the largest subspace which satisfies (i) and (ii) (i.e. contains any other subspace of $\mathbf{R}^n$ which satisfies both (i) and (ii)).

Properties (i) and (ii) imply the existence of a change of coordinates in the state space which induces on the control system (1.1) a decomposition of the form

$$\dot{x}_1 = A_{11}x_1 + A_{12}x_2 + B_1 u$$
$$\dot{x}_2 = A_{22}x_2 + B_2 u$$
$$y = C_2 x_2$$

(in the new coordinates, the elements of W are the points having $x_2 = 0$). This decomposition shows that the set of coordinates denoted by x_1 has no influence on the output y. Thus, any two initial states whose x_2 coordinates are equal, produce identical outputs under any input, i.e. are *indistinguishable*. Actually, since two states whose x_2 coordinates that are equal are such that their difference is an element of W, we deduce that any two states whose difference is an element of W are indeed indistinguishable.

Condition (iii), in turns, guarantees that only the pairs of states characterized in this way (i.e. having a difference in W) are indistinguishable from each other. As a matter of fact, it is known from the linear theory that the condition (iii) is satisfied if and only if

$$W = \ker \begin{bmatrix} C \\ CA \\ \cdots \\ CA^{n-1} \end{bmatrix}$$

(where $\ker(\dots)$ denotes the kernel of a matrix) and, if this is the case, the pair (C_2, A_{22}) is an *observable* pair, i.e. satisfies the condition

$$\text{rank} \begin{bmatrix} C_2 \\ C_2 A_{22} \\ \cdots \\ C_2 A_{22}^{n-d-1} \end{bmatrix} = n - d$$

or, what is the same, has the property that

$$C_2 \exp(A_{22}t)x_2 = 0 \quad \text{for all } t \geq 0 \ \Rightarrow \ x_2 = 0.$$

As a consequence, any two initial states whose difference does not belong to W, are distinguishable from each other, in particular by means of the output produced under zero input.

Again, we may synthesize the above discussion with the following considerations. Given a linear control system, let W be the largest subspace of $\mathbf{R}^n$ satisfying (i) and (ii). Associated with W there is a partition of $\mathbf{R}^n$ into subsets of the form

$$S_p = \{x \in \mathbf{R}^n : x = p + w, w \in W\}$$

characterized by the following property: the set of points indistinguishable from a point p coincides exactly with the element—of the partition—which contains p, i.e. with the set S_p itself. Note that again these sets—as in the previous analysis—are planes parallel to W.

In the following sections of this Chapter and in the following Chapter we shall deduce similar decompositions for nonlinear control systems.

1.2 Notations

Throughout these notes we shall study multivariable nonlinear control systems with m inputs $u_1, \ldots, u_m$ and p outputs $y_1, \ldots, y_p$ described, in state space form, by means of a set of equations of the following type

$$\dot{x} = f(x) + \sum_{i=1}^{m} g_i(x)u_i \tag{2.1a}$$

$$y_i = h_i(x) \qquad 1 \leq i \leq p. \tag{2.1b}$$

The state

$$x = (x_1, \ldots, x_n)$$

is assumed to belong to an open set U of $\mathbf{R}^n$.

The mappings $f, g_1, \ldots, g_m$ which characterize the equation (2.1a) are $\mathbf{R}^n$-valued mappings defined on the open set U; as usual, $f(x), g_1(x), \ldots, g_m(x)$ denote the values they assume at a specific point x of U. Whenever convenient, these mappings may be represented in the form of n-dimensional vectors of real-valued functions of the real variables $x_1, \ldots, x_n$, namely

$$f(x) = \begin{bmatrix} f_1(x_1, \ldots, x_n) \\ f_2(x_1, \ldots, x_n) \\ \ldots \\ f_n(x_1, \ldots, x_n) \end{bmatrix}, \quad g_i(x) = \begin{bmatrix} g_{1i}(x_1, \ldots, x_n) \\ g_{2i}(x_1, \ldots, x_n) \\ \ldots \\ g_{ni}(x_1, \ldots, x_n). \end{bmatrix} \qquad (2.2a)$$

The functions $h_1, \ldots, h_p$ which characterize the equation (2.1b) are real-valued functions also defined on U, and $h_1(x), \ldots, h_p(x)$ denote the values taken at a specific point x. Consistently with the notation (2.2a), these functions may be represented in the form

$$h_i(x) = h_i(x_1, \ldots, x_n) \qquad (2.2b)$$

In what follows, we assume that the mappings $f, g_1, \ldots, g_m$ and the functions $h_1, \ldots, h_p$ are *smooth* in their arguments, i.e. that all entries of (2.2) are real-valued functions of $x_1, \ldots, x_n$ with continuous partial derivatives of any order. Occasionally, this assumption may be replaced by the stronger assumption that the functions in question are *analytic* on their domain of definition.

The class (2.1) describes a large number of physical systems of interest in many engineering applications, including of course linear systems. The latter have exactly the form (2.1), provided that $f(x)$ is a linear function of x, i.e.

$$f(x) = Ax$$

for some $n \times n$ matrix A of real numbers, $g_1(x), \ldots, g_m(x)$ are constant functions of x, i.e.

$$g_i(x) = b_i$$

where $b_1, \ldots, b_m$ are $n \times 1$ vectors of real numbers, and $h_1(x), \ldots, h_p(x)$ are again linear in x, i.e.

$$h_i(x) = c_i x$$

where $c_1, \ldots, c_p$ are $1 \times n$ (i.e. row) vectors of real numbers.

We shall encounter in the sequel many examples of physical control systems that can be modeled by equations of the form (2.1). Note that, as a state space for (2.1), we consider a subset U of $\mathbf{R}^n$ rather than $\mathbf{R}^n$ itself. This limitation may correspond either to a constraint established by the equations themselves (whose solutions may not be free to evolve on the whole of $\mathbf{R}^n$) or to a constraint specifically imposed on the input, for instance to avoid points in the state space where some kind of "singularity" may occurr. We shall be more specific later on. Of course, in many cases one is allowed to set $U = \mathbf{R}^n$.

The mappings $f, g_1, \ldots, g_m$ are smooth mappings assigning to each point x of U a vector of $\mathbf{R}^n$, namely $f(x), g_1(x), \ldots, g_m(x)$. For this reason, they are frequently referred to as smooth *vector fields* defined on U. In many instances, it will be convenient to manipulate—together with vector fields—also *dual* objects called *covector fields*, which are smooth mappings assigning to each point x (of a subset U) an element of the *dual space* $(\mathbf{R}^n)^\star$.

As we will see in a moment, it is quite natural to identify smooth covector fields (defined on a subset U of $\mathbf{R}^n$) with $1 \times n$ (i.e. row) vectors of smooth functions of x. For, recall that the dual space $V^\star$ of a vector space V is the set of all *linear* real-valued functions defined on V. The dual space of an n-dimensional vector space is itself an n-dimensional vector space, whose elements are called *covectors*. Of course, like any linear mapping, an element $w^\star$ of $V^\star$ can be represented by means of a matrix. In particular, since $w^\star$ is a mapping from the n-dimensional space V to the 1-dimensional space $\mathbf{R}$, this representation is a matrix consisting of one row only, i.e. a row vector. On these grounds, one can assimilate $(\mathbf{R}^n)^\star$ with the set of all n-dimensional row vectors, and describe any subspace of $(\mathbf{R}^n)^\star$ as the collection of all linear combinations of some set of n-dimensional row vectors (for instance, the rows of some matrix having n columns). Note also that if

$$v = \begin{bmatrix} v_1 \\ v_2 \\ \dots \\ v_n \end{bmatrix}$$

is the column vector representing an element of V, and

$$w^\star = \begin{bmatrix} w_1 & w_2 & \cdots & w_n \end{bmatrix}$$

is the row vector representing an element of $V^\star$, the "value" of $w^\star$ at v is given by the product

$$w^\star v = \sum_{i=1}^{n} w_i v_i.$$

Most of the times, as often occurring in the literature, the value of $w^\star$ at v will be represented in the form of an *inner product*, writing $\langle w^\star, v \rangle$ instead of simply $w^\star v$.

Suppose now that $\omega_1, \ldots, \omega_n$ are smooth real-valued functions of the real variables $x_1, \ldots, x_n$ defined on an open subset U of $\mathbf{R}^n$, and consider the row vector

$$\omega(x) = \begin{bmatrix} \omega_1(x_1, \ldots, x_n) & \omega_2(x_1, \ldots, x_n) & \ldots & \omega_n(x_1, \ldots, x_n) \end{bmatrix}$$

On the grounds of the previous discussion, it is natural to interpret the latter as a mapping (a smooth one, because the ω_i's are smooth functions) assigning to each point x of a subset U an element $\omega(x)$ of the dual space $(\mathbf{R}^n)^\star$, i.e. exactly the object that was identified as a covector field.

A covector field of special importance is the so-called *differential*, or *gradient*, of a real-valued function λ defined on an open subset U of $\mathbf{R}^n$. This covector field, denoted $d\lambda$, is defined as the $1 \times n$ row vector whose i-th element is the partial derivative of λ with respect to x_i. Its value at a point x is thus

$$d\lambda(x) = [\, \frac{\partial \lambda}{\partial x_1} \quad \frac{\partial \lambda}{\partial x_2} \quad \cdots \quad \frac{\partial \lambda}{\partial x_n} \,]. \tag{2.3a}$$

Note that the right-hand side of this expression is exactly the jacobian matrix of λ, and that the more condensed notation

$$d\lambda(x) = \frac{\partial \lambda}{\partial x} \tag{2.3b}$$

is sometimes preferable. Any covector field having the form (2.3), i.e. the form of the differential of some real-valued function λ, is called an *exact differential*.

We describe now three types of differential operation, involving vector fields and covector fields, that are frequently used in the analysis of nonlinear control systems. The first type of operation involves a real-valued function λ and a vector field f, both defined on a subset U of $\mathbf{R}^n$. From these, a new smooth real-valued function is defined, whose value—at each x in U—is equal to the inner product

$$\langle d\lambda(x), f(x) \rangle = \frac{\partial \lambda}{\partial x} f(x) = \sum_{i=1}^{n} \frac{\partial \lambda}{\partial x_i} f_i(x)$$

This function is sometimes called the *derivative of λ along f* and is often written as $L_f \lambda$. In other words, by definition

$$L_f \lambda(x) = \sum_{i=1}^{n} \frac{\partial \lambda}{\partial x_i} f_i(x)$$

at each x of U.

Of course, repeated use of this operation is possible. Thus, for instance, by taking the derivative of λ first along a vector field f and then along a vector field g one defines the new function

$$L_g L_f \lambda(x) = \frac{\partial (L_f \lambda)}{\partial x} g(x).$$

If λ is being differentiated k times along f, the notation $L_f^k \lambda$ is used; in other words, the function $L_f^k\lambda(x)$ satisfies the recursion

$$L_f^k\lambda(x) = \frac{\partial(L_f^{k-1}\lambda)}{\partial x}f(x)$$

with $L_f^0\lambda(x) = \lambda(x)$.

The second type of operation involves two vector fields f and g, both defined on an open subset U of $\mathbf{R}^n$. From these a new smooth vector field is constructed, noted $[f,g]$ and defined as

$$[f,g](x) = \frac{\partial g}{\partial x}f(x) - \frac{\partial f}{\partial x}g(x)$$

at each x in U. In this expression

$$\frac{\partial g}{\partial x} = \begin{bmatrix} \frac{\partial g_1}{\partial x_1} & \frac{\partial g_1}{\partial x_2} & \cdots & \frac{\partial g_1}{\partial x_n} \\ \frac{\partial g_2}{\partial x_1} & \frac{\partial g_2}{\partial x_2} & \cdots & \frac{\partial g_2}{\partial x_n} \\ . & . & \cdots & . \\ \frac{\partial g_n}{\partial x_1} & \frac{\partial g_n}{\partial x_2} & \cdots & \frac{\partial g_n}{\partial x_n} \end{bmatrix} \qquad \frac{\partial f}{\partial x} = \begin{bmatrix} \frac{\partial f_1}{\partial x_1} & \frac{\partial f_1}{\partial x_2} & \cdots & \frac{\partial f_1}{\partial x_n} \\ \frac{\partial f_2}{\partial x_1} & \frac{\partial f_2}{\partial x_2} & \cdots & \frac{\partial f_2}{\partial x_n} \\ . & . & \cdots & . \\ \frac{\partial f_n}{\partial x_1} & \frac{\partial f_n}{\partial x_2} & \cdots & \frac{\partial f_n}{\partial x_n} \end{bmatrix}$$

denote the Jacobian matrices of the mappings g and f, respectively.

The vector field thus defined is called the *Lie product* (or *bracket*) *of f and g*. Of course, repeated bracketing of a vector field g with the same vector field f is possible. Whenever this is needed, in order to avoid a notation of the form $[f,[f,\ldots,[f,g]]]$, that could generate confusion, it is preferable to define such an operation recursively, as

$$ad_f^k g(x) = [f, ad_f^{k-1}g](x)$$

for any $k \geq 1$, setting $ad_f^0 g(x) = g(x)$.

The Lie product between vector fields is characterized by three basic properties, that are summarized in the following statement. The proof of them is extremely easy and is left as an exercise to the reader.

Proposition 2.1. The Lie product of vector fields has the following properties:

(i) is *bilinear* over $\mathbf{R}$, i.e. if f_1, f_2, g_1, g_2 are vector fields and r_1, r_2 real numbers, then

$$[r_1f_1 + r_2f_2, g_1] = r_1[f_1, g_1] + r_2[f_2, g_1]$$
$$[f_1, r_1g_1 + r_2g_2] = r_1[f_1, g_1] + r_2[f_1, g_2]$$

(ii) is *skew commutative*, i.e.

$$[f,g] = -[g,f]$$

(iii) satisfies the *Jacobi identity*, i.e., if f, g, p are vector fields, then

$$[f, [g, p]] + [g, [p, f]] + [p, [f, g]] = 0.$$

The third type of operation of frequent use involves a covector field ω and a vector field f, both defined on an open subset U of $\mathbf{R}^n$. This operation produces a new covector field, noted $L_f \omega$ and defined as

$$L_f \omega(x) = f^T(x) \left[\frac{\partial \omega^T}{\partial x} \right]^T + \omega(x) \frac{\partial f}{\partial x}$$

at each x of U, where the superscript T denotes transposition. This covector field is called the *derivative of ω along f*.

The operations thus defined are used very frequently in the sequel. For convenience we list in the following statement a series of "rules" of major interest, involving these operations either separately or jointly. Again, proofs are very elementary and left to the reader.

Proposition 2.2. The three types of differential operations introduced so far are such that

(i) if α is a real-valued function, f a vector field and λ a real-valued function, then

$$L_{\alpha f} \lambda(x) = (L_f \lambda(x)) \alpha(x) \tag{2.4}$$

(ii) if α, β are real-valued functions and f, g vector fields, then

$$[\alpha f, \beta g](x) = \alpha(x) \beta(x) [f, g](x) + (L_f \beta(x)) \alpha(x) g(x) - (L_g \alpha(x)) \beta(x) f(x) \tag{2.5}$$

(iii) if f, g are vector fields and λ a real-valued function, then

$$L_{[f,g]} \lambda(x) = L_f L_g \lambda(x) - L_g L_f \lambda(x) \tag{2.6}$$

(iv) if α, β are real-valued functions, f a vector field and ω a covector field, then

$$L_{\alpha f} \beta \omega(x) = \alpha(x) \beta(x) (L_f \omega(x)) + \beta(x) \langle \omega(x), f(x) \rangle d\alpha(x) + \\ + (L_f \beta(x)) \alpha(x) \omega(x) \tag{2.7}$$

(v) if f is a vector field and λ a real-valued function, then

$$L_f d\lambda(x) = dL_f \lambda(x) \tag{2.8}$$

(vi) if f, g are vector fields and ω a covector field, then

$$L_f \langle \omega, g \rangle(x) = \langle L_f \omega(x), g(x) \rangle + \langle \omega(x), [f, g](x) \rangle. \tag{2.9}$$

Example 2.1. As an exercise one can check, for instance, (2.4). By definition, one has

$$L_{\alpha f}\lambda(x) = \sum_{i=1}^{n}\left(\frac{\partial\lambda}{\partial x_i}\right)\left(\alpha(x)f_i(x)\right) = \sum_{i=1}^{n}\left(\frac{\partial\lambda}{\partial x_i}f_i(x)\right)\alpha(x) = \left(L_f\lambda(x)\right)\alpha(x).$$

Or, as far as (2.7) is concerned,

$$\begin{aligned}
[L_{\alpha f}\beta\omega]_i &= \sum_{j=1}^{n}\alpha f_j\frac{\partial\beta\omega_i}{\partial x_j} + \sum_{j=1}^{n}\beta\omega_j\frac{\partial\alpha f_j}{\partial x_i}\\
&= \sum_{j=1}^{n}\alpha f_j\beta\frac{\partial\omega_i}{\partial x_j} + \sum_{j=1}^{n}\alpha f_j\omega_i\frac{\partial\beta}{\partial x_j} + \sum_{j=1}^{n}\beta\omega_j\alpha\frac{\partial f_j}{\partial x_i} + \sum_{j=1}^{n}\beta\omega_j f_j\frac{\partial\alpha}{\partial x_i}\\
&= [\alpha\beta(L_f\omega)]_i + [\alpha(L_f\beta)\omega]_i + [\beta\langle\omega,f\rangle d\alpha]_i. \quad\bullet
\end{aligned}$$

To conclude the section, we illustrate another procedure of frequent use in the analysis of nonlinear control systems, the *change of coordinates* in the state space. As is well known, transforming the coordinates in the state space is often very useful in order to highlight some properties of interest, like e.g. reachability and observability, or to show how certain control problems, like e.g. stabilization or decoupling, can be solved.

In the case of a linear system, only *linear* changes of coordinates are usually considered. This corresponds to the substitution of the original state vector x with a new vector z related to x by a transformation of the form

$$z = Tx$$

where T is a nonsingular $n \times n$ matrix. Accordingly, the original description of the system

$$\dot{x} = Ax + Bu$$
$$y = Cx$$

is replaced by a new description

$$\dot{z} = \bar{A}z + \bar{B}u$$
$$y = \bar{C}z$$

in which

$$\bar{A} = TAT^{-1} \qquad \bar{B} = TB \qquad \bar{C} = CT^{-1}.$$

If the system is nonlinear, it is more meaningful to consider *nonlinear* changes of coordinates. A nonlinear change of coordinates can be described in the form

$$z = \Phi(x)$$

where $\Phi(x)$ represents a $\mathbf{R}^n$-valued function of n variables, i.e.

$$\Phi(x) = \begin{bmatrix} \phi_1(x) \\ \phi_2(x) \\ \cdots \\ \phi_n(x) \end{bmatrix} = \begin{bmatrix} \phi_1(x_1, \ldots, x_n) \\ \phi_2(x_1, \ldots, x_n) \\ \cdots \\ \phi_n(x_1, \ldots, x_n) \end{bmatrix}$$

with the following properties

(i) $\Phi(x)$ is invertible, i.e. there exists a function $\Phi^{-1}(z)$ such that

$$\Phi^{-1}(\Phi(x)) = x$$

for all x in $\mathbf{R}^n$.

(ii) $\Phi(x)$ and $\Phi^{-1}(z)$ are both smooth mappings, i.e. have continuous partial derivatives of any order.

A transformation of this type is called a *global diffeomorphism* on $\mathbf{R}^n$. The first of the two properties is clearly needed in order to have the possibility of reversing the transformation and recovering the original state vector as

$$x = \Phi^{-1}(z)$$

while the second one guarantees that the description of the system in the new coordinates is still a smooth one.

Sometimes, a transformation possessing both these properties and defined for all x is difficult to find and the properties in question are difficult to be checked. Thus, in most cases one rather looks at transformations defined only in a *neighborhood* of a given point. A transformation of this type is called a *local diffeomorphism*. In order to check whether or not a given transformation is a local diffeomorphism, the following result is very useful.

Proposition 2.3. Suppose $\Phi(x)$ is a smooth function defined on some subset U of $\mathbf{R}^n$. Suppose the jacobian matrix of Φ is nonsingular at a point $x = x^\circ$. Then, on a suitable open subset U° of U, containing x°, $\Phi(x)$ defines a local diffeomorphism.

Example 2.2. Consider the function

$$\begin{bmatrix} z_1 \\ z_2 \end{bmatrix} = \Phi(x_1, x_2) = \begin{bmatrix} x_1 + x_2 \\ \sin x_2 \end{bmatrix}$$

which is defined for all (x_1, x_2) in $\mathbf{R}^2$. Its jacobian matrix

$$\frac{\partial \Phi}{\partial x} = \begin{bmatrix} 1 & 1 \\ 0 & \cos x_2 \end{bmatrix}$$

has rank 2 at $x^\circ = (0,0)$. On the subset

$$U^\circ = \{(x_1, x_2) : |x_2| < (\pi/2)\}$$

this function defines a diffeomorphism. Note that on a larger set the function does not anymore define a diffeomorphism because the invertibility property is lost. For, observe that for each number x_2 such that $|x_2| > (\pi/2)$, there exists x_2' such that $|x_2'| < (\pi/2)$ and $\sin x_2 = \sin x_2'$. Any pair $(x_1, x_2), (x_1', x_2')$ such that $x_1 + x_2 = x_1' + x_2'$ yields $\Phi(x_1, x_2) = \Phi(x_1', x_2')$ and thus the function is not injective. •

Example 2.3. Consider the function

$$\begin{bmatrix} z_1 \\ z_2 \end{bmatrix} = \Phi(x_1, x_2) = \begin{bmatrix} x_1 \\ x_2 - \dfrac{1}{x_1 + 1} \end{bmatrix}$$

defined on the set

$$U^\circ = \{(x_1, x_2) : x_1 > -1\}$$

This function is a diffeomorphism (onto its image), because $\Phi(x_1, x_2) = \Phi(x_1', x_2')$ implies necessarily $x_1 = x_1'$ and $x_2 = x_2'$. However, this function is not defined on all $\mathbf{R}^2$. •

The effect of a change of coordinates on the description of a nonlinear system can be analyzed in this way. Set

$$z(t) = \Phi(x(t))$$

and differentiate both sides with respect to time. This yields

$$\dot{z}(t) = \frac{dz}{dt} = \frac{\partial \Phi}{\partial x}\frac{dx}{dt} = \frac{\partial \Phi}{\partial x}\big[f(x(t)) + g(x(t))u(t)\big].$$

Then, expressing $x(t)$ as $x(t) = \Phi^{-1}(z(t))$, one obtains

$$\dot{z}(t) = \bar{f}(z(t)) + \bar{g}(z(t))u(t)$$
$$y(t) = \bar{h}(z(t))$$

where

$$\bar{f}(z) = \left[\frac{\partial \Phi}{\partial x} f(x)\right]_{x=\Phi^{-1}(z)}$$

$$\bar{g}(z) = \left[\frac{\partial \Phi}{\partial x} g(x)\right]_{x=\Phi^{-1}(z)}$$

$$\bar{h}(z) = [h(x)]_{x=\Phi^{-1}(z)}.$$

The latter are the expressions relating the new description of the system to the original one. Note that if the system is linear, and if $\Phi(x)$ is linear as well, i.e. if $\Phi(x) = Tx$, then these formulas reduce to ones recalled before.

1.3 Distributions

We have observed in the previous Section that a smooth vector field f, defined on an open set U of $\mathbf{R}^n$, can be intuitively interpreted as a smooth mapping assigning the n- dimensional vector $f(x)$ to each point x of U. Suppose now that d smooth vector fields $f_1, \ldots, f_d$ are given, all defined on the same open set U and note that, at any fixed point x in U, the vectors $f_1(x), \ldots, f_d(x)$ span a vector space (a subspace of the vector space in which all the $f_i(x)$'s are defined, i.e. a subspace of $\mathbf{R}^n$). Let this vector space, which depends on x, be denoted by $\Delta(x)$, i.e. set

$$\Delta(x) = \mathrm{span}\{f_1(x), \ldots, f_d(x)\}$$

and note that, in doing this, we have essentially *assigned a vector space to each point* x of the set U. Motivated by the fact that the vector fields $f_1, \ldots, f_d$ are smooth vector fields, we can regard this assignment as a *smooth* one.

The object thus characterized, namely the assignement—to each point x of an open set U of $\mathbf{R}^n$—of the subspace spanned by the values at x of some smooth vector fields defined on U, is called a *smooth distribution*. We shall now illustrate a series of properties, concerning the notion of smooth distribution, that are of fundamental importance in all the subsequent analysis.

According to the characterization just given, a distribution is identified by a set of vector fields, say $\{f_1, \ldots, f_d\}$; we will use the notation

$$\Delta = \mathrm{span}\{f_1, \ldots, f_d\}$$

to denote the assignment as a whole, and, as before, $\Delta(x)$ to denote the "value" of Δ at a point x.

Pointwise, a distribution is a vector space, a subspace of $\mathbf{R}^n$. Based on this fact, it is possible to extend to the notion thus introduced a number of elementary concepts related to the notion of vector space. Thus, if Δ_1 and Δ_2 are distributions, their *sum* $\Delta_1 + \Delta_2$ is defined by taking pointwise the sum of the subspaces $\Delta_1(x)$ and $\Delta_2(x)$, namely

$$(\Delta_1 + \Delta_2)(x) = \Delta_1(x) + \Delta_2(x).$$

The *intersection* $\Delta_1 \cap \Delta_2$ is defined as

$$(\Delta_1 \cap \Delta_2)(x) = \Delta_1(x) \cap \Delta_2(x).$$

A distribution Δ_1 *contains* a distribution Δ_2, and is written $\Delta_1 \supset \Delta_2$, if $\Delta_1(x) \supset \Delta_2(x)$ for all x. A vector field f *belongs* to a distribution Δ, and is written $f \in \Delta$, if $f(x) \in \Delta(x)$ for all x. The *dimension* of a distribution at a point x of U is the dimension of the subspace $\Delta(x)$.

If F is a matrix having n rows and whose entries are smooth functions of x, its columns can be considered as smooth vector fields. Thus any matrix of this kind identifies a smooth distribution, the one spanned by its columns. The value of such a distribution at each x is equal to the *image* of the matrix $F(x)$ at this point

$$\Delta(x) = \text{Im}(F(x)).$$

Clearly, if a distribution Δ is spanned by the columns of a matrix F, the dimension of Δ at a point x° is equal to the *rank* of $F(x^\circ)$.

Example 3.1. Let $U = \mathbf{R}^3$, and consider the matrix

$$F(x) = \begin{bmatrix} x_1 & x_1 x_2 & x_1 \\ 1 + x_3 & (1 + x_3) x_2 & x_1 \\ 1 & x_2 & 0 \end{bmatrix}.$$

Note that the second column is proportional to the first one, via the coefficient x_2. Thus this matrix has at most rank 2. The first and third columns are independent (and, accordingly, the matrix F has rank exactly equal to 2) if x_1 is nonzero. Thus, we conclude that the columns of F span the distribution characterized as follows

$$\Delta(x) = \text{span}\left\{ \begin{bmatrix} 0 \\ 1 + x_3 \\ 1 \end{bmatrix} \right\} \qquad \text{if } x_1 = 0$$

$$\Delta(x) = \text{span}\left\{ \begin{bmatrix} x_1 \\ 1 + x_3 \\ 1 \end{bmatrix}, \begin{bmatrix} 1 \\ 1 \\ 0 \end{bmatrix} \right\} \qquad \text{if } x_1 \neq 0$$

The distribution has dimension 2 everywhere except on the plane $x_1 = 0$. •

Note that, by construction, the sum of two smooth distributions is a smooth distribution. In fact, if Δ_1 is spanned by smooth vector fields $f_1, \ldots, f_h$ and Δ_2 is spanned by smooth vector fields $g_1, \ldots, g_k$, then $\Delta_1 + \Delta_2$ is spanned by $f_1, \ldots, f_h, g_1, \ldots, g_k$. However, the intersection of two smooth distributions may fail to be smooth. This may be seen in the following example.

Example 3.2. Consider the two distributions defined on $\mathbf{R}^2$

$$\Delta_1 = \text{span} \left\{ \begin{bmatrix} 1 \\ 1 \end{bmatrix} \right\} \qquad \Delta_2 = \text{span} \left\{ \begin{bmatrix} 1 + x_1 \\ 1 \end{bmatrix} \right\}$$

We have

$$(\Delta_1 \cap \Delta_2)(x) = \{0\} \qquad\qquad \text{if } x_1 \neq 0$$
$$(\Delta_1 \cap \Delta_2)(x) = \Delta_1(x) = \Delta_2(x) \qquad \text{if } x_1 = 0.$$

This distribution is not smooth because it is not possible to find a smooth vector field on $\mathbf{R}^2$ which is zero everywhere but on the line $x_1 = 0$. •

Remark 3.1. The previous example shows that sometimes one may encounter an assignment Δ, of a vector space $\Delta(x)$ to each point x of a set U, which is *not* smooth, in the sense that it is not possible to find a set of *smooth* vector fields $\{f_i : i \in I\}$, defined on U, such that $\Delta(x) = \text{span}\{f_i(x) : i \in I\}$ for all x in U. If this is the case, it is convenient to replace Δ by an appropriate smooth distribution, defined on the basis of the following considerations. Suppose Δ_1 and Δ_2 are two smooth distributions, both contained in Δ. The distribution $\Delta_1 + \Delta_2$, is still smooth and contained in Δ, by construction. From this one concludes that the family of all smooth distributions contained in Δ has a unique maximal element (with respect to distributions inclusion), namely the sum of all members of the family. This distribution, which is the largest smooth distribution contained in Δ, will be denoted by $\text{smt}(\Delta)$ and sometimes used as substitute for the original Δ in some specific calculations. •

Other important concepts associated with the notion of distribution are the ones related to the "behavior" of this object as a "function" of x. We have already seen how it is possible to characterize the quality of being smooth, but there are other properties to be considered. A distribution Δ, defined on a open set U, is *nonsingular* if there exists an integer d such that

$$\dim(\Delta(x)) = d$$

for all x in U. A singular distribution, i.e. a distribution for which the above condition is not satisfied, is sometimes called a distribution of variable dimension. A point x° of U is said to be a *regular* point of a distribution Δ, if there exists a neighborhood U° of x° with the property that Δ is nonsingular on U°. Each point of U which is not a regular point is said to be a *point of singularity*.

Example 3.3. Consider again the distribution defined in the Example 3.1. The distribution in question has dimension 2 at each x such that $x_1 \neq 0$ and dimension 1 at each x such that $x_1 = 0$. The plane $\{x \in \mathbf{R}^3 : x_1 = 0\}$ is the set of points of singularity of Δ. •

In what follows we list some properties related to these notions, whose proofs are rather simple, and either omitted or just sketched.

Lemma 3.2. Let Δ be a smooth distribution and x° a regular point of Δ. Suppose $\dim(\Delta(x^\circ)) = d$. Then, there exist an open neighborhood U° of x° and a set $\{f_1, \ldots, f_d\}$ of smooth vector fields defined on U° with the property that
(i) the vectors $f_1(x), \ldots, f_d(x)$ are linearly independent at each x in U°,
(ii) $\Delta(x) = \mathrm{span}\{f_1(x), \ldots, f_d(x)\}$ at each x in U°.

Moreover, every smooth vector field τ belonging to Δ can be expressed, on U°, as

$$\tau(x) = \sum_{i=1}^{d} c_i(x) f_i(x)$$

where $c_1(x), \ldots, c_d(x)$ are smooth real-valued function of x, defined on U°.

Proof. The existence of exactly d smooth vector fields spanning Δ around x° is a trivial consequence of the assumptions. If τ is a vector field in Δ , then for each x near x°, the $n \times (d+1)$ matrix

$$\begin{bmatrix} f_1(x) & f_2(x) & \ldots & f_d(x) & \tau(x) \end{bmatrix}$$

has rank d. Thus, from elementary linear algebra we deduce the representation above, and the smoothness of the entries of this matrix implies that of the $c_i(x)$'s. •

Lemma 3.3. The set of all regular points of a distribution Δ, defined on U, is an open and dense subset of U.

Lemma 3.4. Let Δ_1 and Δ_2 be two smooth distributions, defined on U, with the property that Δ_2 is nonsingular and $\Delta_1(x) \subset \Delta_2(x)$ at each point x of a dense subset of U. Then $\Delta_1(x) \subset \Delta_2(x)$ at each x in U, i.e. $\Delta_1 \subset \Delta_2$.

Lemma 3.5. Let Δ_1 and Δ_2 be two smooth distributions, defined on U, with the property that Δ_1 is nonsingular, $\Delta_1 \subset \Delta_2$ and $\Delta_1(x) = \Delta_2(x)$ at each point x of a dense subset of U. Then $\Delta_1 = \Delta_2$.

As we have seen before, the intersection of two smooth distributions may fail to be smooth. However, around a regular point this cannot happen, as we see from the following statement.

Lemma 3.6. Let x° be a regular point of Δ_1, Δ_2, and $\Delta_1 \cap \Delta_2$. Then, there exists a neighborhood U° of x° such that the restriction of $\Delta_1 \cap \Delta_2$ to U° is smooth.

Proof. Let d_1 and d_2 denote the dimensions of Δ_1 and Δ_2. By Lemma 3.2, Δ_1 and Δ_2 can be described—around x°—as

$$\Delta_1 = \mathrm{span}\{f_i : 1 \leq i \leq d_1\}, \qquad \Delta_2 = \mathrm{span}\{g_i : 1 \leq i \leq d_2\}.$$

At a given point x, the intersection $\Delta_1(x) \cap \Delta_2(x)$ is found by solving the homogeneous equation

$$\sum_{i=1}^{d_1} a_i f_i(x) - \sum_{i=1}^{d_2} b_i g_i(x) = 0$$

in the unknowns $a_i(x)$, $1 \leq i \leq d_1$, and $b_i(x)$, $1 \leq i \leq d_2$. If $\Delta_1 \cap \Delta_2$ has constant dimension d, the coefficient matrix

$$[f_1(x) \quad \cdots \quad f_{d_1}(x) \quad -g_1(x) \quad \cdots \quad -g_{d_2}(x)]$$

of this equation has constant rank $r = d_1 + d_2 - d$; the space of solutions of this equation has dimension d and is spanned by d vectors of the form

$$\mathrm{col}(a_1(x), \ldots, a_{d_1}(x), b_1(x), \ldots, b_{d_2}(x))$$

which are smooth functions of x. As a consequence, it is easy to conclude that $\Delta_1 \cap \Delta_2$ is spanned—around x°—by d smooth vector fields. •

A distribution Δ is *involutive* if the Lie bracket $[\tau_1, \tau_2]$ of any pair of vector fields τ_1 and τ_2 belonging to Δ is a vector field which belongs to Δ, i.e. if

$$\tau_1 \in \Delta, \tau_2 \in \Delta \Rightarrow [\tau_1, \tau_2] \in \Delta.$$

Remark 3.7. Consider a nonsingular distribution Δ, and recall that, using Lemma 3.2, it is possible to express any two vector fields τ_1 and τ_2 of Δ in the form

$$\tau_1(x) = \sum_{i=1}^{d} c_i(x) f_i(x) \qquad \tau_2(x) = \sum_{i=1}^{d} d_i(x) f_i(x)$$

where $f_1, \ldots, f_d$ are smooth vector fields locally spanning Δ. It is easy to see that Δ is involutive if and only if

$$[f_i, f_j] \in \Delta \qquad \text{for all } 1 \leq i, j \leq d. \tag{3.1}$$

The necessity of this follows trivially from the fact that $f_1, \ldots, f_d$ are vector fields of Δ. For the sufficiency, consider the expansion (see (2.5))

$$[\sum_{i=1}^{d} c_i f_i, \sum_{j=1}^{d} d_j f_j] = \sum_{i=1}^{d}\sum_{j=1}^{d}(c_i d_j [f_i, f_j] + c_i (L_{f_i} d_j) f_j - d_j (L_{f_j} c_i) f_i)$$

and note that all vector fields on the right-hand side are vector fields of Δ.

Because of (3.1), checking whether or not a nonsingular distribution is involutive amounts to check that

$$\text{rank}[f_1(x) \ \ldots \ f_d(x)] = \text{rank}[f_1(x) \ \ldots \ f_d(x) \ [f_i, f_j](x)]$$

for all x and all $1 \leq i, j \leq d$. •

Example 3.4. Consider, on $\mathbf{R}^3$, a distribution

$$\Delta = \text{span}\{f_1, f_2\}$$

with

$$f_1(x) = \begin{bmatrix} 2x_2 \\ 1 \\ 0 \end{bmatrix} \qquad f_2(x) = \begin{bmatrix} 1 \\ 0 \\ x_2 \end{bmatrix}.$$

This distribution has dimension 2 for each $x \in \mathbf{R}^3$. Since

$$[f_1, f_2](x) = \begin{bmatrix} 0 & 0 & 0 \\ 0 & 0 & 0 \\ 0 & 1 & 0 \end{bmatrix} \begin{bmatrix} 2x_2 \\ 1 \\ 0 \end{bmatrix} - \begin{bmatrix} 0 & 2 & 0 \\ 0 & 0 & 0 \\ 0 & 0 & 0 \end{bmatrix} \begin{bmatrix} 1 \\ 0 \\ x_2 \end{bmatrix} = \begin{bmatrix} 0 \\ 0 \\ 1 \end{bmatrix}$$

we see that the matrix

$$[f_1 \quad f_2 \quad [f_1, f_2]] = \begin{bmatrix} 2x_2 & 1 & 0 \\ 1 & 0 & 0 \\ 0 & x_2 & 1 \end{bmatrix}$$

has rank 3 (for all x), and therefore the distribution is not involutive. •

Example 3.5. Consider, on the set $U = \{x \in \mathbf{R}^3 : x_1^2 + x_3^2 \neq 0\}$, a distribution

$$\Delta = \text{span}\{f_1, f_2\}$$

with

$$f_1(x) = \begin{bmatrix} 2x_3 \\ -1 \\ 0 \end{bmatrix} \qquad f_2(x) = \begin{bmatrix} -x_1 \\ -2x_2 \\ x_3 \end{bmatrix}.$$

This distribution has dimension 2 for each $x \in U$. Since

$$[f_1, f_2](x) = \begin{bmatrix} -1 & 0 & 0 \\ 0 & -2 & 0 \\ 0 & 0 & 1 \end{bmatrix} \begin{bmatrix} 2x_3 \\ -1 \\ 0 \end{bmatrix} - \begin{bmatrix} 0 & 0 & 2 \\ 0 & 0 & 0 \\ 0 & 0 & 0 \end{bmatrix} \begin{bmatrix} -x_1 \\ -2x_2 \\ x_3 \end{bmatrix} = \begin{bmatrix} -4x_3 \\ 2 \\ 0 \end{bmatrix}$$

the matrix

$$[f_1 \quad f_2 \quad [f_1, f_2]] = \begin{bmatrix} 2x_3 & -x_1 & -4x_3 \\ -1 & -2x_2 & 2 \\ 0 & x_3 & 0 \end{bmatrix}$$

has rank 2 (for all x), and therefore the distribution is involutive. •

Remark 3.8. Any 1-dimensional distribution is involutive. As a matter of fact, such a distribution is locally spanned by a nonzero vector field f and, since

$$[f, f](x) = \frac{\partial f}{\partial x} f(x) - \frac{\partial f}{\partial x} f(x) = 0$$

the condition of involutivity illustrated in the Remark 3.7 is indeed satisfied. •

The intersection of two involutive distributions Δ_1 and Δ_2 is again an involutive distribution, by construction. However, the sum of two involutive distributions in general is not involutive. This is shown, for instance, in the Example 3.4, if one interprets Δ as $\Delta_1 + \Delta_2$ with

$$\Delta_1 = \text{span}\{f_1\} \qquad \Delta_2 = \text{span}\{f_2\}$$

Δ_1 and Δ_2 are involutive (because both one-dimensional), but $\Delta_1 + \Delta_2$ is not.

Remark 3.9. Sometimes, starting from a distribution Δ which is not involutive, it is useful to construct an appropriate involutive distribution, defined on the basis of the following considerations. Suppose Δ_1 and Δ_2 are two involutive distributions, both containing Δ. The distribution $\Delta_1 \cap \Delta_2$ is still involutive and containing Δ, by construction. From this, one concludes that the family of all involutive distributions containing Δ has a unique minimal element (with respect to distributions inclusion), namely the intersection of all members of the family. This distribution, which is the smallest involutive distribution containing Δ, it is called the *involutive closure* of Δ and will be denoted by $\text{inv}(\Delta)$. •

In many instances, calculations are easier if, instead of distributions, one considers *dual* objects, called *codistributions*, that are defined in the following way. Recall that a smooth covector field ω, defined on an open set U of $\mathbf{R}^n$, can be interpreted as the smooth assignment—to each point x of U—of an element of the dual space $(\mathbf{R}^n)^\star$. With a set $\omega_1, \ldots, \omega_d$ of smooth covector fields, all defined on the same subset U of $\mathbf{R}^n$, one can associate the assignment—to each point x of U—of a subspace of $(\mathbf{R}^n)^\star$, the one spanned by the covectors $\omega_1, \ldots, \omega_d$. Motivated by the fact that the covector fields $\omega_1, \ldots, \omega_d$ are smooth covector fields, one may regard this assignment as a *smooth* one. The object characterized in this way is called a *smooth codistribution*.

Coherently with the notations introduced for distributions, we use

$$\Omega = \mathrm{span}\{\omega_1, \ldots, \omega_d\}$$

to denote the assignment as a whole, and

$$\Omega(x) = \mathrm{span}\{\omega_1(x), \ldots, \omega_d(x)\}$$

to denote the "value" of Ω at a point x of U. Since, pointwise, codistributions are vector spaces (subspaces of $(\mathbf{R}^n)^\star$), one can easily extend the notion of addition, intersection, inclusion. Similarly, one can define the dimension of a codistribution at each point x of U, and distinguish between regular points and points of singularity. If W is a matrix having n columns and whose entries are smooth functions of x, its rows can be regarded as smooth covector fields. Thus, any matrix of this kind identifies a codistribution, the one spanned by its rows.

Sometimes, it is possible to construct codistributions starting from given distributions, and conversely. The natural way to do this is the following one: given a distribution Δ, for each x in U consider the *annihilator* of $\Delta(x)$, that is the set of all covectors which annihilates all vectors in $\Delta(x)$

$$\Delta^\perp(x) = \{w^\star \in (\mathbf{R}^n)^\star : \langle w^\star, v \rangle = 0 \text{ for all } v \in \Delta(x)\}.$$

Since $\Delta^\perp(x)$ is a subspace of $(\mathbf{R}^n)^\star$, this construction identifies exactly a codistribution, in that assigns—to each x of U—a subspace of $(\mathbf{R}^n)^\star$. This codistribution, noted $\Delta^\perp$, is called the *annihilator* of Δ.

Conversely, given a codistribution Ω, one can construct a distribution, noted $\Omega^\perp$ and called the annihilator of Ω, setting at each x in U

$$\Omega^\perp(x) = \{v \in \mathbf{R}^n : \langle w^\star, v \rangle = 0 \text{ for all } w^\star \in \Omega(x)\}$$

Some care is required, for distributions/codistributions constructed in this way, about the quality of being smooth. As a matter of fact, the annihilator of a smooth distribution may fail to be smooth, as the following simple example shows.

Example 3.6. Consider the following distribution defined on $\mathbf{R}^1$

$$\Delta = \mathrm{span}\{x\}.$$

Then

$$\Delta^{\perp}(x) = \{0\} \qquad \text{if } x \neq 0$$
$$\Delta^{\perp}(x) = (\mathbf{R}^1)^{\star} \qquad \text{if } x = 0$$

and we see that $\Delta^{\perp}$ is not smooth because it is not possible to find a smooth covector field on $\mathbf{R}^1$ which is zero everywhere but on the point $x = 0$. •

Or, else, the annihilator of a non-smooth distribution can be a smooth codistribution, as in the following example.

Example 3.7. Consider again the two distributions Δ_1 and Δ_2 described in the Example 3.2. Their intersection is not smooth. The annihilator of $[\Delta_1 \cap \Delta_2]$ is

$$[\Delta_1 \cap \Delta_2]^{\perp}(x) = (\mathbf{R}^2)^{\star} \qquad \text{if } x_1 \neq 0$$
$$[\Delta_1 \cap \Delta_2]^{\perp}(x) = \mathrm{span}\{[\,1 \quad -1\,]\} \qquad \text{if } x_1 = 0.$$

The codistribution thus defined is smooth because is spanned, for instance, by the smooth covector fields

$$\omega_1 = [\,1 \quad -1\,]$$
$$\omega_2 = [\,1 \quad -(1 - x_1)\,]. \ \bullet$$

Distributions and codistributions related in this way possess a number of interesting properties. In particular, the sum of the dimensions of Δ and $\Delta^{\perp}$ is equal to n. The inclusion $\Delta_1 \supset \Delta_2$ is satisfied if and only if the inclusion $\Delta_1^{\perp} \subset \Delta_2^{\perp}$ is satisfied. Finally, the annihilator $[\Delta_1 \cap \Delta_2]^{\perp}$ of an intersection of distributions is equal to the sum $\Delta_1^{\perp} + \Delta_2^{\perp}$. If a distribution Δ is spanned by the columns of a matrix F, whose entries are smooth functions of x, its annihilator is identified, at each x in U, by the set of row vectors $w^{\star}$ satisfying the condition

$$w^{\star} F(x) = 0.$$

Conversely, if a codistribution Ω is spanned by the rows of a matrix W, whose entries are smooth functions of x, its annihilator is identified, at each x, by the set of vectors v satisfying

$$W(x)v = 0.$$

Thus, in this case $\Omega^{\perp}(x)$ is the *kernel* of the matrix W at the point x

$$\Omega^{\perp}(x) = \ker(W(x)).$$

One can easily extend Lemmas 3.2 to 3.6. In particular, if x° is a regular point of a smooth codistribution Ω, and $\dim(\Omega(x^\circ)) = d$, it is possible to find an open neighborhood U° of x° and a set of smooth covector fields $\{\omega_1, \ldots, \omega_d\}$ defined on U°, such that the covectors $\omega_1, \ldots, \omega_d$ are linearly independent at each x in U° and

$$\Omega(x) = \mathrm{span}\{\omega_1(x), \ldots, \omega_d(x)\}$$

at each x in U°. Moreover, every smooth covector field ω belonging to Ω can be expressed, on U°, as

$$\omega(x) = \sum_{i=1}^{d} c_i(x)\omega_i(x)$$

where $c_1, \ldots, c_d$ are smooth real-valued functions of x, defined on U°.

In addition one can easily prove the following result.

Lemma 3.10. Let x° be a regular point of a smooth distribution Δ. Then x° is a regular point of $\Delta^\perp$ and there exists a neighborhood U of x° such that the restriction of $\Delta^\perp$ to U° is a smooth codistribution.

Example 3.8. Let Δ be a distribution spanned by the columns of a matrix F and Ω a codistribution spanned by the rows of a matrix W, and suppose the intersection $\Omega \cap \Delta^\perp$ is to be calculated. By definition, a covector in $\Omega \cap \Delta^\perp(x)$ is an element of $\Omega(x)$ which annihilates all the elements of $\Delta(x)$. A generic element in $\Omega(x)$ has the form $\gamma W(x)$, where γ is a row vector of suitable dimension, and this (covector) annihilates all vectors of $\Delta(x)$ if and only if

$$\gamma W(x)F(x) = 0. \tag{3.2}$$

Thus, in order to evaluate $\Omega \cap \Delta^\perp(x)$ at a point x, one can proceed in the following way: first find a basis (say $\gamma_1, \ldots, \gamma_d$) of the space of the solutions of the linear homogeneous equation (3.2), and then express $\Omega \cap \Delta^\perp(x)$ in the form

$$\Omega \cap \Delta^\perp(x) = \mathrm{span}\{\gamma_i \Omega(x) : 1 \leq i \leq d\}$$

Note that the γ_i's depend on the point x. If $W(x)F(x)$ has constant rank for all x in a neighborhood U, then the space of solutions of (3.2) has constant dimension and the γ_i's depend smoothly on x. As a consequence, the row vectors $\gamma_1 \Omega(x), \ldots, \gamma_d \Omega(x)$ are smooth covector fields spanning $\Omega \cap \Delta^\perp$. •

1.4 Frobenius Theorem

In this Section we shall investigate the solvability of a special system of partial differential equations of the first order, which is of paramount importance in the analysis and design of nonlinear control systems. Later on, in the same

Chapter, we will use the results of this investigation in order to establish a fundamental correspondence between the notion of involutive distribution and the existence of local partitions of $\mathbf{R}^n$ into "lower dimensional" smooth surfaces. Such a correspondence is instrumental in the investigation of the existence of decompositions of the system into "reachable" and "unreachable" parts, as well as "observable" and "unobservable" parts, which very naturally extends to the nonlinear setting the analysis anticipated in Section 1.1. In the subsequent Chapters, we shall encounter again the same system of partial differential equations in several problems related to the synthesis of nonlinear feedback control laws.

Consider a nonsingular distribution Δ, defined on an open set U of $\mathbf{R}^n$, and let d denote its dimension. We know from the analysis developed in the previous Section that, in a neighborhood U° of each point x° of U, there exist d smooth vector fields $f_1, \ldots, f_d$, all defined on U°, which span Δ, i.e. are such that

$$\Delta(x) = \text{span}\{f_1(x), \ldots, f_d(x)\}$$

at each x in U°. We know also that the codistribution $\Omega = \Delta^\perp$ is again smooth and nonsingular, has dimension $n - d$ and, locally around each x°, is spanned by $n - d$ covector fields $\omega_1, \ldots, \omega_{n-d}$. By construction, the covector field ω_j is such that

$$\langle \omega_j(x), f_i(x) \rangle = 0 \qquad \text{for all } 1 \le i \le d, 1 \le j \le n - d$$

for all x in U°, i.e. solves the equation

$$\omega_j(x)F(x) = 0 \tag{4.1}$$

where $F(x)$ is the $n \times d$ matrix

$$F(x) = [f_1(x) \quad \cdots \quad f_d(x)].$$

At any fixed x in U, (4.1) can be simply regarded as a linear homogeneous equation in the unknown $\omega_j(x)$. The rank of the coefficient matrix $F(x)$ is d by assumption and the space of solutions is spanned by $n - d$ linearly independent row vectors. In fact, the row vectors $\omega_1(x), \ldots, \omega_{n-d}(x)$ are exactly a basis of this space. Suppose now that, instead of accepting any solution of (4.1), one seeks only solutions having the form

$$\omega_j = \frac{\partial \lambda_j}{\partial x}$$

for suitable real-valued smooth functions $\lambda_1, \ldots, \lambda_{n-d}$. In other words, suppose one is interested in solving the *partial differential equation*

$$\frac{\partial \lambda_j}{\partial x}[f_1(x) \dots f_d(x)] = \frac{\partial \lambda_j}{\partial x} F(x) = 0 \qquad (4.2)$$

and finding $n - d$ independent solutions. By "independent", we mean that the row vectors

$$\frac{\partial \lambda_1}{\partial x}, \dots, \frac{\partial \lambda_{n-d}}{\partial x}$$

are independent at each x. Observing that these row vectors (more precisely, these covector fields) have the form of differentials of real-valued functions, i.e. exact differentials, the problem of establishing the existence of $n - d$ independent solutions of the equation (4.2) can be rephrased in the following terms: when a nonsingular distribution Δ as an annihilator $\Delta^\perp$ which is spanned by *exact differentials*? This problem will be discussed in the present Section. We begin with some terminology. A nonsingular d-dimensional distribution Δ, defined on an open set U of $\mathbf{R}^n$, is said to be *completely integrable* if, for each point x^0 of U there exist a neighborhood U^0 of x^0, and $n - d$ real-valued smooth functions $\lambda_1, \dots, \lambda_{n-d}$, all defined on U^0, such that

$$\text{span}\{d\lambda_1, \dots, d\lambda_{n-d}\} = \Delta^\perp \qquad (4.3)$$

on U^0 (recall the notation (2.3b)). Thus, "complete integrability of the distribution spanned by the columns of the matrix $F(x)$" is essentially a synonimous for "existence of $n - d$ independent solutions of the differential equation (4.2)" . The following result illustrates necessary and sufficient conditions for complete integrability.

Theorem 4.1 (Frobenius). A nonsingular distribution is completely integrable if and only if it is involutive.

Proof. We shall show first that the property of being involutive is a *necessary* condition, for a distribution to be completely integrable. By assumption, there exist functions $\lambda_1, \dots, \lambda_{n-d}$ such that (4.3), or, what is the same, (4.2) is satisfied. Now, observe that the equation (4.2) can also be rewritten as

$$\frac{\partial \lambda_j}{\partial x} f_i(x) = \langle d\lambda_j(x), f_i(x) \rangle = 0 \qquad \text{for all } 1 \le i \le d, \text{ all } x \in U^0 \qquad (4.4)$$

and that the latter, using a notation established in Section **1.2**, can in turn be rewritten as

$$\langle d\lambda_j(x), f_i(x) \rangle = L_{f_i} \lambda_j(x) = 0 \qquad \text{for all } 1 \le i \le d, \text{ all } x \in U^0. \qquad (4.5)$$

Differentiating the function λ_i along the vector field $[f_i, f_k]$, and using (4.5) and

(2.6), one obtains

$$L_{[f_i, f_k]}\lambda_j(x) = L_{f_i} L_{f_k}\lambda_j(x) - L_{f_k} L_{f_i}\lambda_j(x) = 0$$

Suppose now the same operation is repeated for all the functions $\lambda_1, \ldots, \lambda_{n-d}$. We conclude that

$$\begin{bmatrix} L_{[f_i, f_k]}\lambda_1(x) \\ \cdots \\ L_{[f_i, f_k]}\lambda_{n-d}(x) \end{bmatrix} = \begin{bmatrix} d\lambda_1(x) \\ \cdots \\ d\lambda_{n-d}(x) \end{bmatrix} [f_i, f_k](x) = 0 \qquad \text{for all } x \in U^\circ$$

Since by assumption the differentials $\{d\lambda_1, \ldots, d\lambda_{n-d}\}$ span the distribution $\Delta^\perp$, we deduce from this that the vector field $[f_i, f_k]$ is itself a vector field in Δ. Thus, in view of the condition established in the Remark 3.7, we conclude that the distribution Δ is involutive.

The proof of the *sufficiency* is constructive. Namely, it is shown how a set of $n - d$ functions satisfying (4.3) can be found. Recall, that, since Δ is nonsingular and has dimension d, in a neighborhood U° of each point x° of U there exist d smooth vector fields $f_1, \ldots, f_d$, all defined on U°, which span Δ, i.e. are such that

$$\Delta(x) = \text{span}\{f_1(x), \ldots, f_d(x)\}$$

at each x in U°. Let $f_{d+1}, \ldots, f_n$ be a complementary set of vector fields, still defined on U°, with the property that

$$\text{span}\{f_1(x), \ldots, f_d(x), f_{d+1}(x), \ldots, f_n(x)\} = \mathbf{R}^n$$

at each x in U°.

Let $\Phi_t^f(x)$ denote the *flow* of the vector field f, i.e. the smooth function of t and x with the property that $x(t) = \Phi_t^f(x^\circ)$ solves the ordinary differential equation

$$\dot{x} = f(x)$$

with initial condition $x(0) = x^\circ$. In other words, $\Phi_t^f(x)$ is a smooth function of t and x satisfying

$$\frac{\partial}{\partial t}\Phi_t^f(x) = f(\Phi_t^f(x)) \qquad \Phi_0^f(x) = x.$$

Recall also that, for any fixed x° there is a (sufficiently small) t such that the mapping

$$\Phi_t^f : x \mapsto \Phi_t^f(x)$$

is defined for all x in a neighborhood of x°, is a local diffeomorphism (onto its image), and $[\Phi_t^f]^{-1} = \Phi_{-t}^f$. Moreover, for any (sufficiently small) t, s

$$\Phi_{t+s}^f(x) = \Phi_t^f(\Phi_s^f(x)).$$

We show now that a solution of the partial differential equation (4.2) can be constructed by taking an appropriate composition of the flows associated with the vector fields $f_1, \ldots, f_n$, i.e. of

$$\Phi_{t_1}^{f_1}(x), \ldots, \Phi_{t_n}^{f_n}(x).$$

To this end, consider the mapping

$$
\begin{aligned}
F : U_\varepsilon &\to \mathbf{R}^n \\
(z_1, \ldots, z_n) &\mapsto \Phi_{z_1}^{f_1} \circ \cdots \circ \Phi_{z_n}^{f_n}(x^\circ)
\end{aligned}
\tag{4.6}
$$

where $U_\varepsilon = \{x \in \mathbf{R}^n : |z_i| < \varepsilon\}$ and "o" denotes composition with respect to the argument x. If ε is sufficiently small, this mapping has the following properties:

(i) is defined for all $z = (z_1, \ldots, z_n) \in U_\varepsilon$ and is a diffeomorphism onto its image,

(ii) is such that, for all $z \in U_\varepsilon$, the first d columns of the Jacobian matrix

$$\left[\frac{\partial F}{\partial z} \right]$$

are linearly independent vectors in $\Delta(F(z))$

Before proceeding with the proof of these two properties, it is important to observethat they are sufficient to construct a solution of the partial differential equation (4.2). To this end, let U° denote the image of the mapping F, and observe that U° is indeed an open neighborhood of x°, because x° is exactly the value of F at the point $z = 0$. Since this mapping is a diffeomorphism onto its image (property (i)), the inverse F^{-1} exists and is a smooth mapping, defined on U°. Set

$$
\begin{bmatrix}
\phi_1(x) \\
\cdots \\
\phi_n(x)
\end{bmatrix}
= F^{-1}(x)
$$

where, $\phi_1, \ldots, \phi_n$ are real-valued functions, defined for all x in U°. We claim that the last $n - d$ of these functions are independent solutions of the equation (4.2). For, observe that, by definition

$$\left[\frac{\partial F^{-1}}{\partial x} \right]_{x=F(z)} \left[\frac{\partial F}{\partial z} \right] = I$$

where I is the identity matrix, for all $z \in U_\varepsilon$ (i.e. for all $x \in U^\circ$). By property (ii), the first d columns of the second factor on the left-hand side form a basis of Δ at any point $x = F(z)$ of U°. As a consequence, the differentials

$$d\phi_{d+1}(x) = \frac{\partial \phi_{d+1}}{\partial x}, \ldots, d\phi_n(x) = \frac{\partial \phi_n}{\partial x}$$

are annihilated by the vectors of Δ at each x in U°. These differentials (which are independent by construction) are therefore a solution of (4.3). At this point, to complete the proof of the sufficiency we only have to show that (i) and (ii) hold.

Proof of (i). It is known that, for all $x \in \mathbf{R}^n$ and sufficiently small $|t|$, the flow $\Phi_t^f(x)$ of a vector field f is defined and this renders the mapping F well defined for all $(z_1, \ldots, z_n)$ with $|z_i|$ sufficiently small. Moreover, since a flow is smooth, so is F. We prove that F is a local diffeomorphism by showing that the rank of F at 0 is equal to n. To this purpose, let for simplicity $(M)_\star$ denote the jacobian matrix of a mapping $M(x)$, i.e.

$$(M)_\star = \frac{\partial M}{\partial x}$$

and note that, by the chain rule

$$\frac{\partial F}{\partial z_i} = \left(\Phi_{z_1}^{f_1}\right)_\star \cdots \left(\Phi_{z_{i-1}}^{f_{i-1}}\right)_\star \frac{\partial}{\partial z_i}\left(\Phi_{z_i}^{f_i} \circ \cdots \circ \Phi_{z_n}^{f_n}(x^\circ)\right)$$

$$= \left(\Phi_{z_1}^{f_1}\right)_\star \cdots \left(\Phi_{z_{i-1}}^{f_{i-1}}\right)_\star f_i\left(\Phi_{z_i}^{f_i} \circ \cdots \circ \Phi_{z_n}^{f_n}(x^\circ)\right)$$

$$= \left(\Phi_{z_1}^{f_1}\right)_\star \cdots \left(\Phi_{z_{i-1}}^{f_{i-1}}\right)_\star f_i\left(\Phi_{-z_{i-1}}^{f_{i-1}} \circ \cdots \circ \Phi_{-z_1}^{f_1}(F(z))\right).$$

In particular, at $z = 0$, since $F(0) = x^\circ$

$$\frac{\partial F}{\partial z_i}(0) = f_i(x^\circ).$$

The tangent vectors $f_1(x^\circ), \ldots, f_n(x^\circ)$ are by assumption linearly independent, and this proves that the n columns of $(F)_\star$ are linearly independent at $z = 0$. Thus, the mapping F has rank n at $z = 0$.

Proof of (ii). From the previous computations, we deduce also that, at any $z \in U_\varepsilon$,

$$\left(\Phi_{z_1}^{f_1}\right)_\star \cdots \left(\Phi_{z_{i-1}}^{f_{i-1}}\right)_\star f_i\left(\Phi_{-z_{i-1}}^{f_{i-1}} \circ \cdots \circ \Phi_{-z_1}^{f_1}(x)\right) = \frac{\partial F}{\partial z_i}$$

where $x = F(z)$. If we are able to prove that for all x in a neighborhood of x°, for small $|t|$ and for any two vector fields τ and ϑ belonging to Δ,

$$\left(\Phi_t^\vartheta\right)_\star \tau \circ \Phi_{-t}^\vartheta(x) \in \Delta(x)$$

i.e. that $\left(\Phi_t^\vartheta\right)_* \tau \circ \Phi_{-t}^\vartheta(x)$ is a (locally defined) vector field of Δ, then we easily see that (ii) is true. To prove this, one proceeds as follows. Let ϑ be a vector field of Δ and set

$$V_i(t) = \left(\Phi_{-t}^\vartheta\right)_* f_i \circ \Phi_t^\vartheta(x)$$

for $i = 1, \ldots, d$. Since

$$\frac{d}{dt}\left(\Phi_{-t}^\vartheta\right)_* \circ \Phi_t^\vartheta(x) = -\left(\Phi_{-t}^\vartheta\right)_* \frac{\partial \vartheta}{\partial x} \circ \Phi_t^\vartheta(x)$$

(differentiate the identity $\left(\Phi_{-t}^\vartheta\right)_* \left(\Phi_t^\vartheta\right)_* = I$ with respect to t and interchange d/dt with $\partial/\partial x$) and

$$\frac{d}{dt}(f_i \circ \Phi_t^\vartheta(x)) = \frac{\partial f_i}{\partial x} \vartheta \circ \Phi_t^\vartheta(x)$$

the functions $V_i(t)$ just defined satisfy

$$\frac{dV_i}{dt} = \left(\Phi_{-t}^\vartheta\right)_* [\vartheta, f_i] \circ \Phi_t^\vartheta$$

Since both ϑ and f_i belong to Δ and Δ is involutive, there exist functions λ_{ij} defined locally around x° such that

$$[\vartheta, f_i] = \sum_{j=1}^d \lambda_{ij} f_j$$

and, therefore,

$$\frac{dV_i}{dt} = \left(\Phi_{-t}^\vartheta\right)_* \left(\sum_{j=1}^d \lambda_{ij} f_j\right) \circ \left(\Phi_t^\vartheta(x)\right) = \sum_{j=1}^d \lambda_{ij}\left(\Phi_t^\vartheta(x)\right) V_j(t)$$

The functions $V_i(t)$ are seen as solutions of a linear differential equation and, therefore, it is possible to set

$$[V_1(t) \ldots V_d(t)] = [V_1(0) \ldots V_d(0)] X(t)$$

where $X(t)$ is a $d \times d$ fundamental matrix of solutions. By multiplying on the left both sides of this equality by $\left(\Phi_t^\vartheta\right)_*$ we get

$$\left[f_1\left(\Phi_t^\vartheta(x)\right) \ldots f_d(\Phi_t^\vartheta(x))\right] = \left[\left(\Phi_t^\vartheta\right)_* f_1(x) \ldots \left(\Phi_t^\vartheta\right)_* f_d(x)\right] X(t)$$

and also, by replacing x by $\Phi_{-t}^\vartheta(x)$

$$[f_1(x) \ldots f_d(x)] = \left[\left(\Phi_t^\vartheta\right)_* f_1(x) \circ \Phi_{-t}^\vartheta(x) \ldots \left(\Phi_t^\vartheta\right)_* f_d(x) \circ \Phi_{-t}^\vartheta(x)\right] X(t)$$

Since $X(t)$ is nonsingular for all t, we have that, for $i = 1, \ldots, d$

$$\left(\Phi_t^\vartheta\right)_\star f_i(x) \circ \Phi_t^\vartheta(x) \in \mathrm{span}\{f_1(x), \ldots, f_d(x)\}$$

i.e.

$$\left(\Phi_t^\vartheta\right)_\star f_i(x) \circ \Phi_{-t}^\vartheta(x) \in \Delta(x)$$

This result, bearing in mind the possibility of expressing any vector field τ of Δ in the form

$$\tau = \sum_{i=1}^d c_i f_i$$

complets the proof of (ii). •

The proof of this theorem, in the part concerning the sufficiency, is quite interesting, because it shows that the solution of the partial differential equation (4.2) (or, what is the same, (4.3)) can be reduced to the solution of n ordinary differential equations of the form

$$\dot{x} = f_i(x) \qquad 1 \le i \le n$$

where $f_1, \ldots, f_n$ are linearly independent vector fields, with $f_1, \ldots, f_d$ spanning the distribution Δ. As a matter of fact, if the solutions of these equations are composed to build the mapping F defined by (4.6), a solution of (4.2) can be found by taking the last $n - d$ components of the *inverse mapping* F^{-1}. This procedure is applied in the following examples.

Example 4.1. Consider the distribution, defined on $\mathbf{R}^2$

$$\Delta = \mathrm{span}\left\{\begin{bmatrix} \exp x_2 \\ 1 \end{bmatrix}\right\}.$$

This distribution has dimension 1 for each $x \in \mathbf{R}^2$. Thus, Δ is nonsingular and, being 1-dimensional, is also involutive. Set

$$f_1(x) = \begin{bmatrix} \exp x_2 \\ 1 \end{bmatrix} \qquad f_2(x) = \begin{bmatrix} 1 \\ 0 \end{bmatrix}$$

The calculation of the flows of f_1 and f_2 is rather easy. As far as f_1 is concerned, since

$$\dot{x}_1 = \exp x_2$$
$$\dot{x}_2 = 1$$

is solved by

$$x_1(t) = \exp(x_2^\circ)(\exp(t) - 1) + x_1^\circ$$
$$x_2(t) = t + x_2^\circ$$

we have

$$\Phi_{z_1}^{f_1}(x) = \begin{bmatrix} \exp(x_2)(\exp(z_1) - 1) + x_1 \\ z_1 + x_2 \end{bmatrix}.$$

About f_2, since

$$\dot{x}_1 = 1$$
$$\dot{x}_2 = 0$$

is solved by

$$x_1(t) = t + x_1^\circ$$
$$x_2(t) = x_2^\circ$$

we have

$$\Phi_{z_2}^{f_2}(x) = \begin{bmatrix} z_2 + x_1 \\ x_2 \end{bmatrix}.$$

The mapping F, choosing $x_1^\circ = x_2^\circ = 0$, has the form

$$\begin{bmatrix} x_1 \\ x_2 \end{bmatrix} = F(z_1, z_2) = \begin{bmatrix} \exp(z_1) + z_2 - 1 \\ z_1 \end{bmatrix}$$

and its inverse is given by

$$\begin{bmatrix} z_1 \\ z_2 \end{bmatrix} = F^{-1}(x_1, x_2) = \begin{bmatrix} x_2 \\ x_1 - \exp(x_2) + 1 \end{bmatrix}.$$

The function $z_2(x_1, x_2)$ is a solution of the partial differential equation

$$\frac{\partial z_2}{\partial x} f_1(x) = 0$$

as a straigthforward check also confirms. Note that this function is defined in all $\mathbf{R}^2$. •

Example 4.2. Consider the following distribution, defined on $\mathbf{R}^2$

$$\Delta = \text{span} \left\{ \begin{bmatrix} x_1^2 \\ -1 \end{bmatrix} \right\}.$$

Again, this distribution is 1-dimensional, and therefore completely integrable. In order to integrate it we set

$$f_1(x) = \begin{bmatrix} x_1^2 \\ -1 \end{bmatrix} \qquad f_2(x) = \begin{bmatrix} 1 \\ 0 \end{bmatrix}$$

The calculation of the flow of f_1 is not difficult. Since

$$\dot{x}_1 = x_1^2$$
$$\dot{x}_2 = -1$$

is solved by

$$x_1(t) = \frac{x_1^o}{1 - x_1^o t} \qquad x_2(t) = -t + x_2^o$$

we have

$$\Phi_{z_1}^{f_1}(x) = \begin{bmatrix} \dfrac{x_1}{1 - x_1 z_1} \\ -z_1 + x_2 \end{bmatrix}$$

Note that the flow is not defined for $x_1 z_1 \geq 1$ (i.e. the vector field f_1 is not *complete*).The flow of f_2 is identical to the one calculated in the previous example. The mapping F has the form

$$F(z) = \begin{bmatrix} \dfrac{z_2 + x_1^o}{1 - (z_2 + x_1^o)z_1} \\ -z_1 + x_2^o \end{bmatrix}$$

and its inverse is given by

$$F^{-1}(x) = \begin{bmatrix} z_1 \\ z_2 \end{bmatrix} = \begin{bmatrix} x_2^o - x_2 \\ \dfrac{x_1}{1 + x_1(x_2^o - x_2)} - x_1^o \end{bmatrix}.$$

Note that this mapping is not defined on all $\mathbf{R}^2$. However, provided $|x_2 - x_2^o|$ is sufficiently small, the mapping is well-defined for any x^o. The function $z_2(x_1, x_2)$ is then defined in a neighborhood of any x^o and solves the partial differential equation

$$\frac{\partial z_2}{\partial x} f_1(x) = 0. \quad \bullet$$

Example 4.3. Consider the distribution, defined on $\mathbf{R}^3$

$$\Delta = \mathrm{span} \left\{ \begin{bmatrix} 2x_3 \\ -1 \\ 0 \end{bmatrix}, \begin{bmatrix} -x_1 \\ -2x_2 \\ x_3 \end{bmatrix} \right\}$$

This distribution has dimension 2 at each point of the set

$$U = \{x \in \mathbf{R}^3 : x_1^2 + x_3^2 \neq 0\}.$$

The distribution is also involutive on U, as shown in the Example 3.5. Thus, the distribution is completely integrable on U. Set

$$f_1(x) = \begin{bmatrix} 2x_3 \\ -1 \\ 0 \end{bmatrix} \qquad f_2(x) = \begin{bmatrix} -x_1 \\ -2x_2 \\ x_3 \end{bmatrix} \qquad f_3(x) = \begin{bmatrix} 1 \\ 0 \\ 0 \end{bmatrix}.$$

The calculation of the flows of f_1, f_2 and f_3 yields

$$\Phi_{z_1}^{f_1}(x) = \begin{bmatrix} 2z_1x_3 + x_1 \\ -z_1 + x_2 \\ x_3 \end{bmatrix} \qquad \Phi_{z_2}^{f_2}(x) = \begin{bmatrix} \exp(-z_2)x_1 \\ \exp(-2z_2)x_2 \\ \exp(z_2)x_3 \end{bmatrix} \qquad \Phi_{z_3}^{f_3}(x) = \begin{bmatrix} z_3 + x_1 \\ x_2 \\ x_3 \end{bmatrix}$$

Therefore, the mapping F has the form

$$F(z_1, z_2, z_3) = \begin{bmatrix} 2z_1 \exp(z_2)x_3^0 + \exp(-z_2)(z_3 + x_1^0) \\ -z_1 + \exp(-2z_2)x_2^0 \\ \exp(z_2)x_3^0 \end{bmatrix}.$$

Consider for instance the point $x^0 = (0, 0, 1)$. At this point the mapping F^{-1} is given by

$$\begin{bmatrix} z_1 \\ z_2 \\ z_3 \end{bmatrix} = F^{-1}(x_1, x_2, x_3) = \begin{bmatrix} -x_2 \\ \ln(x_3) \\ (x_1 + 2x_2x_3)x_3 \end{bmatrix}.$$

Thus, the partial differential equation

$$\frac{\partial \lambda}{\partial x} \begin{bmatrix} 2x_3 & -x_1 \\ -1 & -2x_2 \\ 0 & x_3 \end{bmatrix} = \begin{bmatrix} 0 & 0 \end{bmatrix}$$

is solved by

$$\lambda(x_1, x_2, x_3) = z_3(x_1, x_2, x_3) = (x_1 + 2x_2x_3)x_3. \quad \bullet$$

One of the most useful consequences of the notion of complete integrability is related to the possibility of using the functions $\lambda_1, \ldots, \lambda_{n-d}$, which solve the partial differential equation (4.2), in order to define (locally around x^0) a coordinates transformation entailing a particularly simple representation for the vector fields of Δ. For, observe that, by construction, the $(n - d)$ differentials

$$d\lambda_1, \ldots, d\lambda_{n-d} \tag{4.7}$$

are linearly independent at the point x°. Then, it is always possible to choose, in the set of functions

$$x_1(x) = x_1, x_2(x) = x_2, \ldots, x_n(x) = x_n$$

a subset of d functions whose differentials at x°, together with those of the set (4.7), form a set of exactly n linearly independent row vectors. Let $\phi_1, \ldots, \phi_d$ denote the functions thus chosen and set

$$\phi_{d+1}(x) = \lambda_1(x), \ldots, \phi_n(x) = \lambda_{n-d}(x)$$

By construction, the jacobian matrix of the mapping

$$z = \Phi(x) = \mathrm{col}(\phi_1(x), \ldots, \phi_d(x), \phi_{d+1}(x), \ldots, \phi_n(x))$$

has rank n at x° and, therefore, the mapping Φ qualifies as a local diffeomorphism (i.e. a local smooth coordinates transformation) around the point x°. Now, suppose τ is a vector field of Δ. In the new coordinates, this vector field is represented in the form

$$\bar{\tau}(z) = \left[\frac{\partial \Phi}{\partial x} \tau(x) \right]_{x = \Phi^{-1}(z)}.$$

Since, by construction, the last $n-d$ rows of the jacobian matrix of Φ span $\Delta^\perp$, it is immediately deduced that the last $n-d$ entries of the vector on the right-hand side are zero, for all x in the set where the coordinates transformation is defined. We conclude from this that any vector field of Δ, in the new coordinates, has a representation of the form

$$\bar{\tau}(z) = \mathrm{col}(\bar{\tau}_1(z), \ldots, \bar{\tau}_d(z), 0, \ldots, 0) \tag{4.8}$$

We end this Section with an additional result that shows how the notion of integrability can be extended to a *collection* of distributions $\Delta_1, \ldots, \Delta_k$, all defined on an open set U. Suppose each distribution of this collection has constant dimension, say $d_1, \ldots, d_k$. Suppose also that the distributions form a *nested* sequence, i.e. that

$$\Delta_1 \supset \Delta_2 \supset \cdots \supset \Delta_k$$

(so that, in particular, $d_1 > d_2 > \cdots > d_k$). If the distribution Δ_1 is completely integrable, by Frobenius Theorem, in a neighborhood of each point x° there exist functions $\lambda_i, 1 \leq i \leq n - d_1$, such that

$$\text{span}\{d\lambda_1, \ldots, d\lambda_{n-d_1}\} = \Delta_1^\perp.$$

Suppose now also Δ_2 is completely integrable. Then, again, $\Delta_2^\perp$ is locally spanned by differentials of suitable functions $\mu_i, 1 \leq i \leq n - d_2$. However, since

$$\Delta_1^\perp \subset \Delta_2^\perp$$

it is immediate to conclude that one can choose

$$\mu_i = \lambda_i \qquad \text{for all } 1 \leq i \leq n - d_1$$

thus obtaining

$$\text{span}\{d\lambda_1, \ldots, d\lambda_{n-d_1}\} + \text{span}\{d\mu_{n-d_1+1}, \ldots, d\mu_{n-d_2}\} = \Delta_2^\perp.$$

Note also that the sum on the left-hand side of this relation is direct, i.e. the two summands have zero intersection. The construction can be repeated for all other distributions of the sequence, provided they are involutive. Thus, one arrives at the following result.

Corollary 4.2. Let $\Delta_1 \supset \Delta_2 \supset \cdots \supset \Delta_k$ be a collection of nested nonsingular distributions. If and only if each distribution of the collection is involutive then, for each point x° of U, there exists a neighborhood U° of x°, and real-valued smooth functions

$$\lambda_1^1, \ldots, \lambda_{n-d_1}^1, \lambda_1^2, \ldots, \lambda_{d_1-d_2}^2, \ldots, \lambda_1^k, \ldots, \lambda_{d_{k-1}-d_k}^k$$

all defined on U°, such that

$$\Delta_1^\perp = \text{span}\{d\lambda_1^1, \ldots, d\lambda_{n-d}^1\}$$
$$\Delta_i^\perp = \Delta_{i-1}^\perp \oplus \text{span}\{d\lambda_1^i, \ldots, d\lambda_{d_{i-1}-d_i}^i\}$$

for $2 \leq i \leq k$.

Remark 4.3. In order to avoid the problem of using double indices, it is sometimes convenient to state the previous, and similar, results by means of a more condensed notation, defined in the following way. Given a set of p_i real-valued functions

$$\phi_1^i(x), \ldots, \phi_{p_i}^i(x)$$

set

$$d\phi^i = \left(d\phi_1^i, \ldots, d\phi_{p_i}^i\right).$$

In this notation, the last expressions of the previous statement can be clearly rewritten in a form like

$$\Delta_1^\perp = \text{span}\{d\lambda^1\}$$

$$\Delta_i^\perp = \Delta_{i-1}^\perp \oplus \text{span}\{d\lambda^i\} = \text{span}\{d\lambda^1, \ldots, d\lambda^i\}. \quad \bullet$$

1.5 The Differential Geometric Point of View

We present in this Section some additional material related to the notion of distribution and to the property, for a distribution, of being completely integrable. The analysis requires some familiarity with a few basic concepts of differential geometry, like the ones that—for convenience of the reader—are summarized in the Appendix A. This background, as well as the knowledge of the material developed in this Section, is indeed helpful in the understanding the proofs of some later results and is essential in any non-local analysis (like the one presented in Chapter 2), but can be dispensed of in a first reading.

Throughout the whole Section, we consider objects defined on an arbitrary n-dimensional smooth manifold N. This point of view is interesting, for instance, when the natural state space on which a control system is defined is not $\mathbf{R}^n$ nor a set diffeomorphic to $\mathbf{R}^n$, but a more abstract set.

If this is the case, one can still describe the control system in a form like

$$\dot{p} = f(p) + \sum_{i=1}^{m} g_i(p)u_i \tag{5.1a}$$

$$y_i = h_i(p) \qquad 1 \le i \le l \tag{5.1b}$$

where $f, g_1, \ldots, g_m$ are smooth vector fields defined on a smooth manifold N, and $h_1, \ldots, h_l$ are smooth real-valued functions defined on N. The first relation represents a differential equation on N, and $\dot{p}$ stands for the tangent vector, at the point p of N, to the smooth curve which characterizes the solution for some fixed initial condition. For the sake of clearness, we have used here p in order to denote a point in a manifold N, leaving the symbol x to denote the n-vector formed by the local coodinates of the point p in some coordinate chart.

Example 5.1. The most common example in which such a situation occurs is the one describing the control of the orientation of a rigid body around its center of mass, for instance the attitude of a spacecraft. Let $e = (e_1, e_2, e_3)$ denote an inertially fixed triplet of orthonormal vectors (the *reference frame*) and let $a = (a_1, a_2, a_3)$ denote a triplet of orthonormal vectors fixed in the body (the *body frame*), as depicted in Fig. 1.2.

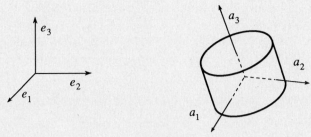

Fig. 1.2

A possible way of defining the attitude of the rigid body in space is to consider the angles between the vectors of a and the vectors of e. Let R be a 3×3 matrix whose element r_{ij} is the cosine of the angle between the vectors a_i and e_j. By definition, then, the elements on the i-th row of R are exactly the coordinates of the vector a_i with respect to the reference frame identified by the triplet e. Since the two triplets are both orthonormal, the matrix R is such that

$$RR^T = I$$

or, what is the same, $R^{-1} = R^T$ (that is, R is an orthogonal matrix); in particular, $\det(R) = 1$. The matrix R completely identifies the orientation of the body frame with respect to the fixed reference frame, and therefore it is possible—and convenient—to use R in order to describe the attitude of the body in space. We shall illustrate now how the equations of the motion of the rigid body and its control can be derived accordingly.

First of all, note that if x_e and x denote the coordinates of an arbitrary vector with respect to e and, respectively, to a, these two sets of coordinates are related by the linear transformation

$$x = Rx_e.$$

Moreover, note that if one associates with a vector

$$w = \text{col}(w_1, w_2, w_3)$$

the 3×3 matrix

$$S(w) = \begin{bmatrix} 0 & w_3 & -w_2 \\ -w_3 & 0 & w_1 \\ w_2 & -w_1 & 0 \end{bmatrix}$$

the usual "vector" product between w and v can be written in the form $w \times v = -S(w)v$.

Suppose the body is rotating with respect to the inertial frame. Let $R(t)$ denote the value at time t of the matrix R describing its attitude, and let $\omega(t)$ (respectively $\omega_e(t)$) denote its angular velocity in the a frame (respectively in the e frame). Consider a point fixed in the body and let x denote its coordinates with respect to the body frame a. Since this frame is fixed with the body, then x is a constant with respect to the time and $dx/dt = 0$. On the other hand the coordinates $x_e(t)$ of the same point with respect to the reference frame e satisfy

$$\dot{x}_e(t) = -S(\omega_e(t))x_e(t).$$

Differentiating $x(t) = R(t)x_e(t)$, and using the identity $RS(\omega_e)x_e = S(\omega)x$, yields

$$0 = \dot{R}x_e + R\dot{x}_e = \dot{R}R^T x - RS(\omega_e)x_e = \dot{R}R^T x - S(\omega)x$$

and, because of the arbitrariness of x,

$$\dot{R}(t) = S(\omega(t))R(t). \qquad (5.2a)$$

This equation, which espresses the relation between the attitude R of the body and its angular velocity (the latter being expressed with respect to a coordinate frame fixed with the body), is commonly known as *kinematic equation*.

Suppose now the body is subject to external torques. If h_e denotes the coordinates of the angular momentum and T_e those of the external torque with respect to the reference frame e, the momentum balance equation yields

$$\dot{h}_e(t) = T_e(t)$$

On the other hand, in the body frame a, the angular momentum can be expressed as

$$h(t) = J\omega(t)$$

where J is a matrix of constants, called the *inertia matrix*. Combining these relations one obtains

$$J\dot{\omega} = \dot{h} = \dot{R}h_e + R\dot{h}_e = S(\omega)Rh_e + RT_e = S(\omega)J\omega + T$$

where $T = RT_e$ is the expression of the external torque in the body frame a. The equation thus obtained, namely

$$J\dot{\omega}(t) = S(\omega(t))J\omega(t) + T(t) \qquad (5.2b)$$

is commonly known as *dynamic equation*.

The equations (5.2a) and (5.2b), describing the control of the attitude of the rigid body, are exactly of the form (5.1a), with

$$p = (R, \omega).$$

In particular, note that R is not *any* 3×3 matrix, but is an *orthogonal* matrix, namely a matrix satisfying $RR^T = I$ (and $\det(R) = 1$). Thus, the natural state space for the system defined by (5.2a)-(5.2b) is not—as one might think just counting the number of equations—$\mathbf{R}^{12}$, but a more abstract set, namely the set of all pairs (R, ω) where R belongs to the set of all orthogonal 3×3 matrices (with determinant equal to 1) and ω belongs to $\mathbf{R}^3$.

The subset of $\mathbf{R}^{3\times 3}$ in which R ranges, namely the set of all 3×3 matrices satisfying $RR^T = I$ and $\det(R) = 1$, is an embedded submanifold of $\mathbf{R}^{3\times 3}$, of dimension 3. In fact, the orthogonality condition $RR^T = I$ can be expressed in the form of 6 equalities

$$\sum_{k=1}^{3} r_{ik}r_{jk} - \delta_{ij} = 0 \qquad 1 \le i \le j \le 3$$

and it is possible to show that the 6 functions on the left hand side of this equality have linearly independent differentials for each nonsingular R (thus, in particular, for any R such that $RR^T = I$). Thus, the set of matrices satisfying this equalities is an embedded 3-dimensional submanifold of $\mathbf{R}^{3\times 3}$, called the *orthogonal group* and noted $O(3)$. Any matrix such that $RR^T = I$ has a determinant which is equal either to 1 or to -1, and therefore $O(3)$ consists of two connected components. The connected component of $O(3)$ in which $\det(R) = 1$ is called the *special orthogonal group* (in $\mathbf{R}^{3\times 3}$) and is denoted by $SO(3)$.

We can conclude that the natural state space of (5.2) is the 6-dimensional smooth manifold

$$N = SO(3) \times \mathbf{R}^3.$$

This is a 6-dimensional smooth manifold, which—however—is *not* diffeomorphic to $\mathbf{R}^6$ (because $SO(3)$ is not diffeomorphic to $\mathbf{R}^3$). •

We begin by showing how the notion of smooth distribution can be rigorously defined in a coordinate-free setting. For, recall that the set of all smooth vector fields defined on N, noted $V(N)$, can be given different algebraic structures. It can be given the structure of a *vector space* over the set R of real numbers, the structure of a *Lie Algebra* (the product of vector fields f_1 and f_2 being defined by their Lie bracket $[f_1, f_2]$) and, also, of a *module* over $C^\infty(N)$, the ring of all smooth real-valued functions defined on N. In the latter structure, the addition $f_1 + f_2$ of vector fields f_1 and f_2 is defined pointwise, i.e. as

$$(f_1 + f_2)(p) = f_1(p) + f_2(p)$$

at each point p of N, and so is the product cf of a vector field f by an element c of $C^\infty(N)$, i.e.

$$(cf)(p) = c(p)f(p)$$

Suppose Δ is a mapping which assigns to each point p of N a subspace, noted

$\Delta(p)$, of the tangent space T_pN to N at p. With Δ it is possible to associate a submodule of $V(N)$, noted $\mathcal{M}_\Delta$, defined as the set of all vector fields in $V(N)$ that pointwise take values in $\Delta(p)$, i.e.

$$\mathcal{M}_\Delta = \{f \in V(N) : f(p) \in \Delta(p) \text{ for all } p \in N\}$$

This set by construction is a *submodule* of $V(N)$. Note, however, that there may be many submodules of $V(N)$ whose vector fields span $\Delta(p)$ at each p; the submodule $\mathcal{M}_\Delta$ thus defined is the largest of them, in the sense that contains any submodule of $V(N)$ consisting of vector fields which span $\Delta(p)$ at each p.

Example 5.2. Suppose $N = \mathbf{R}$, and let Δ be defined in the following way

$$\Delta(x) = 0 \qquad \text{at } x = 0$$
$$\Delta(x) = T_xR \qquad \text{at } x \neq 0.$$

The submodule $\mathcal{M}_\Delta$ is clearly the set of all vector fields of the form

$$f(x) = c(x)\frac{\partial}{\partial x}$$

where c is any element of $C^\infty(\mathbf{R})$ such that $c(0) = 0$. The set $\mathcal{M}'$ of all vector fields of the form

$$f(x) = c(x)x^2\frac{\partial}{\partial x}$$

where c is any element of $C^\infty(\mathbf{R})$, is by construction a submodule of $V(\mathbf{R})$, and its vector fields span Δ at each x. However, $\mathcal{M}'$ does not coincide with $\mathcal{M}_\Delta$, because for instance

$$x\frac{\partial}{\partial x} \notin \mathcal{M}'.$$

In fact, the smooth function x cannot be represented in the form $x = c(x)x^2$ with smooth $c(x)$. •

Conversely, with any submodule $\mathcal{M}$ of $V(N)$, one can associate an assignment, noted $\Delta_\mathcal{M}$, of a subspace of the tangent space T_pN with each point p of N, defining the value of $\Delta_\mathcal{M}$ at p as the set of all the values assumed at p by the vector fields of $\mathcal{M}$, i.e. setting

$$(\Delta_\mathcal{M})(p) = \{v \in T_pN : v = f(p) \text{ with } f \in \mathcal{M}\}$$

This argument shows how two objects of interest: a mapping which assigns to each point p of N a subspace of T_pN and a submodule of $V(N)$, can be related. For consistency reasons, it is desirable that the submodule associated with the mapping $\Delta_\mathcal{M}$ be the module $\mathcal{M}$ itself. For this to be true, it is necessary and sufficient that the submodule $\mathcal{M}$ has the property that, if f is any smooth vector field of $V(N)$ which is pointwise in $\Delta_{\mathcal{M}}$, then f is a vector field of $\mathcal{M}$. If this is the case, the submodule $\mathcal{M}$ is said to be *complete*.

A complete submodule of $V(N)$ is the object that, in a global and coordinate-free setting, replaces the intuitive notion of a smooth distribution introduced in Section 2. Of course, the mapping $\Delta_{\mathcal{M}}$ associated with $\mathcal{M}$ has, locally, smoothness properties which agree with the ones considered so far, i. e. can be (locally) described as the span of a finite set of smooth vector fields.

A similar point of view leads to a coordinate-free notion of codistribution. The latter can be defined, in fact, as a submodule of the module $V^\star(N)$ of all smooth covector fields of N, satisfying a completeness requirement corresponding to the one just discussed.

To the objects thus defined it is possible to extend, quite easily, all the properties discussed in Section 3. There is, however, a specific point that requires a little extra attention: the difference between an *involutive distribution* and a *Lie subalgebra* of the Lie algebra $V(N)$. We recall that an involutive distribution is a distribution Δ having the property that the Lie bracket of any two vector fields of Δ is again a vector field of Δ. In the present setting, we shall say that an involutive distribution is a complete submodule $\mathcal{M}$ having the property that the Lie bracket of any two vector fields of $\mathcal{M}$ is again in $\mathcal{M}$. Since a Lie subalgebra of $V(N)$ is a collection of vector fields having the same property of closure under Lie bracket, one could be led to assimilate the two objects. However, this is not possible, as for instance the following simple example shows.

Example 5.3. Consider the two vector fields of $\mathbf{R}^2$

$$f_1(x) = \frac{\partial}{\partial x_1} \qquad f_2(x) = c(x_1)\frac{\partial}{\partial x_2}$$

where $c(x_1)$ is a C^∞—but not analytic—function vanishing at 0 with all its derivatives and nonzero for $x_1 \neq 0$.

It is easy to check that the Lie algebra $\mathcal{L}\{f_1, f_2\}$ generated by f_1 and f_2 consists of all vector fields having the form

$$\tau(x) = a\frac{\partial}{\partial x_1} + \left(b_0 c(x_1) + b_1\frac{dc}{dx_1} + \cdots + b_k\frac{d^k c}{dx_1^k}\right)\frac{\partial}{\partial x_2}$$

where k is any integer and $a, b_0, \ldots, b_k$ real numbers. Thus, the subspace $\Delta(x)$ of $T_x\mathbf{R}^2$ spanned by the vectors of $\mathcal{L}\{f_1, f_2\}$ at a point x has the following description

$$\Delta(x) = T_x\mathbf{R} \qquad \text{if } x_1 \neq 0$$

$$\Delta(x) = \text{span}\left\{\frac{\partial}{\partial x_1}\right\} \qquad \text{if } x_1 = 0.$$

However, the submodule $\mathcal{M}_\Delta$ consisting of all vector fields of Δ is not involutive, because the Lie bracket of the vector fields f_1 and

$$f_3(x) = x_1 \frac{\partial}{\partial x_2}$$

(both are pointwise in Δ, but f_3 is not in $\mathcal{L}\{f_1, f_2\}$) is the vector field

$$[f_1, f_3] = \frac{\partial}{\partial x_2}$$

which does not belong to Δ at $x_1 = 0$. •

We discuss now an important interpretation of the notion of complete integrability of a distribution. In the previous Section, we defined a nonsingular distribution Δ, of dimension d, to be completely integrable if its annihilator $\Delta^\perp$ is locally spanned by $n - d$ covector fields which are differentials of functions. This definition is still meaningful in a coordinate-free setting, where it requires, for each p° of N, the existence of a neighborhood U° of p° and $n - d$ real-valued smooth functions $\lambda_1, \ldots, \lambda_{n-d}$ defined on U°, such that

$$\text{span}\{d\lambda_1(p), \ldots, d\lambda_{n-d}(p)\} = \Delta^\perp(p) \tag{5.3}$$

for all p in U°. Note that the definition thus given—although in coordinate-free terms—specifies only *local* properties of a distribution. We shall see in the next Chapter a global version of the notion of complete integrability.

By definition, the $n - d$ differentials of the functions $\lambda_1, \ldots, \lambda_{n-d}$ are linearly independent at each point p of the set U° where they are defined. Thus, there exist a neighborhood $U \subset U^\circ$ of p°, and functions $\phi_1, \ldots, \phi_d$ defined on U, that, together with

$$\phi_{d+1} = \lambda_1, \ldots, \phi_n = \lambda_{n-d}$$

define a coordinate chart at p°. Without loss of generality, we may suppose that this is a cubic coordinate chart centered at p°, i.e. that $\phi_i(p^\circ) = 0$ for all $1 \leq i \leq n$ and $\phi_i(U)$ is an open interval of the form $\{x \in \mathbf{R} : |x| < K\}$.

Let $\phi_i = \phi_i(p)$, $1 \leq i \leq n$, denote the i-th coordinate of the point p and recall, at each $p \in U$, the choice of these coordinates induces the choice of a basis in the tangent space $T_p N$, namely

$$\left(\frac{\partial}{\partial \phi_1}\right)_p, \ldots, \left(\frac{\partial}{\partial \phi_n}\right)_p$$

and that of a basis in the cotangent space $T_p^* N$, namely

$$(d\phi_1)_p, \ldots, (d\phi_n)_p.$$

These two bases are dual, i.e. satisfy

$$\langle (d\phi_i)_p, (\frac{\partial}{\partial \phi_j})_p \rangle = \delta_{ij}.$$

The property (5.3) says that the last $n - d$ covectors of the basis of $T_p^* N$ are a basis of the codistribution $\Delta^\perp(p)$, at each $p \in U$. Thus, from the relation of duality, we can conclude that the first d vectors of the basis of $T_p N$ are a basis of $\Delta(p)$, at each $p \in U$. From this argument, one can deduce an alternative characterization of the property, for a distribution, of being completely integrable. A nonsingular distribution Δ, of dimension d, is completely integrable if, at each $p^\circ \in N$, there exists a cubic coordinate chart (U, ϕ), with coordinate functions $\phi_1, \ldots, \phi_n$ such that

$$\Delta(p) = \text{span}\left\{ (\frac{\partial}{\partial \phi_1})_p, \ldots, (\frac{\partial}{\partial \phi_d})_p \right\}$$

for all $p \in U$.

This characterization lends itself to an interesting interpretation. Let p be any point of the cubic coordinate neighborhood U and consider the *slice* of U passing through p consisting of all points whose last $n - d$ coordinates are held constant, i.e. the subset of U

$$S_p = \{q \in U : \phi_{d+1}(q) = \phi_{d+1}(p), \ldots, \phi_n(q) = \phi_n(p)\}. \tag{5.4}$$

This subset, which is a smooth submanifold of U, of dimension d, has the property of having—at each point q—a tangent space that, by construction, is exactly the subspace $\Delta(q)$ of $T_q N$ (Fig. 1.3).

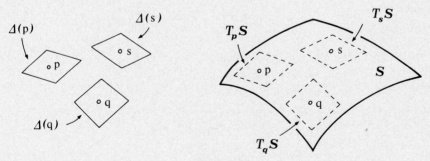

Fig. 1.3

Note that the coordinate neighborhood U is *partitioned* into slices of the form (5.4). Thus, a nonsingular and completely integrable distribution Δ induces, at each point p°, a local partition of N into submanifolds, each one having, at any point, a tangent space which—viewed as a subspace of the tangent space to

N—coincides with the value of Δ at this point.

1.6 Invariant Distributions

The notion of a distribution invariant under a vector field plays, in the theory of nonlinear control systems, a role similar to the one played in the theory of linear systems by the notion of subspace invariant under a linear mapping. A distribution Δ is said to be *invariant* under a vector field f if the Lie bracket $[f, \tau]$ of f with every vector field τ of Δ is again a vector field of Δ, i.e. if

$$\tau \in \Delta \Rightarrow [f, \tau] \in \Delta.$$

In order to represent this condition in a more condensed form, it is convenient to introduce the following notation. We let $[f, \Delta]$ denote the distribution spanned by all the vector fields of the form $[f, \tau]$, with $\tau \in \Delta$, i.e. we set

$$[f, \Delta] = \mathrm{span}\{[f, \tau], \tau \in \Delta\}.$$

Using this notation, it is possible to say that a distribution Δ is invariant under a vector field f if

$$[f, \Delta] \subset \Delta.$$

Remark 6.1. Suppose the distribution Δ is nonsingular (and d-dimensional). Then, using Lemma 3.2, it is possible to express—at least locally—every vector field τ of Δ in the form

$$\tau(x) = \sum_{i=1}^{d} c_i(x)\tau_i(x)$$

where $\tau_1, \ldots, \tau_d$ are vector fields locally spanning Δ. It is easy to see that Δ is invariant under f if and only if

$$[f, \tau_i] \in \Delta \qquad \text{for all } 1 \leq i \leq d.$$

The necessity follows trivially from the fact that $\tau_1, \ldots, \tau_d$ are vector fields of Δ. For the sufficiency, consider the expansion (see (2.5))

$$[f, \tau] = \sum_{i=1}^{d} c_i[f, \tau_i] + \sum_{i=1}^{d} (L_f c_i)\tau_i$$

and note that all the vector fields on the right-hand side are vector fields of Δ.

The previous expression in particular shows that

$$[f, \Delta] \supset \mathrm{span}\{[f, \tau_1], \ldots, [f, \tau_d]\}$$

but note that the distribution on the left-hand side may, in general, be unequal to the one on the right-hand side. However, by adding to both sides the distribution Δ, it is easy to deduce—again from the previous expression—that

$$\Delta + [f, \Delta] = \Delta + \text{span}\{[f, \tau_1], \ldots, [f, \tau_d]\}$$

i.e. that

$$\Delta + [f, \Delta] = \text{span}\{\tau_1, \ldots, \tau_d, [f, \tau_1], \ldots, [f, \tau_d]\}.$$

This property will be utilized in some later developments. •

Remark 6.2. The notion of invariance of a distribution under a vector field incorporates, in some sense, the notion of invariance of a subspace under a linear mapping. In order to see this, consider a subspace V of $\mathbf{R}^n$ invariant under a linear mapping A, i.e. such that $AV \subset V$. Define a distribution Δ_V as

$$\Delta_V(x) = V$$

at each $x \in \mathbf{R}^n$, and a (linear) vector field f_A as

$$f_A(x) = Ax$$

at each $x \in \mathbf{R}^n$. It is easy to prove that the distribution Δ_V is invariant under the vector field f_A, in the sense of the previous definition. On the basis of the previous Remark, all we have to show is that, if $\tau_1, \ldots, \tau_d$ is a set of vector fields locally spanning Δ_V, then $[f, \tau_1], \ldots, [f, \tau_d]$ are again vector fields of Δ_V. To this end, note that if $v_1, \ldots, v_d$ is a basis of V, the vector fields defined as

$$\tau_i(x) = v_i \qquad 1 \leq i \leq d$$

at each $x \in \mathbf{R}^n, 1 \leq i \leq d$, locally span Δ_V. The Lie bracket $[f_A, \tau_i]$ has the expression

$$[f_A, \tau_i](x) = \frac{\partial \tau_i}{\partial x} f_A - \frac{\partial f_A}{\partial x} \tau_i = -Av_i$$

at each $x \in \mathbf{R}^n$. Since, by assumption, Av_i is a vector of V, we conclude that $[f_A, \tau_i]$ is a vector field of Δ_V. •

The notion of invariance under a vector field is particularly useful when referred to completely integrable distributions, because it provides a way of simplifying the local representation of the given vector field.

Lemma 6.3. Let Δ be a nonsingular involutive distribution of dimension d and suppose that Δ is invariant under the vector field f. Then at each point x° there exist a neighborhood U° of x° and a coordinates trasformation $z = \Phi(x)$

defined on U°, in which the vector field f is represented by a vector of the form

$$\bar{f}(z) = \begin{bmatrix} f_1(z_1, \ldots, z_d, z_{d+1}, \ldots, z_n) \\ \cdots \\ f_d(z_1, \ldots, z_d, z_{d+1}, \ldots, z_n) \\ f_{d+1}(z_{d+1}, \ldots, z_n) \\ \cdots \\ f_n(z_{d+1}, \ldots, z_n) \end{bmatrix}. \tag{6.1}$$

Proof. The distribution Δ, being nonsingular and involutive, is also integrable. Therefore, at each point x° there exists a neighborhood U° and a coordinates transformation $z = \Phi(x)$ defined on U° with the property that

$$\mathrm{span}\{d\phi_{d+1}, \ldots, d\phi_n\} = \Delta^\perp.$$

Let $\bar{f}(z)$ denote the representation of the vector field f in the new coordinates. Consider now a vector field

$$\tau(z) = \mathrm{col}(\tau_1(z), \ldots, \tau_n(z))$$

and suppose

$$\tau_k(z) = 0 \qquad \text{for } k \neq i$$
$$\tau_k(z) = 1 \qquad \text{for } k = i.$$

Then

$$[\bar{f}, \tau] = -\frac{\partial \bar{f}}{\partial z}\tau = -\frac{\partial \bar{f}}{\partial z_i}.$$

Recall that (see (4.8)), in the coordinates just chosen, every vector field of Δ is characterized by the property that the last $n - d$ components are vanishing. Thus, if $1 \leq i \leq d$, the vector field τ belongs to Δ. Since Δ is invariant under f, $[f, \tau]$ also belongs to Δ, i.e. its last $n - d$ components must vanish. This yields

$$\frac{\partial \bar{f}_k}{\partial z_i} = 0$$

for all $d + 1 \leq k \leq n, 1 \leq i \leq d$, and proves the assertion. ●

The representation (6.1) is particularly useful in interpreting the notion of invariance of a distribution from a system-theoretic point of view. For, suppose a dynamical system of the form

$$\dot{x} = f(x) \tag{6.2}$$

is given and let Δ be a nonsingular and involutive distribution, invariant under the vector field f. Choose the coordinates as described in the Lemma 6.3 and set

$$\zeta_1 = (z_1, \ldots, z_d)$$
$$\zeta_2 = (z_{d+1}, \ldots, z_n).$$

Then, the system in question is represented by equations of the form

$$\dot{\zeta}_1 = f_1(\zeta_1, \zeta_2) \qquad\qquad (6.3a)$$
$$\dot{\zeta}_2 = f_2(\zeta_2) \qquad\qquad (6.3b)$$

that is exhibits, in the new coordinates, an internal *triangular decomposition*. The block diagram of Fig. 1.4 illustrates this decomposition.

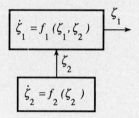

Fig. 1.4

Remark 6.4. Note that, if the vector field f is a *linear* vector field, i.e. if $f(x) = Ax$, the special form on the right-hand side of (6.3) reduces to

$$f(\zeta_1, \zeta_2) = \begin{bmatrix} A_{11} & A_{12} \\ 0 & A_{22} \end{bmatrix} \begin{bmatrix} \zeta_1 \\ \zeta_2 \end{bmatrix}.$$

Thus, we may interpret Lemma 6.3 as an extension of the well known—already recalled in Section **1.1**—result according to which if a subspace V of $\mathbf{R}^n$ is invariant under a matrix A, then choosing an appropriate (linear) change of coordinates the matrix itself can be put into a block upper-triangular form. •

Geometrically, the decomposition described by (6.3) can be interpreted in the following way (see also the end of Section **1.5**). Suppose, without loss of generality, that $\Phi(x^\circ) = 0$ (for, if $\Phi(x^\circ)$ is nonzero, consider the "translated" transformation $z = \Phi'(x) = \Phi(x) - \Phi(x^\circ)$ which still satisfies the requirements of Lemma 6.3 and is such that $\Phi'(x^\circ) = 0$). Suppose also, again without loss of generality, that the neighborhood U° on which the transformation is defined is a neighborhood of the form

$$U^\circ = \{x \in \mathbf{R}^n : |z_i(x)| < \varepsilon\}$$

where ε is a suitable small number. Such a neighborhood U° is called a *cubic neighborhood centered* at x° (Fig. 1.5a). Let x be a point of U°, and consider the subset of U° consisting of all points whose last $n - d$ coordinates (namely the ζ_2 coordinates) coincide with those of x, i.e. the set

$$S_x = \{x' \in U^\circ : \zeta_2(x') = \zeta_2(x)\}. \tag{6.4}$$

This set is called a *slice* of the neighborhood U° (Fig. 1.5b). Note that any set of this type, being the locus (of points of U°) where the smooth coordinates functions $z_{d+1}(x), \ldots, z_n(x)$ assume fixed values, can be regarded as a *smooth surface*, of dimension d. Note also that the collection of all subsets of U° having this form defines a *partition* of U° (Fig. 1.5c).

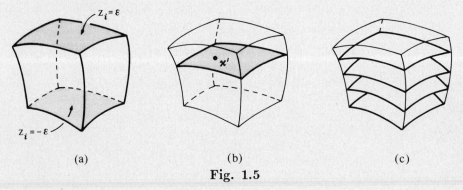

(a)   (b)   (c)

Fig. 1.5

Suppose now that x^a and x^b are two points of U° satisfying the condition

$$\zeta_2(x^a) = \zeta_2(x^b) \tag{6.5}$$

i.e. having the same ζ_2 coordinates but possibly different ζ_1 coordinates. Let $x^a(t)$ and $x^b(t)$ denote the integral curves of the equation (6.2) starting respectively from x^a and x^b at time $t = 0$. Recalling that in the new coordinates the equation (6.2) exhibits the decomposition (6.3), it is easy to conclude that, as far as $x^a(t)$ and $x^b(t)$ are contained in the domain U° of the coordinates transformation $z = \Phi(x)$,

$$\zeta_2(x^a(t)) = \zeta_2(x^b(t)) \tag{6.6}$$

at any time t. As a matter of fact, $\zeta_2(x^a(t))$ and $\zeta_2(x^b(t))$ are both solutions of the same differential equation—namely, the equation (6.3b)—and both satisfy the same initial condition, because

$$\zeta_2(x^a(0)) = \zeta_2(x^a) = \zeta_2(x^b) = \zeta_2(x^b(0)).$$

Two initial conditions x^a and x^b satifying (6.5) belong, by definition, to a slice of the form (6.4). As we have just seen, the two corresponding trajectories $x^a(t)$ and $x^b(t)$ of (6.2) necessarily satisfy (6.6), i.e. at any time t belong necessarily to a slice of the form (6.4). Thus, we can conclude that the flow of (6.2) *carries slices* (of the form (6.4)) *into slices* (Fig. 1.6).

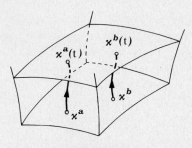

Fig. 1.6

Example 6.1. Consider the 2-dimensional distribution

$$\Delta = \operatorname{span}\{v_1, v_2\}$$

with

$$v_1 = \begin{bmatrix} 1 \\ 0 \\ 0 \\ x_2 \end{bmatrix} \qquad v_2 = \begin{bmatrix} 0 \\ 1 \\ 0 \\ x_1 \end{bmatrix}$$

and the vector field

$$f = \begin{bmatrix} x_2 \\ x_3 \\ x_3 x_4 - x_1 x_2 x_3 \\ \sin x_3 + x_2^2 + x_1 x_3 \end{bmatrix}.$$

A simple calculation shows that

$$[v_1, v_2] = 0$$

and therefore (Remark 3.7) the distribution Δ is involutive. Moreover, since

$$[f, v_1] = 0 \qquad [f, v_2] = -v_1$$

this distribution is also invariant under the vector field f (Remark 6.1).

By Frobenius' Theorem, in a neighborhood of any point x^o there exist functions $\lambda_1(x), \lambda_2(x)$ such that

$$\text{span}\{d\lambda_1, d\lambda_2\} = \Delta^\perp.$$

One can easily verify that, for instance, the functions

$$\lambda_1(x) = x_3$$
$$\lambda_2(x) = -x_1 x_2 + x_4$$

whose differentials have the form

$$d\lambda_1 = [0 \quad 0 \quad 1 \quad 0]$$
$$d\lambda_2 = [-x_2 \quad -x_1 \quad 0 \quad 1]$$

satisfy this condition.

As described in the proof of Lemma, define new (local) coordinates $z_i = \phi_i(x), 1 \le i \le 4$, choosing

$$\phi_3(x) = \lambda_1(x) \qquad \phi_4(x) = \lambda_2(x)$$

and completing, e.g., the set of new coordinates functions with

$$\phi_1(x) = x_1 \qquad \phi_2(x) = x_2.$$

In the new coordinates, the vector field f assumes the form

$$f(z) = \left[\frac{\partial \Phi}{\partial x} f(x)\right]_{x=\Phi^{-1}(z)} = \begin{bmatrix} z_2 \\ z_3 \\ z_3 z_4 \\ \sin z_3 \end{bmatrix}$$

i.e. the form indicated by (6.1), with $\zeta_1 = (z_1, z_2), \zeta_2 = (z_3, z_4)$. •

We discuss now some additional properties related to the notion of invariant distribution, that shall be sometimes used in the sequel.

Lemma 6.5. Let Δ be a distribution invariant under the vector fields f_1 and f_2. Then Δ is also invariant under the vector field $[f_1, f_2]$.

Proof. Suppose τ is a vector field in Δ. From the Jacobi identity, we get

$$[[f_1, f_2], \tau] = [f_1, [f_2, \tau]] - [f_2, [f_1, \tau]].$$

By assumption, $[f_2, \tau] \in \Delta$ and so is $[f_1, [f_2, \tau]]$. For the very same reason

$[f_2, [f_1, \tau]] \in \Delta$, and thus, from the above equality we conclude that $[[f_1, f_2], \tau] \in \Delta$. •

Remark 6.6. Note that the notion of invariance under a given vector field f can be also extended to a (possibly) nonsmooth distribution Δ, by simply requiring that the Lie bracket $[f, \tau]$ of f with every *smooth* vector field τ of Δ be a vector field in Δ, i.e. that

$$[f, \mathrm{smt}(\Delta)] \subset \Delta.$$

Since $[f, \mathrm{smt}(\Delta)]$ is a smooth distribution, this is clearly equivalent to

$$[f, \mathrm{smt}(\Delta)] \subset \mathrm{smt}(\Delta)$$

i.e. to the invariance under f of $\mathrm{smt}(\Delta)$. •

When dealing with codistributions, one can as well introduce the notion of invariance under a vector field in the following way. A codistribution Ω is said to be *invariant* under the vector field f if the derivative $L_f\omega$ of any covector field ω of Ω is again a covector field of Ω, i.e. if

$$\omega \in \Omega \Rightarrow L_f\omega \in \Omega.$$

Using the notation

$$L_f\Omega = \mathrm{span}\{L_f\omega : \omega \in \Omega\}$$

this condition can be rewritten in the form

$$L_f\Omega \subset \Omega.$$

It is easy to prove that the notion thus introduced is the dual version of the notion of invariance of a distribution, as expressed by the following statement.

Lemma 6.7. If a smooth distribution Δ is invariant under the vector field f, then the codistribution $\Omega = \Delta^{\perp}$ is also invariant under f. If a smooth codistribution Ω is invariant under the vector field f, then the distribution $\Delta = \Omega^{\perp}$ is also invariant under f.

Proof. Suppose Δ is invariant under f and let τ be any vector field of Δ. Then $[f, \tau] \in \Delta$. Let ω be any covector field in Ω. Then, by definition

$$\langle \omega, \tau \rangle = 0$$

and also

$$\langle \omega, [f, \tau] \rangle = 0.$$

The identity

$$\langle L_f \omega, \tau \rangle = L_f \langle \omega, \tau \rangle - \langle \omega, [f, \tau] \rangle.$$

yields

$$\langle L_f \omega, \tau \rangle = 0$$

Since Δ is a smooth distribution, given any point x° and any vector v in $\Delta(x^\circ)$, we may find a smooth vector field τ in Δ with the property that $\tau(x^\circ) = v$. Thus, the previous equality shows that

$$\langle L_f \omega(x^\circ), v \rangle = 0$$

for all $v \in \Delta(x^\circ)$, i.e. that $L_f \omega(x^\circ) \in \Omega(x^\circ)$. This proves that $L_f \omega$ is a covector field in Ω, i.e. that Ω is invariant under f. The second part of the statement is proved in the same way. •

Note that, in the previous Lemma, first part, we don't need to assume that the annihilator $\Delta^\perp$ of Δ is smooth nor, in the second part, that the annihilator $\Omega^\perp$ of Ω is smooth. However, if both $\Delta^\perp$ and Δ are smooth, we conclude from the Lemma that the invariance of Δ under f implies and is implied by that of $\Delta^\perp$ under the same vector field f. In view of Lemma 3.10 this is true, in particular, whenever Δ is nonsingular.

Remark 6.8. As an exercise of application of the notion of invariance of a codistribution, and of the previous Lemma, we suggest an alternative proof of Lemma 6.3. First of all, note that if new coordinates are choosen as indicated at the beginning of the proof of Lemma 6.3, any covector field ω of $\Delta^\perp$ has a representation of the form

$$\omega(z) = [0 \ \ldots \ 0 \ \omega_{d+1}(z) \ \ldots \ \omega_n(z)] \tag{6.7}$$

(this is simply because, in these coordinates, Δ is spanned by vectors of the form (4.8)). Observe now that, by construction, the expression of the functions $\phi_1, \ldots, \phi_n$ in the new coordinates is just

$$\phi_i(z) = z_i$$

for all $1 \leq i \leq n$. This implies

$$\frac{\partial \phi_i}{\partial z_j} = \delta_{ij}$$

and, therefore, in the new coordinates all the entries of the differential $d\phi_i$ are zero but the i-th one, which is equal to 1. As a consequence

$$L_f \phi_i(z) = \langle d\phi_i(z), f(z) \rangle = f_i(z)$$

and
$$L_f d\phi_i(z) = dL_f \phi_i(z) = df_i(z).$$

Since Δ is invariant under f and nonsingular, by Lemma 6.7 we have that $L_f \Delta^\perp \subset \Delta^\perp$ and this, since $d\phi_i \in \Delta^\perp$ for $d+1 \le i \le n$, yields

$$L_f d\phi_i = df_i \in \Delta^\perp.$$

The differential df_i, like any covector field of $\Delta^\perp$, must have the form (6.7) and this proves that

$$\frac{\partial f_i}{\partial z_j} = 0$$

if $1 \le j \le d$, $d+1 \le i \le n$.

Remark 6.9. Using (2.7) one can easily prove the dual version of Remark 6.1, namely the fact that if Ω is a nonsingular codistribution of dimension d, spanned by covector fields $\omega_1, \ldots, \omega_d$, then Ω is invariant under f if and only if $L_f \omega_i \in \Omega$ for all $1 \le i \le d$. One also finds that

$$\Omega + L_f \Omega = \text{span}\{\omega_1, \ldots, \omega_d, L_f \omega_1, \ldots, L_f \omega_d\}.$$

1.7 Local Decompositions of Control Systems

In this Section the notion of invariant distribution, and in particular Lemma 6.3, are used in order to obtain, for a nonlinear control system of the form (2.1), namely

$$\dot{x} = f(x) + \sum_{i=1}^{m} g_i(x) u_i \tag{7.1a}$$

$$y_i = h_i(x) \qquad 1 \le i \le p \tag{7.1b}$$

decompositions similar to those described at the beginning of the Chapter.

Proposition 7.1. Let Δ be a nonsingular involutive distribution of dimension d and assume that Δ is invariant under the vector fields $f, g_1, \ldots, g_m$. Moreover, suppose that the distribution $\text{span}\{g_1, \ldots, g_m\}$ is contained in Δ. Then, for each point x° it is possible to find a neighborhood U° of x° and a local coordinates transformation $z = \Phi(x)$ defined on U° such that, in the new coordinates, the control system (7.1a) is represented by equations of the form (see Fig. 1.7a)

$$\dot{\zeta}_1 = f_1(\zeta_1, \zeta_2) + \sum_{i=1}^{m} g_{1i}(\zeta_1, \zeta_2) u_i \tag{7.2a}$$

$$\dot{\zeta}_2 = f_2(\zeta_2) \tag{7.2b}$$

where $\zeta_1 = (z_1, \ldots, z_d)$ and $\zeta_2 = (z_{d+1}, \ldots, z_n)$.

Proof. From Lemma 6.3 it is known that there exists, around each x°, a local coordinates transformation yielding a representation of the form (6.1) for the vector fields $f, g_1, \ldots, g_m$. In the new coordinates the vector fields $g_1, \ldots, g_m$, that by assumption belong to Δ, are represented by vectors whose last $n - d$ components are vanishing (see (4.8)). This proves the Proposition. •

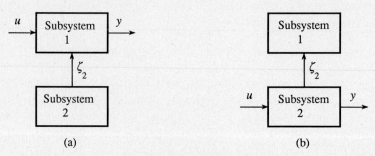

Fig. 1.7

Proposition 7.2.   Let Δ be a nonsingular involutive distribution of dimension d and assume that Δ is invariant under the vector fields $f, g_1, \ldots, g_m$. Moreover, suppose that the codistribution span$\{dh_1, \ldots, dh_p\}$ is contained in the codistribution $\Delta^\perp$. Then, for each point x° it is possible to find a neighborhood U° of x° and a local coordinates transformation $z = \Phi(x)$ defined on U° such that, in the new coordinates, the control system (7.1) is represented by equations of the form (see Fig. 1.7b)

$$\dot\zeta_1 = f_1(\zeta_1, \zeta_2) + \sum_{i=1}^{m} g_{1i}(\zeta_1, \zeta_2)u_i \tag{7.3a}$$

$$\dot\zeta_2 = f_2(\zeta_2) + \sum_{i=1}^{m} g_{2i}(\zeta_2)u_i \tag{7.3b}$$

$$y_i = h_i(\zeta_2) \tag{7.3c}$$

where $\zeta_1 = (z_1, \ldots, z_d)$ and $\zeta_2 = (z_{d+1}, \ldots, z_n)$.

Proof. As before, we know that there exists, around each x°, a coordinates transformation yielding a representation of the form (6.1) for the vector fields $f, g_1, \ldots, g_m$. In the new coordinates, the covector fields $dh_1, \ldots, dh_p$, that by assumption belong to $\Delta^\perp$, must have the form (6.7). Therefore

$$\frac{\partial h_i}{\partial z_j} = 0$$

for all $1 \le j \le d$, $1 \le i \le p$, and this completes the proof. •

The two local decompositions thus obtained are very useful in understanding the input-state and state-output behavior of the control system (7.1). Suppose that the inputs u_i are piecewise constant functions of time, i.e. that there exist real numbers $T_0 = 0 < T_1 < T_2 \ldots$ such that

$$u_i(t) = \bar{u}_i^k \qquad \text{for } T_k \leq t < T_{k+1}.$$

Then, on the time interval $[T_k, T_{k+1})$ the state of the system evolves along the integral curve of the vector field

$$f(x) + g_1(x)\bar{u}_1^k + \cdots + g_m(x)\bar{u}_m^k$$

passing through the point $x(T_k)$. For small values of t, the state $x(t)$ evolves in a neighborhood of the initial point $x(0)$.

Suppose now that the assumptions of the Proposition 7.1 are satisfied, choose a point x°, and set $x(0) = x^\circ$. For small values of t the state evolves on U° and we may use the equations (7.2) to interpret the behavior of the system. From these, we see that the ζ_2 coordinates of $x(t)$ are not affected by the input. In particular, if we denote by $x^\circ(T)$ the point of U° reached at time $t = T$ when no input is imposed (i.e. when $u(t) = 0$ for all $t \in [0, T]$), namely the point

$$x^\circ(T) = \Phi_T^f(x^\circ)$$

($\Phi_t^f(x)$ being the flow of the vector field f), we deduce from the structure of (7.2) that the set of points that can be reached at time T, starting from x°, is a set of points whose ζ_2 coordinates are necessarily equal to the ζ_2 coordinates of $x^\circ(T)$. In other words, *the set of points reachable at time T is necessarily a subset of a slice* of the form (6.4), exactly the one passing through the point $x^\circ(\text{T})$ (see Fig. 1.8).

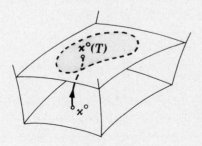

Fig. 1.8

Thus, we conclude that locally the system displays a behavior strictly analogous to the one described in Section 1. The state space can be partitioned into d-dimensional smooth surfaces (the slices of U°) and the states reachable at time T, along trajectories that stay in U° for all $t \in [0, T]$, lie inside the slice passing through the point $x^o(T)$ reached under zero input.

The Proposition 7.2 is useful in studying state-output interactions. Choose a point x^o and take two initial states x^a and x^b belonging to U^o with local coordinates (ζ_1^a, ζ_2^a) and (ζ_1^b, ζ_2^b) such that

$$\zeta_2^a = \zeta_2^b$$

i.e. two initial states belonging to the same *slice* of U^o. Let $x_u^a(t)$ and $x_u^b(t)$ denote the values of the states reached at time t, starting from x^a and x^b, under the action of the same input u. From the equation (7.3b) we see immediately that, if the input u is such that both $x_u^a(t)$ and $x_u^b(t)$ evolve on U^o, the ζ_2 coordinates of $x_u^a(t)$ and $x_u^b(t)$ are the same, no matter what input u we take. As a matter of fact, $\zeta_2(x_u^a(t))$ and $\zeta_2(x_u^b(t))$ are solutions of the same differential equation (the equation (7.3b)) with the same initial condition. If we take into account also the (7.3c), we see that

$$h_i(x_u^a(t)) = h_i(x_u^b(t))$$

for every input u. We may thus conclude that the two states x^a and x^b produce the same output under any input, i.e. are indistinguishable.

Again, we find that locally the state space may be partitioned into d-dimensional smooth surfaces (the slices of U^o) and that *all the initial states on the same slice are indistinguishable*, i.e. produce the same output under any input which keeps the state trajectory evolving on U^o.

In the next Sections we shall reach stronger conclusions, showing that if we add to the hypotheses contained in the Propositions 7.1 and 7.2 the further assumption that the distribution Δ is "minimal" (in the case of Proposition 7.1) or "maximal" (in the case of Proposition 7.2), then from the decompositions (7.2) and (7.3) one may obtain more precise informations about the states reachable from x^o and, respectively, indistinguishable from x^o.

1.8 Local Reachability

In the previous Section we have seen that if there is a nonsingular distribution Δ of dimension d with the properties that

(i) Δ is involutive

(ii) Δ contains the distribution $\text{span}\{g_1, \ldots, g_m\}$

(iii) Δ is invariant under the vector fields $f, g_1, \ldots, g_m$

then at each point $x^o \in U$ it is possible to find a coordinates transformation defined on a neighborhood U^o of x^o and a partition of U^o into slices of dimension d, such that the points reachable at some time T, starting from some initial state $x^o \in U^o$, along trajectories that stay in U^o for all $t \in [0, T]$, lie inside a slice of

U°. Now we want to investigate the actual "thickness" of the subset of points of a slice reached at time T.

The obvious suggestion that comes from the decomposition (7.2) is to look at the "minimal" distribution, if any, that satisfies (ii), (iii) and, then, to examine what can be said about the properties of points which belong to the same slice in the corresponding local decomposition of U. It turns out that this program can be carried out in a rather satisfactory way.

We need first some additional results on invariant distributions. If $\mathcal{D}$ is a family of distributions on U, we define the *smallest* or *minimal* element as the member of $\mathcal{D}$ (when it exists) which is contained in every other element of $\mathcal{D}$.

Lemma 8.1. Let Δ be a given smooth distribution and $\tau_1, \ldots, \tau_q$ a given set of vector fields. The family of all distributions which are invariant under $\tau_1, \ldots, \tau_q$ and contain Δ has a minimal element, which is a smooth distribution.

Proof. The family in question is clearly nonempty. If Δ_1 and Δ_2 are two elements of this family, then it is easily seen that their intersection $\Delta_1 \cap \Delta_2$ contains Δ and, being invariant under $\tau_1, \ldots, \tau_q$, is an element of the same family. This argument shows that the intersection $\hat{\Delta}$ of all elements in the family contains Δ, is invariant under $\tau_1, \ldots, \tau_q$ and is contained in any other element of the family. Thus is its minimal element. $\hat{\Delta}$ must be smooth because otherwise smt($\hat{\Delta}$) would be a smooth distribution containing Δ (because Δ is smooth by assumption), invariant under $\tau_1, \ldots, \tau_q$ (see Remark 6.6) and possibly contained in $\hat{\Delta}$. •

In what follows, the smallest distribution which contains Δ and is invariant under the vector fields $\tau_1, \ldots, \tau_q$ will be denoted by the symbol

$$\langle \tau_1, \ldots, \tau_q | \Delta \rangle.$$

While the existence of a minimal element in the family of distributions which satisfy (ii) and (iii) is always guaranteed, the nonsingularity requires some additional assumptions. We deal with the problem in the following way. Given a distribution Δ and a set $\tau_1, \ldots, \tau_q$ of vector fields we define the nondecreasing sequence of distributions

$$\Delta_0 = \Delta \tag{8.1a}$$

$$\Delta_k = \Delta_{k-1} + \sum_{i=1}^{q} [\tau_i, \Delta_{k-1}]. \tag{8.1b}$$

The sequence of distributions thus defined has the following property.

Lemma 8.2. The distributions Δ_k generated with the algorithm (8.1) are such that

$$\Delta_k \subset \langle \tau_1, \ldots, \tau_q | \Delta \rangle$$

for all k. If there exists an integer k^* such that $\Delta_{k^*} = \Delta_{k^*+1}$, then

$$\Delta_{k^*} = \langle \tau_1, \ldots, \tau_q | \Delta \rangle.$$

Proof. If Δ' is any distribution which contains Δ and is invariant under τ_i, then it is easy to see that $\Delta' \supset \Delta_k$ implies $\Delta' \supset \Delta_{k+1}$. For , we have (recall Remark 6.1)

$$\Delta_{k+1} = \Delta_k + \sum_{i=1}^{q} [\tau_i, \Delta_k] = \Delta_k + \sum_{i=1}^{q} \text{span}\{[\tau_i, \tau] : \tau \in \Delta_k\}$$

$$\subset \Delta_k + \sum_{i=1}^{q} \text{span}\{[\tau_i, \tau] : \tau \in \Delta'\} \subset \Delta'.$$

Since $\Delta' \supset \Delta_0$, by induction we see that $\Delta' \supset \Delta_k$ for all k. If $\Delta_{k^*} = \Delta_{k^*+1}$ for some k^*, we easily see that $\Delta_{k^*} \supset \Delta$ (by definition) and Δ_{k^*} is invariant under $\tau_1, \ldots, \tau_q$ (because $[\tau_i, \Delta_{k^*}] \subset \Delta_{k^*+1} = \Delta_{k^*}$ for all $1 \leq i \leq q$). Thus Δ_{k^*} must coincide with $\langle \tau_1, \ldots, \tau_q | \Delta \rangle$. ●

Remark 8.3. Before proceeding further with the analysis, we want to stress that the recursive construction indicated by (8.1) can be interpreted as a nonlinear analogue of the construction that, in a linear system

$$\dot{x} = Ax + Bu$$
$$y = Cx$$

ends up with the subspace

$$R = \text{Im}(B \ AB \ \ldots \ A^{n-1}B)$$

namely, the smallest subspace of $\mathbf{R}^n$ invariant under A and containing $\text{Im}(B)$. For, suppose the set $\tau_1, \ldots, \tau_q$ consists of only one vector field, namely τ, and set

$$\Delta_0(x) = \text{Im}(B)$$
$$\tau(x) = Ax$$

at each $x \in \mathbf{R}^n$. Observe that any vector field θ of Δ_0 can be locally expressed in the form (see Lemma 3.2)

$$\theta(x) = \sum_{i=1}^{m} c_i(x) b_i$$

where $b_1, \ldots, b_m$ are the columns of B. Thus, in view of a property illustrated in the Remark 6.1.

$$\Delta_1 = \Delta_0 + [\tau, \Delta_0] = \mathrm{span}\{b_1, \ldots, b_m, [\tau, b_1], \ldots, [\tau, b_m]\}.$$

Since

$$[\tau, b_i](x) = [Ax, b_i] = -\frac{\partial(Ax)}{\partial x}b_i = -Ab_i$$

we obtain

$$\Delta_1 = \mathrm{span}\{b_1, \ldots, b_m, Ab_1, \ldots, Ab_m\}$$

i.e.

$$\Delta_1(x) = \mathrm{Im}(B \ AB)$$

at each $x \in \mathbf{R}^n$. Continuing in the same way, we easily deduce that, for any $k \geq 1$,

$$\Delta_k(x) = \mathrm{Im}(B \ AB \ \ldots \ A^k B)$$

Each distribution of the sequence thus constructed is a constant distribution. Since $\Delta_{k+1} \subset \Delta_k$, a dimensionality argument proves that there exists an integer $k^\star < n$ with the property that $\Delta_{k^\star+1} = \Delta_{k^\star}$. Thus Δ_{n-1}, which is indeed the largest distribution of the sequence, by Lemma 8.2 is the smallest distribution invariant under the vector field Ax which contains the distribution $\mathrm{span}\{b_1, \ldots, b_m\}$. At each $x \in \mathbf{R}^n$, this distribution assumes the value

$$\Delta_{n-1}(x) = \mathrm{Im}(B \ AB \ \ldots A^{n-1} B)$$

i.e. that of the smallest subspace of $\mathbf{R}^n$ invariant under A which contains $\mathrm{Im}(B) = \mathrm{span}\{b_1, \ldots, b_m\}$. •

Remark 8.4. Note that, in general, the actual calculation of the distribution Δ_k generated by the algorithm (8.1) can still take advantage of the expression illustrated in the Remark 6.1. For, if Δ_{k-1} is nonsingular and spanned by a set of vector fields $\theta_1, \ldots, \theta_d$, then

$$\Delta_{k-1} + [\tau_i, \Delta_{k-1}] = \mathrm{span}\{\theta_1, \ldots, \theta_d, [\tau_i, \theta_1], \ldots, [\tau_i, \theta_d]\}$$

and therefore

$$\Delta_k = \mathrm{span}\{\theta_s, [\tau_i, \theta_s] : 1 \leq s \leq d, 1 \leq i \leq q\}. \ •$$

We return now to the analysis of the properties of the sequence of distributions generated by (8.1) which, in the nonlinear setting, is quite more elaborate than the one illustrated in the Remark 8.3. The increase of difficulty depends, among the other things, on the fact that—in view of the interest of using Lemma 7.1 for the purpose of obtaining a decomposition of the system—a *nonsingular and involutive* $\langle \tau_1, \ldots, \tau_q | \Delta \rangle$ is sought. First of all, we examine when the stopping condition identified in Lemma 8.2 can be met and then we discuss nonsingularity and involutivity. The simplest practical situation in which the algorithm (8.1) converges in a finite number of steps is when all the distributions of the sequence are nonsingular. In this case, in fact, since by construction

$$\dim \Delta_k \leq \dim \Delta_{k+1} \leq n$$

it is easily seen that there exists an integer $k^\star < n$ such that $\Delta_{k^\star} = \Delta_{k^\star+1}$.

If the distributions Δ_i are singular, one has the following weaker result.

Lemma 8.5. There exists an open and dense subset $U^\star$ of U with the property that at each point $x \in U^\star$

$$\langle \tau_1, \ldots, \tau_q | \Delta \rangle = \Delta_{n-1}(x).$$

Proof. Suppose V is an open set with the property that, for some $k^\star$, $\Delta_{k^\star}(x) = \Delta_{k^\star+1}(x)$ for all $x \in V$. Then, it is possible to show that $\langle \tau_1, \ldots, \tau_q | \Delta \rangle(x) = \Delta_{k^\star}(x)$ for all $x \in V$. For, we already know from Lemma 8.2 that $\langle \tau_1, \ldots, \tau_q | \Delta \rangle \supset \Delta_{k^\star}$. Suppose the inclusion is proper at some $\bar{x} \in V$ and define a new distribution $\bar{\Delta}$ by setting

$$\bar{\Delta}(x) = \Delta_{k^\star}(x) \qquad \text{if } x \in V$$

$$\bar{\Delta}(x) = \langle \tau_1, \ldots, \tau_q | \Delta \rangle(x) \qquad \text{if } x \notin V.$$

This distribution contains Δ and is invariant under $\tau_1, \ldots, \tau_q$. For, if τ is a vector field in $\bar{\Delta}$, then $[\tau_i, \tau] \in \langle \tau_1, \ldots, \tau_q | \Delta \rangle$ (because $\bar{\Delta} \subset \langle \tau_1, \ldots, \tau_q | \Delta \rangle$) and, moreover, $[\tau_i, \tau](x) \in \Delta_{k^\star}(x)$ for all $x \in V$ (because, in a neighborhood of x, $\tau \in \Delta_{k^\star}$ and $[\tau_i, \Delta_{k^\star}] \subset \Delta_{k^\star}$). Since $\bar{\Delta}$ is properly contained in $\langle \tau_1, \ldots, \tau_q | \Delta \rangle$, this would contradict the minimality of $\langle \tau_1, \ldots, \tau_q | \Delta \rangle$. Now let U_k be the set of regular points of Δ_k. This set is an open and dense subset of U (see Lemma 3.3) and so is the set $U^\star = U_0 \cap U_1 \cap \cdots \cap U_{n-1}$. In a neighborhood of every point $x \in U^\star$ the distributions $\Delta_0, \ldots, \Delta_{n-1}$ are nonsingular. This, together with the previous discussion and a dimensionality argument, shows that $\Delta_{n-1} = \langle \tau_1, \ldots, \tau_q | \Delta \rangle$ on $U^\star$ and completes the proof. •

We stress that the equality between Δ_{n-1} and $\langle \tau_1, \ldots, \tau_q | \Delta \rangle$ is true only at points of an open and dense subset of U, and *not* everywhere on U. A simple example in which there are points at which the equality is not true is the following one.

Example 8.1. Let $U = \mathbf{R}^2$, $q = 2$, and set

$$\Delta = \mathrm{span}\{\tau_1\} \qquad \tau_1(x) = \begin{bmatrix} 1 \\ x_1 \end{bmatrix} \qquad \tau_2(x) = \begin{bmatrix} x_2 \\ x_1. \end{bmatrix}$$

Then,

$$\Delta_{n-1} = \Delta_1 = \mathrm{span}\{\tau_1\} + \mathrm{span}\{[\tau_1, \tau_1], [\tau_1, \tau_2]\} = \mathrm{span}\{\tau_1, [\tau_1, \tau_2]\}.$$

Since

$$[\tau_1, \tau_2](x) = \begin{bmatrix} x_1 \\ 1 - x_2 \end{bmatrix}$$

we see that the distribution Δ_1 has dimension 2 at each point of the dense set

$$U^\star = \{x \in \mathbf{R}^2 : 1 - x_2 - x_1^2 \neq 0\}$$

Δ_1 is equal to $\langle \tau_1, \tau_2 | \Delta \rangle$ for all $x \in U^\star$. However $\Delta_1(x) \neq \langle \tau_1, \tau_2 | \Delta \rangle(x)$ at some $x \notin U^\star$. For, note that

$$[\tau_1, [\tau_1, \tau_2]] = \begin{bmatrix} 1 \\ -2x_1 \end{bmatrix}.$$

Thus, Δ_1 is not invariant under τ_1, because this vector is not in $\Delta_1(x)$ if x is such that $x_1 = 1$ and $x_2 = 0$. As a matter of fact, in this case, since

$$[\tau_1, [\tau_1, [\tau_1, \tau_2]]] = \begin{bmatrix} 0 \\ -3 \end{bmatrix} \qquad [\tau_2, [\tau_1, [\tau_1, [\tau_1, \tau_2]]]] = \begin{bmatrix} 3 \\ 0 \end{bmatrix}$$

we have that $\langle \tau_1, \tau_2 | \Delta \rangle(x) = \mathbf{R}^2$ at each $x \in \mathbf{R}^2$. •

We illustrate now a property of $\langle \tau_1, \ldots, \tau_q | \Delta \rangle$ which is instrumental in achieving involutivity.

Lemma 8.6. Suppose Δ is spanned by some of the vector fields of the set $\{\tau_1, \ldots, \tau_q\}$. Then there exists an open and dense submanifold $U^\star$ of U with the following property. For each $x^\circ \in U^\star$ there exist a neighborhood V of x° and d vector fields (with $d = \dim\langle \tau_1, \ldots, \tau_q | \Delta \rangle(x^\circ)$) $\theta_1, \ldots, \theta_d$ of the form

$$\theta_i = [v_r, [v_{r-1}, \ldots, [v_1, v_0]]]$$

where $r \leq n - 1$ is an integer (which may depend on i) and $v_0, \ldots, v_r$ are vector fields in the set $\{\tau_1, \ldots, \tau_q\}$, such that

$$\langle \tau_1, \ldots, \tau_q | \Delta \rangle(x) = \mathrm{span}\{\theta_1(x), \ldots, \theta_d(x)\}$$

for all $x \in V$.

Proof. By induction, using as $U^\star$ the subset of U defined in the proof of Lemma 8.5. Let d_0 denote the dimension of Δ_0 (which may depend on x but is constant locally around x if the latter is a point of $U^\star$). Since, by assumption, Δ_0 is the span of some vector fields in the set $\{\tau_1, \ldots, \tau_q\}$, there exists exactly d_0 vector fields in this set that span Δ_0 locally around x. Let now d_k denote the dimension of Δ_k (constant around x°) and suppose Δ_k is spanned locally around x by d_k vector fields $\theta_1, \ldots, \theta_{d_k}$ of the form

$$\theta_i = [v_r, [v_{r-1}, \ldots, [v_1, v_0]]]$$

where $v_0, \ldots, v_r$ (with $r \leq k$ and possibly depending on i) are vector fields in the set $\{\tau_1, \ldots, \tau_q\}$. Then, a similar result holds for Δ_{k+1}. For, let τ be any vector field in Δ_k. From Lemma 3.2 it is known that there exist real-valued smooth functions $c_1, \ldots, c_{d_k}$ defined locally around x such that τ may be expressed, locally around x, as $\tau = c_1 \theta_1 + \cdots + c_{d_k} \theta_{d_k}$. If τ_j is any vector in the set $\{\tau_1, \ldots, \tau_q\}$ we have

$$[\tau_j, c_1 \theta_1 + \cdots + c_{d_k} \theta_{d_k}] = c_1 [\tau_j, \theta_1] + \cdots + c_{d_k} [\tau_j, \theta_{d_k}] + (L_{\tau_j} c_1) \theta_1 + \cdots + (L_{\tau_j} c_{d_k}) \theta_{d_k}.$$

As a consequence

$$\Delta_{k+1} = \Delta_k + [\tau_1, \Delta_k] + \cdots + [\tau_q, \Delta_k] = \mathrm{span}\{\theta_i, [\tau_1, \theta_i], \ldots, [\tau_q, \theta_i] : 1 \leq i \leq d_k\}.$$

Since Δ_{k+1} is nonsingular around x°, then it is possible to find exactly d_{k+1} vector fields of the form

$$\theta_i = [v_r, [v_{r-1}, \ldots, [v_1, v_0]]]$$

where $v_0, \ldots, v_r$ (with $r \leq k+1$ and possibly depending on i) are vector fields in the set $\{\tau_1, \ldots, \tau_q\}$, which span Δ_{k+1} locally around x. •

On the basis of the previous Lemma it is possible to find conditions under which the distribution $\langle \tau_1, \ldots, \tau_q | \Delta \rangle$ is also involutive.

Lemma 8.7. Suppose Δ is spanned by some of the vector fields $\tau_1, \ldots, \tau_q$ and that $\langle \tau_1, \ldots, \tau_q | \Delta \rangle$ is nonsingular. Then $\langle \tau_1, \ldots, \tau_q | \Delta \rangle$ is involutive.

Proof. We use first the conclusion of Lemma 8.6 to prove that if σ_1 and σ_2 are two vector fields in Δ_{n-1}, then their Lie bracket $[\sigma_1, \sigma_2]$ is such that $[\sigma_1, \sigma_2](x) \in \Delta_{n-1}(x)$ for all $x \in U^\star$. Using again Lemma 3.2 and the previous result we deduce, in fact, that in a neighborhood V of x

$$[\sigma_1, \sigma_2] = \left[\sum_{i=1}^{d} c_i^1 \theta_i, \sum_{j=1}^{d} c_j^2 \theta_j\right] \in \text{span}\{\theta_i, \theta_j, [\theta_i, \theta_j] : 1 \leq i, j \leq d\}$$

where θ_i, θ_j are vector fields of the form described before.

In order to prove the claim, we have only to show that $[\theta_i, \theta_j](x)$ is a tangent vector in $\Delta_{n-1}(x)$. For this purpose, we recall that on $U^\star$ the distribution Δ_{n-1} is invariant under the vector fields $\tau_1, \ldots, \tau_q$ (see Lemma 8.5) and that any distribution invariant under vector fields τ_i and τ_j is also invariant under their Lie bracket $[\tau_i, \tau_j]$ (see Lemma 6.5). Since each θ_i is a repeated Lie bracket of the vector fields $\tau_1, \ldots, \tau_q$, then $[\theta_i, \Delta_{n-1}](x) \subset \Delta_{n-1}(x)$ for all $1 < i < d$ and, thus, in particular $[\theta_i, \theta_j](x)$ is a tangent vector which belongs to $\Delta_{n-1}(x)$.

Thus the Lie bracket of two vector fields σ_1, σ_2 in Δ_{n-1} is such that $[\sigma_1, \sigma_2](x) \in \Delta_{n-1}(x)$. Moreover, it has already been observed that $\langle \tau_1, \ldots, \tau_q | \Delta \rangle = \Delta_{n-1}$ in a neighborhood of x° and, therefore, we conclude that at any point x of $U^\star$ the Lie bracket of any two vector fields σ_1, σ_2 in $\langle \tau_1, \ldots, \tau_q | \Delta \rangle$ is such that $[\sigma_1, \sigma_2](x) \in \langle \tau_1, \ldots, \tau_q | \Delta \rangle(x)$.

Consider now the distribution

$$\bar{\Delta} = \langle \tau_1, \ldots, \tau_q | \Delta \rangle + \text{span}\{[\theta_i, \theta_j] : \theta_i, \theta_j \in \langle \tau_1, \ldots, \tau_q | \Delta \rangle\}$$

which, by construction, is such that

$$\bar{\Delta} \supset \langle \tau_1, \ldots, \tau_q | \Delta \rangle.$$

From the previous result it is seen that $\bar{\Delta}(x) = \langle \tau_1, \ldots, \tau_q | \Delta \rangle(x)$ at each point x of $U^\star$, which is a dense set in U. By assumption, $\langle \tau_1, \ldots, \tau_q | \Delta \rangle$ is nonsingular. So, by Lemma 3.5 we deduce that $\bar{\Delta} = \langle \tau_1, \ldots, \tau_q | \Delta \rangle$, and, therefore, that $[\theta_i, \theta_j] \in \langle \tau_1, \ldots, \tau_q | \Delta \rangle$ for all pairs $\theta_i, \theta_j \in \langle \tau_1, \ldots, \tau_q | \Delta \rangle$. This concludes the proof. •

Lemma 8.8. Suppose Δ is spanned by some of the vector fields $\tau_1, \ldots, \tau_q$ and that Δ_{n-1} is nonsingular. Then $\langle \tau_1, \ldots, \tau_q | \Delta \rangle$ is involutive and

$$\langle \tau_1, \ldots, \tau_q | \Delta \rangle = \Delta_{n-1}.$$

Proof. An immediate consequence of Lemmas 8.5, 8.7 and 3.5. •

We now come back to the original problem of the study of the smallest distribution which contains $\text{span}\{g_1, \ldots, g_m\}$ and is invariant under the vector fields $f, g_1, \ldots, g_m$. From the previous Lemmas it is seen that if $\langle f, g_1, \ldots, g_m | \text{span}\{g_1, \ldots, g_m\}\rangle$ is nonsingular, then it is also involutive and, therefore, the decomposition (7.2) may be performed. We will see later that the minimality of $\langle f, g_1, \ldots, g_m | \text{span}\{g_1, \ldots, g_m\} \rangle$ makes it possible to deduce an interesting topological property of the set of points reached at some fixed time T starting from a given point x°. However, before doing this, it is convenient to illustrate the results obtained so far with the aid of a simple example and to analyze some other characteristics of the decomposition (7.2).

Example 8.2. Consider the system

$$\dot{x} = f(x) + g(x)u$$

with

$$f(x) = \begin{bmatrix} x_1 x_3 + x_2 e^{x_2} \\ x_3 \\ x_4 - x_2 x_3 \\ x_3^2 + x_2 x_4 - x_2^2 x_3 \end{bmatrix} \qquad g(x) = \begin{bmatrix} x_1 \\ 1 \\ 0 \\ x_3 \end{bmatrix}.$$

Computing the sequence (8.1), we find $\Delta_0 = \text{span}\{g\}$, $\Delta_1 = \text{span}\{g, [f, g]\}$, with

$$[f, g](x) = \frac{\partial g}{\partial x} f(x) - \frac{\partial f}{\partial x} g(x) = - \begin{bmatrix} e^{x_2} \\ 0 \\ 0 \\ 0 \end{bmatrix}.$$

Note that the distribution Δ_1 has dimension 2 for all x. Proceeding further, we have clearly

$$\Delta_2 = \Delta_1 + [f, \Delta_1] + [g, \Delta_1] = \Delta_1 + \text{span}\{[f, [f, g]], [g, [f, g]]\}$$

However, in this specific case we have $[f, [f, g]] = [g, [f, g]] = 0$. Therefore, the construction terminates, and

$$\langle f, g | \text{span}\{g\} \rangle = \Delta_1 = \text{span}\{g, [f, g]\}$$

is the smallest distribution invariant under f, g and containing the vector field g. Since this distribution is nonsingular and involutive (Lemma 8.7), we may use it in order to find a decomposition of the form indicated in Section 7. To this end, we have first to integrate this distribution, that is to find 2 real-valued functions λ_1, λ_2 such that $\text{span}\{d\lambda_1, d\lambda_2\} = [\langle f, g | \text{span}\{g\} \rangle]^\perp$. This amounts to solve the partial differential equation

$$
\begin{bmatrix} \dfrac{\partial \lambda_1}{\partial x} \\[2mm] \dfrac{\partial \lambda_2}{\partial x} \end{bmatrix}
\begin{bmatrix} x_1 & e^{x_2} \\ 1 & 0 \\ 0 & 0 \\ x_3 & 0 \end{bmatrix}
= \begin{bmatrix} 0 \\ 0 \end{bmatrix}
$$

i.e. to find 2 independent functions satisfying

$$
\frac{\partial \lambda}{\partial x_1}x_1 + \frac{\partial \lambda}{\partial x_2} + \frac{\partial \lambda}{\partial x_4}x_3 = 0 \quad \text{and} \quad \frac{\partial \lambda}{\partial x_1}e^{x_2} = 0.
$$

Since the latter implies $(\partial \lambda/\partial x_1) = 0$, the former reduces to

$$
\frac{\partial \lambda}{\partial x_2} + \frac{\partial \lambda}{\partial x_4}x_3 = 0.
$$

Two independent solutions of this equation are the functions

$$
\lambda_1 = x_3
$$
$$
\lambda_2 = x_4 - x_2 x_3.
$$

We may now use these functions in order to construct a change of coordinates in the state space, as explained in Section 7, setting

$$
z_3 = \lambda_1(x) = x_3
$$
$$
z_4 = \lambda_2(x) = x_4 - x_2 x_3.
$$

The change of coordinates can be completed by choosing

$$
z_1 = x_1
$$
$$
z_2 = x_2
$$

In the new coordinates, the system becomes

$$
\dot{z} = \begin{bmatrix} z_1 z_3 + z_2 e^{z_2} \\ z_3 \\ z_4 \\ 0 \end{bmatrix} + \begin{bmatrix} z_1 \\ 1 \\ 0 \\ 0 \end{bmatrix} u
$$

i.e. exactly in the form (7.2). •

We introduce now another distribution, which plays an important role in the study of local decompositions of the form (7.2) and is related to $\langle f, g_1, \ldots, g_m | \mathrm{span}\{g_1, \ldots, g_m\}\rangle$. The distribution in question is $\langle f, g_1, \ldots, g_m | \mathrm{span}\{f, g_1, \ldots, g_m\}\rangle$, i.e. the smallest distribution invariant under $f, g_1, \ldots, g_m$, which contains $\mathrm{span}\{g_1, \ldots, g_m\}$ and, *also*, the vector field f. If this distribution is nonsingular, and therefore involutive by Lemma 8.7, it may indeed be used in defining a local decomposition of the control system (7.1) similar to the decomposition (7.2). We are going to see in which way this new decomposition is related to the decomposition (7.2) and why it may be of interest. In order to simplify the notation, we set

$$P = \langle f, g_1, \ldots, g_m | \mathrm{span}\{g_1, \ldots, g_m\}\rangle \tag{8.2}$$
$$R = \langle f, g_1, \ldots, g_m | \mathrm{span}\{f, g_1, \ldots, g_m\}\rangle. \tag{8.3}$$

The relation between P and R is described in the following statement.

Lemma 8.9. The distributions P and R are such that
(a) $P + \mathrm{span}\{f\} \subset R$
(b) if x is a regular point of $P + \mathrm{span}\{f\}$, then $(P + \mathrm{span}\{f\})(x) = R(x)$.

Proof. By definition, $P \subset R$ and $f \in R$, so (a) is true. It is known from the proof of Lemma 8.7 that, around each point of an open dense submanifold U^* of U, R is spanned by vector fields of the form

$$\theta_i = [v_r, \ldots, [v_1, v_0]]$$

where $r \leq n - 1$ is an integer which may depend on i, and $v_r, \ldots, v_0$ are vector fields in the set $\{f, g_1, \ldots, g_m\}$.

It is easy to see that all such vector fields belong to $P + \mathrm{span}\{f\}$. For, if θ_i is just one of the vector fields in the set $\{f, g_1, \ldots, g_m\}$ it either belongs to P (which contains $g_1, \ldots, g_m$) or to $\mathrm{span}\{f\}$. If θ_i has the general form shown above we may, without loss of generality, assume that v_0 is in the set $\{g_1, \ldots, g_m\}$. For, if $v_0 = v_1 = f$, then $\theta_i = 0$. Otherwise, if $v_0 = f$ and $v_1 = g_j$, then $\theta_i = [v_r, \ldots, [f, g_j]]$ has the desired form. Any vector of the form

$$\theta_i = [v_r, \ldots, [v_1, g_j]]$$

with $v_r, \ldots, v_1$ in the set $\{f, g_1, \ldots, g_m\}$ is in P because P contains g_j and is invariant under $f, g_1, \ldots, g_m$ and so the claim is proved.

From this fact we deduce that on an open and dense submanifold U^* of U,

$$R \subset P + \mathrm{span}\{f\}$$

and therefore, since $R \supset P + \text{span}\{f\}$ on U, that on U^*

$$R = P + \text{span}\{f\}.$$

Suppose that $P + \text{span}\{f\}$ has constant dimension on some neighborhood V. Then, from Lemma 3.5 we conclude that the two distributions R and $P+\text{span}\{f\}$ coincide on V. •

Corollary 8.10. If P and $P + \text{span}\{f\}$ are nonsingular, then

$$\dim(R) - \dim(P) \le 1.$$

If P and $P + \text{span}\{f\}$ are both nonsingular, so is R and, by Lemma 8.7, both P and R are involutive. Suppose that P is properly contained in R. Then, using Corollary 4.2, one can find for each $x^\circ \in U^\circ$, a neighborhood U° of x° and a coordinates transformation $z = \Phi(x)$ defined on U° such that

$$\text{span}\{d\phi_{r+1}, \ldots, d\phi_n\} = R^\perp$$
$$\text{span}\{d\phi_{r+1}, \ldots, d\phi_n\} = P^\perp$$

on U°, where $r - 1 = \dim(P)$.

In the new coordinates the control system (7.1a) is represented by equations of the form

$$\dot{z}_1 = f_1(z_1, \ldots, z_n) + \sum_{i=1}^{m} g_{1i}(z_1, \ldots, z_n)u_i$$

$$\cdots$$

$$\dot{z}_{r-1} = f_{r-1}(z_1, \ldots, z_n) + \sum_{i=1}^{m} g_{r-1,i}(z_1, \ldots, z_n)u_i \qquad (8.4)$$

$$\dot{z}_r = f_r(z_r, \ldots, z_n)$$

$$\dot{z}_{r+1} = 0$$

$$\cdots$$

$$\dot{z}_n = 0.$$

Note that this differs from the form (7.2), only in that the last $n-r$ components of the vector field f are vanishing (because, by construction, $f \in R$). If, in particular, $R = P$ then also the r-th component of f vanishes and the corresponding equation for z_r is

$$\dot{z}_r = 0.$$

The decomposition (8.4) lends itself to considerations which, to some extent, refine the analysis presented in Sections **1.6** and **1.7**. We knew from the previous analysis that, in suitable coordinates, a set of components of the state (namely, the last $n - r + 1$ ones) was not affected by the input; we see now that in fact all these coordinates but (at most) one are even *constant* with the time.

If U° is partitioned into slices of the form

$$S_{\bar{x}} = \{x \in U^\circ : \phi_{r+1}(x) = \phi_{r+1}(\bar{x}), \ldots, \phi_n(x) = \phi_n(\bar{x})\}$$

then any trajectory $x(t)$ of the system evolving in U° actually evolves on the slice passing through the initial point x°. This slice, in turn, is partitioned into slices of the form

$$S_{\bar{x}} = \{x \in U^\circ : \phi_r(x) = \phi_r(\bar{x}), \ldots, \phi_n(x) = \phi_n(\bar{x})\}$$

each one including the set of points reached at a specific time T (see Fig. 1.9).

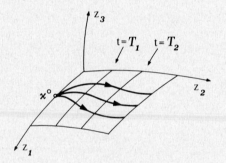

Fig. 1.9

Remark 8.11. A further change of local coordinates makes it possible to better understand the role of the time in the behavior of the control system (8.4). We may assume, without loss of generality, that the initial point x° is such that $\Phi(x^\circ) = 0$. Therefore we have $z_i(t) = 0$ for all $i = r + 1, \ldots, n$ and

$$\dot{z}_r = f_r(z_r, 0, \ldots, 0).$$

Moreover, if we assume that $f \notin P$, then the function f_r is nonzero everywhere on the neighborhood U. Now, let $z_r(t)$ denote the solution of this differential equation, which passes through 0 at $t = 0$. Clearly, the mapping

$$\mu : t \mapsto z_r(t)$$

is a diffeomorphism from an open interval $(-\varepsilon, \varepsilon)$ of the time axis into the open interval of the z_r axis $(z_r(-\varepsilon), z_r(\varepsilon))$. If its inverse μ^{-1} is used as a local coordinates transformation on the z_r axis, one easily sees that, since

$$\mu^{-1}(z_r) = t$$

the *time* t can be taken as new r-th coordinate.

In this way, points on the slice S_{x° passing through the initial state are parametrized by $(z_1, \ldots, z_{r-1}, t)$. In particular, the points reached at time T belong to the $(r-1)$-dimensional slice

$$S'_{x^\circ} = \{x \in U^\circ : \phi_r(x) = T, \phi_{r+1} = 0, \ldots, \phi_n(x) = 0\}. \;\; \bullet$$

Remark 8.12. Note also that, if f is a vector field of P then the local representation (8.4) is such that f_r vanishes on U°. Therefore, starting from a point x° such that $z(x^\circ) = 0$ we shall have $z_i(t) = 0$ for all $i = r, \ldots, n$. $\bullet$

By definition the distribution R is the *smallest* distribution which contains $f, g_1, \ldots, g_m$ and is invariant under $f, g_1, \ldots, g_m$. Thus, we may say that in the associated decomposition (8.4) the dimension r is "minimal", in the sense that it is not possible to find another set of local coordinates $\tilde{z}_1, \ldots, \tilde{z}_s, \ldots, \tilde{z}_n$ with s strictly less than r, with the property that the last $n - s$ coordinates remain constant with the time. We shall now show that, from the point of view of the interaction between input and state, the decomposition (8.4) has even stronger properties. Actually, we are going to prove that the states reachable from the initial state x° fill up at least an open subset of the r-dimensional slice in which they are contained.

Theorem 8.13. Suppose the distribution R (i.e. the smallest distribution invariant under $f, g_1, \ldots, g_m$ which contains $f, g_1, \ldots, g_m$) is nonsingular. Let r denote the dimension of R. Then, for each $x^\circ \in U$ it is possible to find a neighborhood U° of x° and a coordinates transformation $z = \Phi(x)$ defined on U with the following properties

(a) the set $\mathcal{R}(x^\circ)$ of states reachable starting from x° along trajectories entirely contained in U and under the action of piecewise constant input functions is a subset of the slice

$$S_{x^\circ} = \{x \in U^\circ : \phi_{r+1}(x) = \phi_{r+1}(x^\circ), \ldots, \phi_n(x) = \phi_n(x^\circ)\}$$

(b) the set $\mathcal{R}(x^\circ)$ contains an open subset of S_{x°.

Proof. The proof of the statement (a) follows from the previous discussion. We proceed directly to the proof of (b), assuming throughout the proof to operate on the neighborhood U on which the coordinates trasformation $\Phi(x)$ is defined. For convenience, we break up the proof in several steps.

(i) Let $\theta_1, \ldots, \theta_k$ be a set of vector fields, with $k < r$, and let $\Phi_t^1, \ldots, \Phi_t^k$ denote the corresponding flows.

Consider the mapping

$$F : (-\varepsilon, \varepsilon)^k \to U$$
$$(t_1, \ldots, t_k) \mapsto \Phi_{t_k}^1 \circ \cdots \circ \Phi_{t_1}^k (x^\circ)$$

where x° is a point of U and suppose that its differential has rank k at some $s_1, \ldots, s_k$, with $0 < s_i < \varepsilon$ for $1 \leq i \leq k$ (we shall show later that this is true). For ε sufficiently small the mapping

$$\bar{F} : (s_1, \varepsilon) \times \cdots \times (s_k, \varepsilon) \to U$$
$$(t_1, \ldots, t_k) \mapsto F(t_1, \ldots, t_k) \tag{8.5}$$

is an embedding.

Let M denote the image of the mapping (8.5). Consider the slice of U°

$$S_{x^\circ} = \{x \in U^\circ : \phi_i(x) = \phi_i(x^\circ), r+1 \leq i \leq n\}.$$

If the vector fields $\theta_1, \ldots, \theta_k$ have the form

$$\theta_j = f + \sum_{i=1}^m g_i u_i^j$$

with $u_i^j \in \mathbf{R}$ for $1 \leq i \leq m$ and $1 \leq j \leq k$, then M—in view of statement (a)—for small ε is an embedded submanifold of S_{x°. This implies, in particular, that for each $x \in M$

$$T_x M \subset R(x) \tag{8.6}$$

where R, as before, is the smallest distribution invariant under $f, g_1, \ldots, g_m$ which contains $f, g_1, \ldots, g_m$ (recall that $R(x)$ is the tangent space to S_{x° at x).

(ii) Suppose that the vector fields $f, g_1, \ldots, g_m$ are such that

$$f(x) \in T_x M \tag{8.7a}$$
$$g_i(x) \in T_x M \qquad 1 \leq i \leq m \tag{8.7b}$$

for all $x \in M$. We shall show that this contradicts the assumption $k < r$. For, consider the distribution $\bar{\Delta}$ defined by setting

$$\begin{cases} \bar{\Delta}(x) = T_x M & \text{for all } x \in M \\ \bar{\Delta}(x) = R(x) & \text{for all } x \in (U \backslash M). \end{cases}$$

This distribution is contained in R (because of (8.6)) and contains the vector fields $f, g_1, \ldots, g_m$ (because these vector fields are in R and, moreover, it is assumed that (8.7) are true). Let τ be any vector field of $\bar{\Delta}$. Then $\tau \in R$ and since R is invariant under $f, g_1, \ldots, g_m$, then for all $x \in (U \backslash M)$

$$[f, \tau](x) \in \bar{\Delta}(x) \tag{8.8a}$$

$$[g_i, \tau](x) \in \bar{\Delta}(x) \qquad 1 \leq i \leq m. \tag{8.8b}$$

Moreover since $\tau, f, g_1, \ldots, g_m$ are vector fields which are tangent to M at each $x \in M$, we have also that (8.8) hold for all $x \in M$, and therefore all $x \in U$. Having shown $\bar{\Delta}$ is invariant under $f, g_1, \ldots, g_m$ and contains $f, g_1, \ldots, g_m$, we deduce that $\bar{\Delta}$ must coincide with R. But this is a contradiction since at all $x \in M$

$$\dim \bar{\Delta}(x) = k$$
$$\dim R(x) = r > k.$$

(iii) If (8.7) are not true, then it is possible to find m real numbers $u_1^{k+1}, \ldots, u_m^{k+1}$ and a point $\bar{x} \in M$ such that the vector field

$$\theta_{k+1} = f + \sum_{i=1}^{m} g_i u_i^{k+1}$$

satisfies the condition $\theta_{k+1}(\bar{x}) \notin T_{\bar{x}} M$.

Let $\bar{x} = \bar{F}(s_1', \ldots, s_k')$ be this point $(s_i' > s_i, 1 \leq i \leq k)$ and let Φ_t^{k+1} denote the flow of θ_{k+1}. Then the mapping

$$F' : (-\varepsilon, \varepsilon)^{k+1} \to U$$
$$(t_1, \ldots, t_k, t_{k+1}) \mapsto \Phi_{t_{k+1}}^{k+1} \circ F(t_1, \ldots, t_k)$$

at the point $(s_1', \ldots, s_k', 0)$ has rank $k + 1$. For, note that

$$\left[F'_\star \left(\frac{\partial}{\partial t_i} \right) \right]_{(s_1', \ldots, s_k', 0)} = \left[F_\star \left(\frac{\partial}{\partial t_i} \right) \right]_{(s_1', \ldots, s_k')}$$

for $i = 1, \ldots, k$ and that

$$\left[F'_\star \left(\frac{\partial}{\partial t_{k+1}} \right) \right]_{(s_1', \ldots, s_k', 0)} = \theta_{k+1}(\bar{x}).$$

The first k tangent vectors at $\bar{x}$ are linearly independent, because F has rank k at all points of $(s_1, \varepsilon) \times \cdots \times (s_k, \varepsilon)$. The $(k+1)$-th one is independent from the first k by construction and therefore F' has rank $k+1$ at $(s_1', \ldots, s_k', 0)$. We may thus conclude that the mapping F' has rank $k + 1$ at a point $(s_1', \ldots, s_k', s_{k+1}')$, with $0 < s_i' < \varepsilon$ for $1 \leq i \leq k + 1$.

Note that given any real number $T > 0$ it is always possible to choose the point $\bar{x}$ in such a way that

$$(s_1' - s_1) + \cdots + (s_k' - s_k) < T.$$

For, otherwise, we had that any vector field of the form

$$\theta = f + \sum_{i=1}^{m} g_i u_i$$

would be tangent to the image under $\bar{F}$ of the open set

$$\{(t_1, \ldots, t_k) \in (s_1, \varepsilon) \times \cdots \times (s_k, \varepsilon) : (t_1 - s_1) + \cdots + (t_k - s_k) < T\}$$

and this, as in (ii), would be a contradiction.

(iv) We can now construct a sequence of mappings of the form (8.5). Let

$$\theta_1 = f + \sum_{i=1}^{m} g_i u_i^1$$

be a vector field which is not zero at x° (such a vector field can always be found because, otherwise, we would have $R(x^\circ) = \{0\}$) and let M_1 denote the image of the mapping

$$\bar{F}_1 : (0, \varepsilon) \to U$$
$$t_1 \mapsto \Phi_{t_1}^1(x^\circ).$$

Let $\bar{x} = \bar{F}_1(s_1^1)$ be a point of M_1 in which a vector field of the form

$$\theta_2 = f + \sum_{i=1}^{m} g_i u_i^2$$

is such that $\theta_2(\bar{x}) \notin T_{\bar{x}} M_1$. Then we may define the mapping (see Fig. 1.10)

$$\bar{F}_2 : (s_1^1, \varepsilon) \times (0, \varepsilon) \to U$$
$$(t_1, t_2) \mapsto \Phi_{t_2}^2 \circ \Phi_{t_1}^1(x^\circ).$$

Iterating this procedure, at stage k we start with a mapping

$$\bar{F}_k : (s_1^{k-1}, \varepsilon) \times \cdots \times (s_{k-1}^{k-1}, \varepsilon) \times (0, \varepsilon) \to U$$
$$(t_1, \ldots, t_k) \mapsto \Phi_{t_k}^k \circ \ldots \circ \Phi_{t_1}^1(x^\circ)$$

and we find a point $\bar{x} = \bar{F}_k(s_1^k, \ldots, s_k^k)$ of its image M_k and a vector field $\theta_{k+1} = f + \sum_{i=1}^{m} g_i u_i^{k+1}$ such that $\theta_{k+1}(\bar{x}) \notin T_{\bar{x}} M_k$. This makes it possible to define the next mapping $\bar{F}_{k+1}$. Note that $s_i^k > s_i^{k-1}$ for $i = 1, \ldots, k-1$ and

$$s_k^k > 0.$$

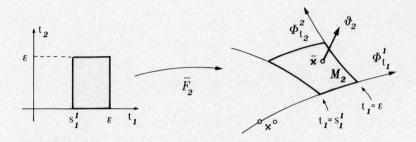

Fig. 1.10

The procedure clearly stops at the stage r, when a mapping $\bar{F}_r$ is defined

$$\bar{F}_r : (s_1^{r-1}, \varepsilon) \times \cdots \times (s_{r-1}^{r-1}, \varepsilon) \times (0, \varepsilon) \to U$$
$$(t_1, \ldots, t_r) \mapsto \Phi_{t_r}^k \circ \ldots \circ \Phi_{t_1}^1 (x^\circ).$$

(v) Observe that a point $x = \bar{F}_r(t_1, \ldots, t_r)$ in the image M_r of the embedding $\bar{F}_r$ can be reached, starting from the state x° at time $t = 0$, under the action of the piecewise constant control defined by

$$u_i(t) = u_i^1 \qquad \text{for } t \in [0, t_1)$$
$$u_i(t) = u_i^k \qquad \text{for } t \in [t_1 + \cdots + t_{k-1}, t_1 + t_2 + \cdots + t_k).$$

We know from our previous discussions that M_r must be contained in the slice of U°

$$S_{x^\circ} = \{x \in U^\circ : \phi_i(x) = \phi_i(x^\circ), r+1 \leq i \leq n\}.$$

The images under $\bar{F}_r$ of the open sets of

$$U_r = (s_1^{r-1}, \varepsilon) \times \cdots \times (s_{r-1}^{r-1}, \varepsilon) \times (0, \varepsilon)$$

are open in the topology of M_r as a subset of U° (because $\bar{F}_r$ is an embedding) and therefore they are also open in the topology of M_r as a subset of S_{x° (because S_{x° is an embedded submanifold of U°). Thus we have that M_r is an embedded submanifold of S_{x° and a dimensionality argument tells us that M_r is actually an open submanifold of S_{x°. •

Theorem 8.14. Suppose the distributions P (i.e. the smallest distribution invariant under $f, g_1, \ldots, g_m$ which contains $g_1, \ldots, g_m$) and $P + \text{span}\{f\}$ are nonsingular. Let p denote the dimension of P. Then, for each $x^\circ \in U$ it is possible to find a neighborhood U° of x° and a coordinates transformation $z = \Phi(x)$ defined on U° with the following properties

(a) the set $\mathcal{R}(x^\circ, T)$ of states reachable at time $t = T$ starting from x° at $t = 0$, along trajectories entirely contained in U and under the action of piecewise constant input functions, is a subset of the slice

$$S_{x^\circ,T} = \{x \in U : \phi_{p+1}(x) = \phi_{p+1}(\Phi_T^f(x^\circ)), \ldots, \phi_n(x) = \phi_n(\Phi_T^f(x^\circ))\}$$

where $\Phi_T^f(x^\circ)$ denotes the state reached at time $t = T$ when $u(t) = 0$ for all $t \in [0, T]$

(b) the set $\mathcal{R}(x^\circ, T)$ contains an open subset of $S_{x^\circ,T}$.

Proof. We know from Lemma 8.9 that R is nonsingular. Therefore one can repeat the construction used to prove the part (b) of Theorem 8.13. Moreover, from Corollary 8.10 it follows that r, the dimension of R, is equal either to $p+1$ or to p. Suppose the first situation occurs. Given any real number $T \in (0, \varepsilon)$, consider the set

$$U_r^T = \{(t_1, \ldots, t_r) \in U_r : t_1 + \cdots + t_r = T\}$$

where U_r is as defined at step (v) in the proof of Theorem 8.13. From the last remark at the step (iii) we know that there exists always a suitable choice of $s_1^{r-1}, \ldots, s_{r-1}^{r-1}$ after which this set is not empty. Clearly the image $\bar{F}_r(U_r^T)$ consists of points reachable at time T and therefore is contained in $\mathcal{R}(x^\circ, T)$. Moreover, using the same arguments as in (v), we deduce that the set $\bar{F}_r(U_r^T)$ is an open subset of $S_{x^\circ,T}$. If $p = r$, i.e. if $P = R$, the proof can be carried out by simply adding an extra state variable satisfying the equation

$$\dot{z}_{n+1} = 1$$

and showing that this reduces the problem to the previous one. The details are left to the reader. •

1.9 Local Observability

We have seen in Section 7 that if there is a nonsingular distribution Δ of dimension d with the properties that
  (i) Δ is involutive
  (ii) Δ is contained in the distribution $(\text{span}\{dh_1, \ldots, dh_p\})^\perp$
  (iii) Δ is invariant under the vector fields $f, g_1, \ldots, g_m$
then, at each point $x^\circ \in U$ it is possible to find a coordinates transformation defined in a neighborhood U° of x° and a partition of U° into slices of dimension d, such that points on each slice produce the same output under any input u which keeps the state trajectory evolving on U°.

We want now to find conditions under which points belonging to different slices of U° produce different outputs, i.e. are distinguishable. In this case we see from the decomposition (7.3) that the right object to look for is now the "largest" distribution which satisfies (ii), (iii). Since the existence of a nonsingular distribution Δ which satisfies (i), (ii), (iii) implies and is implied by the existence of a codistribution Ω (namely $\Delta^\perp$) with the properties that
(i') Ω is spanned, locally around each point $x \in U$, by $n - d$ exact covector fields
(ii') Ω contains the codistribution span $\{dh_1, \ldots, dh_p\}$
(iii') Ω is invariant under the vector fields $f, g_1, \ldots, g_m$
we may as well look for the "smallest" codistribution which satisfies (ii'), (iii').

Like in the previous Section, we need some background material. However, most of the results stated below require proofs which are similar to those of the corresponding results stated before and, for this reason, will be omitted.

Lemma 9.1. Let Ω be a given smooth codistribution and $\tau_1, \ldots, \tau_q$ a given set of vector fields. The family of all codistributions which are invariant under $\tau_1, \ldots, \tau_q$ and contain Ω has a minimal element, which is a smooth codistribution.

We shall use the symbol $\langle \tau_1, \ldots, \tau_q | \Omega \rangle$ to denote the smallest codistribution which contains Ω and is invariant under $\tau_1, \ldots, \tau_q$. Given a codistribution Ω and a set of vector fields $\tau_1, \ldots, \tau_q$ one can consider the following dual version of the algorithm (8.1)

$$\Omega_0 = \Omega \tag{9.1a}$$

$$\Omega_k = \Omega_{k-1} + \sum_{i=1}^{q} L_{\tau_i} \Omega_{k-1} \tag{9.1b}$$

and have the following result.

Lemma 9.2. The codistributions $\Omega_0, \Omega_1, \ldots$ generated with the algorithm (9.1) are such that

$$\Omega_k \subset \langle \tau_1, \ldots, \tau_q | \Omega \rangle$$

for all k. If there exists an integer $k^\star$ such that $\Omega_{k^\star} = \Omega_{k^\star+1}$, then

$$\Omega_{k^\star} = \langle \tau_1, \ldots, \tau_q | \Omega \rangle.$$

Remark 9.3. In the case of the linear system

$$\dot{x} = Ax$$
$$y = Cx$$

the sequence (9.1) can be interpreted as a nonlinear analogue of a sequence leading to the largest subspace of $\mathbf{R}^n$ invariant under A and contained in $\ker(C)$.

Suppose—as in Remark 8.3—that the set $\tau_1, \ldots, \tau_d$ consists only of the vector field τ and set

$$\Omega_0(x) = \text{span}\{c_1, \ldots, c_p\}$$
$$\tau(x) = Ax$$

for each $x \in \mathbf{R}^n$, where $c_1, \ldots, c_p$ denote the rows of C. Since any covector field ω in Ω_0 can be locally expressed as

$$\omega = \sum_{i=1}^{p} c_i \gamma_i(x)$$

where $\gamma_1, \ldots, \gamma_p$ are smooth real-valued functions, it is easy to deduce (see Remark 6.9) that

$$\Omega_1 = \Omega_0 + L_\tau \Omega_0 = \text{span}\{c_1, \ldots, c_p, L_\tau c_1, \ldots, L_\tau c_p\}.$$

Therefore, since

$$L_\tau c_i = L_{Ax} c_i = c_i \frac{\partial(Ax)}{\partial x} = c_i A$$

we have

$$\Omega_1(x) = \text{span}\{c_1, \ldots, c_p, c_1 A, \ldots, c_p A\}$$

at each $x \in \mathbf{R}^n$. Continuing in the same way, we have, for any $k \geq 1$,

$$\Omega_k(x) = \text{span}\{c_1, \ldots, c_p, c_1 A, \ldots, c_p A, \ldots, c_1 A^k, \ldots, c_p A^k\}.$$

Each codistribution of the sequence thus constructed is a constant codistribution. Since $\Omega_{k+1} \supset \Omega_k$, a dimensionality argument proves that there exists an integer $k^\star < n$ with the property that $\Omega_{k^\star+1} = \Omega_{k^\star}$. Thus Ω_{n-1}, which is indeed the largest codistribution of the sequence, by Lemma 9.2 is the smallest codistribution invariant under the vector field Ax which contains the codistribution $\Omega_0 = \text{span}\{c_1, \ldots, c_p\}$.

By duality $\Omega_{n-1}^\perp$ is the largest distribution invariant under the vector field Ax and contained in the distribution $\Omega_0^\perp$. Observe now that, by construction, at each $x \in \mathbf{R}^n$,

$$\Omega_0^\perp(x) = \ker(C)$$

$$\Omega_{n-1}^\perp(x) = \ker \begin{bmatrix} C \\ CA \\ \ldots \\ CA^{n-1} \end{bmatrix}.$$

Thus, we can conclude that the value of $\Omega_{n-1}^\perp$ at each x coincides with that of the largest subspace of $\mathbf{R}^n$ which is invariant under A and is contained in the subspace $\ker(C)$. ●

Returning to the case of nonlinear systems, one may obtain the following dual versions of Lemmas 8.5, 8.6, 8.7, 8.8.

Lemma 9.4. There exists an open and dense subset $U^\star$ of U with the property that at each point $x \in U^\star$

$$\langle \tau_1, \ldots, \tau_q | \Omega \rangle(x) = \Omega_{n-1}(x)$$

Lemma 9.5. Suppose Ω is spanned by a set $d\lambda_1, \ldots, d\lambda_s$ of exact covector fields. Then, there exists an open and dense submanifold $U^\star$ of U with the following property. For each $x^\circ \in U^\star$ there exist a neighborhood U° of x° and d exact covector fields (with $d = \dim\langle \tau_1, \ldots, \tau_q | \Omega \rangle(x^\circ)$) $\omega_1, \ldots, \omega_d$ which have the form

$$\omega_i = dL_{v_r} \ldots L_{v_1} \lambda_j$$

where $r \leq n-1$ is an integer (which may depend on i), $v_1, \ldots, v_r$ are vector fields in the set $\{\tau_1, \ldots, \tau_q\}$ and λ_j is a function in the set $\{\lambda_1, \ldots, \lambda_s\}$, such that

$$\langle \tau_1, \ldots, \tau_q | \Omega \rangle(x) = \operatorname{span}\{\omega_1(x), \ldots, \omega_d(x)\}$$

for all $x \in U^\circ$.

Lemma 9.6. Suppose Ω is spanned by a set $d\lambda_1, \ldots, d\lambda_s$ of exact covector fields and that $\langle \tau_1, \ldots, \tau_q | \Omega \rangle$ is nonsingular. Then $\langle \tau_1, \ldots, \tau_q | \Omega \rangle^\perp$ is involutive.

Proof. From the previous Lemma, it is seen that in a neighborhood of each point x in an open and dense submanifold $U^\star$, the codistribution $\langle \tau_1, \ldots, \tau_q | \Omega \rangle$ is spanned by exact covector fields. Therefore, the Lie bracket of any two vector fields θ_1, θ_2 in $\langle \tau_1, \ldots, \tau_q | \Omega \rangle^\perp$ is such that $[\theta_1, \theta_2](x) \in \langle \tau_1, \ldots, \tau_q \ \Omega \rangle^\perp(x)$ (see Section 1.4) for each $x \in U^\star$. From this result, using again Lemma 3.5 as in the proof of Lemma 8.7, one deduces that $\langle \tau_1, \ldots, \tau_q | \Omega \rangle^\perp$ is involutive. ●

Lemma 9.7. Suppose Ω is spanned by a set $d\lambda_1, \ldots, d\lambda_s$ of exact covector fields and that Ω_{n-1} is nonsingular. Then $\langle \tau_1, \ldots, \tau_q | \Omega \rangle^\perp$ is involutive and

$$\langle \tau_1, \ldots, \tau_q | \Omega \rangle = \Omega_{n-1}.$$

In the study of the state-output interactions for a control system of the form (7.1), we consider the distribution

$$Q = \langle f, g_1, \ldots, g_m | \operatorname{span}\{dh_1, \ldots, dh_p\}\rangle^\perp$$

From Lemma 6.7 we deduce that this distribution is invariant under $f, g_1, \ldots, g_m$ and we also see that, by definition, it is contained in

$(\text{span}\{dh_1, \ldots, dh_p\})^{\perp}$. If nonsingular, then, according to Lemma 9.6, is also involutive.

Invoking Proposition 7.2, this distribution may be used in order to find locally around each $x^{\circ} \in U$ an open neighborhood U° of x° and a coordinates transformation yielding a decomposition of the form (7.3). Let s denote the dimension of Q. Since $Q^{\perp}$ is the smallest codistribution invariant under $f, g_1, \ldots, g_m$ which contains $dh_1, \ldots, dh_p$, then in this case the decomposition we find is maximal, in the sense that it is not possible to find another set of local coordinates $\tilde{z}_1, \ldots, \tilde{z}_{\tilde{r}}, \tilde{z}_{\tilde{r}+1}, \ldots, \tilde{z}_n$ with $\tilde{r}$ strictly larger than s, with the property that only the last $n - \tilde{r}$ coordinates influence the output. We show now that this corresponds to the fact that points belonging to different slices of the neighborhood U° are distinguishable.

Theorem 9.8. Suppose the distribution Q (i.e. the annihilator of the smallest codistribution invariant under $f, g_1, \ldots, g_m$ and which contains $dh_1, \ldots, dh_p$) is nonsingular. Let s denote the dimension of Q. Then, for each $x^{\circ} \in U$ it is possible to find a neighborhood U° of x° and a coordinates transformation $z = \Phi(x)$ defined U° with the following properties

(a) Any two initial states x^a and x^b of U° such that

$$\phi_i(x^a) = \phi_i(x^b) \qquad i = s+1, \ldots, n$$

produce identical output functions under any input which keeps the state trajectories evolving on U°

(b) Any initial state x of U° which cannot be distinguished from x° under piecewise constant input functions belongs to the slice

$$S_{x^{\circ}} = \{x \in U^{\circ} : \phi_i(x) = \phi_i(x^{\circ}), s+1 \leq i \leq n\}.$$

Proof. We need only to prove (b). For simplicity, we break up the proof in various steps.

(i) Consider a piecewise-constant input function

$$\begin{aligned} u_i(t) &= u_i^1 && \text{for } t \in [0, t_1) \\ u_i(t) &= u_i^k && \text{for } t \in [t_1 + \cdots + t_{k-1}, t_1 + \cdots + t_k). \end{aligned}$$

Define the vector field

$$\theta_k = f + \sum_{i=1}^{m} g_i u_i^k$$

and let Φ_t^k denote the corresponding flow. Then, the state reached at time t_k starting from x° at time $t = 0$ under this input may be expressed as

$$x(t_k) = \Phi_{t_k}^k \circ \cdots \circ \Phi_{t_1}^1(x^{\circ})$$

and the corresponding output y as

$$y_i(t_k) = h_i(x(t_k)).$$

Note that this output may be regarded as the value of a mapping

$$F_i^{x^\circ} : (-\varepsilon, \varepsilon)^k \to \mathbf{R}$$
$$(t_1, \ldots, t_k) \mapsto h_i \circ \Phi_{t_k}^k \circ \cdots \circ \Phi_{t_1}^1(x^\circ).$$

If two initial states x^a and x^b are such that they produce two identical outputs for any possible piecewise constant input, we must have

$$F_i^{x^a}(t_1, \ldots, t_k) = F_i^{x^b}(t_1, \ldots, t_k)$$

for all possible $(t_1, \ldots, t_k)$, with $0 \le t_i < \varepsilon$ for $1 \le i \le p$. From this we deduce that

$$\left(\frac{\partial F_i^{x^a}}{\partial t_1 \ldots \partial t_k} \right)_{t_1 = \cdots = t_k = 0} = \left(\frac{\partial F_i^{x^b}}{\partial t_1 \ldots \partial t_k} \right)_{t_1 = \cdots = t_k = 0}.$$

An easy calculation shows that

$$\left(\frac{\partial F_i^{x^\circ}}{\partial t_1 \ldots \partial t_k} \right)_{t_1 = \cdots = t_k = 0} = L_{\theta_1} \ldots L_{\theta_k} h_i(x^\circ)$$

and, therefore, we must have

$$L_{\theta_1} \ldots L_{\theta_k} h_i(x^a) = L_{\theta_1} \ldots L_{\theta_k} h_i(x^b).$$

(ii) Now, remember that θ_j, $j = 1, \ldots, k$, depends on $(u_1^j, \ldots, u_m^j)$ and that the above equality must hold for all possible choices of $(u_1^j, \ldots, u_m^j) \in \mathbf{R}^m$. By appropriately selecting these $(u_1^j, \ldots, u_m^j)$ one easily arrives at an equality of the form

$$L_{v_1} \ldots L_{v_k} h_i(x^a) = L_{v_1} \ldots L_{v_k} h_i(x^b) \tag{9.2}$$

where $v_1, \ldots, v_k$ are vector fields belonging to the set $\{f, g_1, \ldots, g_m\}$. For, set $\gamma_2 = L_{\theta_2} \ldots L_{\theta_k} h$. From the equality $L_{\theta_1} \gamma_2(x^a) = L_{\theta_1} \gamma_2(x^b)$ we obtain

$$L_f \gamma_2(x^a) + \sum_{i=1}^m L_{g_i} \gamma_2(x^a) u_i^1 = L_f \gamma_2(x^b) + \sum_{i=1}^m L_{g_i} \gamma_2(x^b) u_i^1.$$

This, due to the arbitrariness of the $(u_1^1, \ldots, u_m^1)$, implies that

$$L_v \gamma_2(x^a) = L_v \gamma_2(x^b)$$

where v is any vector in the set $\{f, g_1, \ldots, g_m\}$. This procedure can be iterated, by setting $\gamma_3 = L_{\theta_3} \ldots L_{\theta_k} h$. From the above equality one gets

$$L_v L_f \gamma_3(x^a) + \sum_{i=1}^{m} L_v L_{g_i} \gamma_3(x^a) u_i^2 = L_v L_f \gamma_3(x^b) + \sum_{i=1}^{m} L_v L_{g_i} \gamma_3(x^b) u_i^2$$

and, therefore,

$$L_{v_1} L_{v_2} \gamma_3(x^a) = L_{v_1} L_{v_2} \gamma_3(x^b)$$

for all v_1, v_2 belonging to the set $\{f, g_1, \ldots, g_m\}$. Finally, one arrives at (9.2).

(iii) Let U° be a neighborhood of the point x° on which a coordinates transformation $\Phi(x)$ is defined which makes the condition

$$Q(x) = \mathrm{span}\{ \left(\frac{\partial}{\partial \phi_1} \right)_x, \ldots, \left(\frac{\partial}{\partial \phi_s} \right)_x \} \tag{9.3}$$

satisfied for all $x \in U^\circ$. From Lemma 9.5, we know that there exists an open subset $U^\star$ of U°, dense in U°, with the property that, around each $x \in U^\star$ it is possible to find a set of $n - s$ real-valued functions $\lambda_1, \ldots, \lambda_{n-s}$ which have the form

$$\lambda_i = L_{v_r} \ldots L_{v_1} h_j \tag{9.4}$$

with $v_1, \ldots, v_r$ vector fields in $\{f, g_1, \ldots, g_m\}$ and $1 \leq j \leq p$, such that

$$Q^{\perp} = \mathrm{span}\{d\lambda_1, \ldots, d\lambda_{n-s}\}.$$

Suppose $x^\circ \in U^\star$. Since $Q^{\perp}(x^\circ)$ has dimension $n - s$, it follows that the tangent covectors $d\lambda_1(x^\circ), \ldots, d\lambda_{n-s}(x^\circ)$ are linearly independent. In the local coordinates which satisfy (9.3), $\lambda_1, \ldots, \lambda_{n-s}$ are functions only of $z_{s+1}, \ldots, z_n$ (see (6.7)). Therefore, we may deduce that the mapping

$$\Lambda : (z_{s+1}, \ldots, z_n) \mapsto (\lambda_1(z_{s+1}, \ldots, z_n), \ldots, \lambda_{n-s}(z_{s+1}, \ldots, z_n))$$

has a jacobian matrix which is square and nonsingular at $(z_{s+1}(x^\circ), \ldots, z_n(x^\circ))$. In particular, this mapping is locally injective. We may use this property to deduce that, for some suitable neighborhood U' of x°, any other point x of U' such that

$$\lambda_i(x') = \lambda_i(x'')$$

for $1 \leq i \leq n - s$, must be such that

$$\phi_{s+i}(x^\circ) = \phi_{s+i}(x^\circ)$$

for $1 \leq i \leq n - s$, i.e. must belong to the slice of U° passing through x°. This, in view of the results proved in (ii) completes the proof in the case where $x^\circ \in U^\star$.

(iv) Suppose $x^\circ \notin U^\star$. Let $x(x^\circ, T, u)$ denote the state reached at time $t = T$ under the action of the piecewise constant input function u. If T is sufficiently small, $x(x^\circ, T, u)$ is still in U°. Suppose $x(x^\circ, T, u) \in U^\star$. Then, using the conclusions of (iii), we deduce that in some neighborhood U' of $x' = x(x^\circ, T, u)$, the states indistinguishable from x' lie on the slice of U° passing through x'. Now, recall that the mapping

$$\Phi : x^\circ \to x(x^\circ, T, u)$$

is a local diffeomorphism. Thus, there exists a neighborhood $\bar{U}$ of x° whose (diffeomorphic) image under Φ is a neighborhood $U'' \subset U'$ of x'. Let $\bar{x}$ denote a point of $\bar{U}$ indistinguishable from x° under piecewise constant inputs. Then, clearly, also $x'' = x(\bar{x}, T, u)$ is indistinguishable from $x' = x(x^\circ, T, u)$. From the previous discussion we know that x'' and x' belong to the same slice of U°. But this implies also that x° and $\bar{x}$ belong to the same slice of U°. Thus the proof is completed, provided that

$$x(x^\circ, T, u) \in U^\star. \tag{9.5}$$

(v) All we have to show now is that (9.5) can be satisfied. For, suppose $\mathcal{R}(x^\circ)$, the set of states reachable from x° under piecewise constant control along trajectories entirely contained in U°, is such that

$$\mathcal{R}(x^\circ) \cap U^\star = \emptyset. \tag{9.6}$$

If this is true, we know from Theorem 8.13 that it is possible to find an r-dimensional embedded submanifold V of U° entirely contained in $\mathcal{R}(x^\circ)$ and therefore such that $V \cap U^\star = 0$. For any choice of functions $\lambda_1, \ldots, \lambda_{n-s}$, of the form (9.4), at any point $x \in V$ the covectors $d\lambda_1(x), \ldots, d\lambda_{n-s}(x)$ are linearly dependent. Thus, without loss of generality, we may assume that there exist $d < n - s$ functions $\gamma_1, \ldots, \gamma_d$ still of the form (9.4) such that, for some open subset V' of V,

- $\text{span}\{dh_1(x), \ldots, dh_p(x)\} \subset \text{span}\{d\gamma_1(x), \ldots, d\gamma_d(x)\}$ for all $x \in V'$
- $d\gamma_1(x), \ldots, d\gamma_d(x)$ are linearly independent covectors at all $x \in V'$,
- $dL_v\gamma_j \in \text{span}\{d\gamma_1(x), \ldots, d\gamma_d(x)\}$ for all $x \in V'$ and $v \in \{f, g_1, \ldots, g_m\}$.

Now, we define a codistribution on U° as follows: $\Omega(x) = Q^\perp(x)$, for $x \notin V'$, and $\Omega(x) = \text{span}\{d\gamma_1(x), \ldots, d\gamma_d(x)\}$, for $x \in V'$. Using the fact that $f, g_1, \ldots, g_m$ are tangent to V', it is not difficult to verify that this codistribution is invariant under $f, g_1, \ldots, g_m$, contains $\text{span}\{dh_1, \ldots, dh_p\}$ and is smaller than $\langle f, g_1, \ldots, g_m | \text{span}\{dh_1, \ldots, dh_p\}\rangle$. This is a contradiction and therefore (9.6) must be false. $\bullet$

2 Global Decompositions of Control Systems

2.1 Sussmann's Theorem and Global Decompositions

In the previous Chapter, we have shown that a nonsingular and involutive distribution induces a local partition of the state space into lower dimensional submanifolds and we have used this result to obtain local decompositions of control systems. The decompositions thus obtained are very useful to understand the behavior of control systems from the point of view of input-state and, respectively, state-output interaction. However, it must be stressed that the existence of decompositions of this type is strictly related to the assumption that the dimension of the distribution is constant at least over a neighborhood of the point around which we want to investigate the behavior of our control system.

In this Section we shall see that the assumption that Δ is nonsingular can be removed and that global partitions of the state space can be obtained. Since we are interested in establishing results which have a global validity, it is convenient—for more generality—to consider, as anticipated in Section 1.5, the case of control systems whose state space is a *manifold* N. Of course, this more general analysis will cover in particular the case in which $N = U$.

To begin with, we need to introduce a few more concepts. Let Δ be a distribution defined on the manifold N. A submanifold S of N is said to be an *integral submanifold* of the distribution Δ if, for every $p \in S$, the tangent space $T_p S$ to S at p coincides with the subspace $\Delta(p)$ of $T_p N$. A *maximal* integral submanifold of Δ is a connected integral submanifold S of Δ with the property that every other connected integral submanifold of Δ which contains S coincides with S. We see immediately from this definition that any two maximal integral submanifolds of Δ passing through a point $p \in N$ must coincide. Motivated by this, it is said that a distribution Δ on N has the *maximal integral manifolds property* if through every point $p \in N$ passes a maximal integral submanifold of Δ or, in other words, if there exists a *partition* of N into maximal integral submanifolds of Δ.

It is easily seen that this is a global version of the notion of complete integrability for a distribution. As a matter of fact, a nonsingular and completely integrable distribution is such that for each $p \in N$ there exists a neighborhood U of p with the property that Δ restricted to U has the maximal integral manifolds property.

A simple consequence of the previous definitions is the following one.

Lemma 1.1. A distribution Δ which has the maximal integral manifolds property is involutive.

Proof. If τ is a vector field which belongs to a distribution Δ with the maximal integral manifolds property, then τ must be tangent to every maximal integral submanifold S of Δ. As a consequence, the Lie bracket $[\tau_1, \tau_2]$ of two vector fields τ_1 and τ_2 both belonging to Δ must be tangent to every maximal integral submanifold S of Δ. Thus, $[\tau_1, \tau_2]$ belongs to Δ. •

Thus, involutivity is a necessary condition for Δ to have the maximal integral manifolds property but, unlike the notion of complete integrability, this condition is no longer sufficient.

Example 1.1. Let $N = \mathbf{R}^2$ and let Δ be a distribution defined by

$$\Delta(x) = \text{span}\left\{ \left(\frac{\partial}{\partial x_1}\right)_x, \lambda(x_1)\left(\frac{\partial}{\partial x_2}\right)_x \right\}$$

where $\lambda(x_1)$ is a C^∞ function such that $\lambda(x_1) = 0$ for $x_1 \leq 0$ and $\lambda(x_1) > 0$ for $x_1 > 0$. This distribution is involutive and

$$\dim \Delta(x) = 1 \qquad \text{if } x \text{ is such that } x_1 \leq 0$$
$$\dim \Delta(x) = 2 \qquad \text{if } x \text{ is such that } x_1 > 0.$$

Clearly, the open subset of N

$$\{(x_1, x_2) \in \mathbf{R}^2 : x_1 > 0\}$$

is an integral submanifold of Δ (actually a maximal integral submanifold) and so is any subset of the form

$$\{(x_1, x_2) \in \mathbf{R}^2 : x_1 < 0, x_2 = c\}.$$

However, it is not possible to find integral submanifolds of Δ passing through a point $(0, c)$. •

Another important point to be stressed, which emphasizes the difference between the general problem here considered and its local version described in Section 1.4, is that the elements of a global partition of N induced by a distribution which has the integral manifolds property are *immersed* submanifolds. On the contrary, local partitions induced by a nonsingular and completely integrable distribution are always made of slices of a coordinate neighborhood, i.e. of *embedded* submanifolds.

Example 1.2. Consider a torus $T_2 = S_1 \times S_1$, We define a vector field on the torus in the following way. Let τ be a vector field on $\mathbf{R}^2$ defined by setting

$$\tau(x_1, x_2) = -x_2 \left(\frac{\partial}{\partial x_1}\right)_x + x_1 \left(\frac{\partial}{\partial x_2}\right)_x.$$

At each point $(x_1, x_2) \in S_1$ this mapping defines a tangent vector in $T_{(x_1,x_2)}S_1$, and therefore a vector field on S_1 whose flow is given by

$$\Phi_t^\tau(x_1^\circ, x_2^\circ) = (x_1^\circ \cos t - x_2^\circ \sin t, x_1^\circ \sin t - x_2^\circ \cos t).$$

In order to simplify the notation we may represent a point (x_1, x_2) of S_1 with the complex number $z = x_1 + jx_2$, $|z| = 1$, and have $\Phi_t^\tau(z) = e^{jt}z$. Similarly, by setting

$$\theta(x_1, x_2) = -x_2\alpha \left(\frac{\partial}{\partial x_1}\right)_x + x_1\alpha \left(\frac{\partial}{\partial x_2}\right)_x$$

we define another vector field on S_1, whose flow is now given by $\Phi_t^\theta(z) = e^{j\alpha t}z$.

From τ and θ we may define a vector field f on T_2 by setting

$$f(z_1, z_2) = \big(\tau(z_1), \theta(z_2)\big)$$

and we readily see that the flow of f is given by

$$\Phi_t^f(z_1, z_2) = (e^{jt}z_1, e^{j\alpha t}z_2).$$

If α is a rational number, then there exists a T such that $\Phi_t^f = \Phi_{t+kT}^f$ for all $t \in \mathbf{R}$ and all $k \in \mathbf{Z}$. Otherwise , if α is irrational, for each fixed $p = (z_1, z_2) \in T_2$ the mapping $F_p : t \mapsto \Phi_t^f(z_1, z_2)$ is an injective immersion of $\mathbf{R}$ into T_2, and $F_p(\mathbf{R})$ is an immersed submanifold of T_2.

From the vector field f we can define the one-dimensional distribution $\Delta = \mathrm{span}\{f\}$ and see that, if α is irrational, the maximal integral submanifold of Δ passing through a point $p \in T_2$ is exactly $F_p(\mathbf{R})$ and Δ has the maximal integral manifold property.

$F_p(\mathbf{R})$ is an immersed but not an embedded submanifold of T_2. For, it is easily seen that given any point $p \in T_2$ and any open (in the topology of T_2) neighborhood U of p, the intersection $F_p(\mathbf{R}) \cap U$ is dense in U and this excludes the possibility of finding a coordinate cube (U, ϕ) around p whith the property that $F_p(\mathbf{R}) \cap U$ is a slice of U. •

The following Theorem establishes the desired necessary and sufficient condition for a distribution to have the maximal integral manifolds property.

Theorem 1.2 (Sussmann). A distribution Δ has the maximal integral manifolds property if and only if, for every vector field $\tau \in \Delta$ and for every pair $(t, p) \in \mathbf{R} \times N$ such that the flow $\Phi_t^\tau(p)$ of τ is defined, the differential $(\Phi_t^\tau)_*$ at p maps the subspace $\Delta(p)$ into the subspace $\Delta\big(\Phi_t^\tau(p)\big)$. •

We are not going to give the proof of this theorem, that can be found in the literature. Nevertheless, some remarks are in order.

Remark 1.3. An intuitive understanding of the constructions that are behind the statement of Sussmann's theorem may be obtained in this way.

Let $\tau_1, \ldots, \tau_k$ be a collection of vector fields of Δ and let $\Phi_{t_1}^{\tau_1}, \ldots, \Phi_{t_k}^{\tau_k}$ denote the corresponding flows. It is clear that if p is a point of N, and S is an integral manifold of Δ passing through p, then $\Phi_{t_i}^{\tau_i}(p)$ should be a point of S for all values of t_i for which $\Phi_{t_i}^{\tau_i}(p)$ is defined. Thus, S should include all points of N that can be expressed in the form

$$\Phi_{t_k}^{\tau_k} \circ \Phi_{t_{k-1}}^{\tau_{k-1}} \circ \cdots \circ \Phi_{t_1}^{\tau_1}(p). \tag{1.1}$$

In particular, if τ and θ are vector fields of Δ, the smooth curve

$$\sigma : (-\varepsilon, \varepsilon) \to N$$
$$t \mapsto \Phi_{t_1}^{\tau} \circ \Phi_t^{\theta} \circ \Phi_{-t_1}^{\tau}(p)$$

passing through p at $t = 0$, should be contained in S and its tangent vector at p should be contained in $\Delta(p)$. Computing this tangent vector, we obtain

$$(\Phi_{t_1}^{\tau})_* \theta \big(\Phi_{-t_1}^{\tau}(p) \big) \in \Delta(p)$$

i.e. setting $q = \Phi_{-t_1}^{\tau}(p)$

$$(\Phi_{t_1}^{\tau})_* \theta(q) \in \Delta \big(\Phi_{t_1}^{\tau}(q) \big)$$

and this motivates the necessity of Sussmann's condition. •

According to the statement of Theorem 1.2, in order to "test" whether or not a given distribution Δ is integrable, one should check that $(\Phi_t^{\tau})_*$ maps $\Delta(p)$ into $\Delta \big(\Phi_t^{\tau}(p) \big)$ for all vector fields τ in Δ. Actually one could limit oneself to make this test only on some suitable subset of vector fields in Δ because the statement of the Theorem 1.2 can be given the following weaker version, also due to Sussmann.

Theorem 1.4. A distribution Δ has the maximal integral manifolds property if and only if there exists a set of vector fields $\mathcal{T}$, which spans Δ, with theproperty that for every $\tau \in \mathcal{T}$ and every pair $(t, p) \in \mathbf{R} \times N$ such that the flow$\Phi_t^{\tau}(p)$ is defined, the differential $(\Phi_t^{\tau})_*$ at p maps the subspace$\Delta(p)$ into the subspace$\Delta \big(\Phi_t^{\tau}(p) \big)$.

Remark 1.5. It is clear that the proof of the "if" part of Theorem 1.2 is implied by the "if" part of Theorem 1.4 because the set of all vector fields in Δ is indeed a set of vector fields which spans Δ. Conversely, the "only if" part of Theorem 1.4 is implied by the "only if" part of Theorem 1.2. •

We have seen that involutivity is a necessary but not sufficient condition for a distribution Δ to have the maximal integral manifolds property. However, the involutivity is something easier to test—in principle—because it involves only the computation of the Lie bracket of vector fields in Δ whereas the test of the condition stated in the Theorem 1.4 requires the knowledge of the flows Φ_t^τ associated with all vector fields τ of the subset $\mathcal{T}$ which spans Δ. Therefore, one might wish to identify some special *classes* of distributions for which the involutivity becomes a sufficient condition for them to have the maximal integral manifolds property. Actually, this is possible with a relatively little effort.

A set $\mathcal{T}$ of vector fields is *locally finitely generated* if, for every $p \in N$ there exists a neighborhood U of p and a finite set $\{\tau_1, \ldots, \tau_k\}$ of vector fields of $\mathcal{T}$ with the property that every other vector field belonging to $\mathcal{T}$ can be represented on U in the form

$$\tau = \sum_{i=1}^{k} c_i \tau_i \tag{1.2}$$

where each c_i is a real-valued smooth function defined on U.

The class of the distributions which are spanned by locally finitely generated sets of vector fields is actually one of the classes we were looking for, as it will be shown hereafter.

We prove first a slightly different result which will also be used independently.

Lemma 1.6. Let $\mathcal{T}$ be a locally finitely generated set of vector fields which spans Δ and θ another vector field such that $[\theta, \tau] \in \mathcal{T}$ for all $\tau \in \mathcal{T}$. Then, for every pair $(t, p) \in \mathbf{R} \times N$ such that the flow $\Phi_t^\theta(p)$ is defined, the differential $(\Phi_t^\theta)_*$ at p maps the subspace $\Delta(p)$ into the subspace $\Delta\big(\Phi_t^\theta(p)\big)$.

Proof. The reader will have no difficulty in finding that the same arguments usedfor the statement (ii) in the proof of Theorem 1.4.1 can be used. •

Note that in the above statement the vector field θ may not belong to $\mathcal{T}$. If the set $\mathcal{T}$ is involutive, i.e. if the Lie bracket $[\tau_1, \tau_2]$ of any two vector fields $\tau_1 \in \mathcal{T}$, $\tau_2 \in \mathcal{T}$ is again a vector field in $\mathcal{T}$, from the previous Lemma and from Sussmann's Theorem we derive immediately the following result

Theorem 1.7. A distribution Δ spanned by an involutive and locally finitely generated set of vector fields $\mathcal{T}$ has the maximal integral manifolds property.

The existence of an involutive and locally finitely generated set of vector fields appears to be something easier to prove, at least in principle. In particular, there are some classes of distributions in which the existence of a locally finitely generated set of vector fields is automatically guaranteed. This yields the following corollaries of Theorem 1.7.

Corollary 1.8. A nonsingular distribution has the maximal integral manifold property if and only if it is involutive.

Proof. In this case, the set of all vector fields which belong to the distribution is involutive and, as a consequence of Lemma 1.3.2, locally finitely generated. •

Corollary 1.9. An analytic distribution on a real analytic manifold has the maximal integral manifolds property if and only if it is involutive.

Proof. It depends on the fact that any set of analytic vector fields defined on a real analytic manifold is locally finitely generated. •

We conclude this Section with another interesting consequence of the previous results, which will be used later on.

Lemma 1.10. Let Δ be a distribution with the maximal integral manifolds property and let S be a maximal integral submanifold of Δ. Then, given any two points p and q in S, there exist vector fields $\tau_1, \ldots, \tau_k$ in Δ and real numbers $t_1, \ldots, t_k$ such that $q = \Phi_{t_1}^{\tau_1} \circ \cdots \circ \Phi_{t_k}^{\tau_k}(p)$.

Theorem 1.11. Let Δ be an involutive distribution invariant under a complete vector field θ. Suppose the set of all vector fields in Δ is locally finitely generated. Let p_1 and p_2 be two points belonging to the same maximal integral submanifold of Δ. Then, for all T, $\Phi_T^\theta(p_1)$ and $\Phi_T^\theta(p_2)$ belong to the same maximal integral submanifold of Δ.

Proof. Observe, first of all, that Δ has the maximal integral manifolds property (see Theorem 1.7).

Let τ be a vector field in Δ. Then, for ε sufficiently small the mapping

$$\sigma : (-\varepsilon, \varepsilon) \rightarrow N$$
$$t \mapsto \Phi_T^\theta \circ \Phi_t^\tau \circ \Phi_{-T}^\theta(p)$$

defines a smooth curve on N which passes through p at $t = 0$. Computing the tangent vector to this curve at t we get

$$\sigma_\star \left(\frac{d}{dt} \right)_t = (\Phi_T^\theta)_\star \tau \left(\Phi_t^\tau \circ \Phi_{-T}^\theta(p) \right)$$
$$= (\Phi_T^\theta)_\star \tau \left(\Phi_{-T}^\theta(\sigma(t)) \right).$$

But since $\tau \in \Delta$, we know from Lemma 1.6 that, for all q, $(\Phi_T^\theta)_\star \tau \left(\Phi_{-T}^\theta(q) \right) \subseteq \Delta(q)$ and therefore we get

$$\sigma_\star \left(\frac{d}{dt} \right)_t \in \Delta(\sigma(t))$$

for all $t \in (-\varepsilon, \varepsilon)$. This shows that the smooth curve σ lies on an integral submanifold of Δ. Now, let $p_1 = \Phi_{-T}^\theta(p)$ and $p_2 = \Phi_t^\tau(p_1)$. Then p_1 and p_2 are two points belonging to a maximal integral submanifold of Δ, and the previous result shows that $\Phi_T^\theta(p_1)$ and $\Phi_T^\theta(p_2)$ again are two points belonging to a maximal integral submanifold of Δ. Thus the Theorem is proved for points p_1, p_2 such that $p_2 = \Phi_t^\tau(p_1)$. If this is not the case, using Lemma 1.10 we can always find vector fields $\tau_1, \ldots, \tau_k$ of Δ such that $p_2 = \Phi_{t_1}^{\tau_1} \circ \cdots \circ \Phi_{t_k}^{\tau_k}(p_1)$ and use the above result in order to prove the Theorem.

2.2 The Control Lie Algebra

The notions developed in the previous Section are useful in dealing with the study ofinput-state interaction properties from a global point of view. As in Section 1.5, we consider here control systems described by equations of the form

$$\dot{p} = f(p) + \sum_{i=1}^{m} g_i(p) u_i. \tag{2.1}$$

Recall that the local analysis of these properties was based upon the consideration of the smallest distribution, denoted R, invariant under the vector fields $f, g_1, \ldots, g_m$ and which contains $f, g_1, \ldots, g_m$. It was also shown that if this distribution is nonsingular, then it is involutive (Lemma 1.8.7). This property makes it possible to use immediately one of the results discussed in the previous Section and find a global decomposition of the state space N.

Lemma 2.1. Suppose R is nonsingular, then R has the maximal integral manifolds property.

Proof. Just use Corollary 1.8. •

The decomposition of N into maximal integral submanifolds of R has the following intepretation from the point of view of the study of interactions between inputs and states. It is known that each of the vector fields $f, g_1, \ldots, g_m$ is in R, and therefore tangent to each maximal integral submanifold of R. Let S_{p° be the maximal integral submanifold of R passing through p°. From what we have said before, we know that any vector field of the form $\tau = f + \sum_{i=1}^{m} g_i u_i$, where $u_1, \ldots, u_m$ are real numbers, will be tangent to S_{p° and, therefore, that the integral curve of τ passing through p^0 at time $t = 0$ will belong to S_{p°. We conclude that any state trajectory emanating from the point p^0, under the action of a piecewise constant control, will stay in S_{p°.

Putting together this observation with the part (b) of the statement of Theorem 1.8.13, one obtains the following result.

Theorem 2.2. Suppose R is nonsingular. Then there exists a partition of N into maximal integral submanifolds of R, all with the same dimension. Let S_{p° denote the maximal integral submanifold of R passing through p^0. The set $\mathcal{R}(p^\circ)$ of states reachable from p^0 under piecewise constant input functions
(a) is a subset of S_{p°

(b) contains an open subset of S_{p°.

The result might be interpreted as a global version of Theorem 1.8.13. However, there are more general versions, which do not require the assumption that R is nonsingular. Of course, since one is interested in having global decompositions, it is necessary to work with distributions having the maximal integral manifolds property. From the discussions of the previous Section, we see that a reasonable situation is the one in which the distributions are spanned by a set of vector fields which is involutive and locally finitely generated. This motivates the interest in the following considerations.

Let $\{\tau_i : 1 \leq i \leq q\}$ be a finite set of vector fields and $\mathcal{L}_1, \mathcal{L}_2$ two subalgebras of $V(N)$ which both contain the vector fields $\tau_1, \ldots, \tau_q$. Clearly, the intersection $\mathcal{L}_1 \cap \mathcal{L}_2$ is again a subalgebra of $V(N)$ and contains $\tau_1, \ldots, \tau_q$. Thus we conclude that there exists an unique subalgebra $\mathcal{L}$ of $V(N)$ which contains $\tau_1, \ldots, \tau_q$ and has the property of being contained in all the subalgebras of $V(N)$ which contain the vector fields $\tau_1, \ldots, \tau_q$. We refer to this as the *smallest* subalgebra of $V(N)$ which contains the vector fields $\tau_1, \ldots, \tau_q$.

Remark 2.3. One may give a description of the subalgebra $\mathcal{L}$ also in the following terms. Consider the set

$$L_o = \left\{ \tau \in V(N) : \tau = [\tau_{i_k}, [\tau_{i_{k-1}}, \ldots, [\tau_{i_2}, \tau_{i_1}]]]; 1 \leq i_k \leq q, 1 < k < \infty \right\}$$

and let $LC(L_o)$ denote the set of all finite $\mathbf{R}$-linear combinations of elements of L_o. Then, it is possible to see that $\mathcal{L} = LC(L_o)$. For, by construction, every element of L_o is an element of $\mathcal{L}$ because $\mathcal{L}$, being a subalgebra of $V(N)$ which contains $\tau_1, \ldots, \tau_q$, must contain every vector field of the form $[\tau_{i_k}, [\tau_{i_{k-1}}, \ldots, [\tau_{i_2}, \tau_{i_1}]]]$. Therefore $LC(L_o) \subset \mathcal{L}$ and also $\tau_i \in LC(L_o)$ for $1 \leq i \leq q$. To prove that $\mathcal{L} = LC(L_o)$ we only need to show that $LC(L_o)$ is a subalgebra of $V(N)$. This follows from the fact that the Lie bracket of any two vector fields in L_o is a $\mathbf{R}$-linear combination of elements of L_o. •

With the subalgebra $\mathcal{L}$ we may associate a distribution $\Delta_{\mathcal{L}}$ in a natural way, by setting

$$\Delta_{\mathcal{L}} = \text{span}\{\tau : \tau \in \mathcal{L}\}$$

Clearly, $\Delta_{\mathcal{L}}$ need not to be nonsingular. Thus, in order to be able to operate with $\Delta_{\mathcal{L}}$, we have to set explicitly some suitable assumptions. In view of the results discussed at the end of the previous Section we shall assume that $\mathcal{L}$ is locally finitely generated.

An immediate consequence of this assumption is the following one.

Lemma 2.4. If the subalgebra $\mathcal{L}$ is locally finitely generated, the distribution $\Delta_{\mathcal{L}}$ has the maximal integral manifolds property.

Proof. The set $\mathcal{L}$ is involutive by construction (because it is a subalgebra of $V(N)$). Then, using Theorem 1.7 we see that $\Delta_{\mathcal{L}}$ has the maximal integral manifolds property. •

When dealing with control systems of the form (2.1), we take into consideration the smallest subalgebra of $V(N)$ which contains the vector fields $f, g_1, \ldots, g_m$. This subalgebra will be denoted by $\mathcal{C}$ and called *Control Lie Algebra*. With $\mathcal{C}$ we associate the distribution

$$\Delta_{\mathcal{C}} = \text{span}\{\tau : \tau \in \mathcal{C}\}.$$

Remark 2.5. It is not difficult to prove that the codistribution $\Delta_{\mathcal{C}}^{\perp}$ is invariant under the vector fields $f, g_1, \ldots, g_m$. For, let τ be any vector field in $\mathcal{C}$ and ω a covector field in $\Delta_{\mathcal{C}}^{\perp}$. Then $\langle \omega, \tau \rangle = 0$ and $\langle \omega, [f, \tau] \rangle = 0$ because $[f, \tau]$ is again a vector field in $\mathcal{C}$. Therefore, from the equality

$$\langle L_f \omega, \tau \rangle = L_f \langle \omega, \tau \rangle - \langle \omega, [f, \tau] \rangle = 0$$

we deduce that $L_f \omega$ annihilates all vector fields in $\mathcal{C}$. Since $\Delta_{\mathcal{C}}$ is spanned by vector fields in $\mathcal{C}$, it follows that $L_f \omega$ is a covector field in $\Delta_{\mathcal{C}}^{\perp}$, i.e. that $\Delta_{\mathcal{C}}^{\perp}$ is invariant under f. In the same way it is proved that $\Delta_{\mathcal{C}}^{\perp}$ is invariant under $g_1, \ldots, g_m$.

If the codistribution $\Delta_{\mathcal{C}}^{\perp}$ is smooth (e.g. when the codistribution $\Delta_{\mathcal{C}}$ is nonsingular), then using Lemma 1.6.7, one concludes that $\Delta_{\mathcal{C}}$ itself is invariant under $f, g_1, \ldots, g_m$. •

Remark 2.6. The distribution $\Delta_{\mathcal{C}}$, and the distributions P and R introduced in the previous Chapter are related in the following way.
(a) $\Delta_{\mathcal{C}} \subset P + \text{span}\{f\} \subset R$
(b) if p is a regular point of $\Delta_{\mathcal{C}}$, then $\Delta_{\mathcal{C}}(p) = (P + \text{span}\{f\})(p) = R(p)$.

We leave to the reader the proof of this statement. •

The role of the control Lie algebra $\mathcal{C}$ in the study of interactions between input and state depends on the following considerations. Suppose $\Delta_{\mathcal{C}}$ has the maximal integral manifolds property and let S_{p° be the maximal integral submanifold of $\Delta_{\mathcal{C}}$ passing through p°. Since the vector fields $f, g_1, \ldots, g_m$, as well as any vector field τ of the form $\tau = f + \sum_{i=1}^{m} g_i u_i$ with $u_1, \ldots, u_m$ real numbers, are in $\Delta_{\mathcal{C}}$ (and therefore tangent to S_{p°), then any state trajectory of the control system (2.1) passing through p° at $t = 0$, due to the action of a piecewise constant control, will stay in S_{p°.

As a consequence of this we see that, when studying the behavior of a control system initialized at $p^\circ \in N$, we may regard as a natural state space the submanifold S_{p° of N instead of the whole N. Since for all $\hat{p} \in S_{p^\circ}$ the tangent vectors $f(\hat{p}), g_1(\hat{p}), \ldots, g_m(\hat{p})$ are elements of the tangent space to S_{p° at $\hat{p}$, by taking the *restriction* to S_{p° of the original vector fields $f, g_1, \ldots, g_m$ one may define a set of vector fields $\hat{f}, \hat{g}_1, \ldots, \hat{g}_m$ on S_{p° and a control system evolving on S_{p°

$$\dot{\hat{p}} = \hat{f}(\hat{p}) + \sum_{i=1}^{m} \hat{g}_i(\hat{p}) u_i \qquad (2.2)$$

which behaves exactly as the original one.

By construction, the smallest subalgebra $\hat{\mathcal{C}}$ of $V(S_{p^\circ})$ which contains $\hat{f}, \hat{g}_1, \ldots, \hat{g}_m$ spans, at each $\hat{p} \in S_{p^\circ}$, the whole tangent space $T_{\hat{p}} S_{p^\circ}$. This may easily be seen using for $\mathcal{C}$ and $\hat{\mathcal{C}}$ the description illustrated in the Remark 2.3.

Therefore, one may conclude that for the control system (2.2) (which evolves on S_{p°), the dimension of $\Delta_{\hat{\mathcal{C}}}$ is equal to that of S_{p° at each point or, also, that the smallest distribution $\hat{R}$ invariant under $\hat{f}, \hat{g}_1, \ldots, \hat{g}_m$ which contains $\hat{f}, \hat{g}_1, \ldots, \hat{g}_m$ is nonsingular (see Remark 2.6), with a dimension equal to that of S_{p°.

The control system (2.2) is such that the assumptions of Theorem 2.2 are satisfied and this makes it possible to state the following result.

Theorem 2.7. Suppose the distribution $\Delta_{\mathcal{C}}$ has the maximal integral manifolds property. Let S_{p° denote the maximal integral submanifold of $\Delta_{\mathcal{C}}$ passing through p°. The set $\mathcal{R}(p^\circ)$ of states reachable from p° under piecewise constant input functions

(a) is a subset of S_{p°
(b) contains an open subset of S_{p°.

Remark 2.8. Note that, if $\Delta_{\mathcal{C}}$ has the maximal integral manifolds property but is *singular*, then the dimensions of different maximal integral submanifolds of $\Delta_{\mathcal{C}}$ may be different. Thus, it may happen that at two different initial states p^1 and p^2 one obtains two control systems of the form (2.2) which evolve on two manifolds S_{p^1} and S_{p^2} of different dimensions. We will see examples of this in Section 4. •

Remark 2.9. Note that the assumption "the distribution $\Delta_{\mathcal{C}}$ has the maximal integral manifolds property" is implied by the assumption "the distribution $\Delta_{\mathcal{C}}$ is nonsingular". In this case, in fact, $\Delta_{\mathcal{C}} = R$ (see Remark 2.6) and R has the maximal integral manifolds property (Lemma 2.1). •

We conclude this Section by the illustration of some terminology which is frequently used. The control system (2.1) is said to satisfy the *controllability rank condition* at p° if

$$\dim \Delta_{\mathcal{C}}(p^\circ) = n. \qquad (2.3)$$

Clearly, if this is the case, and $\Delta_{\mathcal{C}}$ has the maximal integral manifolds property, then the maximal integral submanifold of $\Delta_{\mathcal{C}}$ passing through p° has dimension

n and, according to Theorem 2.7, the set of states reachable from p° fill up at least an open set of the state space N.

The following Corollary of Theorem 2.7 describes the situation which holds when one is free to choose arbitrarily the initial state p°. A control system of the form (2.1) is said to be *weakly controllable* on N if for every initial state $p^\circ \in N$ the set of states reachable under piecewise constant input functions contains at least an open subset of N.

Corollary 2.10. A sufficient condition for a control system of the form (2.1) to be weakly controllable on N is that

$$\dim \Delta_{\mathcal{C}}(p) = n$$

for all $p \in N$. If the distribution $\Delta_{\mathcal{C}}$ has the maximal integral manifolds property then this condition is also necessary.

Proof. If this condition is satisfied, $\Delta_{\mathcal{C}}$ is nonsingular, involutive and therefore, from the previous discussion, we conclude that the system is weakly controllable. Conversely, if the distribution $\Delta_{\mathcal{C}}$ has the maximal integral manifolds property and $\dim \Delta_{\mathcal{C}}(p^\circ) < n$ at some $p^\circ \in N$ then the set of states reachable from p° belongs to a submanifold of N whose dimension is strictly less than n (Theorem 2.7). So this set cannot contain an open subset of N. •

2.3 The Observation Space

In this Section we study state-output interaction properties from a global point of view, for a system described by equations of the form (2.1), together with an output map

$$y = h(p). \tag{3.1}$$

The presentation will be closely analogue to the one given in the previous Section. First of all, recall that the local analysis carried out in Section **1.9** was based upon the consideration of the smallest codistribution invariant under the vector fields $f, g_1, \ldots, g_m$ and containing the covector fields $dh_1, \ldots, dh_l$. If the annihilator Q of this codistribution is nonsingular, then it is also involutive (Lemma **1.9.6**) and may be used to perform a global decomposition of the state space. Parallel to Lemma 2.1 we have the following result.

Lemma 3.1. Suppose Q is nonsingular. Then Q has the maximal integral manifolds property.

The role of this decomposition in describing the state-output interaction may be explained as follows. Observe that Q, being nonsingular and involutive, satisfies the assumptions of Theorem 1.11 (because the set of all vector fields in a nonsingular distribution is locally finitely generated). Let S be any maximal integral submanifold of Q. Since Q is invariant under $f, g_1, \ldots, g_m$ and also under any vector field of the form $\tau = f + \sum_{i=1}^{m} g_i u_i$, where $u_1, \ldots, u_m$ are real numbers, using Theorem 1.11 we deduce that given any two points p^a and p^b in S and any vector field of the form $\tau = f + \sum_{i=1}^{m} g_i u_i$, the points $\Phi_t^\tau(p^a)$ and $\Phi_t^\tau(p^b)$ for all t belong to the same maximal integral submanifold of Q. In other words, we see that from any two initial states on some maximal integral submanifold of Q, under the action of the same piecewise constant control one obtains two trajectories which, at any time, pass through the same maximal integral submanifold of Q.

Moreover, it is easily seen that the functions $h_1, \ldots, h_l$ are constant on each maximal integral submanifold of Q. For, let S be any of these submanifolds and let $\hat{h}_i$ denote the restriction of h_i to S. At each point p of S the derivative of $\hat{h}_i$ along any vector v of $T_p S$ is zero because $Q \subset (\mathrm{span}\{dh_i\})^\perp$, and therefore the function $\hat{h}_i$ is a constant.

As a conclusion, we immediately see that if p^a and p^b are two initial states belonging to the same integral manifold of Q then under the action of the same piecewise constant control one obtains two trajectories which, at any time, produce identical values on each component of the output, e.g. are indistinguishable.

These considerations enable us to state the following global version of Theorem 1.9.8.

Theorem 3.2. Suppose Q is nonsingular. Then there exists a partition of N into maximal integral submanifolds of Q, all with the same dimension. Let S_{p° denote the maximal integral submanifold of Q passing through p°. Then

(a) no other point of S_{p° can be distinguished from p° under piecewise constant input functions

(b) there exists an open neighborhood U of p° in N with the property that any point $p \in U$ which cannot be distinguished from p° under piecewise constant input functions necessarily belongs to $U \cap S_{p^\circ}$.

Proof. The statement (a) has already been proved. The statement (b) requires some remark. Since Q is nonsingular, we know that around any point p° we can find a neighborhood U and a partition of U into slices each of which is clearly an integral submanifold of Q. But also the intersection of S_{p° with U, which is a nonempty open subset of S_{p° is an integral submanifold of Q. Therefore, since S_{p° is maximal, we deduce that the slice of U passing through p° is contained into $U \cap S_{p^\circ}$. From the statement (b) of Theorem 1.9.8 we deduce that any other state of U which cannot be distinguished from p° under piecewise constant inputs belongs to the slice of U passing through p° and therefore to $U \cap S_{p^\circ}$. •

If the distribution Q is singular, one may approach the problem on the basis of the following considerations. Let $\{\lambda_i : 1 \leq i \leq l\}$ be a finite set of real-valued functions and $\{\tau_i : 1 \leq i \leq q\}$ be a finite set of vector fields. Let S_1 and S_2 be two subspaces of $C^\infty(N)$ which both contain the functions $\lambda_1, \ldots, \lambda_l$ and have the property that, for all $\lambda \in S_i$ and for all $1 \leq j \leq q, L_{\tau_j}\lambda \in S_i, i = 1, 2$. Clearly the intersection $S_1 \cap S_2$ is again a subspace of $C^\infty(N)$ which contains $\lambda_1, \ldots, \lambda_l$ and is such that, for all $\lambda \in S_1 \cap S_2$ and for all $1 \leq j \leq q, L_{\tau_j}\lambda \in S_1 \cap S_2$. Thus we conclude that there exists a unique subspace S of $C^\infty(N)$ which contains $\lambda_1, \ldots, \lambda_l$ and is such that, for all $\lambda \in S$ and for all $1 \leq j \leq q, L_{\tau_j}\lambda \in S$. This is the smallest subspace of C^∞ which contains $\lambda_1, \ldots, \lambda_l$ and is closed under differentiation along $\tau_1, \ldots, \tau_q$.

Remark 3.3. The subspace S may be described as follows. Consider the set

$$S_0 = \{\lambda \in C^\infty(N) : \lambda = \lambda_j \text{ or } \lambda = L_{\tau_{i_1}} \ldots L_{\tau_{i_k}} \lambda_j; 1 \leq j \leq l, 1 \leq i_k \leq q,$$
$$1 \leq k < \infty\}$$

and let $LC(S_0)$ denote the set of all **R**-linear combinations of elements of S_0. Then, $LC(S_0) = S$. As a matter of fact, it is easily checked that every element of $LC(S_0)$ is an element of S, so $LC(S_0) \subset S$, that $\lambda_j \in LC(S_0)$ for $1 \leq j \leq l$ and that $LC(S_0)$ is closed under differentiation along $\tau_1, \ldots, \tau_q$. •

With the subspace S we may associate a codistribution Ω_S, in a natural way, by setting

$$\Omega_S = \text{span}\{d\lambda : \lambda \in S\}.$$

The codistribution Ω_S is smooth by construction, but—as we know—the distribution $\Omega_S^\perp$ may fail to be so. Since we are interested in smooth distributions because we use them to partition the state space into maximal integral submanifolds, we should rather be looking at the distribution $\text{smt}(\Omega_S^\perp)$ (see Remark 1.3.1).

The following result is important when looking at $\text{smt}(\Omega_S^\perp)$ for the purpose of finding global decompositions of N.

Lemma 3.4. Suppose the set of all vector fields in $\text{smt}(\Omega_S^\perp)$ is locally finitely generated. Then $\text{smt}(\Omega_S^\perp)$ has the maximal integral manifolds property.

Proof. In view of Theorem 1.7, we have only to show that $\text{smt}(\Omega_S^\perp)$ is involutive. Let τ_1 and τ_2 be two vector fields in $\text{smt}(\Omega_S^\perp)$ and λ any function in S. Since $\langle d\lambda, \tau_1 \rangle = 0$ and $\langle d\lambda, \tau_2 \rangle = 0$ we have

$$\langle d\lambda, [\tau_1, \tau_2] \rangle = L_{\tau_1} \langle d\lambda, \tau_2 \rangle - L_{\tau_2} \langle d\lambda, \tau_1 \rangle = 0.$$

The vector field $[\tau_1, \tau_2]$ is thus in $\Omega_S^\perp$. But $[\tau_1, \tau_2]$, being smooth, is also in $\text{smt}(\Omega_S^\perp)$. •

In order to study the observability we consider the smallest subspace of $C^\infty(N)$ which contains the functions $h_1, \ldots, h_l$ and is closed under differentiation along the vector fields $f, g_1, \ldots, g_m$. This subspace will be denoted by $\mathcal{O}$ and called the *Observation Space*. Moreover, with $\mathcal{O}$ we associate the codistribution

$$\Omega_{\mathcal{O}} = \operatorname{span}\{d\lambda : \lambda \in \mathcal{O}\}.$$

Remark 3.5. It is possible to prove that the distribution $\Omega_{\mathcal{O}}^{\perp}$ is invariant under the vector fields $f, g_1, \ldots, g_m$. For, let λ be any function in $\mathcal{O}$ and τ a vector field in $\Omega_{\mathcal{O}}^{\perp}$. Then $\langle d\lambda, \tau \rangle = 0$ and $\langle dL_f\lambda, \tau \rangle = 0$ because $L_f\lambda$ is again a function in $\mathcal{O}$. Therefore, from the equality

$$\langle d\lambda, [f, \tau] \rangle = L_f \langle d\lambda, \tau \rangle - \langle dL_f\lambda, \tau \rangle = 0$$

we deduce that $[f, \tau]$ annihilates all functions in $\mathcal{O}$. Since $\Omega_{\mathcal{O}}$ is spanned by differentials of functions in $\mathcal{O}$, it follows that $[f, \tau]$ is a vector field in $\Omega_{\mathcal{O}}^{\perp}$. In the same way one proves the invariance under $g_1, \ldots, g_m$.

If the distribution $\Omega_{\mathcal{O}}^{\perp}$ is smooth (e.g. when the codistribution $\Omega_{\mathcal{O}}$ is nonsingular) the using Lemma **1.6.7** one concludes that $\Omega_{\mathcal{O}}$ itself is invariant under $f, g_1, \ldots, g_m$. •

Remark 3.6 The distribution $\Omega_{\mathcal{O}}^{\perp}$ and the distribution Q introduced in the previous Chapter are related in the following way
(a) $\Omega_{\mathcal{O}}^{\perp} \supset Q$
(b) if p is a regular point of $\Omega_{\mathcal{O}}$, then $\Omega_{\mathcal{O}}^{\perp}(p) = Q(p)$.

We leave to the reader the proof of this statement. •

From the previous Remark 3.5 and from Remark **1.6.6** it is deduced that the distribution $\operatorname{smt}(\Omega_{\mathcal{O}}^{\perp})$ is invariant under the vector fields $f, g_1, \ldots, g_m$ and so under any vector field τ of the form $\tau = f + \sum_{i=1}^{m} g_i u_i$, where $u_1, \ldots, u_m$ are real numbers. Now suppose that the set of all vector fields in $\operatorname{smt}(\Omega_{\mathcal{O}}^{\perp})$ is locally finitely generated, so that $\operatorname{smt}(\Omega_{\mathcal{O}}^{\perp})$ has the maximal integral manifolds property. Using Theorem **1.11**, as we did before in the case of nonsingular Q, we may conclude that from any two states on the same integral submanifold of $\operatorname{smt}(\Omega_{\mathcal{O}}^{\perp})$, under the action of the same piecewise constant control one obtains two trajectories that at any time lie on the same maximal integral submanifold of $\operatorname{smt}(\Omega_{\mathcal{O}}^{\perp})$. Observe now that $\operatorname{smt}(\Omega_{\mathcal{O}}^{\perp})$ is also contained in $(\operatorname{span}\{dh_i\})^{\perp}, 1 \leq i \leq l$, because every tangent vector in $\operatorname{smt}(\Omega_{\mathcal{O}}^{\perp})(p)$ is also in $\Omega_{\mathcal{O}}^{\perp}(p)$ and every tangent vector v in $\Omega_{\mathcal{O}}^{\perp}(x)$ is such that $\langle dh_i(p), v \rangle = 0$. Therefore one may deduce that the functions h_i are constant on each maximal integral submanifold of $\operatorname{smt}(\Omega_{\mathcal{O}}^{\perp})$.

This, together with the previous observations, shows that any two initial states p^a and p^b on the same maximal integral submanifold of $\operatorname{smt}(\Omega_{\mathcal{O}}^{\perp})$ are indistinguishable under piecewise constant inputs. This extends the statement (a) of Theorem 3.2. As for the statement (b), some regularity is required, as it is seen hereafter.

Theorem 3.7. Suppose the set of all vector fields contained in $\mathrm{smt}(\Omega_{\mathcal{O}}^{\perp})$ is locally finitely generated. Let $S_{p^{\circ}}$ denote the maximal integral submanifold of $\mathrm{smt}(\Omega_{\mathcal{O}}^{\perp})$ passing through p°. Then
(a) No other point of $S_{p^{\circ}}$ can be distinguished from p° under piecewise constant inputs
(b) If p° is a regular point of $\Omega_{\mathcal{O}}$, then there exists an open neighborhood U of p° in N with the property that any point $p \in U$ which cannot be distinguished from p° under piecewise constant inputs necessarily belongs to $U \cap S_{p^{\circ}}$.

Proof. The statement (a) has already been proved. The statement (b) is proved essentially in the same way as the statement (b) of Theorem 3.2. ●

The following example illustrates the need for the "regularity" assumption in the statement (b) of the previous Theorem.

Example 3.1. Consider the following system with $N = \mathbf{R}$ and

$$\dot{x} = 0$$
$$y = h(x)$$

where $h(x)$ is defined as

$$h(x) = \exp\left(-\frac{1}{x^2}\right) \sin\left(\frac{1}{x}\right) \quad \text{for } x \neq 0$$
$$h(0) = 0.$$

For this system, two states x^a and x^b are indistinguishable if and only if $h(x^a) = h(x^b)$. In particular, the set of states which are indistinguishable from the state $x = 0$ concides with the set of the roots of the equation $h(x) = 0$. Each point in this set is isolated but the point $x = 0$. Thus, no matter how small we choose an open neighborhood U of $x = 0$, U contains points indistinguishable from $x = 0$.

It is also seen that the codistribution $\Omega_{\mathcal{O}} = \mathrm{span}\{dh\}$ has dimension 1 everywhere but at the points x in which $dh/dx = 0$ where its dimension is 0. Thus, any smooth vector field belonging to $\Omega_{\mathcal{O}}^{\perp}$ must vanish identically on $\mathbf{R}$ and $\mathrm{smt}(\Omega_{\mathcal{O}}^{\perp}) = \{0\}$. The maximal integral submanifold of $\mathrm{smt}(\Omega_{\mathcal{O}}^{\perp})$ passing through x is the point x itself.

At the point $x = 0$, which is not a regular point of $\Omega_{\mathcal{O}}$, we have that $U \cap S_{0} = \{0\}$ for all U, whereas we know there are other points of U indistinguishable from $x = 0$. ●

We conclude this Section with some global considerations. The control system

(2.1)-(3.1) is said to satisfy the *observability rank condition* at p° if

$$\dim \Omega_{\mathcal{O}}(p^\circ) = n \tag{3.2}$$

Clearly, if this is the case then p° is a regular point of $\Omega_{\mathcal{O}}$ and from the previous discussion it is seen that any point p in a suitable neighborhood U of p° can be distinguished under piecewise constant inputs. A control system of the form (2.1)-(3.1) is said to be *locally observable* on N if for every state p° there is a neighborhood U of p° in which every point can be distinguished from p° under piecewise constant inputs.

Corollary 3.8. A sufficient condition for a control system of the form (2.1)-(3.1) to be locally observable on N is that

$$\dim \Omega_{\mathcal{O}}(p) = n$$

for all $p \in N$.

2.4 Linear Systems and Bilinear Systems

In this Section we describe some elementary examples, in order to make the reader more familiar with the ideas introduced so far.

As a first application, we shall compute the Lie algebra $\mathcal{C}$ and the distribution $\Delta_{\mathcal{C}}$ for a *linear system*

$$\dot{x} = Ax + Bu$$
$$y = Cx$$

Recall (see also Section 1.2) that this is indeed a system of the form (2.2)-(3.2), with $N = \mathbf{R}^n$ and

$$f(x) = Ax$$
$$g_i(x) = b_i \qquad 1 \le i \le m$$

where b_i is the i-th column of the the matrix B and

$$h_i(x) = c_i x \qquad 1 \le i \le l$$

where c_i is the i-th row of the matrix C.

We want to prove first that the control Lie algebra $\mathcal{C}$ is the subspace of $V(N)$ consisting of all vector fields which are $\mathbf{R}$-linear combinations of the vector fields in the set

$$\{Ax\} \cup \{A^k b_i : 1 \le i \le m, 0 \le k \le n-1\} \tag{4.1}$$

For, observe that this set contains the vector fields Ax and $b_1, \ldots, b_m$ (i.e. the vector fields $f, g_1, \ldots, g_m$) and also that this set is contained in $\mathcal{C}$, because any of its elements is a repeated Lie bracket of f and $g_1, \ldots, g_m$. As a matter of fact,

$$A^k b_i = (-1)^k a d_f^k g_i.$$

Moreover, it is easy to see that the set

$$LC(\{Ax\} \cup \{A^k b_i : 1 \le i \le m, 0 \le k \le n - 1\}) \qquad (4.2)$$

of all **R**-linear combinations of vector fields in the set (4.1) is already a Lie subalgebra, i.e. is closed under Lie bracketing.

For, one easily sees that if $\tau_1(x)$ and $\tau_2(x)$ are vector fields of the form

$$\tau_1(x) = A^k b_i$$
$$\tau_2(x) = A^h b_j$$

then $[\tau_1, \tau_2](x) = 0$. On the other hand, if

$$\tau_1(x) = A^k b_i$$
$$\tau_2(x) = Ax$$

then

$$[\tau_1, \tau_2] = A^{k+1} b_i.$$

If $k < n - 1$, this vector field is in the set (4.1) and, if $k = n - 1$, this vector field is an **R**-linear combination of vector fields in the set (4.1) (by Cayley-Hamilton Theorem).

It τ_1 and τ_2 are **R**-linear combinations of vector fields of (4.1) then their Lie bracket is still an **R**-linear combination of vector fields of (4.1), and this proves that the set (4.2) is a Lie subalgebra.

The set (4.2) is a Lie algebra which contains $f, g_1, \ldots, g_m$ and is contained in $\mathcal{C}$, the smallest Lie subalgebra which contains $f, g_1, \ldots, g_m$. Then, the set (4.2) coincides with $\mathcal{C}$.

Evaluating the distribution $\Delta_{\mathcal{C}}$ we get, at a point $x \in \mathbf{R}^n$,

$$\Delta_{\mathcal{C}}(x) = \text{span}\{Ax\} + \text{span}\{A^k b_i : 1 \le i \le m, 0 \le k \le n - 1\}$$
$$= \text{span}\{Ax\} + \sum_{k=0}^{n-1} \text{Im}(A^k B). \qquad (4.3)$$

We are now interested in the distribution P, the smallest distribution which contains $g_1, \ldots, g_m$ and is invariant under $f, g_1, \ldots, g_m$. By means of calculations similar to the ones in Remark **1.8.3**, it is easy to check that at any point $x \in \mathbf{R}^n$

$$P(x) = \mathrm{span}\{A^k b_i : 1 \leq i \leq m, 0 \leq k \leq n-1\}. \tag{4.4}$$

Thus, we see that

$$\Delta_C = \mathrm{span}\{f\} + P.$$

The distribution Δ_C is spanned by a set of vector fields which is locally finitely generated (because any vector field in C is *analytic* on $\mathbf{R}^n$), and therefore—by Lemma **2.4**—the distribution Δ_C has the maximal integral manifolds property. The distribution P is nonsingular and involutive and thus—by Corollary **1.8**—it also has the maximal integral manifolds property.

The maximal integral submanifolds of P, all of the same dimension, have the form $x + V$, where

$$V = \mathrm{Im}(B \ AB \ \ldots \ A^{n-1}B).$$

The maximal integral submanifolds of Δ_C may have different dimensions, because Δ_C may have singularities.

If, at some point $x \in \mathbf{R}^n$, $f(x) \in P(x)$, then the maximal integral submanifold of Δ_C passing through x coincides with the one of the distribution P, i.e. is a subset of the form $x + V$. Otherwise, if such a condition is not verified, the maximal integral submanifold of Δ_C is a submanifold whose dimension exceeds by 1 that of P and this submanifold, in turn, is partitioned into subsets of the form $x' + V$.

Example 4.1. The following simple example illustrates the case of a singular Δ_C. Let the system be described by

$$\dot{x} = \begin{bmatrix} 1 & 0 & 0 \\ 0 & -1 & 0 \\ 0 & 0 & 1 \end{bmatrix} x + \begin{bmatrix} 1 \\ 0 \\ 0 \end{bmatrix} u.$$

Then we easily see that

$$V = \{x \in \mathbf{R}^3 : x_2 = x_3 = 0\}$$

and that

$$P = \mathrm{span}\left\{\frac{\partial}{\partial x_1}\right\}.$$

The tangent vector $f(x)$ belongs to P only at those x in which $x_2 = x_3 = 0$, i.e. only on V. Thus, the maximal integral submanifolds of Δ_C will have dimension 2 everywhere but on V. A direct computation shows that these submanifolds may be described in the following way (Fig. 2.1)

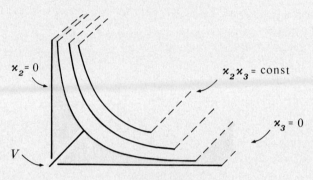

Fig. 2.1

(i) if x° is such that $x_2^\circ = 0$ (resp. $x_3^\circ = 0$) then the maximal submanifold passing through x° is the half open plane

$$\{x \in \mathbf{R}^n : x_2 = 0 \text{ and } \mathrm{sgn}(x_3) = \mathrm{sgn}(x_3^\circ)\}$$
$$(\text{resp. } \{x \in \mathbf{R}^n : x_3 = 0 \text{ and } \mathrm{sgn}(x_2) = \mathrm{sgn}(x_2^\circ)\})$$

(ii) if x° is such that both $x_2^\circ \neq 0$ and $x_3^\circ \neq 0$, then the maximal submanifold passing through x° is the surface

$$\{x \in \mathbf{R}^n : x_2 x_3 = x_2^\circ x_3^\circ\}. \quad \bullet$$

We turn now to the computation of the subspace $\mathcal{O}$ and the codistribution $\Omega_{\mathcal{O}}$. It is easy to prove that $\mathcal{O}$ is the subspace of $C^\infty(N)$ consisting of all $\mathbf{R}$-linear combinations of functions of the form $c_i A^k x$ or $c_i A^k b_j$, namely that

$$\mathcal{O} = LC\{\lambda \in C^\infty(N) : \lambda(x) = c_i A^k x \quad \text{or} \\ \lambda(x) = c_i A^k b_j; \ 1 \leq i \leq l, 1 \leq j \leq m, 0 \leq k \leq n-1\}. \tag{4.5}$$

For, note that functions of the form $c_i A^k x$ or $c_i A^k b_j$ are such that

$$c_i A^k x = L_f^k h_i(x)$$
$$c_i A^k b_j = L_{g_j} L_f^k h_i(x)$$

and this implies that the right-hand-side of (4.5) is contained in $\mathcal{O}$. Moreover, the functions $h_1, \ldots, h_l$ are elements of the right-hand-side of (4.5). Then, the proof of (4.5) is completed as soon as we show that its right-hand-side is closed under differentiation along $f, g_1, \ldots, g_m$.

If $\lambda(x) = c_i A^k x$, then $L_f \lambda = c_i A^{k+1} x$ and $L_{g_j} \lambda(x) = c_i A^k b_j$. If $\lambda(x) = c_i A^k b_j$, then $L_f \lambda(x) = L_{g_j} \lambda(x) = 0$. Thus, using again Cayley-Hamilton Theorem, it is easily seen that the right-hand-side of (4.5) is closed under differentiation along $f, g_1, \ldots, g_m$.

At each point x, the codistribution $\Omega_{\mathcal{O}}$ is given by $\Omega_{\mathcal{O}}(x) = \mathrm{span}\{c_i A^k : 1 \leq i \leq l, 0 \leq k \leq n-1\}$. Thus, $\Omega_{\mathcal{O}}$ is nonsingular and, in view of Remark 3.6 (part (b))

$$\Omega_{\mathcal{O}}^{\perp}(x) = \bigcap_{k=0}^{n-1} \ker(CA^k) = Q(x)$$

(see also Remark 1.9.3). The maximal integral submanifolds of Q have now the form $x + W$ where

$$W = \bigcap_{i=0}^{n-1} \ker(CA^i).$$

As a second application we consider a *bilinear system*, i.e. a system described by equations of the form

$$\dot{x} = Ax + \sum_{i=1}^{m}(N_i x)u_i$$

$$y = Cx.$$

Here also the manifold on which the system evolves is the whole of $\mathbf{R}^n$; f and $h_1, \ldots, h_l$ are the same as before, and

$$g_i(x) = N_i x \qquad 1 \leq i \leq m.$$

In order to compute the subalgebra $\mathcal{C}$ we note first that any vector field τ in the set $\{f, g_1, \ldots, g_m\}$ has the form $\tau(x) = Tx$, where T is an $n \times n$ matrix. If we want to take the Lie bracket of two vector fields τ_1, τ_2 of the form

$$\tau_1(x) = T_1 x \qquad \tau_2(x) = T_2 x$$

we have

$$[\tau_1, \tau_2](x) = (T_2 T_1 - T_1 T_2)x = [T_1, T_2]x$$

where $[T_1, T_2] = (T_2 T_1 - T_1 T_2)$ is the *commutator* of T_1 and T_2.

On the basis of this observation, it is easy to set up a recursive procedure yielding the smallest Lie subalgebra which contains a set of vector fields of the form $\tau_1(x) = T_1 x, \ldots, \tau_r(x) = T_r x$.

Lemma 4.1. Consider the nondecreasing sequence of subspaces of $\mathbf{R}^{n \times n}$, the space of all $n \times n$ matrices of real numbers, defined by setting

$$M_0 = \mathrm{span}\{T_1, \ldots, T_r\}$$
$$M_k = M_{k-1} + \mathrm{span}\{[T_1, T], \ldots, [T_r, T] : T \in M_{k-1}\}.$$

Then, there exists an integer $k^\star$ such that

$$M_k = M_{k^\star}$$

for all $k > k^\star$. The set of vector fields

$$\mathcal{L} = \{\tau \in V(\mathbf{R}^n) : \tau(x) = Tx, T \in M_{k^\star}\}$$

is the smallest Lie subalgebra of vector fields which contains $\tau_1(x) = T_1 x, \ldots, \tau_r(x) = T_r x$.

Proof. The proof is rather simple and consists in the following steps. A dimensionality argument proves the existence of the integer $k^\star$ such that $M_k = M_{k^\star}$ for all $k > k^\star$. Then, one checks that the subspace $M_{k^\star}$ contains $T_1, \ldots, T_r$ and any repeated commutator of the form $[T_{i_1}, \ldots, [T_{i_{h-1}}, T_{i_h}]]$ and is such that $[P, Q] \in M_{k^\star}$ for all $P \in M_{k^\star}$ and $Q \in M_{k^\star}$. From these properties, it is straightforward to deduce that $\mathcal{L}$ is the desired Lie algebra. •

Based on this result, it is easy to construct the Lie algebra $\mathcal{C}$ by simply initializing the algorithm described in the above Lemma with the matrices $A, N_1, \ldots, N_m$.

In this case, unlike the previous one, we cannot anymore give a simple expression of $\Delta_{\mathcal{C}}(x)$ and/or its maximal integral submanifolds. In some special situations, however, like the one illustrated in the following example, a rather satisfactory analysis is possible.

Example 4.2. Consider the system

$$\dot{x} = Ax + Nxu$$

where $x \in \mathbf{R}^3$ and

$$A = \begin{bmatrix} 0 & 1 & 0 \\ -1 & 0 & 0 \\ 0 & 0 & 0 \end{bmatrix} \qquad N = \begin{bmatrix} 0 & 0 & 1 \\ 0 & 0 & 0 \\ -1 & 0 & 0 \end{bmatrix}.$$

An easy calculation shows that

$$[A, N] = \begin{bmatrix} 0 & 0 & 0 \\ 0 & 0 & 1 \\ 0 & -1 & 0 \end{bmatrix}$$

$$[N, [A, N]] = A \qquad [A, [A, N]] = -N.$$

Therefore, we have

$$\mathcal{C} = \left\{ \tau \in V(\mathbf{R}^3) : \tau(x) = Tx, T \in \operatorname{span}\{A, N, [A, N]\} \right\}.$$

To compute the dimension of $\Delta_\mathcal{C}$ we evaluate the rank of the matrix

$$\left(Ax, Nx, [A, N]x \right) = \begin{bmatrix} x_2 & x_3 & 0 \\ -x_1 & 0 & x_3 \\ 0 & -x_1 & -x_2 \end{bmatrix}$$

and we find the following result

$$\dim \Delta_\mathcal{C}(x) = 0 \quad \text{if } x = 0$$
$$\dim \Delta_\mathcal{C}(x) = 2 \quad \text{if } x \neq 0.$$

A direct computation shows that the maximal integral submanifold of $\Delta_\mathcal{C}$ passing through x° is the set

$$\left\{ x \in \mathbf{R}^3 : x_1^2 + x_2^2 + x_3^2 = (x_1^\circ)^2 + (x_2^\circ)^2 + (x_3^\circ)^2 \right\}$$

i.e. the sphere centered at the origin passing through x°.

Therefore, we can say that the state of the system is not free to evolve on the whole of $\mathbf{R}^n$, but rather on the sphere centered at the origin which passes through the initial state.

Around any point $x \neq 0$ the distribution $\Delta_\mathcal{C}$ is nonsingular, so we can obtain locally a decomposition of the form (1.8.4), by means of a suitable coordinates transformation. To this end, we may make use of the construction introduced in the proof of Theorem 1.4.1 and find a set of three vector fields τ_1, τ_2, τ_3 with the property that τ_1 and τ_2 belong to $\Delta_\mathcal{C}$ and $\tau_1(x^\circ)$, $\tau_2(x^\circ)$ and $\tau_3(x^\circ)$ are linearly independent. If we consider an initial point on the line

$$\{x \in \mathbf{R}^3 : x_1 = x_2 = 0\}$$

we may take the vector fields

$$\tau_1(x) = (Nx)$$
$$\tau_2(x) = ([A, N]x)$$
$$\tau_3 = \begin{bmatrix} 0 \\ 0 \\ 1 \end{bmatrix}.$$

Accordingly we get

$$\Phi_t^1(x) = \begin{bmatrix} (\cos t)x_1 + (\sin t)x_3 \\ x_2 \\ -(\sin t)x_1 + (\cos t)x_3 \end{bmatrix}$$

$$\Phi_t^2(x) = \begin{bmatrix} x_1 \\ (\cos t)x_2 + (\sin t)x_3 \\ -(\sin t)x_2 + (\cos t)x_3 \end{bmatrix}$$

$$\Phi_t^3(x) = \begin{bmatrix} x_1 \\ x_2 \\ t + x_3 \end{bmatrix}.$$

The local coordinate chart around the point x^o is given by the inverse of the function

$$F : (z_1, z_2, z_3) \mapsto \Phi_{z_1}^1 \circ \Phi_{z_2}^2 \circ \Phi_{z_3}^3 (x^o).$$

For $x_1^o = x_2^o = 0$ and $x_3^o = a$ we have

$$F(z_1, z_2, z_3) = \begin{bmatrix} (\sin z_1)(\cos z_2)(z_3 + a) \\ (\sin z_2)(z_3 + a) \\ (\cos z_1)(\cos z_2)(z_3 + a) \end{bmatrix}.$$

The local representations of the vector fields f and g in the new coordinate chart are given by

$$\tilde{f}(z) = (F_\star)^{-1} f(F(z)) = (F_\star)^{-1} A F(z)$$
$$\tilde{g}(z) = (F_\star)^{-1} g(F(z)) = (F_\star)^{-1} N F(z).$$

A simple but tedious computation yields

$$\tilde{f}(z) = \begin{bmatrix} \cos z_1 \tan z_2 \\ -\sin z_1 \\ 0 \end{bmatrix} \qquad \tilde{g}(z) = \begin{bmatrix} 1 \\ 0 \\ 0 \end{bmatrix}.$$

We conclude that around x^o the system, in the z coordinates, is described by the equations

$$\dot{z}_1 = \cos z_1 \tan z_2 + u$$
$$\dot{z}_2 = -\sin z_1$$
$$\dot{z}_3 = 0. \quad \bullet$$

The study of the observability of a bilinear system is much simpler. By means of arguments similar to those used in the case of linear systems it is easy to prove that $\mathcal{O}$ is given by

$$\mathcal{O} = \{\lambda \in C^\infty(N) : \lambda(x) = c_i x \text{ or } \lambda(x) = c_i N_{j_1} \dots N_{j_k} x;$$

$$1 \leq i \leq l, 1 \leq k \leq n-1, 0 \leq j_1, \dots, j_k \leq m\}$$

(with $N_0 = A$). Therefore

$$\Omega_{\mathcal{O}}^{\perp} = \bigcap_{k=0}^{n-1} \bigcap_{j_1,\dots,j_k=0}^{m} \ker(CN_{j_1} \dots N_{j_k}).$$

The distribution $\Omega_{\mathcal{O}}^{\perp} = Q$ is nonsingular and its maximal integral submanifolds have the form $x + W$, where now

$$W = \bigcap_{k=0}^{n-1} \bigcap_{j_1,\dots,j_k=0}^{m} \ker(CN_{j_1} \dots N_{j_k}).$$

It may be worth observing that the subspace W thus defined is invariant under $A, N_1, \dots, N_m$, is contained in $\ker(C)$ and is the largest subspace of $\mathbf{R}^n$ having these properties. From linear algebra we know that by taking a suitable change of coordinates in $\mathbf{R}^n$ (see e.g. Section 1.1) the matrices $A, N_1, \dots, N_m$ become block triangular and, therefore, the dynamics of the system become described by equations of the form

$$\dot{x}_1 = A_{11}x_1 + A_{12}x_2 + \sum_{i=1}^{m}(N_{i,11}x_1 + N_{i,12}x_2)u_i$$

$$\dot{x}_2 = A_{22}x_2 + \sum_{i=1}^{m} N_{i,22}x_2 u_i.$$

Moreover, the output y depends only on the x_2 coordinates, $y = C_2 x_2$

The above equations are exactly of the form (1.7.3), this time obtained by means of standard linear algebra arguments.

2.5 Examples

In this Section we discuss an example of application of the theories illustrated in the Chapter to a control system whose state space is a manifold N not diffeormorphic to $\mathbf{R}^n$. More precisely, we study the system—already introduced in Section 5.1—which describes the control of the attitude of a spacecraft by means of equations of the form

$$J\dot{\omega} = S(\omega)J\omega + T \tag{5.1}$$
$$\dot{R} = S(\omega)R \tag{5.2}$$

with state $(\omega, R) \in \mathbf{R}^3 \times SO(3)$ and input $T \in \mathbf{R}^3$. The orthogonal matrix R represents the orientation of the spacecraft with respect to an inertially fixed reference frame, the vector ω its angular velocity, and the vector T represents the external torque. The matrix J is the so-called inertia matrix of the spacecraft, and $S(\omega)$ is the skew-simmetric matrix

$$S(\omega) = \begin{bmatrix} 0 & \omega_3 & -\omega_2 \\ -\omega_3 & 0 & \omega_1 \\ \omega_2 & -\omega_1 & 0 \end{bmatrix}.$$

If we suppose the external torque T generated by a set of r independent pairs of gas jets (*thrusters*), it is possible to set

$$T = b_1 u_1 + \cdots + b_r u_r$$

where $b_1, \ldots, b_r \in \mathbf{R}^3$ represent the vectors of direction cosines—with respect to the body frame—of the axes about which the control torques are applied and $u_1, \ldots, u_r$ the corresponding magnitudes. Of course, we assume the set $\{b_1, \ldots, b_r\}$ is a linearly independent set (and thus $r \leq 3$).

We want to analyze the partition induced by the distribution $\Delta_{\mathcal{C}}$, in the two cases $r = 3$ and $r = 2$. For convenience, we begin by discussing the dynamic equation (5.1) only. Note that, setting

$$x = J\omega$$

and using the property

$$S(w)v = -S(v)w$$

(which holds for any pair of vectors $v, w \in \mathbf{R}^3$) the equation in question can be rewritten in the form

$$\dot{x} = -S(x)J^{-1}x + Bu$$

where $B = (b_1 \ldots b_r)$, i.e.

$$\dot{x} = f(x) + g_1(x)u_1 + \cdots + g_r(x)u_r$$

with

$$f(x) = -S(x)J^{-1}x \qquad g_i(x) = b_i \qquad 1 \leq i \leq r.$$

The case in which $r = 3$ is rather trivial. In fact, since the control Lie algebra $\mathcal{C}$ contains, by definition, the three vector fields $g_1(x), g_2(x), g_3(x)$, and these vector fields—which are constant—are by assumption linearly independent at each $x \in \mathbf{R}^3$, we have immediately

$$\Delta_{\mathcal{C}}(x) = T_x \mathbf{R}^3 \qquad \text{for all } x \in \mathbf{R}^3.$$

In other words, the controllability rank condition (2.3) is satisfied at each x, and the partition of $\mathbf{R}^3$ induced by $\Delta_{\mathcal{C}}$ degenerates into one single element, namely $\mathbf{R}^3$ itself.

The case $r = 2$ is more interesting (at least from the point of view of the analysis). In this case, to obtain meaningful informations, one has to compute a few Lie brackets between $f(x)$ and the $g_i(x)$'s. Let c_1 and c_2 be real numbers and consider the (constant) vector field

$$g(x) = c_1 g_1(x) + c_2 g_2(x)$$

Since, as a straightforward calculation shows,

$$\frac{\partial f}{\partial x} = -S(x)J^{-1} + S(J^{-1}x)$$

then, setting $b = c_1 b_1 + c_2 b_2$, it is immediate to see that

$$[f, g](x) = S(x)J^{-1}b - S(J^{-1}x)b$$
$$[[f, g], g](x) = -2S(b)J^{-1}b.$$

By definition, the control Lie algebra contains the three (constant) vector fields $g_1(x), g_2(x), [[f, g], g](x)$. Thus, if

$$\text{rank}(b_1 \ b_2 \ S(b)J^{-1}b) = 3 \tag{5.3}$$

we again obtain that $\Delta_{\mathcal{C}}(x)$ has dimension 3 at each x as before, and the associated partition of $\mathbf{R}^3$ degenerates into one single element.

Note that the vector b is an arbitrary vector in the image if the matrix

$$B = (b_1 \ b_2)$$

and, therefore, the possibility of having the condition (5.3) fulfilled can be restated in the following terms

$$S(b)J^{-1}b \notin \text{Im}(B) \qquad \text{for some } b \in \text{Im}(B) \tag{5.4}.$$

We show now that, if the condition (5.4) is not satisfied, then $\Delta_{\mathcal{C}}$ has dimension 2 at all points of a certain plane in $\mathbf{R}^3$, and dimension 3 everywhere else. For, note that the control Lie algebra contains the three vector fields $g_1(x), g_2(x), [f, g_i](x)$, whose corresponding tangent vectors span $T_x\mathbf{R}^3$ at each x such that

$$\det(b_1 \ b_2 \ [f, g_i](x)) \neq 0$$

This determinant, being linear in x, is either identically vanishing or zero only at points of a plane in $\mathbf{R}^3$. It is easy to check that—for an appropriate value of i—the first situation cannot occurr. In fact, as x ranges over $\mathbf{R}^3$, the vector $[f, g_i](x)$ cannot be contained into $\mathrm{Im}(B)$ for both values of i, because b_1 and b_2 are independent. This proves that the distribution Δ_C has dimension 3 everywhere but possibly at points of a plane. We will see now that there *is* a plane where the dimension of Δ_C is less than 3.

For, let γ denote a (nonzero) row vector satisfying

$$\gamma B = 0$$

and suppose a linear (thus globally defined) coordinates transformation is performed, changing x into $z = Tx$, with

$$z_1(x) = \gamma x.$$

By definition of γ, we have

$$\dot{z}_1 = \gamma \dot{x} = \gamma(-S(x)J^{-1}x + Bu) = -\gamma S(x)J^{-1}x. \qquad (5.5)$$

If the condition (5.4) does not hold, at each point x of B, $S(x)J^{-1}x$ is a vector in $\mathrm{Im}(B)$. Since

$$x \in \mathrm{Im}(B) \Leftrightarrow \gamma x = 0 \Leftrightarrow z_1(x) = 0$$

we see from (5.5) that, if the condition (5.4) does not hold, at each point where $z_1 = 0$, then $\dot{z}_1 = 0$ also. This means that any trajectory of the system (5.1) starting in the plane

$$M = \{x \in \mathbf{R}^3 : \gamma x = 0\}$$

remains in this plane for all times. As a consequence, in view of the results established in Section 2, we deduce that necessarily Δ_C has at most dimension 2 at each point of M (in fact, it *has* dimension 2 because b_1 and b_2 are independent).

Summarizing, we have seen that if the condition (5.4) does not hold, the distribution Δ_C has dimension 2 at all points of M and dimension 3 everywhere else. As a result, the state space of (5.1) is partitioned by Δ_C into three maximal integral manifolds: the plane M and the two (open) half-spaces separated by M.

We study now the same kind of problems for the full system (5.1)-(5.2), whose state space is the manifold

$$N = \mathbf{R}^3 \times SO(3).$$

To this end, a few preliminary remarks about the structure of the tangent space to $SO(3)$ are in order.

Recall that $SO(3)$ is an embedded 3-dimensional submanifold of the manifold $\mathbf{R}^{3\times3}$. As a consequence, the tangent space to $SO(3)$ at R can be viewed as a(3-dimensional) subspace of the tangent space $T_R \mathbf{R}^{3\times3}$. Let x_{ij} denotes the (i,j) element of a matrix $X \in \mathbf{R}^{3\times3}$, and choose the natural (globally defined on $\mathbf{R}^{3\times3}$) coordinate functions

$$\{\phi_{ij}(X) = x_{ij} : 1 \leq i,j \leq 3\}$$

This choice induces, at each X, a choice of a basis for $T_X \mathbf{R}^{3\times3}$, namely the set of tangent vectors

$$\{\frac{\partial}{\partial x_{ij}} : 1 \leq i,j \leq 3\}. \tag{5.6}$$

Using this basis, any tangent vector v at a point X of $\mathbf{R}^{3\times3}$ will be represented in the form

$$v = \sum_{i,j=1}^{3} v_{ij} \frac{\partial}{\partial x_{ij}}$$

where v_{ij} denotes the (i,j) element of a 3×3 matrix V.

Now, consider the three matrices

$$A_1 = \begin{bmatrix} 0 & 1 & 0 \\ -1 & 0 & 0 \\ 0 & 0 & 0 \end{bmatrix} \qquad A_2 = \begin{bmatrix} 0 & 0 & 1 \\ 0 & 0 & 0 \\ -1 & 0 & 0 \end{bmatrix} \qquad A_3 = \begin{bmatrix} 0 & 0 & 0 \\ 0 & 0 & 1 \\ 0 & -1 & 0 \end{bmatrix}$$

and the corresponding exponentials $\exp(A_1 t), \exp(A_2 t), \exp(A_3 t)$, with $t \in \mathbf{R}$. An easy calculation shows that , for each $1 \leq k \leq 3$, $\exp(A_k t)$ is an orthogonal matrix, with determinant equal to 1. Thus, $\exp(A_k t) \in SO(3)$. Consider now the mapping

$$P_k : \mathbf{R} \to SO(3)$$
$$t \mapsto (\exp(A_k t))R$$

where R is an element of $SO(3)$. By construction, $P(t)$ is a smooth curve on $SO(3)$, passing through R at $t = 0$. Its tangent vector at $t = 0$, in the basis (5.6), is represented by the matrix

$$[\dot{P}_k(t)]_{t=0} = A_k R$$

i.e. has the form

$$v_k = \sum_{i,j=1}^{3} (A_k R)_{ij} \frac{\partial}{\partial x_{ij}}.$$

Since the three matrices A_1, A_2, A_3 are linearly independent, so are the three corresponding tangent vectors $\{v_1, v_2, v_3\}$. Moreover, each v_k is an element of $T_R SO(3)$ by construction, and $T_R SO(3)$ is 3-dimensional. As a consequence, we can conclude that the set $\{v_1, v_2, v_3\}$ is actually a *basis* of $T_R SO(3)$.

In particular we see that, in the basis (5.6), any vector of $T_R SO(3)$ can be represented by means of a matrix of the form

$$c_1 A_1 R + c_2 A_2 R + c_3 A_3 R$$

where c_1, c_2, c_3 are real numbers, i.e.—in view of the special structure of A_1, A_2, A_3—in the form

$$\begin{bmatrix} 0 & c_1 & c_2 \\ -c_1 & 0 & c_3 \\ -c_2 & -c_3 & 0 \end{bmatrix} R = S(c)R$$

where $c = \text{col}(c_3, -c_2, c_1)$.

We return now to the problem of discussing the partition of the state space of system (5.1)-(5.2) induced by the distribution Δ_C. The system in question has the form

$$\dot{p} = f(p) + g_1(p)u_1 + \cdots + g_r(p)u_r$$

with

$$p = (x, R) \in N = \mathbf{R}^3 \times SO(3)$$
$$\dot{p} \in T_p N = T_x \mathbf{R}^3 \times T_R SO(3)$$
$$f(p) = (-S(x)J^{-1}x, S(J^{-1}x)R)$$
$$g_i(p) = (b_i, 0) \qquad 1 \le i \le r$$

(recall that we set $J\omega = x$).

Suppose $r = 3$ and note that, by definition, the control Lie algebra $\mathcal{C}$ contains the six vector fields $g_i(p), [f, g_i](p), 1 \le i \le 3$. An easy calculation shows that

$$[f, g_i](x, R) = ((\frac{\partial S(x)J^{-1}x}{\partial x})b_i, -S(J^{-1}b_i)R).$$

Note that the b_i's are linearly independent vectors, and so are the vectors $J^{-1}b_i$ and the matrices $S(J^{-1}b_i)R, 1 \le i \le 3$. Thus, in view of the previous discussion, we deduce that the matrices $S(J^{-1}b_i)R, 1 \le i \le 3$ represent, in the basis (5.6), three independent tangent vectors that span $T_R SO(3)$, for each $R \in SO(3)$.

On the other hand, the vectors $b_i, 1 \leq i \leq 3$, span $T_x\mathbf{R}^3$. Therefore, we can conclude that the set of six vectors $g_i(p), [f, g_i](p), 1 \leq i \leq 3$ span the tangent space $T_x\mathbf{R}^3 \times T_R SO(3)$ at each $(x, R) \in N$. The controllability rank condition (2.3) is satisfied at each point, and the partition of N induced by Δ_C degenerates into one single element, namely N itself.

The study of the case in which $r = 2$ can be carried out in the same way, using the condition (5.4) to show that the matrices $S(J^{-1}b_1)R, S(J^{-1}b_2)R$, and

$$S(J^{-1}S(b)J^{-1}b)R$$

(with $b = c_1b_1 + c_2b_2$) are linearly independent, and proving that T_pN is spanned by $g_i(p), [f, g_i](p), 1 \leq i \leq 2, [[f, g], g](p)$, and $[f, [[f, g], g]](p)$, with $g(p) = c_1g_1(p) + c_2g_2(p)$.

3 Input-Output Maps and Realization Theory

3.1 Fliess Functional Expansions

The purpose of this section and of the following one is to describe representations of the input-output behavior of a nonlinear system. We consider, as usual, systems described by differential equations of the form

$$\dot{x} = f(x) + \sum_{i=1}^{m} g_i(x) u_i \tag{1.1a}$$

$$y_j = h_j(x) \qquad 1 \leq j \leq p. \tag{1.1b}$$

This system, as in Chapter 1, is assumed to be defined on an open set U of $\mathbf{R}^n$. Moreover, throughout the present Chapter, we constantly suppose also that the vector fields $f, g_1, \ldots, g_m$ are *analytic* vector fields defined on U. Likewise, the output functions $h_1, \ldots, h_p$ are analytic functions defined on U.

For the sake of notational convenience most of the times we represent the output of the system as a vector-valued function

$$y = h(x) = \operatorname{col}(h_1(x), \ldots, h_p(x)).$$

We require first some combinatorial notations. Consider the set of $m + 1$ indexes $I = \{0, 1, \ldots, m\}$ (we represent here, as usual, indexes with integer numbers, but we could as well represent the $m + 1$ indexes with elements of any set Z with $card(Z) = m + 1$). Let I_k be the set of all sequences $(i_k \ldots i_1)$ of k elements $i_k, \ldots, i_1$ of I. An element of this set I_k will be called a multiindex of length k. For consistency we define also a set I_0 whose unique element is the empty sequence (i.e. a multiindex of length 0), denoted $\emptyset$. Finally, let

$$I^\star = \bigcup_{k \geq 0} I_k.$$

It is easily seen that the set $I^\star$ can be given a structure of free monoid with composition rule

$$(i_k \ldots i_1)(j_h \ldots j_1) \mapsto (i_k \ldots i_1 j_h \ldots j_1)$$

with neutral element $\emptyset$.

A *formal power series* in $m + 1$ noncommutative indeterminates and coefficients in $\mathbf{R}$ is a mapping

$$c : I^\star \mapsto \mathbf{R}.$$

In what follows we represent the value of c at some element $i_k \ldots i_0$ of $I^\star$ with the symbol $c(i_k \ldots i_0)$.

The second relevant object we have to introduce is called an *iterated integral* of a given set of functions and is defined in the following way. Let T be a fixed value of the time and suppose $u_1 \ldots u_m$ are real-valued piecewise continuous functions defined on $[0, T]$. For each multiindex $(i_k \ldots i_0)$ the corresponding iterated integral is a real-valued function of t

$$E_{i_k \ldots i_1 i_0}(t) = \int_0^t d\xi_{i_k} \ldots d\xi_{i_1} d\xi_{i_0}$$

defined for $0 \leq t \leq T$ by recurrence on the length, setting

$$\xi_0(t) = t$$

$$\xi_i(t) = \int_0^t u_i(\tau) \, d\tau \qquad \text{for } 1 \leq i \leq m$$

and

$$\int_0^t d\xi_{i_k} \ldots d\xi_{i_0} = \int_0^t d\xi_{i_k}(\tau) \int_{k-1}^\tau d\xi_{i_{k-1}} \ldots d\xi_{i_0}.$$

The iterated integral corresponding to the multiindex $\emptyset$ is the real number 1.

Example 1.1. Just for convenience, let us compute the first few iterated integrals, in a case where $m = 1$.

$$\int_0^t d\xi_0 = t, \qquad \int_0^t d\xi_1 = \int_0^t u_1(\tau) \, d\tau$$

$$\int_0^t d\xi_0 d\xi_0 = \frac{t^2}{2!}, \qquad \int_0^t d\xi_0 d\xi_1 = \int_0^t \int_0^\tau u_1(\theta) \, d\theta d\tau$$

$$\int_0^t d\xi_1 d\xi_0 = \int_0^t u_1(\tau) \tau d\tau, \qquad \int_0^t d\xi_1 d\xi_1 = \int_0^t u_1(\tau) \int_0^\tau u_1(\theta) \, d\theta d\tau, \qquad \cdots \;\; \bullet$$

Given a formal power series in $m + 1$ non-commutative indeterminates, it is possible to associate with this series a functional of $u_1, \ldots, u_m$ by taking the sum over $I^\star$ of all the products of the form

$$c(i_k \ldots i_0) \int_0^t d\xi_{i_k} \ldots d\xi_{i_0}.$$

The convergence of a sum of this kind is guaranteed by some growth condition on the "coefficients" $c(i_k \ldots i_0)$ as stated below.

Lemma 1.1. Suppose there exist real numbers $K > 0, M > 0$ such that

$$|c(i_k \ldots i_0)| < K(k+1)! M^{k+1} \tag{1.2}$$

for all $k \geq 0$ and all multiindexes $i_k \ldots i_0$.

Then, there exists a real number $T > 0$ such that, for each $0 \leq t \leq T$ and each set of piecewise continuous functions $u_1, \ldots, u_m$ defined on $[0, T]$ and subject to the constraint

$$\max_{0 \leq \tau \leq T} |u_i(\tau)| < 1, \tag{1.3}$$

the series

$$y(t) = c(\emptyset) + \sum_{k=0}^{\infty} \sum_{i_0, \ldots, i_k = 0}^{m} c(i_k \ldots i_0) \int_0^t d\xi_{i_k} \ldots d\xi_{i_0} \tag{1.4}$$

is absolutely and uniformly convergent.

Proof. It is easy to see, from the definition of iterated integral that, if the functions $u_1 \ldots u_m$ satisfy the constraint (1.3) then

$$\int_0^t d\xi_{i_k} \ldots d\xi_{i_0} \leq \frac{t^{k+1}}{(k+1)!}.$$

If the growth condition is satisfied, then

$$\left| \sum_{i_0, \ldots, i_k = 0}^{m} c(i_k \ldots i_0) \int_0^t d\xi_{i_k} \ldots d\xi_{i_0} \right| \leq K[M(m+1)t]^{k+1}.$$

As a consequence, if T is sufficiently small, the series (1.4) converges absolutely and uniformly on $[0, T]$. ●

The expression (1.4) clearly defines a functional of $u_1 \ldots u_m$. This functional is *causal*, in the sense that $y(t)$ depends only on the restrictions of $u_1, \ldots, u_m$ to the time interval $[0, t]$.

A representation of the form (1.4) is unique.

Lemma 1.2. Let c^a and c^b be two formal power series in $m + 1$ non-commutative indeterminates and let the associated functionals of the form (1.4) be defined on the same interval $[0, T]$. Then the two functionals coincides if and only if $c^a = c^b$.

Proof. Let c^a, c^b be two formal power series and $y^a(t), y^b(t)$ the associated functionals of the form (1.4). Note that

$$y(t) = y^a(t) - y^b(t)$$

is still a functional of the form (1.4) associated with a formal power series c whose coefficients are defined as differences between the corresponding coefficients of c^a and c^b. To prove the lemma, all we need is to show that if $y(t) = 0$ for all $t \in [0, T]$ and for all input functions, all the coefficients of the series c vanish.

If, in particular, $u_1 = \cdots = u_m = 0$ on $[0, T]$ then $y(t) = 0$ for all $t \in [0, T]$ implies

$$c(\emptyset) + c(0)t + c(00)\frac{t^2}{2!} + \cdots = 0$$

for all $t \in [0, T]$, i.e.

$$c(\emptyset) = 0$$

$$c(\underbrace{0 \ldots 0}_{k \text{ times}}) = 0 \qquad 1 \le k \le \infty.$$

Taking the derivative of (1.4) with respect to time and evaluating it at $t = 0$, one obtains

$$\left(\frac{dy}{dt}\right)_{t=0} = \sum_{i=1}^{m} c(i)u_i(0).$$

Therefore, $(dy/dt)_{t=0} = 0$ for all $u_1(0), \ldots, u_m(0)$ implies

$$c(i) = 0 \qquad 1 \le i \le m$$

Continuing this way, one may compute the second derivative of $y(t)$ at $t = 0$ and get

$$\left(\frac{d^2 y}{dt^2}\right)_{t=0} = \sum_{i_0, i_1 = 1}^{m} c(i_1 i_0)u_{i_1}(0)u_{i_0}(0) + \sum_{i=1}^{m} (c(0i) + c(i0))u_i(0).$$

If this is zero for all $u_1(0), \ldots, u_m(0)$, then

$$c(i_1 i_0) = 0 \qquad 1 \le i_1, i_0 \le m$$

$$c(0i) = -c(i0) \qquad 1 \le i \le m.$$

In the third derivative, the contribution of terms

$$\sum_{i=1}^{m} \left(c(0i) \int_0^t d\xi_0 d\xi_i + c(i0) \int_0^t d\xi_i d\xi_0\right)$$

is

$$\sum_{i=1}^{m} [\frac{1}{6}c(0i) + \frac{1}{3}c(i0)] \left(\frac{du_i}{dt}\right)_{t=0}.$$

If this is zero for all $(du_i/dt)_{t=0}$, then $c(0i) = -2c(i0)$ which, together with the previous equality $c(0i) = -c(i0)$ implies

$$c(0i) = 0 \qquad 1 \le i \le m.$$

Continuing in the same way, one may complete the proof. •

We are now going to show that the output $y(t)$ of the nonlinear system (1.1) can be represented as a functional of the inputs $u_1, \ldots, u_m$ in the form (1.4). To this end we need some preliminary results.

Lemma 1.3. Let $g_0, \ldots, g_m$ be a set of analytic vector fields and λ a real-valued analytic function defined on U. Given a point $x^\circ \in U$, consider the formal power series defined by

$$\begin{aligned} c(\emptyset) &= \lambda(x^\circ) \\ c(i_k \ldots i_0) &= L_{g_{i_0}} \ldots L_{g_{i_k}} \lambda(x^\circ). \end{aligned} \qquad (1.5)$$

Then, there exist real numbers $K > 0$ and $M > 0$ such that the growth condition (1.2) is satisfied.

Proof. The reader is referred to the literature. •

In view of this result and of Lemma 1.1, one may associate with $g_0, \ldots, g_m$ and λ the functional

$$v(t) = \lambda(x^\circ) + \sum_{k=0}^{m} \sum_{i_0, \ldots, i_k = 0}^{m} L_{g_{i_0}} \ldots L_{g_{i_k}} \lambda(x^\circ) \int_0^t d\xi_{i_k} \ldots d\xi_{i_0}. \qquad (1.6)$$

Lemma 1.4. Let $g_0, \ldots, g_m$ be as in the previous Lemma and let $\lambda_1, \ldots, \lambda_l$ be real-valued analytic functions defined on U. Moreover, let γ be a real-valued analytic function defined on $\mathbf{R}^l$. Let $v_1(t), \ldots, v_l(t)$ denote the functionals defined by setting, in a functional of the form (1.6), $\lambda = \lambda_1, \ldots, \lambda = \lambda_l$. The composition $\gamma(v_1(t), \ldots, v_l(t))$ is again a functional of the form(1.6), corresponding to the setting $\lambda = \gamma(\lambda_1, \ldots, \lambda_l)$.

Proof. We will only give a trace to the reader for the proof. Let c_1, c_2 denote the formal power series defined by setting, in (1.5), $\lambda = \lambda_1$ and respectively $\lambda = \lambda_2$, and let $v_1(t), v_2(t)$ denote the associated functionals (1.6). Then, it is immediately seen that with the formal power series defined by setting $\lambda = \alpha_1 \lambda_1 + \alpha_2 \lambda_2$, where α_1 and α_2 are real numbers, there is associated the functional $\alpha_1 v_1(t) + \alpha_2 v_2(t)$.

With a little work, it is also seen that with the formal power series defined by setting $\lambda = \lambda_1 \lambda_2$, there is associated the functional $v_1(t) v_2(t)$. We show only the very first computations needed for that. For, consider the product

$$v_1(t)v_2(t) = (\lambda_1 + L_{g_0}\lambda_1 \int_0^t d\xi_0 + L_{g_1}\lambda_1 \int_0^t d\xi_1 + L_{g_0}L_{g_0}\lambda_1 \int_0^t d\xi_0 d\xi_0 + \cdots)$$

$$(\lambda_2 + L_{g_0}\lambda_2 \int_0^t d\xi_0 + L_{g_1}\lambda_2 \int_0^t d\xi_1 + L_{g_0}L_{g_0}\lambda_2 \int_0^t d\xi_0 d\xi_0 + \cdots)$$

where, for simplicity, we have omitted specifying that the values of all the functions of x are to be taken at $x = x^0$. Multiplying term-by-term we have

$$v_1(t)v_2(t) = \lambda_1\lambda_2$$

$$+ (\lambda_1 L_{g_0}\lambda_2 + \lambda_2 L_{g_0}\lambda_1) \int_0^t d\xi_0 + (\lambda_1 L_{g_1}\lambda_2 + \lambda_2 L_{g_1}\lambda_1) \int_0^t d\xi_1$$

$$+ (\lambda_1 L_{g_0}L_{g_0}\lambda_2 + \lambda_2 L_{g_0}L_{g_0}\lambda_1) \int_0^t d\xi_0 d\xi_0$$

$$+ (L_{g_0}\lambda_1)(L_{g_0}\lambda_2) \int_0^t d\xi_0 \int_0^t d\xi_0 + \cdots$$

The factors that multiply $\int_0^t d\xi_0$ and $\int_0^t d\xi_1$ are clearly $L_{g_0}\lambda_1\lambda_2$ and respectively $L_{g_1}\lambda_1\lambda_2$. For the other three, we have

$$L_{g_0}L_{g_0}\lambda_1\lambda_2 = \lambda_1 L_{g_0}L_{g_0}\lambda_2 + \lambda_2 L_{g_0}L_{g_0}\lambda_1 + 2(L_{g_0}\lambda_1)(L_{g_0}\lambda_2)$$

but also

$$\int_0^t d\xi_0 \int_0^t d\xi_0 = 2 \int_0^t d\xi_0 d\xi_0$$

so that the three terms in question give exactly

$$L_{g_0}L_{g_0}\lambda_1\lambda_2 \int_0^t d\xi_0 d\xi_0.$$

It is not difficult to set up a recursive formalism which makes it possible to completely verify the claim.

If now γ is any real-valued analytic function defined on $\mathbf{R}^l$, we may take its Taylor series expansion at the origin and use recursively the previous results in order to show that the composition $\gamma(v_1(t), \ldots, v_l(t))$ may be represented as a series like the (1.6) with λ replaced by the Taylor series expansion of $\gamma(\lambda_1, \ldots, \lambda_l)$. ●

At this point, it is easy to obtain the desired representation of $y(t)$ as a functional of the form (1.6).

Theorem 1.5. Suppose the inputs $u_1, \ldots, u_m$ of the control system (1.1) satisfy the constraint (1.3). If T is sufficiently small, then for all $0 \leq t \leq T$ the

j-th output $y_j(t)$ of the system (1.1) may be expanded in the following way

$$y_j(t) = h_j(x^\circ) + \sum_{k=0}^{\infty} \sum_{i_0,\ldots,i_k=0}^{m} L_{g_{i_0}} \ldots L_{g_{i_k}} h_j(x^\circ) \int_0^t d\xi_{i_k} \ldots d\xi_{i_0} \qquad (1.7)$$

where $g_0 = f$.

Proof. We first show that the j-th component of the solution of the differential equation (1.1a) may be expressed as

$$x_j(t) = x_j(x^\circ) + \sum_{k=0}^{\infty} \sum_{i_0,\ldots,i_k=0}^{m} L_{g_{i_0}} \ldots L_{g_{i_k}} x_j(x^\circ) \int_0^t d\xi_{i_k} \ldots d\xi_{i_0} \qquad (1.8)$$

where the function $x_j(x)$ stands for

$$x_j : (x_1,\ldots,x_n) \mapsto x_j.$$

Note that, by definition of iterated integral

$$\frac{d}{dt} \int_0^t d\xi_0 d\xi_{i_{k-1}} \ldots d\xi_{i_0} = \int_0^t d\xi_{i_{k-1}} \ldots d\xi_{i_0}$$

and

$$\frac{d}{dt} \int_0^t d\xi_i d\xi_{i_{k-1}} \ldots d\xi_{i_0} = u_i(t) \int_0^t d\xi_{i_{k-1}} \ldots d\xi_{i_0}$$

for $1 \leq i \leq m$. Then, taking the derivative of the right-hand-side of (1.8) with respect to the time and rearranging the terms we have

$$\dot{x}_j(t) = L_f x_j(x^\circ) + \sum_{k=0}^{\infty} \sum_{i_0,\ldots,i_k=0}^{m} L_{g_{i_0}} \ldots L_{g_{i_k}} L_f x_j(x^\circ) \int_0^t d\xi_{i_{k-1}} \ldots d\xi_{i_0}$$

$$+ \sum_{i=1}^{m} \left[L_{g_i} x_j(x^\circ) + \sum_{k=0}^{\infty} \sum_{i_0,\ldots,i_k=0}^{m} L_{g_{i_0}} \ldots L_{g_{i_k}} L_{g_i} x_j(x^\circ) \int_0^t d\xi_{i_k} \ldots d\xi_{i_0} \right] u_i(t).$$

Now, let f_j and g_{ij} denote the j-th components of f and g_i, $1 \leq j \leq n, 1 \leq i \leq m$ and observe that

$$L_f x_j = f_j(x_1,\ldots,x_n).$$

Therefore, on the basis of Lemma 1.4, we may write

$$L_f x_j(x^\circ) + \sum_{k=0}^{\infty} \sum_{i_0,\dots,i_k=0}^{m} L_{g_{i_0}} \dots L_{g_{i_k}} L_f x_j(x^\circ) \int_0^t d\xi_{i_k} \dots d\xi_{i_0} =$$

$$= f_j(x^\circ) + \sum_{k=0}^{\infty} \sum_{i_0,\dots,i_k=0}^{m} L_{g_{i_0}} \dots L_{g_{i_k}} f_j(x^\circ) \int_0^t d\xi_{i_k} \dots d\xi_{i_0}$$

$$= f_j(x_1(t), \dots, x_n(t)).$$

A similar substitution can be performed on the other terms thus yielding

$$\dot{x}_j(t) = f_j(x_1(t), \dots, x_n(t)) + \sum_{i=1}^{m} g_{ij}(x_1(t), \dots, x_n(t)) u_i(t).$$

Moreover, the $x_j(t)$ satisfy the condition

$$x_j(0) = x_j^\circ$$

and therefore are the components of the solution $x(t)$ of the differential equation (1.1a).

A further application of Lemma (1.4) shows that the output (1.1b) can be expressed in the form (1.7). •

The development (1.7) will be from now on referred to as the *fundamental formula*, or *Fliess' functional expansion* of $y_j(t)$. Obviously, one may deal directly with the case of a vector-valued output with the same formalism, by just replacing the real-valued function $h_j(x)$ with the vector-valued function $h(x)$. We stress that, from Lemma 1.1, it is known that the series (1.7) converges absolutely and uniformly on $[0, T]$.

Remark 1.6 The reader will immediately observe that the functions $h_j(x)$ and $L_{g_{i_0}} \dots L_{g_{i_k}} h_j(x)$, with $1 \leq j \leq p$ and $(i_k \dots i_0) \in (I^\star \backslash I_0)$, whose values at x° characterize the functional (1.7), span the so called *observation space* $\mathcal{O}$. The latter, in fact, was characterized—in Section **2.3**—as the space $\mathcal{O}$ of all **R**-linear combinations of functions of the form $h_j(x)$ and $L_{g_{i_0}} \dots L_{g_{i_k}} h_j(x)$ with $1 \leq j \leq p, 0 \leq i_k \leq m, 1 \leq k \leq \infty$. •

Example 1.2. In the case of a linear system, the formal power series which characterizes the functional (1.7) is such that $c(\emptyset) = c_j x^\circ$,

$$c(i_k \dots i_0) = \begin{cases} c_j A^{k+1} x^\circ & \text{if } i_0 = \dots = i_k = 0 \\ c_j A^k b_{i_0} & \text{if } i_0 \neq i_1 = \dots = i_k = 0 \end{cases}$$

and $c(i_k \dots i_0) = 0$ elsewhere.

In the case of a bilinear system, the formal power series which characterizes the functional (1.7) takes the form

$$c(\emptyset) = c_j x^\circ$$
$$c(i_k \ldots i_0) = c_j N_{i_k} \ldots N_{i_0} x^\circ$$

where $N_0 = A$. ●

3.2 Volterra Series Expansions

The input-output behavior of a nonlinear system of the form (1.1) may also be represented by means of a series of *generalized convolution integrals*. A generalized convolution integral of order k is defined as follows. Let $(i_k \ldots i_1)$ be a multiindex of length k, with $i_k, \ldots, i_1$ elements of the set $\{1, \ldots, m\}$. With this multiindex there is associated a real-valued continuous function $w_{i_k \ldots i_1}$, defined on the subset of $\mathbf{R}^{k+1}$

$$S_k = \left\{ (t, \tau_k, \ldots, \tau_1) \in \mathbf{R}^{k+1} : T \geq t \geq \tau_k \geq \cdots \geq \tau_1 \geq 0 \right\}$$

where T is a fixed number. If $u_1, \ldots, u_m$ are real-valued piecewise continuous functions defined on $[0, T]$, the generalized convolution integral of order k of $u_1, \ldots, u_k$ with kernel $w_{i_k \ldots i_1}$ is defined as

$$\int_0^t \int_0^{\tau_k} \cdots \int_0^{\tau_2} w_{i_k \ldots i_1}(t, \tau_k, \ldots, \tau_1) u_{i_k}(\tau_k) \ldots u_{i_1}(\tau_1) \, d\tau_1 \ldots d\tau_k$$

for $0 \leq t \leq T$.

For consistency, if $k = 0$, rather than a generalized convolution integral, one considers simply a continuous real-valued function w_0 defined on the set

$$S_0 = \{ t \in \mathbf{R} : T \geq t \geq 0 \}.$$

The sum of a series of generalized convolution integrals may describe a functional of $u_1, \ldots, u_m$, under the conditions stated below.

Lemma 2.1. Suppose there exist real numbers $K > 0$, $M > 0$ such that

$$\left| w_{i_k \ldots i_1}(t, \tau_k, \ldots, \tau_1) \right| < K(k)! M^k \tag{2.1}$$

for all $k > 0$, for all multiindexes $(i_k \ldots i_1)$, and all $(t, \tau_k \ldots \tau_1) \in S_k$.

Then, there exists a real number $T > 0$ such that, for each $0 \leq t \leq T$ and each set of piecewise continuous functions $u_1, \ldots, u_m$ defined on $[0, T]$ and subject to the constraint

$$\max_{0 \leq \tau \leq T} \left| u_i(\tau) \right| < 1, \tag{2.2}$$

the series

$$y(t) = w_0(t)+$$

$$\sum_{k=1}^{\infty} \sum_{i_1,\dots,i_k=1}^{m} \int_0^t\!\!\int_0^{\tau_k}\!\!\cdots\int_0^{\tau_2} w_{i_k\dots i_1}(t,\tau_k,\dots,\tau_1)u_{i_k}(\tau_k)\dots u_{i_1}(\tau_1)d\tau_1\dots d\tau_k$$

$$(2.3)$$

is absolutely and uniformly convergent.

Proof. It is similar to that of Lemma 1.1. •

The expression (2.3) clearly defines a functional of $u_1,\dots,u_m$, which is causal, and is called a *Volterra series expansion.*

As in the previous Section, we are interested in the possibility of using an expansion of the form (2.3) for the output of the nonlinear system (1.1). The existence of such an expansion and the expressions of the kernels may be described in the following way.

Lemma 2.2. Let $f,g_1,\dots,g_m$ be a set of analytic vector fields and λ a real-valued analytic function defined on U. Let Φ_t^f denote the flow of f. For each pair $(t,x) \in \mathbf{R} \times U$ for which the flow $\Phi_t^f(x)$ is defined, let $Q_t(x)$ denote the function

$$Q_t(x) = \lambda \circ \Phi_t^f(x) \tag{2.4}$$

and $P_t^1(x),\dots,P_t^m(x)$ the vector fields

$$P_t^i(x) = (\Phi_{-t}^f)_* g_i \circ \Phi_t^f(x) \tag{2.5}$$

$1 \leq i \leq m$. Moreover, let

$$w_0(t) = Q_t(x^\circ) \tag{2.6a}$$

$$w_{i_k\dots i_1}(t,\tau_k,\dots,\tau_1) = L_{P_{\tau_1}^{i_1}} \dots L_{P_{\tau_k}^{i_k}} Q_t(x^\circ). \tag{2.6b}$$

Then, there exist real numbers $K > 0$ and $M > 0$ such that the condition (2.1) is satisfied.

From this result it is easy to obtain the desired representation of $y(t)$ in the form of a Volterra series expansion.

Theorem 2.3. Suppose the inputs $u_1,\dots,u_m$ of the control system (1.1) satisfy the constraint (2.2). If T is sufficiently small, then for all $0 \leq t \leq T$ the output $y_j(t)$ of the system (1.1) may be expanded in the form of a Volterra series, with kernels (2.6), where $Q_t(x)$ and $P_t^i(x)$ are as in (2.4)-(2.5) and $\lambda = h_j$.

This result may be proved either directly, by showing that the Volterra series in question satisfies the equations (1.1), or indirectly, after establishing a correspondence between the functional expansion described at the beginning of the previous Section and the Volterra series expansion. We take the second way.

For, observe that for all $(i_k \ldots i_1)$ the kernel $w_{i_k \ldots i_1}(t, \tau_k, \ldots, \tau_1)$ is analytic in a neighborhood of the origin, and consider the Taylor series expansion of this kernel as a function of the variables $t - \tau_k, \tau_k - \tau_{k-1}, \ldots, \tau_2 - \tau_1, \tau_1$. This expansion has clearly the form

$$
w_{i_k \ldots i_1}(t, \tau_k, \ldots, \tau_1) = \sum_{n_0, \ldots, n_k = 0}^{\infty} c_{i_k \ldots i_1}^{n_0 \ldots n_k} \frac{(t - \tau_k)^{n_k} \ldots (\tau_2 - \tau_1)^{n_1} \tau_1^{n_0}}{n_k! \ldots n_1! n_0!}
$$

where

$$
c_{i_k \ldots i_1}^{n_0 \ldots n_k} = \left[\frac{\partial^{n_0 + \cdots + n_k} w_{i_k \ldots i_1}}{\partial(t - \tau_k)^{n_k} \ldots \partial(\tau_2 - \tau_1)^{n_1} \partial \tau_1^{n_0}} \right]_{t - \tau_k = \cdots = \tau_2 - \tau_1 = \tau_1 = 0}.
$$

If we substitute this expression in the convolution integral associated with $w_{i_k \ldots i_1}$, we obtain an integral of the form

$$
\sum_{n_0, \ldots, n_k = 0}^{\infty} c_{i_k \ldots i_1}^{n_0 \ldots n_k} \int_0^t \int_0^{\tau_k} \cdots \int_0^{\tau_2} \frac{(t - \tau_k)^{n_k}}{n_k!} u_{i_k}(\tau_k) \ldots \frac{(\tau_2 - \tau_1)^{n_1}}{n_1!} u_{i_1}(\tau_1) \frac{\tau_1^{n_0}}{n_0!} d\tau_k \ldots d\tau_1.
$$

The integral which appears in this expression is actually an iterated integral of $u_1, \ldots, u_m$, and precisely the integral

$$
\int_0^t (d\xi_0)^{n_k} d\xi_{i_k} \ldots (d\xi_0)^{n_1} d\xi_{i_1} (d\xi_0)^{n_0} \tag{2.7}
$$

(where $(d\xi_0)^n$ stands for n-times $d\xi_0$).

Thus, the expansion (2.3) may be replaced with the expansion

$$
y(t) = \sum_{n=0}^{\infty} c_0^n \int_0^t (d\xi_0)^n +
$$
$$
\sum_{k=1}^{\infty} \sum_{i_1, \ldots, i_k = 1}^{m} \sum_{n_0, \ldots, n_k = 0}^{\infty} c_{i_k \ldots i_1}^{n_0 \ldots n_k} \int_0^t (d\xi_0)^{n_k} d\xi_{i_k} \ldots (d\xi_0)^{n_1} d\xi_{i_1} (d\xi_0)^{n_0} \tag{2.8}
$$

which is clearly an expansion of the form (1.4). Of course, one could rearrange the terms and establish a correspondence between the coefficients $c_0^n, c_{i_k \ldots i_1}^{n_0 \ldots n_k}$ (i.e. the values of the derivatives of w_0 and $w_{i_k \ldots i_1}$ at $t - \tau_k = \cdots = \tau_2 - \tau_1 = \tau_1 = 0$) and the coefficients $c(\emptyset)$, $c(i_k \ldots i_0)$ of the espansion (1.4), but this is not needed at this point.

On the basis of these considerations it is very easy to find Taylor series expansions of the kernels which characterize the Volterra series expansion of

$y_j(t)$. We see from (2.8) that the coefficient $c_{i_k\ldots i_1}^{n_0\ldots n_k}$ of the Taylor series expansion of $w_{i_k\ldots i_1}$ coincides with the coefficient of the iterated integral (2.7) in the expansion (1.4), but we know also from (1.7) that the coefficient of the iterated integral (2.7) has the form

$$L_f^{n_0} L_{g_{i_1}} L_f^{n_1} \ldots L_f^{n_{k-1}} L_{g_{i_k}} L_f^{n_k} h_j(x^\circ).$$

This makes it possible to write down immediately the expression of the Taylor series expansions of all the kernels which characterize the Volterra series expansion of $y_j(t)$

$$w_0(t) = \sum_{n=0}^{\infty} L_f^n h_j(x^\circ) \frac{t^n}{n!} \tag{2.9a}$$

$$w_i(t, \tau_1) = \sum_{n_1=0}^{\infty} \sum_{n_0=0}^{\infty} L_f^{n_0} L_{g_i} L_f^{n_1} h_j(x^\circ) \frac{(t-\tau_1)^{n_1}}{n_1!} \frac{\tau_1^{n_0}}{n_0!} \tag{2.9b}$$

$$w_{i_2 i_1}(t, \tau_2, \tau_1) = \sum_{n_2=0}^{\infty} \sum_{n_1=0}^{\infty} \sum_{n_0=0}^{\infty} L_f^{n_0} L_{g_{i_1}} L_f^{n_1} L_{g_{i_2}} L_f^{n_2} h_j(x^\circ)$$
$$\frac{(t-\tau_2)^{n_2}(\tau_2-\tau_1)^{n_1}\tau_1^{n_0}}{n_2!n_1!n_0!} \tag{2.9c}$$

and so on.

The last step needed in order to prove Theorem 2.3 is to show that the Taylor series expansions of the kernels (2.6), with $Q_t(x)$ and $P_t^i(x)$ defined as in (2.4), (2.5) for $\lambda = h_j(x)$ coincide with the expansions (2.9)

This is only a routine computation, which may be carried out with a little effort by keeping in mind the well-known Campbell-Baker-Hausdorff formula, which provides a Taylor series expansion of $P_t^i(x)$. According to this formula it is possible to expand $P_t^i(x)$ in the following way

$$P_t^i(x) = (\Phi_{-t}^f)_* g_i \circ \Phi_t^f(x) = \sum_{n=0}^{\infty} ad_f^n g_i(x) \frac{t^n}{n!}$$

where, as usual, $ad_f^n g = [f, ad_f^{n-1}g]$ and $ad_f^0 g = g$.

Example 2.1. In the case of bilinear systems, the flow Φ_t^f may be clearly given the following closed form expression

$$\Phi_t^f(x) = (\exp At)x.$$

From this it is easy to find the expressions of the kernels of the Volterra series expansion of $y_i(t)$. In this case

$$Q_t(x) = c_j(\exp At)x$$
$$P_t^i(x) = (\exp(-At))N_i(\exp At)x$$

and, therefore,

$$w_0(t) = c_j(\exp At)x^\circ$$
$$w_i(t, \tau_1) = c_j(\exp A(t - \tau_1))N_i(\exp A\tau_1)x^\circ$$
$$w_{i_2 i_1}(t, \tau_2, \tau_1) = c_j(\exp A(t - \tau_2))N_{i_2}(\exp A(\tau_2 - \tau_1))N_{i_1}(\exp A\tau_1)x^\circ$$

and so on.

3.3 Output Invariance

In this Section we want to find the conditions under which the output is not affected by the input. These conditions will be used later on in the next Chapters when dealing with the disturbance decoupling or with the noninteracting control.

Consider again a system of the form

$$\dot{x} = f(x) + \sum_{i=1}^{m} g_i(x)u_i$$
$$y_j = h_j(x) \qquad 1 \le j \le p$$

and let

$$y_j(t; x^\circ; u_1, \ldots, u_m)$$

denote the value at time t of the j-th output, corresponding to an initial state x° and to a set of input functions $u_1, \ldots, u_m$. We say that the output y_j is *unaffected* by (or *invariant* under) the input u_i, if for every initial state $x^\circ \in U$, for every set of input functions $u_1, \ldots, u_{i-1}, u_{i+1}, \ldots, u_m$ and for all t

$$y_j(t; x^\circ; u_1, \ldots, u_{i-1}, v^a, u_{i+1}, \ldots, u_m) = y_j(t; x^\circ; u_1, \ldots, u_{i-1}, v^b, u_{i+1}, \ldots, u_m)$$
$$(3.1)$$

for every pair of functions v^a and v^b.

There is a simple test that identifies the system having the output y_j unaffected by the input u_i.

Lemma 3.1. The output y_j is unaffected by the input u_i if and only if, for all $r \ge 1$ and for any choice of vector fields $\tau_1, \ldots, \tau_r$ in the set $\{f, g_1, \ldots, g_m\}$

$$L_{g_i} h_j(x) = 0$$
$$L_{g_i} L_{\tau_1} \ldots L_{\tau_r} h_j(x) = 0$$
$$(3.2)$$

for all $x \in U$.

Proof. Suppose the above condition is satisfied. Then, one easily sees that the function

$$L_{\tau_1} \ldots L_{\tau_r} h_j(x) \tag{3.3}$$

is identically zero whenever at least one of the vector fields $\tau_1, \ldots, \tau_r$ coincides with g_i. If we now look, for instance, at the Fliess expansion of $y_j(t)$, we observe that under these circumstances

$$c(i_k \ldots i_0) = 0$$

whenever one of the indexes $i_0, \ldots, i_k$ is equal to i, and this, in turn, implies that any iterated integral which involves the input function u_i is multiplied by a zero factor. Thus the condition (3.1) is satisfied and the output y_j is decoupled from the input u_i.

Conversely, suppose the condition (3.1) is satisfied, for every $x^\circ \in U$, for every set of inputs $u_1, \ldots, u_{i-1}, u_{i+1}, \ldots, u_m$ and every pair of functions v^a and v^b. Take in particular $v^a(t) = 0$ for all t. Then in the Fliess expansion of $y_j(t; x^\circ; u_1, \ldots, u_{i-1}, v^a, u_{i+1}, \ldots, u_m)$ an iterated integral of the form

$$\int_0^t d\xi_{i_k} \ldots d\xi_{i_0}$$

will be zero whenever one of the indexes $i_0, \ldots, i_k$ is equal to i. All other iterated integrals of this expansion (i.e. the ones in which none of the indexes $i_0, \ldots, i_k$ is equal to i) will be equal to the corresponding iterated integrals in the expansion of $y_j(t; x^\circ; u_1, \ldots, u_{i-1}, v^b, u_{i+1}, \ldots, u_m)$ because the inputs $u_1, \ldots, u_{i-1}, u_{i+1}, \ldots, u_m$ are the same. Therefore, we deduce that the difference between the right-hand-side and the left-hand-side of (3.1) is a series of the form

$$\sum_{k=0}^{\infty} \sum_{i_0, \ldots, i_k = 0}^{m} c(i_k \ldots i_0) \int_0^t d\xi_{i_k} \ldots d\xi_{i_0}$$

in which the only nonzero coefficients are those with at least one of the indexes $i_0, \ldots, i_k$ equal to i. The sum of this series is zero for every input $u_1, \ldots, u_{i-1}, v^b, u_{i+1}, \ldots, u_m$. Therefore, according to Lemma 1.2, all its coefficients must vanish, for all $x^\circ \in U$. •

The condition (3.2) can be given other formulations, in geometric terms. Remember (Remark 1.6) that we have already observed that the coefficients of the Fliess expansion of $y(t)$ coincide with the values at x° of the functions that span the observation space $\mathcal{O}$. The differentials of these functions span, by definition, the codistribution

$$\Omega_{\mathcal{O}} = \text{span}\{d\lambda : \lambda \in \mathcal{O}\}.$$

If we fix our attention only on the j-th output, we may in particular define an observation space $\mathcal{O}_j$ as the space of all R-linear combinations of functions of the form h_j and $L_{g_{i_0}} \ldots L_{g_{i_k}} h_j$, $0 \leq i_k \leq m$, $0 \leq k < \infty$. The set of differentials $dh_j, dL_{g_{i_0}} \ldots L_{g_{i_k}} h_j$ with $i_k, \ldots, i_0 \in I$ and j fixed spans the codistribution

$$\Omega_{\mathcal{O}_j} = \text{span}\{d\lambda : \lambda \in \mathcal{O}_j\}.$$

Now, observe that the condition (3.2) can be written as

$$\langle dh_j, g_i \rangle(x) = 0$$
$$\langle dL_{g_{i_0}} \ldots L_{g_{i_k}} h_j, g_i \rangle(x) = 0$$

for all $k \geq 0$ and for all $i_k, \ldots, i_0 \in I$. From the above discussion we conclude that the condition stated in Lemma 3.1 is equivalent to the condition

$$g_i \in \Omega_{\mathcal{O}_j}^{\perp}. \qquad (3.4)$$

Other formulations are possible. For, remember that we have shown in Section 2.3 that the distribution $\Omega_{\mathcal{O}}^{\perp}$ is invariant under the vector fields $f, g_1, \ldots, g_m$. For the same reasons, also the distribution $\Omega_{\mathcal{O}_j}^{\perp}$ is invariant under $f, g_1, \ldots, g_m$.

Now, let $\langle f, g_1, \ldots, g_m | \text{span}\{g_i\} \rangle$ denote, as usual, the smallest distribution invariant under $f, g_1, \ldots, g_m$ which contains $\text{span}\{g_i\}$. If (3.4) is true, then, since $\Omega_{\mathcal{O}_j}^{\perp}$ is invariant under $f, g_1, \ldots, g_m$, we must have

$$\langle f, g_1, \ldots, g_m | \text{span}\{g_i\} \rangle \subset \Omega_{\mathcal{O}_j}^{\perp}. \qquad (3.5)$$

Moreover, since

$$\Omega_{\mathcal{O}_j}^{\perp} \subset (\text{span}\{dh_j\})^{\perp}$$

we see also that if (3.5) is true, we must have

$$\langle f, g_1, \ldots, g_m | \text{span}\{g_i\} \rangle \subset (\text{span}\{dh_j\})^{\perp}. \qquad (3.6)$$

Thus, we have seen that (3.4) implies (3.5) and this, in turn, implies (3.6). We will show now that (3.6) implies (3.4) thus proving that the three conditions are in fact equivalent.

For, observe that any vector field of the form $[\tau, g_i]$ with $\tau \in \{f, g_1, \ldots, g_m\}$ is by definition in the left-hand-side of (3.6). Therefore, if (3.6) is true,

$$0 = \langle dh_j, [\tau, g_i] \rangle = L_{\tau} L_{g_i} h_j - L_{g_i} L_{\tau} h_j.$$

But, again from (3.6), $g_i \in \left(\mathrm{span}\{dh_j\}\right)^\perp$ so we can conclude

$$L_{g_i} L_\tau h_j = 0$$

i.e.

$$g_i \in \left(\mathrm{span}\{dL_\tau h_j\}\right)^\perp.$$

By iterating this argument it is easily seen that if $\tau_k, \ldots, \tau_1$ is any set of k vector fields belonging to the set $\{f, g_1, \ldots, g_m\}$, then

$$g_i \in \left(\mathrm{span}\{dL_{\tau_k} \ldots L_{\tau_1} h_j\}\right)^\perp \tag{3.7}$$

We know that $\mathcal{O}_j$ consists of $\mathbf{R}$-linear combinations of functions of the form h_j or $L_{\tau_k} \ldots L_{\tau_1} h_j$, with $\tau_i \in \{f, g_1, \ldots, g_m\}$, $1 \leq i \leq k$, $1 \leq k < \infty$. Thus, from (3.7) we deduce that g_i annihilates the differential of any function in $\mathcal{O}_j$, i.e. that (3.4) is satisfied.

Summing up we may state the following result.

Theorem 3.2. The output y_j is unaffected by the input u_i if and only if any one of the following (equivalent) conditions is satisfied

(i) $$g_i \in \Omega_{\mathcal{O}_j}^\perp$$

(ii) $$\langle f, g_1, \ldots, g_m | \mathrm{span}\{g_i\} \rangle \subset \left(\mathrm{span}\{dh_j\}\right)^\perp$$

(iii) $$\langle f, g_1, \ldots, g_m | \mathrm{span}\{g_i\} \rangle \subset \Omega_{\mathcal{O}_j}^\perp.$$

Remark 3.3. It is clear that the statement of Lemma 3.1 can be slightly modified (and weakened) by asking that

$$L_{g_i} h_j(x) = 0$$
$$L_{g_i} L_{\tau_1} \ldots L_{\tau_r} h_j(x) = 0$$

for all $r \geq 1$ and any choice of vector fields $\tau_1, \ldots, \tau_r$ in the set $\{f, g_1, \ldots, g_{i-1}, g_{i+1}, \ldots, g_m\}$. Consistently, instead of $\mathcal{O}_j$, one should consider the subspace of all $\mathbf{R}$-linear combinations of h_j and $L_{\tau_1} \ldots L_{\tau_r} h_j$, with $\tau_1, \ldots, \tau_k$ vector fields in the set $\{f, g_1, \ldots, g_{i-1}, g_{i+1}, \ldots, g_m\}$.

Remark 3.4. Suppose $\langle f, g_1, \ldots, g_m | \mathrm{span}\{g_i\} \rangle$ and $\Omega_{\mathcal{O}_j}^\perp$ are nonsingular. The distributions are also involutive (see Lemmas 1.8.7, 1.9.6 and Remark 2.3.6). If the condition (iii) of Theorem 3.2 is satisfied, then around each point $x \in U$ it is possible to find a coordinate neighborhood on which the nonlinear system is locally represented by equations of the form

$$\dot{x}_1 = f_1(x_1, x_2) + \sum_{\substack{k=1 \\ k \neq i}}^{m} g_{1k}(x_1, x_2)u_k + g_{1i}(x_1, x_2)u_i$$

$$\dot{x}_2 = f_2(x_2) + \sum_{\substack{k=1 \\ k \neq i}}^{m} g_{2k}(x_2)u_k$$

$$y_j = h_j(x_2)$$

from which one sees that the input u_i has no influence on the output y_j. •

Suppose there is a distribution Δ which is invariant under the vector fields $f, g_1, \ldots, g_m$, contains the vector field g_i and is contained in the distribution $(\text{span}\{dh_j\})^{\perp}$. Then

$$\langle f, g_1, \ldots, g_m | \text{span}\{g_i\} \rangle \subset \Delta \subset (\text{span}\{dh_j\})^{\perp}.$$

We conclude from the above inequality that the condition (ii) of Theorem 3.2 is satisfied. Conversely, if condition (i) of Theorem 3.2 is satisfied, we have a distribution, $\Omega_{\bar{\mathcal{O}}_j}^{\perp}$, which is invariant under the vector fields $f, g_1, \ldots, g_m$, contains g_i and is contained in $(\text{span}\{dh_j\})^{\perp}$. Therefore we may give another different and useful formulation to the invariance condition.

Theorem 3.5. The output y_j is unaffected by the input u_i if and only if there exists a distribution Δ with the following properties

(i) $\qquad\qquad\qquad\qquad \Delta$ is invariant under $f, g_1, \ldots, g_m$

(ii) $\qquad\qquad\qquad\qquad g_i \in \Delta \subset (\text{span}\{dh_j\})^{\perp}.$

Remark 3.6. Again the condition (i) may be weakened by simply asking that

(i') $\qquad\qquad\qquad \Delta$ is invariant under $f, g_1, \ldots, g_{i-1}, g_{i+1}, \ldots, g_m.$

Note that this implies that if there exists a distribution Δ with the properties (i') and (ii) then there exists another distribution Δ with the properties (i) and (ii). •

We leave to the reader the task of extending the previuos result to the situation in which it is required that a specified set of outputs $y_{j_1}, \ldots, y_{j_r}$ has to be unaffected by a given set of inputs $u_{i_1}, \ldots, u_{i_s}$. The conditions stated in Lemma 3.1 remain formally the same, while the ones stated in Theorems 3.2 and 3.5 require appropriate modifications.

Example 3.1. In concluding this Section it may be worth observing that in case the system in question reduces to a linear system of the form

$$\dot{x} = Ax + \sum_{i=1}^{m} b_i u_i$$

$$y_j = c_j x \qquad 1 \le j \le p$$

then the condition (3.2) becomes

$$c_j A^k b_i = 0 \qquad \text{for all } k \ge 0.$$

The conditions (i),(ii),(iii) of Theorem 3.2 become respectively

$$b_i \in \bigcap_{k=0}^{n-1} \ker\big(c_j A^k\big)$$

$$\sum_{k=0}^{n-1} \operatorname{Im}\big(A^k b_i\big) \subset \bigcap_{k=0}^{n-1} \ker\big(c_j A^k\big)$$

$$\sum_{k=0}^{n-1} \operatorname{Im}\big(A^k b_i\big) \subset \ker\big(c_j\big).$$

These clearly imply and are implied by the existence of a subspace V invariant under A and such that

$$b_i \subset V \in \ker(c_j). \quad \bullet$$

3.4 Realization Theory

The problem of "realizing" a given input-output behavior is generally known as the problem of finding a dynamical system with inputs and outputs able to reproduce, when initialized in a suitable state, the given input-output behavior. The dynamical system is thus said to "realize", from the chosen initial state, the prescribed input-output map.

Usually, the search for dynamical systems which realize the input-output map is restricted to special classes in the universe of all dynamical systems, depending on the structure and/or properties of the given map. For example, when this map may be represented as a convolution integral of the form

$$y(t) = \int_0^t w(t - \tau)u(\tau)\, d\tau$$

where w is a prescribed function of t defined for $t \ge 0$, then one usually looks for a linear dynamical system

$$\dot{x} = Ax + Bu$$
$$y = Cx$$

able to reproduce, when initialized in $x^\circ = 0$, the given behavior. For this to be true, the matrices A, B, C must be such that

$$C\exp(At)B = w(t).$$

We will now describe the fundamentals of the realization theory for the (rather general) class of input-output maps which can be represented like functionals of the form (1.4). In view of the results of the previous Sections, the search for "realizations" of this kind of maps will be restricted to the class of dynamical system of the form (1.1).

From a formal point of view, the problem is stated in the following way. Given a formal power series in $m + 1$ noncommutative indeterminates with coefficients in $\mathbf{R}^p$, find an integer n, an element x° of $\mathbf{R}^n$, $m + 1$ analytic vector fields $g_0, \ldots, g_m$ and an analytic p-vector valued function h defined on a neighborhood U of x° such that

$$h(x^\circ) = c(\emptyset)$$
$$L_{g_{i_0}} \ldots L_{g_{i_k}} h(x^\circ) = c(i_k \ldots i_0).$$

If these conditions are satisfied, then it is clear that the dynamical system

$$\dot{x} = g_0(x) + \sum_{i=1}^m g_i(x)u_i$$
$$y = h(x)$$

initialized in $x^\circ \in \mathbf{R}^n$ produces an input-output behavior of the form

$$y(t) = c(\emptyset) + \sum_{k=0}^\infty \sum_{j_0,\ldots,j_k=0}^m c(j_k \ldots j_0) \int_0^t d\xi_{j_k} \ldots d\xi_{j_0}.$$

In view of this, the set $\{g_0, \ldots, g_m, h, x^\circ\}$ will be called a *realization* of the formal power series c.

In order to present the basic results of the relization theory, we need first to develop some notations and describe some simple algebraic concepts related to the formal power series. In view of the need of dealing with sets of series and defining certain operations on these sets, it is useful to represent each series as a *formal* infinite *sum* of "monomials". Let $z_0, \ldots, z_m$ denote a set of $m + 1$ abstract noncommutative indeterminates and let $Z = \{z_0, \ldots, z_m\}$. With each multiindex $(i_k \ldots i_0)$ we associate the monomial $(z_{i_k} \ldots z_{i_0})$ and we represent the series in the form

$$c = c(\emptyset) + \sum_{k=0}^{\infty} \sum_{i_0,\dots,i_k=0}^{m} c(i_k \dots i_0) z_{i_k} \dots z_{i_0}. \tag{4.1}$$

The set of all *power series* in $m+1$ noncommutative indeterminates (or, in other words, in the noncommutative indeterminates $z_0, \dots, z_m$) and coefficients in $\mathbf{R}^p$ is denoted with the symbol $\mathbf{R}^p \langle\!\langle Z \rangle\!\rangle$. A special subset of $\mathbf{R}^p \langle\!\langle Z \rangle\!\rangle$ is the set of all those series in which the number of nonzero coefficients (i.e. the number of nonzero terms in the sum (4.1)) is finite. A series of this type is a *polynomial* in $m+1$ noncommutative indeterminates and the set of all such polynomials is denoted with the symbol $\mathbf{R}^p \langle Z \rangle$. In particular $\mathbf{R}\langle Z \rangle$ is the set of all polynomials in the $m+1$ noncommutative indeterminates $z_0, \dots, z_m$ and coefficients in $\mathbf{R}$.

An element of $\mathbf{R}\langle Z \rangle$ may be represented in the form

$$p = p(\emptyset) + \sum_{k=0}^{d} \sum_{i_0,\dots,i_k=0}^{m} p(i_0 \dots i_k) z_{i_k} \dots z_{i_0} \tag{4.2}$$

where d is an integer which depends on p and $p(\emptyset), p(i_0 \dots i_k)$ are real numbers.

The sets $\mathbf{R}\langle Z \rangle$ and $\mathbf{R}^p \langle\!\langle Z \rangle\!\rangle$ may be given different algebraic structures. They can clearly be regarded as $\mathbf{R}$-vector *spaces*, by letting $\mathbf{R}$-linear combinations of polynomials and/or series be defined coefficient-wise. The set $\mathbf{R}\langle Z \rangle$ may also be given a *ring* structure, by letting the operation of sum of polinomials be defined coefficient-wise (with the neutral element given by the polinomial whose coefficients are all zero) and the operation of product of polynomials defined through the customary product of the corresponding representations (4.2) (in which case the neutral element is the polynomial whose coefficients are all zero but $p(\emptyset)$ is equal to 1). Later on, in the proof of Theorem 4.4 we shall also endow $\mathbf{R}\langle Z \rangle$ and $\mathbf{R}\langle\!\langle Z \rangle\!\rangle$ with structures of *modules* over the ring $\mathbf{R}\langle Z \rangle$, but, for the moment, those additional structures are not required.

What is important at this point is to know that the set $\mathbf{R}\langle Z \rangle$ can also be given a structure of *Lie algebra*, by taking the above-mentioned $\mathbf{R}$-vector space structure and defining a Lie bracket of two polynomials p_1, p_2 by setting $[p_1, p_2] = p_2 p_1 - p_1 p_2$. The smallest sub-algebra of $\mathbf{R}\langle Z \rangle$ which contains the monomials $z_0, \dots, z_m$ will be denoted by $\mathcal{L}(Z)$. Clearly, $\mathcal{L}(Z)$ may be viewed as a subspace of the $\mathbf{R}$-vector space $\mathbf{R}\langle Z \rangle$, which contains $z_0, \dots, z_m$ and is closed under Lie bracketing with $z_0, \dots, z_m$. Actually, it is not difficult to see that $\mathcal{L}(Z)$ is the smallest subspace of $\mathbf{R}\langle Z \rangle$ which has these properties.

Now we return to the problem of realizing an input-output map represented by a functional of the form (1.4). As expected, the existence of realizations will be characterized as a property of the formal power series which specifies the functional. We associate with the formal power series c two integers which will be called, following Fliess, the *Hankel rank* and the *Lie rank* of c. This is done in the following manner. We use the given formal power series c to define a mapping

$$F_c : \mathbf{R}\langle Z \rangle \to \mathbf{R}^p \langle\langle Z \rangle\rangle$$

in the following way:

(a) The image under F_c of any polynomial in the set $Z^\star = \{z_{j_k} \ldots z_{j_0} \in \mathbf{R}\langle Z \rangle : (j_k \ldots j_0) \in I^\star\}$ (by definition, the polynomial associated with the multiindex $\emptyset \in I^\star$ will be the polynomial in which all coefficients are zero but $p(\emptyset)$ which is equal to 1, i.e. the unit of $\mathbf{R}\langle Z \rangle$) is a formal power series defined by setting

$$\left[F_c(z_{j_k} \ldots z_{j_0}) \right](i_r \ldots i_0) = c(i_r \ldots i_0 j_k \ldots j_0)$$

for all $j_k \ldots j_0 \in I^\star$.

(b) The map F_c is an $\mathbf{R}$-vector space morphism of $\mathbf{R}\langle Z \rangle$ into $\mathbf{R}^p \langle\langle Z \rangle\rangle$.

Note that any polynomial in $\mathbf{R}\langle Z \rangle$ may be expressed as an $\mathbf{R}$-linear combination of elements of $Z^\star$ and, therefore, the prescriptions (a) and (b) completely specify the mapping F_c.

Looking at F_c as a morphism of $\mathbf{R}$-vector spaces, we define the *Hankel rank* $\rho_H(c)$ of c as the rank of F_c, i.e. the dimension of the subspace

$$F_c(\mathbf{R}\langle Z \rangle) \subset \mathbf{R}^p \langle\langle Z \rangle\rangle.$$

Moreover, we define the *Lie rank* $\rho_L(c)$ of c as the dimension of the subspace

$$F_c(\mathcal{L}(Z)) \subset \mathbf{R}^p \langle\langle Z \rangle\rangle$$

i.e. the rank of the mapping $F_c|_{\mathcal{L}(Z)}$.

Remark 4.1. It is easy to get a matrix representation of the mapping F_c. For, suppose we represent an element p of $\mathbf{R}\langle Z \rangle$ with an infinite column vector of real numbers whose entries are indexed by the elements of $I^\star$ and the entry indexed by $j_k \ldots j_0$ is exactly $p(j_k \ldots j_0)$. Of course, p being a polynomial, only finitely many elements of this vector are nonzero. In the same way, we may represent an element c of $\mathbf{R}^p \langle\langle Z \rangle\rangle$ with an infinite column vector whose entries are p-vectors of real numbers, indexed by the elements of $I^\star$ and such that the entry indexed $i_r \ldots i_0$ is $c(i_r \ldots i_0)$. Then, any $\mathbf{R}$-vector space morphism defined on $\mathbf{R}\langle Z \rangle$ with values in $\mathbf{R}^p \langle\langle Z \rangle\rangle$ will be represented by an infinite matrix, whose columns are indexed by elements of $I^\star$ and in which each block of p-rows of index $(i_r \ldots i_0)$ on the column of index $(j_k \ldots j_0)$ is exactly the coefficient

$$c(i_r \ldots i_0 \ j_k \ldots j_0)$$

of c. We leave to the reader the elementary check of this statement. •

The matrix H_c is called the *Hankel matrix* of the series c. It is clear from the above definition that the rank of the matrix H_c coincides with the Hankel rank of c.

Example 4.1 If the set I consists of only one element, then it is easily seen that I^* can be identified with the set $\mathbf{Z}^+$ of the non-negative integer numbers. A formal power series in one indeterminate with coefficients in $\mathbf{R}$, i.e. a mapping

$$c : \mathbf{Z}^+ \to \mathbf{R}$$

may be represented, like in (5.1), as an infinite sum

$$c = \sum_{k=0}^{\infty} c_k z^k$$

and the Hankel matrix associated with the mapping F_c coincides with the classical Hankel matrix associated with the sequence $c_0, c_1, \ldots$

$$H_c = \begin{bmatrix} c_0 & c_1 & c_2 & \cdots \\ c_1 & c_2 & c_3 & \cdots \\ c_2 & c_3 & c_4 & \cdots \\ \vdots & \vdots & \vdots & \ddots \end{bmatrix}. \quad \bullet$$

The importance of the Hankel and Lie ranks of the mapping F_c depends on the following basic results.

Lemma 4.2. Let $f, g_1, \ldots, g_m, h$ and a point $x^\circ \in \mathbf{R}^n$ be given. Let Δ_C be the distribution associated with the control Lie algebra C and Ω_O the codistribution associated with the observation space O. Let $K(x^\circ)$ denote the subset of vectors of $\Delta_C(x^\circ)$ which annihilate $\Omega_O(x^\circ)$ i.e. the subspace of $T_{x^\circ}\mathbf{R}^n$ defined by

$$K(x^\circ) = \Delta_C(x^\circ) \cap \Omega_O^\perp(x^\circ) = \{v \in \Delta_C(x^\circ) : \langle d\lambda(x^\circ), v \rangle = 0 \ \forall \lambda \in O \}.$$

Finally, let c be the formal power series defined by

$$c(\emptyset) = h(x^\circ) \tag{4.3a}$$

$$c(i_k \ldots i_0) = L_{g_{i_0}} \ldots L_{g_{i_k}} h(x^\circ) \tag{4.3b}$$

with $g_0 = f$. Then the Lie rank of c has the value

$$\rho_L(c) = \dim \Delta_C(x^\circ) - \dim K(x^\circ) = \dim \frac{\Delta_C(x^\circ)}{\Delta_C(x^\circ) \cap \Omega_{\mathcal{O}}^\perp(x^\circ)}$$

Proof. Define a morphism of Lie algebras

$$\mu : \mathcal{L}(Z) \to V(\mathbf{R}^n)$$

by setting

$$\mu(z_i) = g_i \qquad 0 \leq i \leq m.$$

Then, it is easy to check that if p is a polynomial in $\mathcal{L}(Z)$ the $(i_k \ldots i_0)$-th coefficient of $F_c(p)$ is $L_{\mu(p)} L_{g_{i_0}} \ldots L_{g_{i_k}} h(x^\circ)$, Thus, the series $F_c(p)$ has the expression

$$F_c(p) = L_{\mu(p)} h(x^\circ) + \sum_{k=0}^{\infty} \sum_{i_0,\ldots,i_k=0}^{m} L_{\mu(p)} L_{g_{i_0}} \ldots L_{g_{i_k}} h(x^\circ) z_{i_k} \ldots z_{i_0}.$$

If we let v denote the value of the vector field $\mu(p)$ at x°, the above can be rewritten as

$$F_c(p) = \langle dh(x^\circ), v \rangle + \sum_{k=0}^{\infty} \sum_{i_0,\ldots,i_k=0}^{m} \langle dL_{g_{i_0}} \ldots L_{g_{i_k}} h(x^\circ), v \rangle z_{i_k} \ldots z_{i_0}.$$

When p ranges over $\mathcal{L}(Z)$, the tangent vector v takes any value in $\Delta_C(x^\circ)$. Moreover, the covectors $dh(x^\circ), \ldots, dL_{g_{i_0}} \ldots L_{g_{i_k}} h(x^\circ), \ldots$ span $\Omega_{\mathcal{O}}(x^\circ)$. This implies that the number of $\mathbf{R}$-linearly independent power series in $F_c(\mathcal{L}(Z))$ is exactly equal to

$$\dim \Delta_C(x^\circ) - \dim \Delta_C(x^\circ) \cap \Omega_{\mathcal{O}}^\perp(x^\circ)$$

and this, in view of the definition of the Lie rank of F_c, proves the claim. •

We immediately see from this that if an input-output functional of the form (1.4) is realized by a dynamical system of dimension n, then necessarily the Lie rank of the formal power series which specifies the functional is bounded by n. In other words, the *finiteness* of the Lie rank $\rho_L(c)$ is a necessary condition for the existence of finite-dimensional realizations. We shall see later on that this condition is also sufficient. For the moment, we wish to investigate the role of the finiteness of the other rank associated with F_c i.e. the Hankel rank. It comes from the definition that

$$\rho_L(c) \leq \rho_H(c)$$

so that the Hankel rank may be infinite when the Lie rank is finite. However, there are special cases in which $\rho_H(c)$ is finite.

Lemma 4.3. Suppose $f, g_1, \ldots, g_m, h$ are linear in x, i.e. that

$$f(x) = Ax \quad g_1(x) = N_1 x, \ldots, g_m(x) = N_m x \quad h(x) = Cx$$

for suitable matrices $A, N_1, \ldots, N_m, C$. Let x° be a point of $\mathbf{R}^n$. Let V denote the smallest subspace of $\mathbf{R}^n$ which contains x° and is invariant under $A, N_1, \ldots, N_m$. Let W denote the largest subspace of $\mathbf{R}^n$ which is contained in $\ker(C)$ and is invariant under $A, N_1, \ldots, N_m$. The Hankel rank of the formal power series (4.3) has the value

$$\rho_H(c) = \dim V - \dim W \cap V = \dim \frac{V}{W \cap V}.$$

Proof. We have already seen, in Section 2.4, that the subspace W may be expressed in the following way

$$W = (\ker C) \cap \left[\bigcap_{r=0}^{\infty} \bigcap_{i_0, \ldots, i_r = 0}^{m} \ker(C N_{i_r} \ldots N_{i_0}) \right]$$

with $N_0 = A$. With the same kind of arguments one proves that the subspace V may be expressed as

$$V = \text{span}\{x^\circ\} + \sum_{k=0}^{\infty} \sum_{j_0, \ldots, j_k = 0}^{m} \text{span}\{N_{j_k} \ldots N_{j_0} x^\circ\}.$$

In the present case the Hankel matrix of F_c is such that the block of p rows of index $(i_r \ldots i_0)$ on the column of index $(j_k \ldots j_0)$, i.e. the coefficient $c(i_r \ldots i_0 j_k \ldots j_0)$ of c has the expression

$$C N_{i_r} \ldots N_{i_0} N_{j_k} \ldots N_{j_0} x^\circ.$$

By factoring out this expression in the form

$$(C N_{i_r} \ldots N_{i_0})(N_{j_k} \ldots N_{j_0} x^\circ)$$

it is seen that the Hankel matrix can be factored as the product of two matrices, of which the one on the left-hand-side has a kernel equal to the subspace W, while the one on the right-hand-side has an image equal to the subspace V. From this the claimed result follows immediately. ●

Thus, it is seen from this Lemma that if an input-output functional of the form (1.4) is realized by a dynamical system of dimension n described by equations of the form

$$\dot{x} = Ax + \sum_{i=1}^{m} N_i x u_i$$

$$y = Cx$$

i.e. by a bilinear dynamical system of dimension n, then the Hankel rank of the formal power series which specifies the functional is bounded by n. The finiteness of the Hankel rank $\rho_H(c)$ is a necessary condition for the existence of bilinear realizations.

We turn now to the problem of showing the sufficiency of the above two conditions. We treat first the case of bilinear realizations, which is simpler. In analogy with the definition given at the beginning of the Section, we say that the set $\{N_0, \ldots, N_m, C, x^\circ\}$, where $x^\circ \in \mathbf{R}^n$, $N_i \in \mathbf{R}^{n \times n}$ for $0 \le i \le m$ and $C \in \mathbf{R}^{p \times n}$ is a *bilinear realization* of the formal power series c if the set $\{g_0, \ldots, g_m, h, x^\circ)$ defined by

$$g_0(x) = N_0 x, \ g_1(x) = N_1 x, \ \ldots, \ g_m(x) = N_m x$$
$$h(x) = Cx$$

is a realization of c.

Theorem 4.4. Let c be a formal power series in $m + 1$ noncommutative indeterminates and coefficients in $\mathbf{R}^p$. There exists a bilinear realization of c if and only if the Hankel rank of c is finite.

Proof. We need only to prove the "if" part. For, consider again the mapping F_c. The sets $\mathbf{R}^p \langle Z \rangle$ and $\mathbf{R}^p \langle\!\langle Z \rangle\!\rangle$ will now be endowed with structures of modules. The ring $\mathbf{R} \langle Z \rangle$ is regarded as a module over itself. $\mathbf{R}^p \langle\!\langle Z \rangle\!\rangle$ is given an $\mathbf{R} \langle Z \rangle$-module structure by letting the operation of sum of power series be defined coefficient-wise and the product $p \cdot s$ of a polynomial $p \in \mathbf{R} \langle Z \rangle$ by a series $s \in \mathbf{R}^p \langle\!\langle Z \rangle\!\rangle$ be defined in the following way
(a) $1 \cdot s = s$
(b) for all $0 \le i \le m$ the series $z_i \cdot s$ is given by

$$(z_i \cdot s)(i_r \ldots i_0) = s(i_r \ldots i_0 i)$$

(c) for all $p_1, p_2 \in \mathbf{R} \langle Z \rangle$ and $\alpha_1, \alpha_2 \in \mathbf{R}$

$$(\alpha_1 p_1 + \alpha_2 p_2) \cdot s = \alpha_1 (p_1 \cdot s) + \alpha_2 (p_2 \cdot s).$$

Note that from a) and b) we have that for all $j_k \ldots j_0 \in I^\star$

$$(z_{j_k} \ldots z_{j_0} \cdot s)(i_r \ldots i_0) = s(i_r \ldots i_0 j_k \ldots j_0)$$

Note also that since the ring $\mathbf{R} \langle Z \rangle$ is not commutative, the order in which the products are performed is essential.

We leave to the reader the simple proof that the map F_c previously defined becomes an $\mathbf{R}\langle Z\rangle$-module morphism when $\mathbf{R}^p\langle\langle Z\rangle\rangle$ is endowed with this kind of $\mathbf{R}\langle Z\rangle$-module structure. As a matter of fact, it is trivial to check that $F_c(p) = p \cdot c$.

Now consider the canonical factorization of F_c

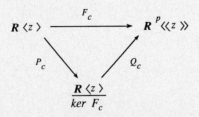

in which, as usual, P_c denotes the canonical projection $p \mapsto (p + \ker F_c)$ and Q_c the injection $(p + \ker F_c) \mapsto F_c(p)$. P_c and Q_c are $\mathbf{R}$-vector space morphisms, but there is also a canonical $\mathbf{R}\langle Z\rangle$-module structure on $\mathbf{R}\langle Z\rangle/\ker F_c$ which makes P_c and Q_c $\mathbf{R}\langle Z\rangle$-module morphisms.

Since, by definition, $\mathbf{R}\langle Z\rangle/\ker(F_c)$ is isomorphic to the image of F_c, we have that the dimension of $\mathbf{R}\langle Z\rangle/\ker(F_c)$ as an $\mathbf{R}$-vector space is equal to the Hankel rank $\rho_H(c)$ of the formal power series c. Let, for simplicity, denote

$$X = \frac{\mathbf{R}\langle Z\rangle}{\ker(F_c)}$$

But X is also an $\mathbf{R}\langle Z\rangle$-module, so to each of the indeterminates $z_0,\ldots,z_m$ we may associate mappings

$$M_i : X \to X$$

$$x \mapsto z_i \cdot x.$$

The mappings M_i are clearly $\mathbf{R}$-vector space morphisms. We also define an $\mathbf{R}$-vector space morphism

$$H : X \to \mathbf{R}^p$$

by taking

$$Hx = \big[Q_c(x)\big](\emptyset).$$

With the notation on the right-hand side we mean the coefficient with empty index in the series $Q_c(x)$.

Finally, let x° be the element of X

$$x^\circ = P_c(1)$$

where 1 is the unit polynomial in $\mathbf{R}\langle Z\rangle$.

We claim that

$$c(\emptyset) = Hx^\circ \qquad (4.4.a)$$

$$c(i_k \ldots i_0) = H M_{i_k} \ldots M_{i_0} x^\circ. \qquad (4.4.b)$$

For, it is seen immediately that

$$c = F_c(1) = Q_c P_c(1) = Q_c(x^\circ) \qquad (4.5a)$$

Moreover, suppose that

$$F_c(z_{i_k} \ldots z_{i_0}) = Q_c M_{i_k} \ldots M_{i_0} x^\circ \qquad (4.5b)$$

then we have

$$F_c(z_i z_{i_k} \ldots z_{i_0}) = z_i \cdot F_c(z_{i_k} \ldots z_{i_0}) = z_i(Q_c M_{i_k} \ldots M_{i_0} x^\circ)$$
$$= Q_c(z_i \cdot M_{i_k} \ldots M_{i_0} x^\circ) = Q_c M_i M_{i_k} \ldots M_{i_0} x^\circ$$

for $0 \leq i \leq m$. Thus (4.5b) is true for all $(i_k \ldots i_0) \in I^\star$.

Now, keeping in mind the definition of F_c, one has

$$\left[F_c(z_{i_k} \ldots z_{i_0}) \right](\emptyset) = c(z_{i_k} \ldots z_{i_0})$$

and, therefore, in view of the definition of the mapping H, (4.4) are proved.

Take now a basis in the $\rho_H(c)$-dimensional vector space X. The mappings $M_0, \ldots, M_m$ and H will be represented by matrices $N_0, \ldots, N_m$ and C; x° will be represented by a vector $\hat{x}^\circ$. These quantities are such that

$$c(i_k \ldots i_0) = C N_{i_k} \ldots N_{i_0} \hat{x}^\circ$$

for all $(i_k \ldots i_0) \in I^\star$. This shows that the set $\{C, N_0, \ldots, N_m, \hat{x}^\circ\}$ is a bilinear realization for our series. •

The result which follows presents a necessary and sufficient condition for the existence of realizations of an input-output functional of the form (1.4), provided that the coefficients of the power series which characterize the functional are suitably bounded.

Theorem 4.5. Let c be a formal power series whose coefficients satisfy the condition

$$\| c(i_k \ldots i_0) \| \leq C(k+1)! r^{(k+1)} \qquad (4.6)$$

for all $(i_k \ldots i_0) \in I^\star$, for some pair of real numbers $C > 0$ and $r > 0$. Then there exists a realization of c if and only if the Lie rank of c is finite.

Proof. Some more machinery is required. For each polynomial $p \in \mathbf{R}\langle Z \rangle$ we define a mapping $S_p : \mathbf{R}^p \langle\!\langle Z \rangle\!\rangle \to \mathbf{R}^p \langle\!\langle Z \rangle\!\rangle$ in the following way

(a) if $p \in Z^\star = \{z_{j_k} \ldots z_{j_0} \in \mathbf{R}\langle Z \rangle : (j_k \ldots j_0) \in I^\star\}$ then $S_p(c)$is a formal power series defined by setting

$$[S_{z_{j_k} \ldots z_{j_0}}(c)](i_r \ldots i_0) = c(j_k \ldots j_0 i_r \ldots i_0)$$

(b) if $\alpha_1, \alpha_2 \in \mathbf{R}$ and $p_1, p_2 \in \mathbf{R}\langle Z \rangle$ then

$$S_p(c) = \alpha_1 S_{p_1}(c) + \alpha_2 S_{p_2}(c).$$

Moreover, suppose that, given a formal power series $s_1 \in \mathbf{R}\langle\!\langle Z \rangle\!\rangle$ and a formal power series $s_2 \in \mathbf{R}\langle Z \rangle$, the sum of the numerical series

$$s_1(\emptyset)s_2(\emptyset) + \sum_{k=0}^{\infty} \sum_{i_0,\ldots,i_k=0}^{m} s_1(i_k \ldots i_0)s_2(i_k \ldots i_0) \tag{4.7}$$

exists. If this is the case, the sum of this series will be denoted by $\langle s_1, s_2 \rangle$.

We now turn our attention to the problem of finding a realization of c. In order to simplify the notation, we assume $p = 1$ (i.e. we consider the case of a single-output system). By definition, there exist n polynomials in $\mathcal{L}(Z)$, denoted $p_1, \ldots, p_n$, with the property that the formal power series $F_c(p_1), \ldots, F_c(p_n)$ are $\mathbf{R}$-linearly independent.

With the polynomials $p_1, \ldots, p_n$ we associate a formal power series

$$w = \exp\left(\sum_{i=1}^{n} x_i p_i\right) = 1 + \sum_{k=1}^{\infty} \frac{1}{k!}\left(\sum_{i=1}^{n} x_i p_i\right)^k \tag{4.8}$$

where $x_1, \ldots, x_n$ are real variables.

The series c which is to be realized and the series w thus defined are used in order to construct a set of analytic functions of $x_1, \ldots, x_n$, defined in a neighborhood of 0 and indexed by the elements of $I^\star$, in the following way

$$h(x) = \langle c, w \rangle$$
$$h_{i_k \ldots i_0}(x) = \langle S_{z_{i_k} \ldots z_{i_0}}(c), w \rangle$$

The growth condition (4.6) guarantees the convergence of the series on the right-hand-side for all x in a neighborhood of $x = 0$.

It will be shown now that there exist $m + 1$ vector fields, $g_0(x), \ldots, g_m(x)$, defined in a neighborhood of 0, with the property

$$L_{g_i} h_{i_k \ldots i_0}(x) = h_{i_k \ldots i_0 i}(x) \tag{4.9}$$

for all $(i_k \ldots i_0) \in I^\star$. This will be actually enough to prove the Theorem because, at $x = 0$, the functions $h_{i_k \ldots i_0}(x)$ by construction are such that

$$h(0) = c(\emptyset)$$
$$h_{i_k \ldots i_0}(0) = c(i_k \ldots i_0)$$

and this shows that the set $\{h, g_0, \ldots, g_m\}$ together with the initial state $x = 0$ is a realization of c.

To find the vector fields $g_0, \ldots, g_m$ one proceeds as follows. Since the n series $F_c(p_1), \ldots, F_c(p_n)$ are $\mathbf{R}$-linear independent, it is easily seen that there exists n monomials $m_1, \ldots, m_n$ in the set $Z^\star$ with the property that the $(n \times n)$ matrix of real numbers

$$\begin{bmatrix} [F_c(p_1)](m_1) & \cdots & [F_c(p_n)](m_1) \\ \vdots & \ddots & \vdots \\ [F_c(p_1)](m_n) & \cdots & [F_c(p_n)](m_n) \end{bmatrix} \tag{4.10}$$

has rank n. It is easy to see that

$$[F_c(p_i)](m_j) = \left(\frac{\partial}{\partial x_i} \langle S_{m_j}(c), w \rangle \right)_{x=0}.$$

For, if $p_i \in Z^\star$, then by definition

$$[F_c(p_i)](m_j) = c(m_j p_i) = [S_{m_j}(c)](p_i) = \left(\frac{\partial}{\partial x_i} \langle S_{m_j}(c), w \rangle \right)_{x=0}$$

From this, using linearity, one concludes that the above expression is true also in the (general) case where p_i is an $\mathbf{R}$-linear combination of elements of $Z^\star$.

Using this property, we conclude that the j-th row of the matrix (4.10) coincides with the value at 0 of the differential of one of the functions $h_{i_k \ldots i_0}$, the one whose multiindex corresponds to the monomial m_j.

Consider now the system of linear equations

$$\begin{bmatrix} \frac{\partial}{\partial x} \langle S_{m_1}(c), w \rangle \\ \vdots \\ \frac{\partial}{\partial x} \langle S_{m_n}(c), w \rangle \end{bmatrix} g_k(x) = \begin{bmatrix} \langle S_{m_1 z_k}(c), w \rangle \\ \vdots \\ \langle S_{m_n z_k}(c), w \rangle \end{bmatrix}$$

in the unknown vector $g_k(x)$. The coefficient matrix is nonsingular for all x in a neighborhood of 0 (because at $x = 0$ it coincides—as we have seen—with the matrix (4.10)). Thus, in a neighborhood of 0 it is possible to find a vector field $g_k(x)$ such that

$$L_{g_k} \langle S_{m_i}(c), w \rangle = \langle S_{m_i z_k}(c), w \rangle$$

and this proves that (4.9) can be satisfied, at least for those $h_{i_k \ldots i_0}$ whose multiindexes correspond to the monomials $m_1, \ldots, m_n$.

The proof that (4.9) holds for all other functions $h_{i_k \ldots i_0}(x)$ depends on the fact that every formal power series in $F_c(\mathcal{L}(Z))$ is an $\mathbf{R}$-linear combination of $F_c(p_1), \ldots, F_c(p_n)$, and is left to the reader. •

It is seen from the above Theorem that if a formal power series c has finite Lie rank, and its coefficients satisfy the growth condition (4.6), then it is possible to find a dynamical system of dimension $\rho_L(c)$ which realizes the series.

This fact, together with the result stated before in Lemma 4.2, induces to some further remarks. A realization $\{f, g_1, \ldots, g_m, x^\circ\}$ of a formal power series c is *minimal* if its dimension, i.e. the dimension of the underlying manifold on which $f, g_1, \ldots, g_m$ are defined, is less than or equal to the dimension of any other realization of c. Thus, from Lemma 4.2 we immediately deduce the following corollaries.

Corollary 4.6. A realization $\{f, g_1, \ldots, g_m, x^\circ\}$ of a formal power series c is minimal if and only if its dimension is equal to the Lie rank $\rho_L(c)$.

Corollary 4.7. A realization $\{f, g_1, \ldots, g_m, x^\circ\}$ of a formal power series c is minimal if and only if

$$\dim \Delta_C(x^\circ) = \dim \Omega_\mathcal{O}(x^\circ) = n$$

or, which is the same, the realization satisfies the controllability rank condition and the observability rank condition at x°.

3.5 Uniqueness of Minimal Realizations

In this Section we prove an interesting uniqueness result, by showing that any two minimal realizations of a formal power series are locally "diffeomorphic".

Theorem 5.1. Let c be a formal power series and let n denote its Lie rank. Let $\{g_0^a, \ldots, g_m^a, h^a, x^a\}$ and $\{g_0^b, \ldots, g_m^b, h^b, x^b\}$ be two minimal, i.e. n-dimensional, realizations of c. Let g_i^a, $0 \le i \le m$, and h^a be defined on a neighborhood U^a of x^a in $\mathbf{R}^n$ and g_i^b, $0 \le i \le m$, and h^b be defined on a neighborhood U^b of x^b in $\mathbf{R}^n$. Then, there exist open subsets $V^a \subset U^a$ and $V^b \subset U^b$ and a diffeomorphism $F : V^a \to V^b$ such that

$$g_i^b(x) = F_\star g_i^a \circ F^{-1}(x) \qquad 0 \le i \le m \tag{5.1}$$

$$h^b(x) = h^a \circ F^{-1}(x) \tag{5.2}$$

for all $x \in V^b$.

Proof. We break up the proof in several steps.

(i) Recall that a minimal realization $\{f, g_1, \ldots, g_m, x^\circ\}$ of c satisfies the observability rank condition at x° (Corollary 4.7). From the definitions of $\mathcal{O}$ and $\Omega_{\mathcal{O}}$, one deduces that there exist n real-valued functions $\lambda_1, \ldots, \lambda_n$, defined in a neighborhood U of x°, having the form

$$\lambda_i(x) = L_{v_r} \ldots L_{v_1} h_j(x)$$

with $v_1, \ldots, v_r$ vector fields in the set $\{f, g_1, \ldots, g_m\}$, r (possibly) depending on i and $1 \leq j \leq p$ such that the covectors $d\lambda_1(x^\circ), \ldots, d\lambda_n(x^\circ)$ are linearly independent (i.e. span the cotangent space $T_{x^\circ}^\star U$). From this property, using the inverse function theorem, it is deduced that there exists a neighborhood $U_H \subset U$ of x° such that the mapping

$$H : x \mapsto \big(\lambda_1(x), \ldots, \lambda_n(x)\big)$$

is a diffeomorphism of U_H onto its image $H(U_H)$.

From any two minimal realizations, labeled "a" and "b", we will construct two of such mappings, denoted H^a and respectively H^b.

(ii) Let $\theta_1, \ldots, \theta_n$ be a set of vector fields, defined in a neighborhood U of x°, having the form

$$\theta_i = f + \sum_{j=1}^{m} g_j \bar{u}_j^i$$

with $\bar{u}_j^i \in \mathbf{R}$ for $1 \leq j \leq m$. Let Φ_t^i denote the flow of θ_i and G denote the mapping

$$G : (t_1, \ldots, t_n) \mapsto \Phi_{t_n}^n \circ \cdots \circ \Phi_{t_1}^1(x^\circ)$$

defined on a neighborhood $(-\varepsilon, \varepsilon)^n$ of 0.

From any two minimal realizations, labeled "a" and "b", we will construct two of such mappings, denoted G^a and G^b (the same set of $\bar{u}_j^i$'s being used in both G^a and G^b).

Recall that a minimal realization $\{f^a, g_1^a, \ldots, g_m^a, x^a\}$ satisfies the controllability rank condition at x^a (Corollary 4.7). From the properties of Δ_c and R (see Remark 2.2.6), one deduces that the distribution R is nonsingular and n-dimensional around x^a. Then, using the same arguments as the ones used in the proof of Theorem 1.8.13, it is possible to see that there exists a choice of $\bar{u}_j^i$'s and an open subset W of $(0, \varepsilon)^n$ such that the restriction of G^a to W is a diffeomorphism of W onto its image $G^a(W)$.

(iii) It is easily proved that if $\{f^a, g_0^a, \ldots, g_m^a, h^a, x^a\}$ and $\{f^b, g_0^b, \ldots, g_m^b, h^b, x^b\}$ are two realizations of the same formal power series c, then, for all $0 < t_i < \varepsilon$, $1 \leq i \leq n$, with sufficiently small ε,

$$H^a \circ G^a(t_1, \ldots, t_n) = H^b \circ G^b(t_1, \ldots, t_n). \tag{5.3}$$

As a matter of fact, if ε is small then $G(t_1, \ldots, t_n)$ is a point of U_H reached from x^0 under the piecewise constant control defined by

$$u_j(t) = \bar{u}_j^i \quad \text{for} \quad t \in [t_1 + \cdots + t_{i-1}, t_1 + \cdots + t_i).$$

Moreover the values of the components of H (i.e. the values of the functions $\lambda_1, \ldots, \lambda_n$) at a point were shown to coincide with the values of certain derivatives, at time $t = 0$, of some components of an output function $y(t)$ obtained under suitable piecewise constant controls (see proof of Theorem 1.9.8). So, one may interpret the components of $H \circ G(t_1, \ldots, t_n)$ as the values at time $t = t_1 + \cdots + t_n$ of certain derivatives of an output function $y(t)$ obtained under suitable piecewise constant controls.   Two minimal realizations of the same power series c characterize two systems which by definition display the same input-output behavior. These two systems, initialized respectively in x^a and x^b, under any piecewise constant control produce two identical output functions. Thus the two sides of (5.3) must concide.

(iv) Recall that, if the realization "a" is minimal, if $(t_1, \ldots, t_n) \in W$ and ε is sufficiently small, the mapping $H^a \circ G^a$ is composition of diffeomorphisms. If also the realization "b" is minimal, H^b is indeed a diffeomorphism, but also G^b must be a diffeomorphism of W onto its image, because of the equality (5.3) and of the fact that the left-hand-side is itself a diffeomorphism. The following diagram

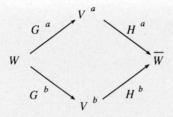

where $V^a = G^a(W)$, $V^b = G^b(W)$, $V^a \subset U_H^a$, $V^b \subset U_H^b$ and $\bar{W} = H^a \circ G^a(W) = H^b \circ G^b(W)$, is a commutative diagram of diffeomorphisms. Thus, we may define a diffeomorphism

$$F : V^a \to V^b$$

as

$$F = (H^b)^{-1} \circ H^a \tag{5.4.a}$$

whose inverse may also be expressed as

$$F^{-1} = G^a \circ (G^b)^{-1}. \tag{5.4b}$$

(v) By means of the same arguments as the ones already used in (iii) one may easily prove a more general version of (5.3). More precisely, setting

$$\theta^a = f^a + \sum_{i=1}^m g_i^a v_i \qquad \theta^b = f^b + \sum_{i=1}^m g_i^b v_i$$

one may deduce that, for sufficiently small t

$$H^a \circ \Phi_t^{\theta^a} \circ G^a(t_1, \dots, t_n) = H^b \circ \Phi_t^{\theta^b} \circ G^b(t_1, \dots, t_n).$$

Differentiating this one with respect to t and setting $t = 0$ one obtains

$$(H^a)_* \theta^a \circ G^a(t_1, \dots, t_n) = (H^b)_* \theta^b \circ G^b(t_1, \dots, t_n).$$

Because of the arbitrariness of $v_1, \dots, v_m$ one has then

$$(H^a)_* g_i^a \circ G^a(t_1, \dots, t_n) = (H^b)_* g_i^b \circ G^b(t_1, \dots, t_n)$$

for all $0 \le i \le m$. But these ones, in view of the definitions (5.4) may be rewritten as

$$g_i^b(x) = F_* g_i^a \circ F^{-1}(x) \qquad 0 \le i \le m$$

for all $x \in V^b$, thus proving (5.1).

(vi) Again, using the same arguments already used in (ii) one may easily see that

$$h^a \circ G^a(t_1, \dots, t_n) = h^b \circ G^b(t_1, \dots, t_n)$$

i.e. that

$$h^b(x) = h^a \circ F^{-1}(x)$$

for all $x \in V^b$, thus proving also (5.2). $\bullet$

4 Elementary Theory of Nonlinear Feedback for Single-Input Single-Output Systems

4.1 Local coordinates transformations

Beginning from this Chapter, we will study—in order of increasing complexity—a series of problems concerned with the synthesis of *feedback control laws* for nonlinear systems of the form (1.2.1). We will discuss first the case of single-input single-output systems, whose simple structure lends itself to a rather elementary analysis, and then—in the next Chapter—a special class of multivariable systems, in which a straightforward extension of most of the theory developed for single-input single-output systems is possible. Finally—in the last two Chapters—we will present a set of more powerful tools for the analysis and the design of rather more general classes of multivariable control systems.

The purpose of this introductory Section is to show how single-input single-output nonlinear systems can be locally given, by means of a suitable change of coordinates in the state space, a "normal form" of special interest, on which several important properties can be elucidated.

The point of departure of the whole analysis is the notion of relative degree of the system, which is formally described in the following way. The single-input single-output nonlinear system

$$\dot{x} = f(x) + g(x)u \qquad (1.1a)$$

$$y = h(x) \qquad (1.1b)$$

is said to have *relative degree* r at a point x° if

(i) $L_g L_f^k h(x) = 0$ for all x in a neighborhood of x° and all $k < r - 1$

(ii) $L_g L_f^{r-1} h(x^\circ) \neq 0$

Note that there may be points where a relative degree cannot be defined. This occurs, in fact, when the first function of the sequence

$$L_g h(x), L_g L_f h(x), \ldots, L_g L_f^k h(x), \ldots$$

which is not identically zero (in a neighborhood of x°) has a zero exactly at the point $x = x^\circ$. However, the set of points where a relative degree can be defined is clearly an open and dense subset of the set U where the system (1.1) is defined.

Example 1.1. Consider the equations describing a controlled Van der Pol oscillator in state space form

$$\dot{x} = f(x) + g(x)u = \begin{bmatrix} x_2 \\ 2\omega\zeta(1 - \mu x_1^2)x_2 - \omega^2 x_1 \end{bmatrix} + \begin{bmatrix} 0 \\ 1 \end{bmatrix} u.$$

Suppose the output function is chosen as

$$y = h(x) = x_1.$$

In this case we have

$$L_g h(x) = \frac{\partial h}{\partial x} g(x) = \begin{bmatrix} 1 & 0 \end{bmatrix} \begin{bmatrix} 0 \\ 1 \end{bmatrix} = 0$$

and

$$L_f h(x) = \frac{\partial h}{\partial x} f(x) = \begin{bmatrix} 1 & 0 \end{bmatrix} \begin{bmatrix} x_2 \\ 2\omega\zeta(1 - \mu x_1^2)x_2 - \omega^2 x_1 \end{bmatrix} = x_2.$$

Moreover

$$L_g L_f h(x) = \frac{\partial (L_f h)}{\partial x} g(x) = \begin{bmatrix} 0 & 1 \end{bmatrix} \begin{bmatrix} 0 \\ 1 \end{bmatrix} = 1$$

and thus we see that the system thus considered has relative degree 2 at any point x°.

However, if the output function is, for instance

$$y = h(x) = \sin x_2$$

then $L_g h(x) = \cos x_2$. The system has relative degree 1 at any point x°, provided that $(x^\circ)_2 \neq (2k+1)\pi/2$. If the point x° is such that this condition is violated, no relative degree can be defined. ●

Remark 1.1. In order to compare the notion thus introduced with a familiar concept, let us calculate the relative degree of a linear system

$$\dot{x} = Ax + Bu$$
$$y = Cx.$$

In this case, since $f(x) = Ax$, $g(x) = B$, $h(x) = Cx$, it easily seen that

$$L_f^k h(x) = CA^k x$$

and therefore

$$L_g L_f^k h(x) = CA^k B.$$

Thus, the integer r is characterized by the conditions

$$CA^k B = 0 \qquad \text{for all } k < r - 1$$
$$CA^{r-1} B \neq 0.$$

It is well-known that the integer satisfying these conditions is exactly equal to the *difference* between the degree of the denominator polynomial and the degree of the numerator polynomial of the transfer function

$$H(s) = C(sI - A)^{-1} B$$

of the system. •

We illustrate now a simple interpretation of the notion of relative degree, which is not restricted to the assumption of linearity considered in the previous Remark. Assume the system at some time t° is in the state $x(t^\circ) = x^\circ$ and suppose we wish to calculate the value of the output $y(t)$ and of its derivatives with respect to time $y^{(k)}(t)$, for $k = 1, 2, \ldots$, at $t = t^\circ$. We obtain

$$y(t^\circ) = h(x(t^\circ)) = h(x^\circ)$$
$$y^{(1)}(t) = \frac{\partial h}{\partial x} \frac{dx}{dt} = \frac{\partial h}{\partial x}(f(x(t)) + g(x(t))u(t))$$
$$= L_f h(x(t)) + L_g h(x(t))u(t).$$

If the relative degree r is larger than 1, for all t such that $x(t)$ is near x°, i.e. for all t near t°, we have $L_g h(x(t)) = 0$ and therefore

$$y^{(1)}(t) = L_f h(x(t)).$$

This yields

$$y^{(2)}(t) = \frac{\partial L_f h}{\partial x} \frac{dx}{dt} = \frac{\partial L_f h}{\partial x}(f(x(t)) + g(x(t))u(t))$$
$$= L_f^2 h(x(t)) + L_g L_f h(x(t))u(t).$$

Again, if the relative degree is larger than 2, for all t near t° we have $L_g L_f h(x(t)) = 0$ and

$$y^{(2)}(t) = L_f^2 h(x(t)).$$

Continuing in this way, we get

$$y^{(k)}(t) = L_f^k h(x(t)) \qquad \text{for all } k < r \text{ and all } t \text{ near } t^\circ$$
$$y^{(r)}(t^\circ) = L_f^r h(x^\circ) + L_g L_f^{r-1} h(x^\circ)u(t^\circ).$$

Thus, the relative degree r is exactly equal to the number of times one has to differentiate the output $y(t)$ at time $t = t°$ in order to have the value $u(t°)$ of the input explicitly appearing.

Note also that if

$$L_g L_f^k h(x) = 0 \qquad \text{for all } x \text{ in a neighborhood of } x° \text{ and all } k \geq 0$$

(in which case no relative degree can be defined at any point around $x°$) then the output of the system is not affected by the input, for all t near $t°$. As a matter of fact, if this is the case, the previous calculations show that the Taylor series expansion of $y(t)$ at the point $t = t°$ has the form

$$y(t) = \sum_{k=0}^{\infty} L_f^k h(x°) \frac{(t - t°)^k}{k!}$$

i.e. that $y(t)$ is a function depending only on the initial state and not on the input.

These calculations suggest that the functions $h(x)$, $L_f h(x), \ldots, L_f^{r-1} h(x)$ must have a special importance. As a matter of fact, it is possible to show that they can be used in order to define, at least partially, a local coordinates transformation around $x°$ (recall that $x°$ is a point where $L_g L_f^{r-1} h(x°) \neq 0$). This fact is based on the following property.

Lemma 1.2. The row vectors

$$dh(x°), dL_f h(x°), \ldots, dL_f^{r-1} h(x°)$$

are linearly independent.

In order to prove this Lemma, we illustrate first another property, which will also be used several other times in the sequel.

Lemma 1.3. Let ϕ be a real-valued function and f, g vector fields, all defined in an open set U of $\mathbf{R}^n$. Then, for any choice of $s, k, r > 0$,

$$\langle dL_f^s \phi(x), ad_f^{k+r} g(x) \rangle = \sum_{i=0}^{r} (-1)^i \binom{r}{i} L_f^{r-i} \langle dL_f^{s+i} \phi(x), ad_f^k g(x) \rangle. \qquad (1.2)$$

As a consequence, the two sets of conditions

$$(i) \ L_g \phi(x) = L_g L_f \phi(x) = \ldots = L_g L_f^k \phi(x) = 0 \qquad \text{for all } x \in U \quad (1.3a)$$

$$(ii) \ L_g \phi(x) = L_{ad_f g} \phi(x) = \cdots = L_{ad_f^k g} \phi(x) = 0 \qquad \text{for all } x \in U \quad (1.3b)$$

are equivalent.

Proof of Lemma 1.3. The proof of (1.2) is easily obtained by induction on r, in view of the fact that

$$\langle dL_f^s \phi(x), ad_f^{k+r+1} g(x) \rangle = \langle dL_f^s \phi(x), [f, ad_f^{k+r} g(x)] \rangle$$
$$= L_f \langle dL_f^s \phi(x), ad_f^{k+r} g(x) \rangle - \langle dL_f^{s+1} \phi(x), ad_f^{k+r} g(x) \rangle.$$

The equivalence of (1.3a) and (1.3b) is a straightforward consequence of (1.2). ●

Proof of Lemma 1.2. Observe that by definition of relative degree, using (1.2) we obtain for all i, j such that $i + j \leq r - 2$

$$\langle dL_f^j h(x), ad_f^i g(x) \rangle = 0 \qquad \text{for all } x \text{ around } x^o$$

and

$$\langle dL_f^j h(x^o), ad_f^i g(x^o) \rangle = (-1)^{r-1-j} L_g L_f^{r-1} h(x^o) \neq 0$$

for all i, j such that $i + j = r - 1$.

The above conditions, all together, show that the matrix

$$
\begin{bmatrix}
dh(x^o) \\
dL_f h(x^o) \\
\cdots \\
dL_f^{r-1} h(x^o)
\end{bmatrix}
[g(x^o) \quad ad_f g(x^o) \quad \cdots \quad ad_f^{r-1} g(x^o)] =
$$

$$
= \begin{bmatrix}
0 & & \cdots & \langle dh(x^o), ad_f^{r-1} g(x^o) \rangle \\
0 & & \cdots & \star \\
\cdots & & \cdots & \star \\
\langle dL_f^{r-1} h(x^o), g(x^o) \rangle & \star & & \star
\end{bmatrix} \qquad (1.4)
$$

has rank r and, thus, that the row vectors $dh(x^o)$, $dL_f h(x^o), \ldots, dL_f^{r-1} h(x^o)$ are linearly independent. ●

Lemma 1.2 shows that necessarily $r \leq n$ and that the r functions $h(x), L_f h(x), \ldots, L_f^{r-1} h(x)$ qualify as a partial set of new coordinate functions around the point x^o (recall Proposition **1**.2.3). As we shall see in a moment, the choice of these new coordinates entails a particularly simple structure in the equations describing the system. However, before doing this, it is convenient to summarize the results discussed so far in a formal statement, that also illustrates a way in which the set of new coordinates can be completed in case the relative degree r is strictly less than n.

Proposition 1.4. Suppose the system has relative degree r at x^o. Then

$r \leq n$. Set

$$\phi_1(x) = h(x)$$
$$\phi_2(x) = L_f h(x)$$
$$\cdots$$
$$\phi_r(x) = L_f^{r-1} h(x).$$

If r is strictly less than n, it is always possible to find $n - r$ more functions $\phi_{r+1}(x), \ldots, \phi_n(x)$ such that the mapping

$$\Phi(x) = \begin{bmatrix} \phi_1(x) \\ \cdots \\ \phi_n(x) \end{bmatrix}$$

has a jacobian matrix which is nonsingular at x° and therefore qualifies as a local coordinates transformation in a neighborhood of x°. The value at x° of these additional functions can be fixed arbitrarily. Moreover, it is always possible to choose $\phi_{r+1}(x), \ldots, \phi_n(x)$ in such a way that

$$L_g \phi_i(x) = 0 \qquad \text{for all } r + 1 \leq i \leq n \text{ and all } x \text{ around } x^\circ.$$

Proof. By definition of relative degree, the vector $g(x^\circ)$ is nonzero, and, thus, the distribution $G = \mathrm{span}\{g\}$ is nonsingular around x°. Being 1-dimensional, this distribution is also involutive. Therefore, by Frobenius' Theorem, we deduce the existence of $n - 1$ real-valued functions, $\lambda_1(x), \ldots, \lambda_{n-1}(x)$, defined in a neighborhood of x°, such that

$$\mathrm{span}\{d\lambda_1, \ldots, d\lambda_{n-1}\} = G^\perp. \tag{1.5}$$

It is easy to showthat

$$\dim(G^\perp + \mathrm{span}\{dh, dL_f h, \ldots, dL_f^{r-1} h\}) = n \tag{1.6}$$

at x°. For, suppose this is false. Then

$$G(x^\circ) \cap (\mathrm{span}\{dh, dL_f h, \ldots, dL_f^{r-1} h\})^\perp(x^\circ) \neq 0$$

i. e. the vector $g(x^\circ)$ annihilates all the covectors in

$$\mathrm{span}\{dh, dL_f h, \ldots, dL_f^{r-1} h\}(x^\circ)$$

But this is a contradiction, because by definition $\langle dL_f^{r-1} h(x^\circ), g(x^\circ) \rangle$ is nonzero. From (1.5), (1.6) and from the fact that $\mathrm{span}\{dh, dL_f h, \ldots, dL_f^{r-1} h\}$ has dimension r, by Lemma 1.2, we conclude that in the set $\{\lambda_1, \ldots, \lambda_{n-1}\}$ it is possible to find $n - r$ functions, without loss of generality $\lambda_1, \ldots, \lambda_{n-r}$, with the property that the n differentials $dh, dL_f h, \ldots, dL_f^{r-1} h, d\lambda_1, \ldots, d\lambda_{n-r}$, are linearly independent at x°. Since by construction the functions $\lambda_1, \ldots, \lambda_{n-r}$ are such that

$$\langle d\lambda_i(x), g(x) \rangle = L_g \lambda_i(x) = 0 \qquad \text{for all } x \text{ near } x^\circ \text{ and all } 1 \leq i \leq n - r$$

this establishes the required result. Note that any other set of functions of the form $\lambda_i'(x) = \lambda_i(x) + c_i$, where c_i is a constant, satisfies the same conditions, thus showing that the value of these functions at x° can be chosen arbitrarily. •

The description of the system in the new coordinates $z_i = \phi_i(x)$, $1 \leq i \leq n$, is found very easily. Looking at the calculations already carried out at the beginning, we obtain for $z_1, \ldots, z_r$

$$\frac{dz_1}{dt} = \frac{\partial \phi_1}{\partial x} \frac{dx}{dt} = \frac{\partial h}{\partial x} \frac{dx}{dt} = L_f h(x(t)) = \phi_2(x(t)) = z_2(t)$$

$$\ldots$$

$$\frac{dz_{r-1}}{dt} = \frac{\partial \phi_{r-1}}{\partial x} \frac{dx}{dt} = \frac{\partial (L_f^{r-2} h)}{\partial x} \frac{dx}{dt} = L_f^{r-1} h(x(t)) = \phi_r(x(t)) = z_r(t).$$

For z_r we obtain

$$\frac{dz_r}{dt} = L_f^r h(x(t)) + L_g L_f^{r-1} h(x(t)) u(t).$$

On the right-hand side of this equation we must now replace $x(t)$ with its expression as a function of $z(t)$, i.e. $x(t) = \Phi^{-1}(z(t))$. Thus, setting

$$a(z) = L_g L_f^{r-1} h(\Phi^{-1}(z))$$

$$b(z) = L_f^r h(\Phi^{-1}(z))$$

the equation in question can be rewritten as

$$\frac{dz_r}{dt} = b(z(t)) + a(z(t)) u(t).$$

Note that at the point $z^\circ = \Phi(x^\circ)$, $a(z^\circ) \neq 0$ by definition. Thus, the coefficient $a(z)$ is nonzero for all z in a neighborhood of z°.

As far as the other new coordinates are concerned, we cannot expect any special structure for the corresponding equations, if nothing else has been specified. However, if $\phi_{r+1}(x), \ldots, \phi_n(x)$ have been chosen in such a way that

$L_g \phi_i(x) = 0$, then

$$\frac{dz_i}{dt} = \frac{\partial \phi_i}{\partial x}(f(x(t)) + g(x(t))u(t)) = L_f \phi_i(x(t)) + L_g \phi_i(x(t))u(t) = L_f \phi_i(x(t)).$$

Setting

$$q_i(z) = L_f \phi_i(\Phi^{-1}(z)) \qquad \text{for all } r+1 \leq i \leq n$$

the latter can be rewritten as

$$\frac{dz_i}{dt} = q_i(z(t)).$$

Thus, in summary, the state-space description of the system in the new coordinates will be as follows

$$\begin{aligned}
\dot{z}_1 &= z_2 \\
\dot{z}_2 &= z_3 \\
&\cdots \\
\dot{z}_{r-1} &= z_r \\
\dot{z}_r &= b(z) + a(z)u \\
\dot{z}_{r+1} &= q_{r+1}(z) \\
&\cdots \\
\dot{z}_n &= q_n(z).
\end{aligned} \tag{1.7a}$$

In addition to these equations one has to specify how the output of the system is related to the new state variables. But, being $y = h(x)$, it is immediately seen that

$$y = z_1. \tag{1.7b}$$

The structure of these equations is best illustrated in the block diagram depicted in Fig. 4.1. The equations thus defined are said to be in *normal form*. We will find them useful in understanding how certain control problems can be solved.

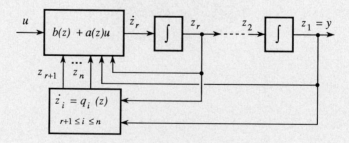

Fig. 4.1

Remark 1.5. Note that sometimes it is not easy to construct $n - r$ functions $\phi_{r+1}(x), \ldots, \phi_n(x)$ such that $L_g \phi_i(x) = 0$, because this, as shown in the proof of Proposition 1.4, amounts to solve a system of $n - r$ partial differential equations. Usually, it is much simpler to find functions $\phi_{r+1}(x), \ldots, \phi_n(x)$ with the only property that the jacobian matrix of $\Phi(x)$ is nonsingular at x^o, and this is sufficient to define a coordinates trasformation. Using a trasformation constructed in this way, one gets the same structure for the first r equations, i. e.

$$\dot{z}_1 = z_2$$
$$\dot{z}_2 = z_3$$
$$\cdots$$
$$\dot{z}_{r-1} = z_r$$
$$\dot{z}_r = b(z) + a(z)u$$

but it is not possible to obtain anything special for the last $n - r$ ones, that therefore will appear in a form like

$$\dot{z}_{r+1} = q_{r+1}(z) + p_{r+1}(z)u$$
$$\cdots$$
$$\dot{z}_n = q_n(z) + p_n(z)u$$

with the input u explicitly present. •

Example 1.2. Consider the system

$$\dot{x} = \begin{bmatrix} -x_1 \\ x_1 x_2 \\ x_2 \end{bmatrix} + \begin{bmatrix} \exp(x_2) \\ 1 \\ 0 \end{bmatrix} u$$
$$y = h(x) = x_3.$$

For this system we have

$$\frac{\partial h}{\partial x} = [0 \quad 0 \quad 1], \qquad L_g h(x) = 0, \qquad L_f h(x) = x_2$$
$$\frac{\partial (L_f h)}{\partial x} = [0 \quad 1 \quad 0], \qquad L_g L_f h(x) = 1.$$

In order to find the normal form, we set

$$z_1 = \phi_1(x) = h(x) = x_3$$
$$z_2 = \phi_2(x) = L_f h(x) = x_2$$

and we seek for a function $\phi_3(x)$ such that

$$\frac{\partial \phi_3}{\partial x} g(x) = \frac{\partial \phi_3}{\partial x_1} \exp(x_2) + \frac{\partial \phi_3}{\partial x_2} = 0.$$

It is easily seen that the function

$$\phi_3(x) = 1 + x_1 - \exp(x_2)$$

satisfies this condition. This and the previous two functions define a trasformation $z = \Phi(x)$ whose jacobian matrix

$$\frac{\partial \Phi}{\partial x} = \begin{bmatrix} 0 & 0 & 1 \\ 0 & 1 & 0 \\ 1 & -\exp(x_2) & 0 \end{bmatrix}$$

is nonsingular for all x. The inverse transformation is given by

$$x_1 = -1 + z_3 + \exp(z_2)$$
$$x_2 = z_2$$
$$x_3 = z_1.$$

Note also that $\Phi(0) = 0$. In the new coordinates the system is described by

$$\dot{z}_1 = z_2$$
$$\dot{z}_2 = (-1 + z_3 + \exp(z_2))z_2 + u$$
$$\dot{z}_3 = (1 - z_3 - \exp(z_2))(1 + z_2 \exp(z_2)).$$

These equations are globally valid because the transformation we considered was a global coordinates transformation. •

Example 1.3. Consider the system

$$\dot{x} = \begin{bmatrix} x_1 x_2 - x_1^3 \\ x_1 \\ -x_3 \\ x_1^2 + x_2 \end{bmatrix} + \begin{bmatrix} 0 \\ 2 + 2x_3 \\ 1 \\ 0 \end{bmatrix} u$$

$$y = h(x) = x_4.$$

For this system we have

$$\frac{\partial h}{\partial x} = [0 \quad 0 \quad 0 \quad 1], \qquad L_g h(x) = 0, \qquad L_f h(x) = x_1^2 + x_2$$

$$\frac{\partial (L_f h)}{\partial x} = [2x_1 \quad 1 \quad 0 \quad 0], \qquad L_g L_f h(x) = 2(1 + x_3).$$

Note that $L_g L_f h(x) \neq 0$ if $x_3 \neq -1$. This means that we shall be able to find a normal form only locally, away from any point such that $x_3 = -1$.

In order to find this normal form, we have to set first of all

$$z_1 = \phi_1(x) = h(x) = x_4$$
$$z_2 = \phi_2(x) = L_f h(x) = x_2 + x_1^2$$

and then find $\phi_3(x)$, $\phi_4(x)$ which complete the transformation and are such that $L_g \phi_3(x) = L_g \phi_4(x) = 0$.

Suppose we do not want to search for these particular functions and we just take any choice of $\phi_3(x)$, $\phi_4(x)$ which completes the transformation. This can be done, e.g. by taking

$$z_3 = \phi_3(x) = x_3$$
$$z_4 = \phi_4(x) = x_1.$$

The jacobian matrix of the transformation thus defined

$$\frac{\partial \Phi}{\partial x} = \begin{bmatrix} 0 & 0 & 0 & 1 \\ 2x_1 & 1 & 0 & 0 \\ 0 & 0 & 1 & 0 \\ 1 & 0 & 0 & 0 \end{bmatrix}$$

is nonsingular for all x, and the inverse transformation is given by

$$x_1 = z_4$$
$$x_2 = z_2 - z_4^2$$
$$x_3 = z_3$$
$$x_4 = z_1.$$

Note also that $\Phi(0) = 0$. In these new coordinates the system is described by

$$\dot{z}_1 = z_2$$
$$\dot{z}_2 = z_4 + 2z_4(z_4(z_2 - z_4^2) - z_4^3) + (2 + 2z_3)u$$
$$\dot{z}_3 = -z_3 + u$$
$$\dot{z}_4 = -2z_4^3 + z_2 z_4.$$

These equations are valid globally (because the trasformation we considered was a global coordinates transformation), but they are not in normal form because of the presence of the input u in the equation for z_3.

If one wants to get rid of u in this equation, it is necessary to use a different function $\phi_3(x)$, making sure that

$$\frac{\partial \phi_3}{\partial x} g(x) = \frac{\partial \phi_3}{\partial x_2}(2 + 2x_3) + \frac{\partial \phi_3}{\partial x_3} = 0.$$

An easy calculation shows that the function

$$\phi_3(x) = x_2 - 2x_3 - x_3^2$$

satisfies this equation. Using this new function and still taking $\phi_4(x) = x_1$ one finds a transformation (whose domain of definition does not include the points at which $x_3 = -1$) yielding the required normal form. •

4.2 Exact Linearization Via Feedback

As we anticipated at the beginning of the previous Section, one of the main purposes of these notes is the analysis and the design of feedback control laws for nonlinear systems. In almost all situations, we assume the state x of the system being available for measurements, and we let the input of the system to depend on this state and, possibly, on external reference signals. If the value of the control at time t depends only on the values, at the same instant of time, of the state x and of the external reference input, the control is said to be a *Static* (or *Memoryless*) *State Feedback Control Mode*. Otherwise, if the control depends also on a set of additional state variables, i.e. if this control is itself the output of an appropriate dynamical system having its own internal state, driven by x and by the external reference input, we say that a *Dynamic State Feedback Control Mode* is implemented.

In a single-input single-output system, the most convenient structure for a Static State Feedback Control is the one in which the input variable u is set equal to

$$u = \alpha(x) + \beta(x)v \tag{2.1}$$

where v is the external reference input (see Fig. 4.2). In fact, the composition of this control with a system of the form

$$\dot{x} = f(x) + g(x)u$$
$$y = h(x)$$

yields a closed loop characterized by the similar structure

$$\dot{x} = f(x) + g(x)\alpha(x) + g(x)\beta(x)v$$
$$y = h(x)$$

The functions $\alpha(x)$ and $\beta(x)$ that characterize the control (2.1) are defined on a suitable open set of $\mathbf{R}^n$. For obvious reasons, $\beta(x)$ is assumed to be nonzero for all x in this set.

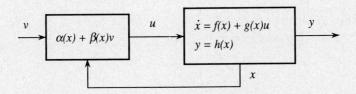

$$v \longrightarrow \boxed{\alpha(x) + \beta(x)v} \xrightarrow{\ u\ } \boxed{\begin{array}{l} \dot{x} = f(x) + g(x)u \\ y = h(x) \end{array}} \xrightarrow{\ y\ }$$

Fig. 4.2

The first application that will be discussed is the use of state feedback (and change of coordinates in the state-space) to the purpose of transforming a given nonlinear system into a linear and controllable one. The point of departure of this study will be the normal form developed and illustrated in the previous Section.

Consider a nonlinear system having relative degree $r = n$, i.e. exactly equal to the dimension of the state space, at some point $x = x^\circ$. In this case the change of coordinates required to construct the normal form is given exactly by

$$\Phi(x) = \begin{bmatrix} \phi_1(x) \\ \phi_2(x) \\ \cdots \\ \phi_n(x) \end{bmatrix} = \begin{bmatrix} h(x) \\ L_f h(x) \\ \cdots \\ L_f^{n-1} h(x) \end{bmatrix}$$

i.e. by the function $h(x)$ and its first $n-1$ derivatives along $f(x)$. No extra functions are needed in order to complete the transformation. In the new coordinates

$$z_i = \phi_i(x) = L_f^{i-1}(x) \qquad 1 \le i \le n$$

the system will appear described by equations of the form

$$\dot{z}_1 = z_2$$
$$\dot{z}_2 = z_3$$
$$\cdots$$
$$\dot{z}_{n-1} = z_n$$
$$\dot{z}_n = b(z) + a(z)u$$

where $z = (z_1, \ldots, z_n)$. Recall also that at the point $z^\circ = \Phi(x^\circ)$, and thus at all z in a neighborhood of z°, the function $a(z)$ is nonzero.

Suppose now the following state feedback control law is chosen

$$u = \frac{1}{a(z)}(-b(z) + v) \tag{2.2}$$

which indeed exists and is well-defined in a neighborhood of z^o. The resulting closed loop system is governed by the equations (Fig. 4.3)

$$\dot{z}_1 = z_2$$
$$\dot{z}_2 = z_3$$
$$\cdots$$
$$\dot{z}_{n-1} = z_n$$
$$\dot{z}_n = v$$

i.e. is *linear* and *controllable*. Thus we conclude that any nonlinear system with relative degree n at some point x^o can be transformed into a system which, in a neighborhood of the point $z^o = \Phi(x^o)$, is linear and controllable. It is important to stress that the trasformation in question consists of two basic ingredients
(i) a change of coordinates, defined locally around the point x^o
(ii) a state feedback, also defined locally around the point x^o.

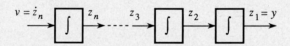

Fig. 4.3

Remark 2.1. It is easily checked that the two transformations used in order to obtain the linear form can be interchanged. One can first use a feedback and then change the coordinates in the state space, and the result is the same. The feedback needed to this purpose is exactly the same feedback just used, but now expressed in the x coordinates, i.e. as

$$u = \frac{1}{a(\Phi(x))}(-b(\Phi(x)) + v).$$

Comparing this with the expressions for $a(z)$ and $b(z)$ given in the previous Section, one realizes that this feedback—expressed in terms of the functions

$f(x)$, $g(x)$, $h(x)$ which characterize the original system—has the form

$$u = \frac{1}{L_g L_f^{n-1} h(x)} (-L_f^n h(x) + v). \tag{2.3}$$

An immediate calculation shows that this feedback, together with the same change of coordinates used so far, exactly yields the same linear and controllable system already obtained. •

Remark 2.2. Note that if x° is an equilibrium point for the original nonlinear system, i.e. if $f(x^\circ) = 0$, and if also $h(x^\circ) = 0$, then $z^\circ = \Phi(x^\circ) = 0$. As a matter of fact

$$\phi_1(x^\circ) = h(x^\circ) = 0$$

$$\phi_i(x^\circ) = \frac{\partial(L_f^{i-2} h)}{\partial x} f(x^\circ) = 0 \qquad \text{for all } 2 \le i \le n.$$

Note also that a condition like $h(x^\circ) = 0$ can always be satisfied, by means of a suitable translation of the origin of the output space.

Thus, we conclude that if x° is an equilibrium point for the original system, and this system has relative degree n at x°, there is a feedback control law (defined in a neighborhood of x°) and a coordinates transformation (also defined in a neighborhood of x°) changing the system into a linear and controllable one, defined in a neighborhood of 0. •

Remark 2.3. On the linear system thus obtained one can impose new feedback controls, like for instance

$$v = Kz$$

where

$$K = (c_0 \ldots c_{n-1})$$

can be chosen e.g. in order to assign a specific set of eigenvalues, or to satisfy an optimality criterion, etc. Recalling the expression of the z_i's as functions of x, the feedback in question can be rewritten as

$$v = c_0 h(x) + c_1 L_f h(x) + \cdots + c_{n-1} L_f^{n-1} h(x) \tag{2.4}$$

i.e. in the form of a nonlinear feedback from the state x of the original description of the system. Note that the composition of (2.3) and (2.4) is again a state feedback, having the form

$$u = \frac{-L_f^n h(x) + \sum_{i=0}^{n-1} c_i L_f^i h(x)}{L_g L_f^{n-1} h(x)}. \quad •$$

Example 2.1. Consider the system

$$\dot{x} = \begin{bmatrix} 0 \\ x_1 + x_2^2 \\ x_1 - x_2 \end{bmatrix} + \begin{bmatrix} \exp(x_2) \\ \exp(x_2) \\ 0 \end{bmatrix} u$$

$$y = x_3.$$

For this system we have

$$L_g h(x) = 0, \qquad L_f h(x) = x_1 - x_2,$$
$$L_g L_f h(x) = 0, \qquad L_f^2 h(x) = -x_1 - x_2^2,$$
$$L_g L_f^2 h(x) = -(1 + 2x_2)\exp(x_2),$$
$$L_f^3 h(x) = -2x_2(x_1 + x_2^2).$$

Thus, we see that the system has relative degree 3 (i.e. equal to n) at each point such that $1 + 2x_2 \neq 0$. Around any of such points, for instance around $x = 0$, the system can be transformed into a linear and controllable system by means of the feedback control

$$u = \frac{-2x_2(x_1 + x_2^2)}{(1 + 2x_2)\exp(x_2)} - \frac{1}{(1 + 2x_2)\exp(x_2)} v$$

and the change of coordinates

$$z_1 = h(x) = x_3$$
$$z_2 = L_f h(x) = x_1 - x_2$$
$$z_3 = L_f^2 h(x) = -x_1 - x_2^2.$$

Note that both the feedback and the change of coordinates are defined only locally around $x = 0$. In particular, the feedback u is not defined at points x such that $1 + 2x_2 = 0$ and the jacobian matrix of the coordinates transformation is singular at these points.

In the new coordinates, the system appears as

$$\dot{z} = \begin{bmatrix} 0 & 1 & 0 \\ 0 & 0 & 1 \\ 0 & 0 & 0 \end{bmatrix} z + \begin{bmatrix} 0 \\ 0 \\ 1 \end{bmatrix} v$$

which is linear and controllable. ●

Of course, the basic feature of the system that made it possible to change it into a linear and controllable one was the existence of an "output" function $h(x)$ for which the system had relative degree exactly n (at x°). We shall see now that the existence of such a function is not only a sufficient—as the previous discussion shows—but also a necessary condition for the existence of a state feedback and a change of coordinates transforming a given system into a linear and controllable one.

More precisely, consider a system (without output)

$$\dot{x} = f(x) + g(x)u$$

and suppose the following problem is set: given a point x° find (if possible), a neighborhood U of x°, a feedback

$$u = \alpha(x) + \beta(x)v$$

defined on U, and a coordinates transformation $z = \Phi(x)$ also defined on U, such that the corresponding closed loop

$$\dot{x} = f(x) + g(x)\alpha(x) + g(x)\beta(x)v$$

in the coordinates $z = \Phi(x)$, is linear and controllable, i.e. such that

$$\left[\frac{\partial \Phi}{\partial x}(f(x) + g(x)\alpha(x))\right]_{x=\Phi^{-1}(z)} = Az \tag{2.5}$$

$$\left[\frac{\partial \Phi}{\partial x}(g(x)\beta(x))\right]_{x=\Phi^{-1}(z)} = B \tag{2.6}$$

for some suitable matrix $A \in \mathbf{R}^{n \times n}$ and vector $B \in \mathbf{R}^n$ satisfying the condition

$$\text{rank}(B \ AB \ \ldots \ A^{n-1}B) = n.$$

This problem is the "single-input" version of the so-called *State Space Exact Linearization Problem*. The previous analysis has already established a sufficient condition for the existence of a solution; we show now that this condition is also necessary.

Lemma 2.4. The State Space Exact Linearization Problem is solvable if and only if there exists a neighborhood U of x° and a real-valued function $\lambda(x)$, defined on U, such that the system

$$\dot{x} = f(x) + g(x)u$$
$$y = \lambda(x)$$

has relative degree n at x°.

Proof. Clearly, we only have to show that the condition is necessary. We begin by showing an interesting feature of the notion of relative degree, namely that the latter is invariant under coordinates transformations and feedback. For, let $z = \Phi(x)$ be a coordinates transformation, and set

$$\bar{f}(z) = \left[\frac{\partial \Phi}{\partial x} f(x)\right]_{x=\Phi^{-1}(z)} \qquad \bar{g}(z) = \left[\frac{\partial \Phi}{\partial x} g(x)\right]_{x=\Phi^{-1}(z)} \qquad \bar{h}(z) = h(\Phi^{-1}(z)).$$

Then

$$L_{\bar{f}}\bar{h}(z) = \frac{\partial \bar{h}}{\partial z} \bar{f}(z) = \left[\frac{\partial h}{\partial x}\right]_{x=\Phi^{-1}(z)} \left[\frac{\partial \Phi^{-1}}{\partial z}\right]\left[\frac{\partial \Phi}{\partial x} f(x)\right]_{x=\Phi^{-1}(z)}$$

$$= \left[\frac{\partial h}{\partial x} f(x)\right]_{x=\Phi^{-1}(z)} = [L_f h(x)]_{x=\Phi^{-1}(z)}.$$

Iterated calculations of this kind show that

$$L_{\bar{g}}L_{\bar{f}}^k\bar{h}(z) = [L_g L_f^k h(x)]_{x=\Phi^{-1}(z)}$$

from which it is easily concluded that the relative degree is invariant under coordinates transformation. As far as the feedback is concerned, note that

$$L_{f+g\alpha}^k h(x) = L_f^k h(x) \qquad \text{for all } 0 \le k \le r - 1 \tag{2.7}$$

As a matter of fact, this equality is trivially true for $k = 0$. By induction, suppose is true for some $0 < k < r - 1$. Then

$$L_{f+g\alpha}^{k+1} h(x) = L_{f+g\alpha} L_f^k h(x) = L_f^{k+1} h(x) + L_g L_f^k h(x)\alpha(x) = L_f^{k+1} h(x)$$

thus showing that the equality in question holds for $k + 1$. From (2.7), one deduces that

$$L_{g\beta}L_{f+g\alpha}^k h(x) = 0 \qquad \text{for all } 0 \le k < r - 1$$

and that, if $\beta(x^\circ) \ne 0$

$$L_{g\beta} L_{f+g\alpha}^{r-1} h(x^\circ) \ne 0.$$

This shows that r is invariant under feedback.

Now, let (A, B) be a reachable pair. Then, it is well-known from the theory of linear systems that there exists a nonsingular $n \times n$ matrix T and a $1 \times n$ vector

k such that

$$T(A+Bk)T^{-1} = \begin{bmatrix} 0 & 1 & 0 & \cdots & 0 \\ 0 & 0 & 1 & \cdots & 0 \\ \vdots & \vdots & \vdots & \ddots & \vdots \\ 0 & 0 & 0 & \cdots & 1 \\ 0 & 0 & 0 & \cdots & 0 \end{bmatrix} \qquad TB = \begin{bmatrix} 0 \\ 0 \\ \vdots \\ 0 \\ 1 \end{bmatrix}. \qquad (2.8)$$

Suppose (2.5) and (2.6) hold and set

$$\bar{z} = \bar{\Phi}(x) = T\Phi(x)$$
$$\bar{\alpha}(x) = \alpha(x) + \beta(x)k\Phi(x).$$

Then, it is easily seen that

$$\left[\frac{\partial\bar{\Phi}}{\partial x}(f(x)+g(x)\bar{\alpha}(x))\right]_{x=\bar{\Phi}^{-1}(\bar{z})} = \begin{bmatrix} 0 & 1 & 0 & \cdots & 0 \\ 0 & 0 & 1 & \cdots & 0 \\ \vdots & \vdots & \vdots & \ddots & \vdots \\ 0 & 0 & 0 & \cdots & 1 \\ 0 & 0 & 0 & \cdots & 0 \end{bmatrix} \bar{z}$$

$$\left[\frac{\partial\bar{\Phi}}{\partial x}(g(x)\beta(x))\right]_{x=\bar{\Phi}^{-1}(\bar{z})} = \begin{bmatrix} 0 \\ \cdots \\ 0 \\ 1 \end{bmatrix}.$$

From this, it is deduced that there is no loss of generality in assuming that the pair (A, B) which renders the (2.5)-(2.6) satisfied has the form indicated in the right-hand-sides of (2.8).

Define now the "output" function

$$y = (1 \ 0 \ \ldots \ 0)\bar{z}.$$

A straightforward calculation shows that the linear system with A and B in the form of the right-hand-sides of (2.8) and with this output function has exactly relative degree n. Thus, since the relative degree is invariant under feedback and coordinates transformation, the proof is complete. •

The problem of finding a function $\lambda(x)$ such that the relative degree of the system at x° is exactly n, namely a function such that

$$L_g\lambda(x) = L_gL_f\lambda(x) = \ldots = L_gL_f^{n-2}\lambda(x) = 0 \quad \text{for all } x \text{ near } x^\circ \quad (2.9)$$
$$L_gL_f^{n-1}\lambda(x^\circ) \neq 0 \quad (2.10)$$

is apparently a problem involving the solution of a system of partial differential equations (namely the equations (2.9)), in which the unknown function $\lambda(x)$ is differentiated up to $n-1$ times, together with a condition (namely the condition (2.10)) which singles-out trivial solutions like e.g. $\lambda(x) = 0$. However, thanks to Lemma 1.3, this system is in fact equivalent to a system of first order partial differential equations, of a rather simple form. As a matter of fact, this Lemma exactly shows that the equations (2.9) are equivalent to

$$L_g\lambda(x) = L_{ad_fg}\lambda(x) = \ldots = L_{ad_f^{n-2}g}\lambda(x) = 0 \quad (2.11)$$

and that the nontriviality condition (2.10) is equivalent to

$$L_{ad_f^{n-1}g}\lambda(x^\circ) \neq 0. \quad (2.12)$$

The existence of a function satisfying these relations is an easy consequence of Frobenius' Theorem, as it can be seen in the proof of the following result.

Lemma 2.5.   There exists a real-valued function $\lambda(x)$ defined in a neighborhood U of x° solving the partial differential equations (2.11), and satisfying the nontriviality condition (2.12), if and only if
(i) the matrix $[g(x^\circ)\ ad_fg(x^\circ)\ \ldots\ ad_f^{n-2}g(x^\circ)\ ad_f^{n-1}g(x^\circ)]$ has rank n,
(ii) the distribution $D = \text{span}\{g, ad_fg, \ldots, ad_f^{n-2}g\}$ is involutive in a neighborhood of x°.

Proof. Suppose a function $\lambda(x)$ satisfying (2.11) and (2.12) exists. Then, from the proof of Lemma 1.2, in particular from the nonsingularity of the matrix (1.4), we deduce that the n vectors

$$g(x^\circ), ad_fg(x^\circ), \ldots, ad_f^{n-2}g(x^\circ), ad_f^{n-1}g(x^\circ)$$

are linearly independent. This proves the necessity of (i). If (i) holds then the distribution D is nonsingular and $(n-1)$-dimensional around x°. The equations (2.11), that can be rewritten as

$$d\lambda(x)[g(x^\circ)\ ad_fg(x^\circ)\ \ldots\ ad_f^{n-2}g(x^\circ)] = 0 \quad (2.13)$$

tell us that the differential $d\lambda(x)$ is a basis of the 1-dimensional codistribution $D^\perp$ around x°. So, by Frobenius' Theorem, the distribution D is involutive, and this proves the necessity of (ii). Conversely, suppose (i) holds. Then the distribution D is nonsingular and $(n-1)$-dimensional around x°. If also (ii) holds, by Frobenius' Theorem we know there exists a real-valued function $\lambda(x)$, defined in a neighborhood U of x° whose differential $d\lambda(x)$ spans $D^\perp$, i.e. solves the partial differential equation (2.11). Moreover, $d\lambda(x)$ also satisfies (2.12), because otherwise $d\lambda(x)$ would be annihilated by a set of n linearly independent vectors, i.e. a contradiction. $\bullet$

We can at this point summarize theresults established so far in the following formal statement

Theorem 2.6. Suppose a system

$$\dot{x} = f(x) + g(x)u$$

is given. The State Space Exact Linearization Problem is solvable near a point x° (i.e. there exists an "output" function $\lambda(x)$ for which the system has relative degree n at x°) if and only if the following conditions are satisfied
 (i) the matrix $[g(x^\circ)\ ad_f g(x^\circ)\ \ldots\ ad_f^{n-2}g(x^\circ)\ ad_f^{n-1}g(x^\circ)]$ has rank n,
 (ii) the distribution $D = \mathrm{span}\{g, ad_f g, \ldots, ad_f^{n-2}g\}$ is involutive near x°.

On the basis of the previous discussion, it is now clear that the procedure leading to the construction of a feedback $u = \alpha(x) + \beta(x)v$ and of a coordinates transformation $z = \Phi(x)$ solving the State Space Exact Linearzation problem consists of the following steps
 - from $f(x)$ and $g(x)$, construct the vector fields

$$g(x), ad_f g(x), \ldots, ad_f^{n-2}g(x), ad_f^{n-1}g(x)$$

 and check the conditions (i) and (ii),
 - if both are satisfied, solve for $\lambda(x)$ the partial differential equation (2.11),
 - set

$$\alpha(x) = \frac{-L_f^n \lambda(x)}{L_g L_f^{n-1}\lambda(x)} \qquad \beta(x) = \frac{1}{L_g L_f^{n-1}\lambda(x)} \qquad (2.14)$$

 - set

$$\Phi(x) = \mathrm{col}(\lambda(x), L_f \lambda(x), \ldots, L_f^{n-1}\lambda(x)) \qquad (2.15)$$

The feedback defined by the functions (2.14) is called the *linearizing feedback* and the new coordinates defined by (2.15) are called the *linearizing coordinates*. We illustrate now the whole Exact Linearization procedure in a simple example.

Example 2.2. Consider the system

$$\dot{x} = \begin{bmatrix} x_3(1+x_2) \\ x_1 \\ x_2(1+x_1) \end{bmatrix} + \begin{bmatrix} 0 \\ 1+x_2 \\ -x_3 \end{bmatrix} u.$$

In order to check whether or not this system can be transformed into a linear and controllable system via state feedback and coordinates transformation, we have to compute the functions $ad_f g(x)$ and $ad_f^2 g(x)$ and test the conditions of Theorem 2.6.

Appropriate calculations show that

$$ad_f g(x) = \begin{bmatrix} 0 & 0 & 0 \\ 0 & 1 & 0 \\ 0 & 0 & -1 \end{bmatrix} \begin{bmatrix} x_3(1+x_2) \\ x_1 \\ x_2(1+x_1) \end{bmatrix} - \begin{bmatrix} 0 & x_3 & 1+x_2 \\ 1 & 0 & 0 \\ x_2 & 1+x_1 & 0 \end{bmatrix} \begin{bmatrix} 0 \\ 1+x_2 \\ -x_3 \end{bmatrix}$$

$$= \begin{bmatrix} 0 \\ x_1 \\ -(1+x_1)(1+2x_2) \end{bmatrix}$$

and that

$$ad_f^2 g(x) = \begin{bmatrix} (1+x_2)(1+2x_2)(1+x_1) - x_3 x_1 \\ x_3(1+x_2) \\ -x_3(1+x_2)(1+2x_2) - 3x_1(1+x_1) \end{bmatrix}.$$

At $x = 0$, the matrix

$$[g(x) \quad ad_f g(x) \quad ad_f^2 g(x)]_{x=0} = \begin{bmatrix} 0 & 0 & 1 \\ 1 & 0 & 0 \\ 0 & -1 & 0 \end{bmatrix}$$

has rank 3 and therefore the condition (i) is satisfied. It is also easily checked that the product $[g, ad_f g](x)$ has a form

$$[g, ad_f g](x) = \begin{bmatrix} 0 \\ \star \\ \star \end{bmatrix}$$

and therefore also the condition (ii) is satisfied, because the matrix

$$[g(x) \quad ad_f g(x) \quad [g, ad_f g](x)]$$

has rank 2 for all x near $x = 0$.

In the present case, it is easily seen that a function $\lambda(x)$ that solves the equation

$$\frac{\partial \lambda}{\partial x}[g(x) \quad ad_f g(x)] = 0$$

is given by

$$\lambda(x) = x_1.$$

From our previous discussion, we know that considering this as "output" will yield a system having relative degree 3 (i.e. equal to n) at the point $x = 0$. We double-check and observe that

$$L_g\lambda(x) = 0, \quad L_g L_f \lambda(x) = 0, \quad L_g L_f^2 \lambda(x) = (1 + x_1)(1 + x_2)(1 + 2x_2) - x_1 x_3$$

and $L_g L_f^2 \lambda(0) = 1$. Locally around $x = 0$, the system will be transformed into a linear and controllable one by means of the state feedback

$$u = \frac{-L_f^3 \lambda(x) + v}{L_g L_f^2 \lambda(x)}$$
$$= \frac{-x_3^2(1 + x_2) - x_2 x_3(1 + x_2)^2 - x_1(1 + x_1)(1 + 2x_2) - x_1 x_2(1 + x_1) + v}{(1 + x_1)(1 + x_2)(1 + 2x_2) - x_1 x_3}$$

and the coordinates trasformation

$$z_1 = \lambda(x) = x_1$$
$$z_2 = L_f \lambda(x) = x_3(1 + x_2)$$
$$z_3 = L_f^2 \lambda(x) = x_3 x_1 + (1 + x_1)(1 + x_2)x_2. \quad \bullet$$

Remark 2.7. Using the above result, it is easily seen that any nonlinear system whose state space has dimension $n = 2$ can be transformed into a linear system, via state feedback and change of coordinates, around a point x° if and only if the matrix

$$[g(x^\circ) \quad ad_f g(x^\circ)]$$

has rank 2. As a matter of fact, this is exactly the condition (i) of the previous Theorem, and condition (ii) is always satisfied, because $D = \text{span}\{g\}$ is 1-dimensional. In this case it is always possible to find a function $\lambda(x) = \lambda(x_1, x_2)$, defined locally around x°, such that

$$\frac{\partial \lambda}{\partial x}g(x) = \frac{\partial \lambda}{\partial x_1}g_1(x_1, x_2) + \frac{\partial \lambda}{\partial x_2}g_2(x_1, x_2) = 0. \quad \bullet$$

Remark 2.8. The condition (i) of Theorem 2.6 has the following interesting interpretation. Suppose the vector field $f(x)$ has an equilibrium at $x^\circ = 0$, i.e.

$f(0) = 0$, and consider for $f(x)$ an expansion of the form

$$f(x) = Ax + f_2(x)$$

with

$$A = \left[\frac{\partial f}{\partial x}\right]_{x=0} \quad \text{and} \quad \left[\frac{\partial f_2}{\partial x}\right]_{x=0} = 0$$

which separates the linear approximation Ax from the higher-order term $f_2(x)$. Consider also for $g(x)$ an expansion of the form

$$g(x) = B + g_1(x)$$

with $B = g(0)$. These expansions characterize the *linear approximation* of the system at $x = 0$, which is defined as

$$\dot{x} = Ax + Bu.$$

An easy calculation shows that the vector fields $ad_f^k g(x)$ can be expanded in the following way

$$ad_f^k g(x) = (-1)^k A^k B + p_k(x)$$

where $p_k(x)$ is a function such that $p_k(0) = 0$. As a matter of fact, the expansion in question is trivially true for $k = 0$. By induction, suppose is true for some k. Then, by definition

$$
\begin{aligned}
ad_f^{k+1} g(x) &= \frac{\partial(ad_f^k g)}{\partial x} f(x) - \frac{\partial f}{\partial x} ad_f^k g(x) \\
&= \frac{\partial p_k}{\partial x}(Ax + f_2(x)) - (A + \frac{\partial f_2}{\partial x})((-1)^k A^k B + p_k(x)) \\
&= (-1)^{k+1} A^{k+1} B + p_{k+1}(x)
\end{aligned}
$$

where $p_{k+1}(x)$, by construction, is zero at $x = 0$.

From this, we see that the condition (i) of Theorem 2.6 (written at $x^\circ = 0$) is equivalent to the condition

$$\text{rank}(B\ AB\ \ldots\ A^{n-1}B) = n$$

i. e. to the condition that *the linear approximation of the system at x=0 is controllable.*

In other words, we conclude that the controllability of the linear approximation of the system at $x = x^\circ$ is a necessary condition for the solvability of the State Space Exact Linearization Problem. •

Remark 2.9. It is interesting to observe that the conditions (i) and (ii) of

Theorem 2.6 imply the involutivity of the distribution

$$D_k = \text{span}\{g, ad_f g, \ldots, ad_f^k g\}$$

for any $1 \leq k \leq n - 3$. As a matter of fact, since (i) and (ii) imply the existence of a $\lambda(x)$ such that (2.9) and (2.10) hold, from Lemma 1.3 it follows that

$$d\lambda(x)[g(x) \quad ad_f g(x) \quad \cdots \quad ad_f^k g(x)] = 0$$
$$dL_f \lambda(x)[g(x) \quad ad_f g(x) \quad \cdots \quad ad_f^k g(x)] = 0$$
$$\cdots$$
$$dL_f^{n-k-2}\lambda(x)[g(x) \quad ad_f g(x) \quad \cdots \quad ad_f^k g(x)] = 0.$$

These equalities show that

$$\text{span}\{d\lambda, dL_f \lambda, \ldots, dL_f^{n-k-2}\lambda\} \subset D_k^{\perp}$$

Moreover, since (Lemma 1.2) the differentials $d\lambda, dL_f \lambda, \ldots, dL_f^{n-k-2}\lambda$ are linearly independent around x° and $D_k^{\perp}$ has dimension $n - k - 1$ around x° (as a consequence of assumption (i)), it is concluded that $D_k^{\perp}$ is spanned by exact differentials. Then, by Frobenius' Theorem, D_k is involutive.

We see from this property that the involutivity of all the distributions D_k, $1 \leq k \leq n - 2$, is a *necessary* condition for the solvability of the Exact State Space Linearization Problem. ●

Remark 2.10. Note that if the State Space Exact Linearization Problem is solved by means of a feedback and a coordinates transformation $z = \Phi(x)$ defined in a neighborhood U of x°, the corresponding linear system is defined on the open set $\Phi(U)$. For obvious reasons, it is interesting to have $\Phi(U)$ containing the origin of $\mathbf{R}^n$, and in particular to have $\Phi(x^{\circ}) = 0$. In this case, in fact, one could for instance use linear systems theory concepts in order to asymptotically stabilize at $z = 0$ the transformed system and then use the stabilizer thus found in a composite loop to the purpose of stabilizing the nonlinear system at $x = x^{\circ}$ (see Remark 2.3).

This is indeed the case when x° is an equilibrium of the vector field $f(x)$. In this case, in fact, choosing the solution $\lambda(x)$ of the differential equation with the additional constraint $\lambda(x^{\circ}) = 0$, as is always possible, one gets $\Phi(x^{\circ}) = 0$, as already shown at the beginning of the Section (see Remark 2.2).

If x° is not an equilibrium of the vector field $f(x)$, one can manage to have this happening by means of feedback. As a matter of fact, the condition $\Phi(x^{\circ}) = 0$, replaced into (2.5), necessarily yields

$$\left[\frac{\partial \Phi}{\partial x}(f(x) + g(x)\alpha(x))\right]_{x=x^{\circ}} = 0$$

i.e.

$$f(x^\circ) + g(x^\circ)\alpha(x^\circ) = 0.$$

This clearly expresses the fact that the point x° is an equilibrium of the vector field $f(x) + g(x)\alpha(x)$, and can be obtained if and only if the vectors $f(x^\circ)$ and $g(x^\circ)$ are such that

$$f(x^\circ) = cg(x^\circ)$$

where c is a real number. If this is the case, an easy calculation shows that the linearizing coordinates are still zero at x° (if $\lambda(x)$ is such), because, for all $2 \leq i \leq n$

$$L_f^{i-1}\lambda(x^\circ) = cL_g L_f^{i-2}\lambda(x^\circ) = 0.$$

Moreover, the linearizing feedback $\alpha(x)$ is such that

$$\alpha(x^\circ) = -\frac{L_f^n \lambda(x^\circ)}{L_g L_f^{n-1}\lambda(x^\circ)} = -c$$

as expected. •

Remark 2.11. Note that a nonlinear system

$$\dot{x} = f(x) + g(x)u$$
$$y = h(x)$$

having relative degree strictly less than n could as well meet the requirements (i) and (ii) of Theorem 2.6. If this is the case, there will be a *different* "output" function, say $\lambda(x)$, with respect to which the system will have relative degree exactly n. Starting from this new function it will be possible to construct a feedback $u = \alpha(x) + \beta(x)v$ and a change of coordinates $z = \Phi(x)$, that will transform the state space equation

$$\dot{x} = f(x) + g(x)u$$

into a linear and controllable one. However, the real output of the system, in the new coordinates

$$y = h(\Phi^{-1}(z))$$

will in general continue to be a *nonlinear* function of the state z. •

If the system has relative degree $r < n$, for some given output $h(x)$, and either the conditions of Lemma 2.5—for the existence of another output for which the relative degree is equal to n—are not satisfied, or more simply one doesn't like to embark oneself in the solution of the partial differential equation (2.11) yielding such an output, it is still possible to obtain—by means of state feedback—a system which is *partially* linear. As a matter of fact, setting again

$$u = \frac{1}{a(z)}(-b(z) + v) \tag{2.16}$$

on the normal form of the equations, one obtains, if $r < n$, a system like

$$\dot{z}_1 = z_2$$
$$\dot{z}_2 = z_3$$
$$\cdots$$
$$\dot{z}_{r-1} = z_r$$
$$\dot{z}_r = v \tag{2.17}$$
$$\dot{z}_{r+1} = q_{r+1}(z)$$
$$\cdots$$
$$\dot{z}_n = q_n(z)$$
$$y = z_1$$

This system clearly appears decomposed into a *linear subsystem*, of dimension r, which is the only one responsible for the input-output behavior, and a possibly nonlinear system, of dimension $n - r$, whose behavior however does not affect the output (Fig. 4.4).

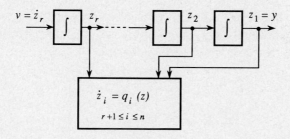

Fig. 4.4

We summarize this result for convenience in a formal statement where, for more generality, the linearizing feedback is specified in terms of the functions $f(x)$, $g(x)$ and $h(x)$ characterizing the original description of the system.

Proposition 2.12. Consider a nonlinear system having relative degree r at a point x°. The state feedback

$$u = \frac{1}{L_g L_f^{r-1} h(x)}(-L_f^r h(x) + v) \tag{2.18}$$

transforms this system into a system whose input-output behavior is identical to that of a linear system having a transfer function

$$H(s) = \frac{1}{s^r}.$$

4.3 The Zero Dynamics

In this Section we introduce and discuss an important concept, that in many circumstances plays a role exactly similar to that of the "zeros" of the transfer function in a linear system. We have already seen that the relative degree r of a linear system can be interpreted as the difference between the number of poles and the number of zeros in the transfer function. In particular, any linear system in which r is strictly less than n has zeros in its transfer function. On the contrary, if $r = n$ the transfer function has no zeros; thus, the systems considered at the beginning of the previous Section are in some sense analogue to linear systems without zeros. We shall see in this Section that this kind of analogy can be pushed much further.

Consider a nonlinear system with r strictly less than n and look at its normal form. In order to write the equations in a slightly more compact manner, we introduce a suitable vector notation. In particular, since there is no specific need to keep track individually of each one of the last $n - r$ components of the state vector, we shall represent all of them together as

$$\eta = \begin{bmatrix} z_{r+1} \\ \cdots \\ z_n \end{bmatrix}.$$

Sometimes, whenever convenient and not otherwise required, we shall represent also the first r components together, as

$$\xi = \begin{bmatrix} z_1 \\ \cdots \\ z_r \end{bmatrix}.$$

With the help of these notations, the normal form of a single-input single-output nonlinear system having $r < n$ (at some point of interest x°, e.g. an equilibrium point) can be rewritten as

$$\dot{z}_1 = z_2$$
$$\dot{z}_2 = z_3$$
$$\cdots$$
$$\dot{z}_{r-1} = z_r$$
$$\dot{z}_r = b(\xi, \eta) + a(\xi, \eta)u$$
$$\dot{\eta} = q(\xi, \eta).$$

Recall that, if x° is such that $f(x^\circ) = 0$ and $h(x^\circ) = 0$, then necessarily the first r new coordinates z_i are 0 at x°. Note also that it is always possible to choose arbitrarily the value at x° of the last $n - r$ new coordinates, thus in particular being 0 at x°. Therefore, without loss of generality, one can assume that $\xi = 0$ and $\eta = 0$ at x°. Thus, if x° was an equilibrium for the system in the original coordinates, its corresponding point $(\xi, \eta) = (0, 0)$ is an equilibrium for the system in the new coordinates and from this we deduce that

$$b(\xi, \eta) = 0 \qquad \text{at } (\xi, \eta) = (0, 0)$$
$$q(\xi, \eta) = 0 \qquad \text{at } (\xi, \eta) = (0, 0).$$

Suppose now we want to analyze the following problem, called the *Problem of Zeroing the Output*. Find, if any, pairs consisting of an initial state x° and of an input function $u^\circ(t)$, defined for all t in a neighborhood of $t = 0$, such that the corresponding output $y(t)$ of the system is identically zero for all t in a neighborhood of $t = 0$. Of course, we are interested in finding *all* such pairs (x°, u°) and not simply in the trivial pair $x^\circ = 0, u^\circ = 0$ (corresponding to the situation in which the system is initially at rest and no input is applied). We perform this analysis on the normal form of the system.

Recalling that in the normal form

$$y(t) = z_1(t)$$

we see that the constraint $y(t) = 0$ for all t entails the following ones

$$\dot{z}_1(t) = \dot{z}_2(t) = \ldots = \dot{z}_r(t) = 0$$

that is $\xi(t) = 0$ for all times.

Thus, we see that when the output of the system is identically zero, its state is constrained to evolve in such a way that also $\xi(t)$ is identically zero. In addition, the input $u(t)$ must necessarily be the unique solution of the equation

$$0 = b(0, \eta(t)) + a(0, \eta(t))u(t)$$

(recall that $a(0, \eta(t)) \neq 0$ if $\eta(t)$ is close to 0). As far as the variable $\eta(t)$ is concerned, it is clear that, being $\xi(t)$ identically zero, its behavior is governed by the differential equation

$$\dot{\eta}(t) = q(0, \eta(t)). \tag{3.1}$$

From this analysis we deduce the following facts. If the output $y(t)$ has to be zero, then necessarily the initial state of the system must be set to a value such that $\xi(0) = 0$, whereas $\eta(0) = \eta^\circ$ can be chosen arbitrarily. According to the value of η°, the input must be set as

$$u(t) = -\frac{b(0, \eta(t))}{a(0, \eta(t))}$$

where $\eta(t)$ denotes the solution of the differential equation

$$\dot{\eta}(t) = q(0, \eta(t)) \qquad \text{with initial condition } \eta(0) = \eta^\circ.$$

Note also that for each set of initial data $\xi = 0$ and $\eta = \eta^\circ$ the input thus defined is the *unique* input capable to keep $y(t)$ identically zero for all times.

The dynamics of (3.1) correspond to the dynamics describing the "internal" behavior of the system when input and initial conditions have been chosen in such a way as to constrain the output to remain identically zero. These dynamics, which are rather important in many of our developments, are called the *zero dynamics* of the system.

Remark 3.1. In order to understand why we used the terminology "zero" dynamics in dealing with the dynamical system (3.1), it is convenient to examine how these dynamics are related to the zeros of the transfer function in a linear system. Let

$$H(s) = K \frac{b_0 + b_1 s + \cdots + b_{n-r-1} s^{n-r-1} + s^{n-r}}{a_0 + a_1 s + \cdots + a_{n-1} s^{n-1} + s^n}$$

denote the transfer function of a linear system (where r characterizes, as expected, the relative degree). Suppose the numerator and denominator polynomials are relatively prime and consider a minimal realization of $H(s)$

$$\dot{x} = Ax + Bu$$
$$y = Cx$$

with

$$A = \begin{bmatrix} 0 & 1 & 0 & \cdots & 0 \\ 0 & 0 & 1 & \cdots & 0 \\ \vdots & \vdots & \vdots & \ddots & \vdots \\ 0 & 0 & 0 & \cdots & 1 \\ -a_0 & -a_1 & -a_2 & \cdots & -a_{n-1} \end{bmatrix} \qquad B = \begin{bmatrix} 0 \\ 0 \\ \vdots \\ 0 \\ K \end{bmatrix}$$

$$C = [b_0 \quad b_1 \quad \cdots \quad b_{n-r-1} \quad 1 \quad 0 \quad \cdots \quad 0].$$

Let us now calculate its normal form. For the first r new coordinates we know we have to take

$$z_1 = Cx = b_0 x_1 + b_1 x_2 + \cdots + b_{n-r-1} x_{n-r} + x_{n-r+1}$$
$$z_2 = CAx = b_0 x_2 + b_1 x_3 + \cdots + b_{n-r-1} x_{n-r+1} + x_{n-r+2}$$
$$\cdots$$
$$z_r = CA^{r-1}x = b_0 x_r + b_1 x_{r+1} + \cdots + b_{n-r-1} x_{n-1} + x_n.$$

For the other $n - r$ new coordinates we have some freedom of choice (provided that the conditions stated in Proposition 1.4 are satisfied), but the simplest one is

$$z_{r+1} = x_1$$
$$z_{r+2} = x_2$$
$$\cdots$$
$$z_n = x_{n-r}.$$

This is indeed an admissible choice because the corresponding coordinates transformation $z = \Phi(x)$ has a jacobian matrix with the following structure

$$\frac{\partial \Phi}{\partial x} = \begin{bmatrix} [\cdots] & \begin{bmatrix} 1 & 0 & \cdots & 0 \\ \star & 1 & \cdots & 0 \\ \vdots & \vdots & \ddots & \vdots \\ \star & \star & \cdots & 1 \end{bmatrix} \\ \begin{bmatrix} 1 & 0 & \cdots & 0 \\ 0 & 1 & \cdots & 0 \\ \vdots & \vdots & \ddots & \vdots \\ 0 & 0 & \cdots & 1 \end{bmatrix} & \begin{bmatrix} 0 & 0 & \cdots & 0 \\ 0 & 0 & \cdots & 0 \\ \vdots & \vdots & \ddots & \vdots \\ 0 & 0 & \cdots & 0 \end{bmatrix} \end{bmatrix}$$

and thus nonsingular.

In the new coordinates we obtain equations in normal form, which, because of the linearity of the system, have the following structure

$$\dot{z}_1 = z_2$$
$$\dot{z}_2 = z_3$$
$$\cdots$$
$$\dot{z}_{r-1} = z_r$$
$$\dot{z}_r = R\xi + S\eta + Ku$$
$$\dot{\eta} = P\xi + Q\eta$$

where R and S are row vectors and P and Q matrices, of suitable dimensions. The zero dynamics of this system, according to our previous definition, are those of

$$\dot{\eta} = Q\eta.$$

The particular choice of the last $n - r$ new coordinates (i.e. of the elements of η) entails a particularly simple structure for the matrix Q. As a matter of fact, is easily checked that

$$\frac{dz_{r+1}}{dt} = \frac{dx_1}{dt} = x_2(t) = z_{r+2}(t)$$
$$\cdots$$
$$\frac{dz_{n-1}}{dt} = \frac{dx_{n-r-1}}{dt} = x_{n-r}(t) = z_n(t)$$
$$\frac{dz_n}{dt} = \frac{dx_{n-r}}{dt} = x_{n-r+1}(t) = -b_0 x_1(t) - \cdots - b_{n-r-1} x_{n-r}(t) + z_1(t)$$
$$= -b_0 z_{r+1}(t) - \cdots - b_{n-r-1} z_n(t) + z_1(t)$$

from which we deduce that

$$Q = \begin{bmatrix} 0 & 1 & 0 & \cdots & 0 \\ 0 & 0 & 1 & \cdots & 0 \\ \vdots & \vdots & \vdots & \ddots & \vdots \\ 0 & 0 & 0 & \cdots & 1 \\ -b_0 & -b_1 & -b_2 & \cdots & -b_{n-r-1} \end{bmatrix}.$$

From the particular form of this matrix, it is evident that the eigenvalues of Q coincide with the zeros of the numerator polynomial of $H(s)$ that is with the zeros of the transfer function. Thus we conclude that in a linear system the zero dynamics are linear dynamics with eigenvalues coinciding with the zeros of the transfer function of the system. •

Remark 3.2. The calculations carried out in the previous Remark are also useful in showing that the linear approximation of the zero dynamics of the system at $\eta = 0$ coincides with the zero dynamics of the linear approximation of the entire system at $x = 0$, i. e. that the operations of taking the linear approximation and calculating the zero dynamics essentially commute.

In order to check this, all we have to show is that the linear approximation of equations in normal form coincides with the normal form of the linear approximation of the original description of the system and this amounts only to show that the relative degree of the system and that of its linear approximation are the same. For, suppose that the system has relative degree r at $x = 0$. Consider the expansions already introduced in Remark 2.8

$$f(x) = Ax + f_2(x)$$
$$g(x) = B + g_1(x)$$

and, in addition, expand $h(x)$ (which is 0 at $x = 0$) as

$$h(x) = Cx + h_2(x)$$

where

$$C = \left[\frac{\partial h}{\partial x}\right]_{x=0} \quad \text{and} \quad \left[\frac{\partial h_2}{\partial x}\right]_{x=0} = 0.$$

An easy calculation shows, by induction, that

$$L_f^k h(x) = CA^k x + d_k(x)$$

where $d_k(x)$ is a function such that

$$\left[\frac{\partial d_k}{\partial x}\right]_{x=0} = 0.$$

From this one deduces that

$$CA^k B = L_g L_f^k h(0) = 0 \qquad \text{for all } k < r - 1$$
$$CA^{r-1} B = L_g L_f^{r-1} h(0) \neq 0$$

i.e. that the relative degree of the linear approximation of the system at $x = 0$ is exactly r.

From this fact, we conclude that taking the linear approximation of equations in normal form, based on expansions of the form

$$b(\xi, \eta) = R\xi + S\eta + b_2(\xi, \eta)$$
$$a(\xi, \eta) = K + a_1(\xi, \eta)$$
$$q(\xi, \eta) = P\xi + Q\eta + q_2(\xi, \eta)$$

a linear system in normal form is obtained. Thus, the jacobian matrix

$$Q = \left[\frac{\partial q}{\partial \eta}\right]_{(\xi,\eta)=0}$$

which describes the linear approximation at $\eta = 0$ of the zero dynamics of the original nonlinear system has eigenvalues which coincide with the zeros of the transfer function of the linear approximation at $x = 0$ of the entire system. •

Example 3.1. Suppose we want to calculate the zero dynamics of the system already analyzed in the Example 1.2. The only thing we have to do is to set $z_1 = z_2 = 0$ in the last equation of the normal form of the equations and get

$$\dot{z}_3 = -z_3.$$

This are the zero dynamics of the system. •

Example 3.2. Suppose we want to analyze the zero dynamics of the system

$$\dot{x} = \begin{bmatrix} x_3 - x_2^3 \\ -x_2 \\ x_1^2 - x_3 \end{bmatrix} + \begin{bmatrix} 0 \\ -1 \\ 1 \end{bmatrix} u$$

$$y = x_1.$$

For this system we have

$$L_g h(x) = 0 \qquad L_f h(x) = x_3 - x_2^3 \qquad L_g L_f h(x) = 1 + 3x_2^2.$$

We can calculate a normal form by taking

$$z_1 = x_1$$
$$z_2 = x_3 - x_2^3$$
$$z_3 = x_2 + x_3$$

which is a globally defined coordinates transformation. Using these new coordinates we obtain equations of the following form

$$\dot{z}_1 = z_2$$
$$\dot{z}_2 = b(z_1, z_2, z_3) + a(z_1, z_2, z_3)u$$
$$\dot{z}_3 = z_1^2 - z_3.$$

The constraint $y(t) = 0$ for all t imposes $z_1(t) = z_2(t) = 0$ for all t, and shows that when the output is identically zero the state must necessarily evolve on the curve (see Fig 4.5)

$$M = \{x \in \mathbf{R}^3 : x_1 = 0 \text{ and } x_3 = x_2^3\}$$

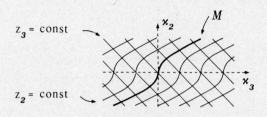

Fig. 4.5

On this curve, the motion of the system is governed by its zero dynamics

$$\dot{z}_3 = -z_3. \ \bullet$$

Although all the properties illustrated so far were discovered and discussed using the normal form, it is not difficult to arrive at similar conclusions starting from equations in different forms. If, for instance, one has not been able to find exactly the normal form because of the difficulty in constructing functions $\phi_{r+1}(x), \ldots, \phi_n(x)$ with the property that $L_g\phi_i(x) = 0$ (see Remark 1.5), one can still identify the zero dynamics of the system working on equations of the form

$$\dot{z}_1 = z_2$$
$$\dot{z}_2 = z_3$$
$$\cdots$$
$$\dot{z}_{r-1} = z_r$$
$$\dot{z}_r = b(\xi, \eta) + a(\xi, \eta)u$$
$$\dot{\eta} = q(\xi, \eta) + p(\xi, \eta)u.$$

As a matter of fact, having seen that the zero dynamics of the system describe its behavior when the output is forced to be zero, we impose this condition on the equations above. We obtain, as before, $\xi(t) = 0$ and

$$0 = b(0, \eta(t)) + a(0, \eta(t))u(t).$$

Replacing $u(t)$ from this equation into the last one, yields a differential equation for $\eta(t)$

$$\dot{\eta} = q(0, \eta) - p(0, \eta)\frac{b(0, \eta)}{a(0, \eta)}$$

which describes the zero dynamics in the new coordinates chosen.

Example 3.3. Suppose we want to calculate the zero dynamics of the system already analyzed in the Example 1.3. In this case we don't have the normal form, but the calculation of the zero dynamics is still very easy. Setting $z_1 = z_2 = 0$ in the second equation yields

$$u = -\frac{z_4 - 4z_4^4}{2 + 2z_3}.$$

Replacing this, and $z_1 = z_2 = 0$, in the third and fourth equation yields

$$\dot{z}_3 = -z_3 - \frac{z_4 - 4z_4^4}{2 + 2z_3}$$
$$\dot{z}_4 = -2z_4^3$$

which describes the zero dynamics of the system. •

The problem of zeroing the output could also have been analyzed directly on the original form of the equations. Keeping in mind the calculations already done at the beginning of Section 4.1, it is easy to deduce that $y^{(i-1)}(t) = 0$ implies $L_f^{i-1}h(x(t)) = 0$, for all $1 \le i \le r$. Thus, as expected, the system has to evolve on the subset

$$Z^\star = \{x \in \mathbf{R}^n : h(x) = L_f h(x) = \cdots = L_f^{r-1}h(x) = 0\}$$

that, locally around x^o, is exactly the set of points whose new coordinates $z_1, \ldots, z_r$ are 0 (see Fig. 4.6). If one writes the additional constraint

$$0 = y^{(r)}(t) = L_f^r h(x(t)) + L_g L_f^{r-1} h(x(t))u(t)$$

this turns out to be exactly the same constraint previously obtained for $u(t)$, but now expressed in terms of the functions which characterize the original equations.

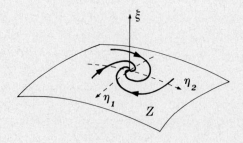

Fig. 4.6

Note that, since the differentials $dL_f^i h(x)$, $0 \leq i \leq r - 1$, are linearly independent at x^0 (Lemma 1.2), the set Z^* is a *smooth manifold* of dimension $n - r$, near x^0. The state feedback

$$u^*(x) = \frac{-L_f^r h(x)}{L_g L_f^{r-1} h(x)}$$

by construction is such that

$$
\begin{bmatrix} dh(x) \\ dL_f h(x) \\ \vdots \\ dL_f^{r-1}h(x) \end{bmatrix} (f(x)+g(x)u^*(x)) = \begin{bmatrix} L_f h(x)+L_g h(x)u^*(x) \\ L_f^2 h(x)+L_g L_f h(x)u^*(x) \\ \vdots \\ L_f^r h(x)+L_g L_f^{r-1}h(x)u^*(x) \end{bmatrix} = \begin{bmatrix} L_f h(x) \\ L_f^2 h(x) \\ \vdots \\ L_f^{r-1}h(x) \\ 0 \end{bmatrix}.
$$

Thus

$$
\begin{bmatrix} dh(x) \\ dL_f h(x) \\ \cdots \\ dL_f^{r-1}h(x) \end{bmatrix} (f(x) + g(x)u^*(x)) = 0
$$

for all $x \in Z^*$ (because $h(x) = L_f h(x) = \cdots = L_f^{r-1}h(x) = 0$ if $x \in Z^*$) and therefore the vector field

$$f^*(x) = f(x) + g(x)u^*(x)$$

is *tangent* to Z^*. As a consequence, any trajectory of the closed loop system

$$\dot{x} = f^*(x)$$

starting at a point of Z^* remains in Z^* (for small values of t). The restriction $f^*(x)|_{Z^*}$ of $f^*(x)$ to Z^* is a well-defined vector field of Z^*, which exactly describes—in a coordinate-free setting—the zero dynamics of the system.

We will illustrate in the sequel a series of relevant issues in which the notion of zero dynamics, and in particular its asymptotic properties, plays an important role. For the time being we can show, for instance, how the zero dynamics are naturally imposed as internal dynamics of a closed loop system whose input-output behavior has been rendered linear by means of state feedback. For, consider again a system in normal form and suppose the feedback control law (2.16) is imposed, under which the input-output behavior becomes identical with that of a linear system consisting of a string of r integrators between input and output (see Fig. 4.4). The closed loop system thus obtained is described by the equations (2.17), that can be rewritten in the form

$$\dot{\xi} = A\xi + Bv$$
$$\dot{\eta} = q(\xi, \eta)$$
$$y = C\xi$$

with

$$A = \begin{bmatrix} 0 & 1 & 0 & \cdots & 0 \\ 0 & 0 & 1 & \cdots & 0 \\ \vdots & \vdots & \vdots & \ddots & \vdots \\ 0 & 0 & 0 & \cdots & 0 \end{bmatrix} \qquad B = \begin{bmatrix} 0 \\ 0 \\ \vdots \\ 0 \\ 1 \end{bmatrix}$$

$$C = [1 \quad 0 \quad \cdots \quad 0].$$

If the linear subsystem is initially at rest and no input is applied, then $y(t) = 0$ for all values of t, and the corresponding internal dynamics of the whole (closed loop) system are exactly those of (3.1), namely the zero dynamics of the open loop system.

We conclude the Section by showing that the interpretation of

$$\dot{\eta}(t) = q(0, \eta(t))$$

as of the dynamics describing the internal behavior of the system when the output is forced to track exactly the output $y(t) = 0$, can easily be extended to the case in which the output to be tracked is any arbitrary function. Consider the following problem, which is called the *Problem of Reproducing the Reference Output $y_R(t)$*. Find, if any, pairs consisting of an initial state x° and of an input function $u^\circ(t)$, defined for all t in a neighborhood of $t = 0$, such that the corresponding output $y(t)$ of the system coincides exactly with $y_R(t)$ for all t in a neighborhood of $t = 0$. Again, we are interested in finding *all* such pairs (x°, u°). Proceeding as before, we deduce that $y(t) = y_R(t)$ necessarily implies

$$z_i(t) = y_R^{(i-1)}(t) \qquad \text{for all } t \text{ and all } 1 \leq i \leq r.$$

Setting

$$\xi_R(t) = \text{col}(y_R(t), y_R^{(1)}(t), \ldots, y_R^{(r-1)}(t)) \qquad (3.2)$$

we then see that the input $u(t)$ must necessarily satisfy

$$y_R^{(r)}(t) = b(\xi_R(t), \eta(t)) + a(\xi_R(t), \eta(t))u(t)$$

where $\eta(t)$ is a solution of the differential equation

$$\dot{\eta}(t) = q(\xi_R(t), \eta(t)). \tag{3.3}$$

Thus, if the ouput $y(t)$ has to track exactly $y_R(t)$, then necessarily the initial state of the system must be set to a value such that $\xi(0) = \xi_R(0)$, whereas $\eta(0) = \eta^\circ$ can be chosen arbitrarily. According to the value of η°, the input must be set as

$$u(t) = \frac{y_R^{(r)}(t) - b(\xi_R(t), \eta(t))}{a(\xi_R(t), \eta(t))} \tag{3.4}$$

where $\eta(t)$ denotes the solution of the differential equation (3.3) with initial condition $\eta(0) = \eta^\circ$. Note also that for each set of initial data $\xi(0) = \xi_R(0)$ and $\eta(0) = \eta^\circ$ the input thus defined is the *unique* input capable of keeping $y(t) = y_R(t)$ for all times.

The (forced) dynamics (3.3) clearly correspond to the dynamics describing the "internal" behavior of the system when input and initial conditions have been chosen in such a way as to constrain the output to track exactly $y_R(t)$. Note that the relations (3.3) and (3.4) describe a system with input $\xi_R(t)$, output $u(t)$ and state $\eta(t)$ that can be interpreted as a *realization* of the *inverse* of the original system.

4.4 Local Asymptotic Stabilization

In this Section we illustrate how the notion of zero dynamics can be helpful in dealing with the problem of asymptotically stabilizing a nonlinear system at a given equilibrium point. Suppose, as usual, a nonlinear system of the form

$$\dot{x} = f(x) + g(x)u$$

is given, with $f(x)$ having an equilibrium point at x° that, without loss of generality, we assume to be $x^\circ = 0$. The problem we want to discuss is the one of finding a smooth state feedback

$$u = \alpha(x)$$

defined locally around the point $x^\circ = 0$ and preserving the equilibrium, i.e. such that $\alpha(0) = 0$, with the property that the corresponding closed loop

$$\dot{x} = f(x) + g(x)\alpha(x)$$

is locally asymptotically stable at $x = 0$. We shall refer to it as to the *Local Asymptotic Stabilization Problem*.

First of all, we review a rather well-known property, by discussing to what extent the possibility of solving the problem in question depends on the properties of the linear approximation of the system near $x^\circ = 0$. To this end, recall that the linear approximation of a system having an equilibrium at $x^\circ = 0$ is defined by expanding $f(x)$ and $g(x)$ as (see Remark 2.8)

$$f(x) = Ax + f_2(x)$$
$$g(x) = B + g_1(x)$$

with

$$A = \left[\frac{\partial f}{\partial x}\right]_{x=0} \qquad \text{and} \qquad B = g(0)$$

From the point of view of stability theory, the importance of the linear approximation is essentially related to the following property.

Proposition 4.1. Suppose the linear approximation is asymptotically stabilizable, i.e. either the pair (A, B) is controllable or—in case the pair (A, B) is not controllable—the uncontrollable modes correspond to eigenvalues with negative real part. Then, any linear feedback which asymptotically stabilizes the linear approximation is also able to asymptotically stabilize the original nonlinear system, at least locally. If the pair (A, B) is not controllable and there exist uncontrollable modes associated with eigenvalues with positive real part, the original nonlinear system cannot be stabilized at all.

Proof. Suppose the linear approximation is asymptotically stabilizable. Let F be any matrix such that $(A+BF)$ has all the eigenvalues in the left-half complex plane, and set

$$u = Fx$$

on the nonlinear system. The resulting closed loop

$$\dot{x} = f(x) + g(x)Fx = (A + BF)x + f_2(x) + g_1(x)Fx$$

has a linear approximation having all the eigenvalues in the left-half complex plane. Thus, the Principle of Stability in the First Approximation proves that the nonlinear closed loop system is locally asymptotically stable.

Conversely, suppose the linear approximation has uncontrollable modes associated with eigenvalues having positive real part. Let $u = \alpha(x)$ be any smooth state feedback. The correspondig closed loop has a linear approximation of the form (recall that $\alpha(0) = 0$)

$$\dot{x} = \left[\frac{\partial[f(x) + g(x)\alpha(x)]}{\partial x}\right]_{x=0} x = \left(A + B\left[\frac{\partial \alpha}{\partial x}\right]_{x=0}\right)x$$

which has eigenvalues in the right-half complex plane, irrespectively of what α is. Thus, again by the Principle of Stability in the First Approximation, the nonlinear closed-loop system is unstable. •

Note that the previous result does not cover the whole spectrum of cases. As a matter of fact, if the pair (A, B) is not controllable and there are uncontrollable modes associated with eigenvalues on the imaginary axis (but none of them is in the right-half complex plane), nothing can be said from the linear approximation, in the sense that the nonlinear system might be locally asymptotically stabilizable—by means of a nonlinear feedback—even though its linear approximation is not. Problems in which this situation occurs are said to be *critical problems* of local asymptotic stabilization.

We show now in which way the notion of zero dynamics is useful when dealing with critical problems of local asymptotic stabilization. Consider again a system in normal form

$$
\begin{aligned}
\dot{z}_1 &= z_2 \\
\dot{z}_2 &= z_3 \\
&\cdots \\
\dot{z}_{r-1} &= z_r \\
\dot{z}_r &= b(\xi, \eta) + a(\xi, \eta)u \\
\dot{\eta} &= q(\xi, \eta)
\end{aligned}
$$

where

$$
\xi = \mathrm{col}(z_1, \ldots, z_r)
$$

and, without loss of generality, assume that $(\xi, \eta) = (0, 0)$ is an equilibrium point. Impose a feedback of the form

$$
u = \frac{1}{a(\xi, \eta)}(-b(\xi, \eta) - c_0 z_1 - c_1 z_2 - \cdots - c_{r-1} z_r) \tag{4.1}
$$

where $c_0, \ldots, c_{r-1}$ are real numbers.

This choice of feedback yields a closed loop system

$$
\begin{aligned}
\dot{\xi} &= A\xi \\
\dot{\eta} &= q(\xi, \eta)
\end{aligned} \tag{4.2}
$$

with

$$A = \begin{bmatrix} 0 & 1 & 0 & \cdots & 0 \\ 0 & 0 & 1 & \cdots & 0 \\ \vdots & \vdots & \vdots & \ddots & \vdots \\ 0 & 0 & 0 & \cdots & 1 \\ -c_0 & -c_1 & -c_2 & \cdots & -c_{r-1} \end{bmatrix}.$$

In particular, the matrix A has a characteristic polynomial

$$p(s) = c_0 + c_1 s + \cdots + c_{r-1} s^{r-1} + s^r.$$

From the form of the equations describing the closed loop we deduce the following interesting property.

Proposition 4.2. If the zero dynamics of the system are asymptotically stable at $\eta = 0$, and if the roots of the polynomial $p(s)$ are in the left-half complex plane, the feedback (4.1) asymptotically stabilizes the system at $(\xi, \eta) = (0, 0)$.

Proof. We only need to use the first Lemma of Section **B**.2. For, note that the closed loop system has the form (**B**.2.1) and that, by assumption, the subsystem

$$\dot{\eta} = q(0, \eta)$$

is asymptotically stable at $\eta = 0$. •

Note that the matrix

$$Q = \left[\frac{\partial q(\xi, \eta)}{\partial \eta} \right]_{(\xi, \eta) = (0, 0)}$$

characterizes the linear approximation of the zero dynamics at $\eta = 0$ (see Remark 3.2). If this matrix had all the eigenvalues in the left-half complex plane, then the result stated in the previous Proposition would be a trivial consequence of the Principle of Stability in the First Approximation, because the linear approximation of (4.2) has in fact the form

$$\begin{bmatrix} \dot{\xi} \\ \dot{\eta} \end{bmatrix} = \begin{bmatrix} A & 0 \\ \star & Q \end{bmatrix} \begin{bmatrix} \xi \\ \eta \end{bmatrix}.$$

However, Proposition 4.2 establishes a stronger result, because it relies upon the assumption that the zero dynamics are *just* asymptotically stable at $\eta = 0$, and this—as is well known—does not necessarily require, for nonlinear dynamics, asymptotic stability of the linear approximation (that is, eigenvalues of Q in the open left-half complex plane). In other words, the result in question may also hold in the presence of some eigenvalue of Q with zero real part.

In order to design the stabilizing control law there is no need to know explicitly the expression of the system in normal form, but only to know *the fact* that the system has an asymptotically stable zero dynamics. Recalling how the coordinates $z_1, \ldots, z_r$ and the functions $a(\xi, \eta)$ and $b(\xi, \eta)$ are related to the original description of the system, it is easily seen that, in the original coordinates, the stabilizing control law assumes the form

$$u = \frac{1}{L_g L_f^{r-1} h(x)} \left(-L_f^r h(x) - c_0 h(x) - c_1 L_f h(x) - \cdots - c_{r-1} L_f^{r-1} h(x) \right)$$

which is particularly interesting, because expressed in terms of quantities that can be immediately calculated from the original data.

By this method we can asymptotically stabilize *also* systems whose linear approximation has uncontrollable modes corresponding to eigenvalues on the imaginary axis, i.e. we can solve critical problems of local asymptotic stabilization, provided we know that for some choice of output the system has an asymptotically stable zero dynamics.

Example 4.1. Consider the system already discussed in the Example 1.3. Its linear approximation at $x = 0$ is described by matrices A and B of the form

$$A = \left[\frac{\partial f}{\partial x}\right]_{x=0} = \begin{bmatrix} 0 & 0 & 0 & 0 \\ 1 & 0 & 0 & 0 \\ 0 & 0 & -1 & 0 \\ 0 & 1 & 0 & 0 \end{bmatrix} \qquad B = g(0) = \begin{bmatrix} 0 \\ 2 \\ 1 \\ 0 \end{bmatrix}$$

and has exactly one uncontrollable mode corresponding to the eigenvalue $\lambda = 0$. However, its zero dynamics (see Example 3.3)

$$\dot{z}_3 = -z_3 - \frac{z_4 - 4z_4^4}{2 + 2z_3}$$

$$\dot{z}_4 = -2z_4^3$$

are asymptotically stable. Thus, from our previous discussion, we conclude that a control law of the form

$$u = \frac{1}{L_g L_f h(x)} \left(-L_f^2 h(x) - c_0 h(x) - c_1 L_f h(x) \right)$$

will eventually stabilize the system. •

If an output function is not defined, the zero dynamics are not defined as well. However, it may happen that one is able to *design* a suitable dummy output whose associated zero dynamics are asymptotically stable. In this case a control law of the form discussed before will guarantee asymptotic stability. This procedure is illustrated in the following simple example.

Example 4.2. Consider the system

$$\dot{x}_1 = x_1^2 x_2^3$$
$$\dot{x}_2 = x_2 + u$$

whose linear approximation at $x = 0$ has an uncontrollable mode corresponding to the eigenvalue $\lambda = 0$. Suppose one is able to find a function $\gamma(x_1)$ such that

$$\dot{x}_1 = x_1^2 [\gamma(x_1)]^3$$

is asymptotically stable. Then, setting

$$y = h(x) = \gamma(x_1) - x_2$$

a system with an asymptotically stable zero dynamics is obtained. As a matter of fact, we know that the zero dynamics correspond to the constraint $y(t) = 0$ for all t, and this, in the present case, imposes

$$\gamma(x_1) = x_2.$$

Thus, the zero dynamics will evolve exactly according to

$$\dot{x}_1 = x_1^2 [\gamma(x_1)]^3$$

and the system will be stabilizable by means of the procedure discussed above. In our case, a suitable choice of $\gamma(x_1)$ will be, e.g.

$$\gamma(x_1) = -x_1.$$

Accordingly, a locally stabilizing feedback is the one given by

$$\alpha(x) = \frac{1}{L_g h(x)} \left(-L_f h(x) - ch(x) \right) = -cx_1 - (1 + c)x_2 - x_1^2 x_2^3$$

with $c > 0$. •

Remark 4.3. It is not difficult to see that the eigenvalues associated with uncontrollable modes of the linear approximation of the system, if any, correspond necessarily to eigenvalues of the jacobian matrix Q, that is of the linear approximation of the zero dynamics. For, observe that the linear approximation of the equations in normal form reads as

$$\dot{z}_1 = z_2$$
$$\dot{z}_2 = z_3$$
$$\cdots$$
$$\dot{z}_{r-1} = z_r$$
$$\dot{z}_r = R\xi + S\eta + Ku$$
$$\dot{\eta} = P\xi + Q\eta$$

where

$$R = \left[\frac{\partial b}{\partial \xi}\right]_{(\xi,\eta)=(0,0)} \qquad S = \left[\frac{\partial b}{\partial \eta}\right]_{(\xi,\eta)=(0,0)}$$
$$P = \left[\frac{\partial q}{\partial \xi}\right]_{(\xi,\eta)=(0,0)} \qquad Q = \left[\frac{\partial q}{\partial \eta}\right]_{(\xi,\eta)=(0,0)}$$

and $K = a(0,0)$. Suppose the linear approximation is uncontrollable. Then for some complex number λ, the matrix

$$\left[\begin{array}{ccccc} \lambda & -1 & 0 & \cdots & 0 \\ 0 & \lambda & -1 & \cdots & 0 \\ \vdots & \vdots & & \ddots & \vdots \\ 0 & 0 & 0 & \cdots & -1 \\ -r_1 & -r_2 & -r_3 & \cdots & \lambda - r_r \end{array}\right. \left.\begin{array}{c} 0 \\ 0 \\ \cdots \\ 0 \\ -S \end{array}\right. \left.\begin{array}{c} 0 \\ 0 \\ \cdots \\ 0 \\ K \end{array}\right]$$
$$\quad -P \qquad\qquad\qquad \lambda I - Q \qquad 0$$

where

$$\begin{bmatrix} r_1 & r_2 & \cdots & r_r \end{bmatrix} = R$$

has rank less than n. More specifically, the values of λ such that this matrix has rank less than n are exactly the eigenvalues associated with the uncontrollable modes. From the structure of the matrix in question it is easily seen that, since K is nonzero, its rank can be less than n only if λ annihilates the determinant of $(\lambda I - Q)$, i.e. if λ is an eigenvalue of the linear approximation of the zero dynamics.

Thus, if the system has a linear approximation with uncontrollable modes corresponding to eigenvalues on the imaginary axis and an output is defined such that the zero dynamics (on the nonlinear system) are asymptotically stable, the latter are not asymptotically stable in the first approximation. However, the system can still be stabilized by the method described before because, as observed, the asymptotic stability in the first approximation of the zero dynamics is not an issue. ●

Remark 4.4. If, instead of the feedback (4.1) one imposes a control

$$u = \frac{1}{a(\xi,\eta)}(-b(\xi,\eta) - c_0 z_1 - \ldots - c_{r-1} z_r + v)$$

in which v is an additional reference input, a closed loop system of the form

$$\dot{\xi} = A\xi + Bv$$
$$\dot{\eta} = q(\xi,\eta)$$

(4.3)

is obtained, with

$$B = \text{col}(0,\ldots,0,1).$$

This, of course, for $v = 0$ reduces to the system (4.2). If the latter is asymptotically stable, then for sufficiently small v the trajectories of (4.3) are *bounded*. More precisely, using the results of Section B.2, it is possible to conclude that for each $\varepsilon > 0$ there exist $\delta > 0$ and $K > 0$ such that

$$\| x(0) \| < \delta \qquad \text{and} \qquad |v(t)| < K \qquad \text{for all } t \geq 0$$

imply $\| x(t) \| < \varepsilon$ for all $t \geq 0$. •

4.5 Asymptotic Output Tracking

In Section 4.3, we have established conditions under which a prescribed reference output function $y_R(t)$ can be *exactly* reproduced. As we have seen, for this to be possible, certain components of the state of the system at time $t = 0$ must fit with the values at this time of the desired output $y_R(t)$ and of its first $r - 1$ derivatives. However, the possibility of presetting the initial state to a prescribed value is quite unusual in practice and, in addition, one cannot neglect the event of unexpected perturbations causing the initial state to be different from the desired one. More realistically, one is then led to investigate the problem of producing an output that, irrespectively of what the initial state of the system is, *converges asymptotically* to the prescribed reference function $y_R(t)$. This problem is called the *Problem of Tracking the Output $y_R(t)$*. Again, an elementary analysis of the problem in question (although not the most general one, as we will observe later in Remark 5.3) is made possible by an appropriate use of the results developed in Sections 4.1 and 4.3.

Consider again a system in normal form

$$\dot{z}_1 = z_2$$
$$\dot{z}_2 = z_3$$
$$\cdots$$
$$\dot{z}_{r-1} = z_r$$
$$\dot{z}_r = b(\xi, \eta) + a(\xi, \eta)u$$
$$\dot{\eta} = q(\xi, \eta)$$
$$y = z_1$$

and choose

$$u = \frac{1}{a(\xi, \eta)}\left(-b(\xi, \eta) + y_R^{(r)} - \sum_{i=1}^{r} c_{i-1}(z_i - y_R^{(i-1)})\right) \tag{5.1}$$

where $c_0, \ldots, c_{r-1}$, are real numbers.

Define an "error" $e(t)$, as the difference between the real output $y(t)$ and the reference ouput $y_R(t)$

$$e(t) = y(t) - y_R(t).$$

Then, since by construction $z_i = y^{(i-1)}(t)$ for $1 \leq i \leq r$, it is immediately seen that the input (5.1) has the expression

$$u = \frac{1}{a(\xi, \eta)}\left(-b(\xi, \eta) + y_R^{(r)} - \sum_{i=1}^{r} c_{i-1}e^{(i-1)}\right).$$

Note that the input in question, if $e(t) = 0$ for all t, reduces exactly to the input needed in order to have precisely $y_R(t)$ as an output (see end of Section 3). For the sake of completeness, note also that the input (5.1), in the original coordinates has the expression

$$u = \frac{1}{L_g L_f^{r-1}h(x)}\left(-L_f^r h(x) + y_R^{(r)} - \sum_{i=1}^{r} c_{i-1}(L_f^{(i-1)}h(x) - y_R^{(i-1)})\right). \tag{5.2}$$

Imposing the input (5.1) implies

$$\dot{z}_r = y^{(r)} = y_R^{(r)} - c_{r-1}e^{(r-1)} - \cdots - c_1 e^{(1)} - c_0 e$$

i.e.

$$e^{(r)} + c_{r-1}e^{(r-1)} + \cdots + c_1 e^{(1)} + c_0 e = 0 \tag{5.3}$$

The error function $e(t)$ satisfies a linear differential equation, of order r, whose coefficients can be arbitrarily preset. The roots of the characteristic equation associated with (5.3) can be arbitrarily assigned, and we see, then, that under the effect of an input of the form (5.1) the output of the system "tracks" the desired output $y_R(t)$, with an error that can be made to converge to zero, as $t \to \infty$, with arbitrarily fast exponential decay.

Of course, an ever present concern in the design of control laws, is that the variables representing the internal behavior of the system remain *bounded* when a specific control law is imposed. In the present situation, the asymptotic analysis of the internal behavior of the system obtained by imposing the control law (5.2) on (1.1) can be carried out in the following way.

First of all, note that if we consider, as we implicitly did, the reference output $y_R(t)$ to be a *fixed* function of time, then the system (1.1) driven by the input (5.2) can be interpreted as a *time-varying* nonlinear system. In particular, looking at the behavior of the state variables in the coodinates used for the normal form, it is easy to check that $z_1, \ldots, z_{r_1}$, satisfy the identities

$$z_i = y_R^{(i-1)} + e^{(i-1)}$$

whereas η satisfies a differential equation of the form

$$\dot{\eta} = q(\xi_R(t) + \chi(t), \eta) \tag{5.4}$$

where, as in (3.2),

$$\xi_R(t) = \mathrm{col}(y_R(t), y_R^{(1)}(t), \ldots, y_R^{(r-1)}(t))$$

and

$$\chi(t) = \mathrm{col}(e(t), e^{(1)}(t), \ldots, e^{(r-1)}(t)).$$

Equation (5.4), in view of the remarks at the end of Section 4.3, can be seen as an equation describing the "response" of the inverse system "driven" by the function $y_R(t) + e(t)$.

Sufficient conditions for the boundedness of the $z_i(t)$'s and $\eta(t)$ are expressed in the following statement.

Proposition 5.1. Suppose $y_R(t), y_R^{(1)}(t), \ldots, y_R^{(r-1)}(t)$ are defined for all $t \geq 0$ and bounded. Let $\eta_R(t)$ denote the solution of

$$\dot{\eta} = q(\xi_R(t), \eta) \tag{5.5}$$

satisfying $\eta_R(0) = 0$. Suppose this solution is defined for all $t \geq 0$, bounded and *uniformly asymptotically stable*. Finally, suppose the roots of the polynomial

$$s^r + c_{r-1}s^{r-1} + \cdots + c_1 s + c_0 = 0$$

all have negative real part. Then, for sufficiently small $a > 0$, if

$$|z_i(t^\circ) - y_R^{(i-1)}(t^\circ)| < a, 1 \le i \le r, \qquad \|\eta(t^\circ) - \eta_R(t^\circ)\| < a$$

the corresponding response $z_i(t)$, $\eta(t)$, $t \ge t^\circ \ge 0$, of the closed loop system (1.1)-(5.2) is bounded. More precisely, for all $\varepsilon > 0$ there exists $\delta > 0$ such that

$$|z_i(t^\circ) - y_R^{(i-1)}(t^\circ)| < \delta \Rightarrow |z_i(t) - y_R^{(i)}(t)| < \varepsilon \qquad \text{for all } t \ge t^\circ \ge 0$$
$$\|\eta(t^\circ) - \eta_R(t^\circ)\| < \delta \Rightarrow \|\eta(t) - \eta_R(t)\| < \varepsilon \qquad \text{for all } t \ge t^\circ \ge 0.$$

Proof. Observe that the system (1.1)-(5.2) can be rewritten in the form

$$\dot\chi = K\chi \qquad\qquad (5.6a)$$
$$\dot\eta = q(\xi_R(t) + \chi, \eta) \qquad\qquad (5.6b)$$

where K is a matrix in companion form, whose characteristic equation is that of (5.3). Set $w = \eta - \eta_R(t)$ and $F(w,\chi,t) = q(\xi_R(t)+\chi, \eta_R(t)+w) - q(\xi_R(t), \eta_R(t))$. Then, the system
$$\dot w = F(w,\chi,t)$$
$$\dot\chi = K\chi$$
is in the form (B.2.5) and $(w,\chi) = (0,0)$ is an equilibrium. Note that, since $q(\xi,\eta)$ is smooth and $\xi_R(t)$, $\eta_R(t)$ are bounded, $F(w,\chi,t)$ is locally Lipschitzian in (w,χ) uniformly with respect to t. The solution $w = 0$ of $\dot w = F(w,0,t)$ is uniformly asymptotically stable and K has eigenvalues with negative real part. Thus, from Section B.2, we deduce that $(0, \eta_R(t))$ is an uniformly stable solution of (5.6), and the indicated estimates follow. ●

Remark 5.2. Note that the solution $\eta_R(t)$ of (5.5) need not to be an equilibrium solution (even in a linear system this is not necessarily the case, as the reader may easily check). The assumption that the solution $\eta_R(t)$ of (5.5) is uniformly asymptotically stable can be interpreted as the—rather natural—requirement that, in the conditions in which $y_R(t)$ is *exactly reproduced* (in this case $\eta(t)$ obeys exactly the equation (5.5)), the internal behavior of the system is that of an uniformly asymptotically stable system. ●

Remark 5.3. It is important to observe that the approach presented so far is not the unique possible, and the assumption for $\eta_R(t)$ of being uniformly asymptotically stable solution of (5.5) is *not necessary* for having bounded response in the state variables of a closed loop system which solves the tracking problem. As a matter of fact, one might have the feeling that this assumption is somehow necessary because it naturally comes together with the imposition of the control law (5.1) which, in turn, was naturally suggested as a adaptation—to the case of mismatched initial state—of the control law (3.4) that was proved to be *necessary* for exact reproduction of $y_R(t)$. The approach followed here, that is the choice of the control law (5.1), incorporates the property that, in the closed loop system, if $e = 0$ at time $t = 0$, then $e(t) = 0$ for all t. However, as we shall see in more detail in Chapter 7 (where we will present a more general approach to the problem), there is no need in principle for such a requirement if one wants a closed loop system whose output tracks a reference output, with the internal variables being bounded. •

Sometimes the reference output is not just a fixed function of time, but the output of a *reference model*, which in turn is subject to some input w, for instance a linear model, described by equations of the form

$$\dot{\zeta} = A\zeta + Bw \tag{5.7a}$$

$$y_R = C\zeta. \tag{5.7b}$$

If this is the case, one may pose the problem of finding a feedback control which, irrespectively of what the initial states of the system and of the model are, causes—for every input $w(t)$ to the model—an output $y(t)$ asymptotically converging to the corresponding output $y_R(t)$ produced by the model under the effect of $w(t)$. This problem is commonly known as *asymptotic model matching*.

In order to solve this problem one could, in principle, think to use the same input (5.2) considered at the beginning, with $y_R(t)$ and its first r derivatives replaced by the ones calculated from the reference model (5.7). However, since

$$y_R^{(i)}(t) = CA^i\zeta(t) + CA^{i-1}Bw(t) + \cdots + CABw^{(i-2)}(t) + CBw^{(i-1)}(t)$$

the control law thus obtained would depend on the first $r - 1$ derivatives of the input $w(t)$ to the reference model, a situation which is not desirable if the control law in question has to be realized by a device which receives $w(t)$ as an input and produces $u(t)$ as an output. In fact, the differentiation of $w(t)$ would indeed boost the effect of unavoidable additive noise.

If we suppose that

$$CB = CAB = \ldots = CA^{r-2}B = 0 \tag{5.8}$$

i.e., that the model has a relative degree equal to or possibly larger than the relative degree r of the system, we have that

$$y_R^{(i)}(t) = CA^i\zeta(t) \qquad \text{for all } 0 \leq i \leq r - 1$$

$$y_R^{(r)}(t) = CA^r\zeta(t) + CA^{r-1}Bw(t).$$

The first $r - 1$ derivatives of $y_R(t)$ do not depend explicitly on $w(t)$, and the r-th one depends explicitly on $w(t)$ but not on its derivatives. Replacing this in the expression (5.2) of u yields

$$u = \frac{1}{L_g L_f^{r-1} h(x)} (-L_f^r h(x) + CA^r \zeta(t) + CA^{r-1} Bw$$

$$- \sum_{i=1}^{r} c_{i-1} (L_f^{(i-1)} h(x) - CA^{(i-1)} \zeta)). \qquad (5.9)$$

By construction the system (1.1), subject to an input of this form—if the coefficients $c_0, \ldots, c_{r-1}$ are appropriately chosen—will produce an output asymptotically converging to the output $y_R(t)$ of the model. Since the latter has the form

$$y_R(t) = Ce^{At} \zeta(0) + \int_0^t Ce^{A(t-s)} Bw(s) \, ds$$

we can conclude that the output of the closed loop system (1.1)-(5.9) (see Fig. 4.7) will be of the form

$$y(t) = e(t) + Ce^{At} \zeta(0) + \int_0^t Ce^{A(t-s)} Bw(s) \, ds$$

with $e(t)$ solution of the differential equation (5.3).

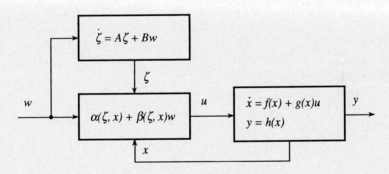

Fig. 4.7

Note that the input (5.9) depends explicitly—at each time t—on the state $x(t)$ of the system, on the input $w(t)$ of the model, and on the state $\zeta(t)$ of the model, which in turn obeys the differential equation (5.7a). Thus, $u(t)$ can be regarded as the "output" of a dynamical system of the form

$$\dot{\zeta} = \gamma(\zeta, x) + \delta(\zeta, x)w \qquad (5.10a)$$

$$u = \alpha(\zeta, x) + \beta(\zeta, x)w \qquad (5.10b)$$

with internal state ζ, driven by the "inputs" w and x. As a matter of fact, the first one of these two equations can be identified with (5.7a) and the second one with (5.9). We see then that the solution of the problem of asymptotically tracking the output of a reference model entails the use of a more general type of state feedback than the one considered so far, in that includes also an internal dynamics. A feedback of this form is called a *dynamic* state feedback.

Summarizing the whole discussion, we can conclude that, if the relative degree of the model (5.7) is larger than or equal to the relative degree of the system, there exists a dynamic feedback of the form (5.10) yielding an output $y(t)$ that converges asymptotically to the output $y_R(t)$ of the model, for every possible input $w(t)$, and for every possible initial state $x(0)$, $\zeta(0)$.

The analysis of the internal asymptotic properties of the system thus obtained is quite similar to the one developed earlier for the system (1.1)-(5.2). As a matter of fact, it is immediate to check that, in the present case, the closed loop system can be described in proper coordinates by equations of the form

$$\dot{\zeta} = A\zeta + Bw$$
$$\dot{\chi} = K\chi$$
$$\dot{\eta} = q(Q\zeta + \chi, \eta)$$

with $Q = \mathrm{col}(C, CA, \ldots, CA^{r-1})$. The first one of these equations describes the dynamics of the model (driven by its own input), the second one the dynamics of the error (which is an autonomous equation), the latter the dynamics of the inverse system driven by the function $C\zeta + \chi_1$.

4.6 Disturbance Decoupling

The normal form introduced in Section 4.1 is also useful in understanding how the output response of a given system can be protected from disturbances affecting the state. Consider a system

$$\dot{x} = f(x) + g(x)u + p(x)w$$
$$y = h(x).$$

in which w represents an undesired input, or disturbance. We want to examine under what conditions there exists a static state feedback control

$$u = \alpha(x) + \beta(x)v$$

yielding a closed loop in which the output y is completely independent of, or *decoupled* from, the disturbance w. This problem is commonly known as *Disturbance Decoupling Problem*.

As usual, we discuss first the solution looking at the normal form of the equations. Let the system have relative degree r at x°, and suppose the vector field $p(x)$ which multiplies the disturbance in the state equation being such that

$$L_p L_f^i h(x) = 0 \qquad \text{for all } 0 \le i \le r-1 \qquad \text{and all } x \text{ near } x^\circ.$$

If we write the state space equations choosing the same coordinates used before to describe the normal form, we obtain

$$
\begin{aligned}
\frac{dz_1}{dt} &= \frac{\partial z_1}{\partial x}\frac{dx}{dt} \\
&= \frac{\partial h}{\partial x}\frac{dx}{dt} \\
&= L_f h(x(t)) + L_g h(x(t))u(t) + L_p h(x(t))w(t) \\
&= L_f h(x(t)) = z_2(t)
\end{aligned}
$$

because, by assumption, $L_p h(x) = 0$ for all t such that $x(t)$ is near x°. A similar situation happens for all other subsequent equations, and thus we get

$$\frac{dz_2}{dt} = z_3(t)$$

$$\cdots$$

$$\frac{dz_{r-1}}{dt} = z_r(t).$$

For z_r we still obtain

$$\frac{dz_r}{dt} = L_f^r h(x(t)) + L_g L_f^{r-1} h(x(t))u(t)$$

because, again, $L_p L_f^{r-1} h(x) = 0$. Thus, the first r equations are exactly the same as those of a normal form of a system without disturbance. This is not anymore the case for the remaining ones, that will now appear depending also on the disturbance w.

Using, as in the previous Sections, a vector notation, we can rewrite the system in the form

$$
\begin{aligned}
\dot{z}_1 &= z_2 \\
\dot{z}_2 &= z_3 \\
&\cdots \\
\dot{z}_{r-1} &= z_r \\
\dot{z}_r &= b(\xi, \eta) + a(\xi, \eta)u \\
\dot{\eta} &= q(\xi, \eta) + k(\xi, \eta)w.
\end{aligned}
$$

In addition we have, as usual

$$y = z_1$$

Suppose now the following state feedback is chosen

$$u = -\frac{b(\xi,\eta)}{a(\xi,\eta)} + \frac{v}{a(\xi,\eta)}.$$

This feedback yields a system which is described by the equations

$$\dot{z}_1 = z_2$$
$$\dot{z}_2 = z_3$$
$$\dots$$
$$\dot{z}_{r-1} = z_r$$
$$\dot{z}_r = v$$
$$\dot{\eta} = q(\xi,\eta) + k(\xi,\eta)w$$

from which it is easily seen that the output, i.e. the state variable z_1, has been completely decoupled from the disturbance w.

The block-diagram interpretation of the closed loop system thus obtained, described in Fig. 4.8, clearly explains what happened. The effect of the input has been that of isolating from the output that part of the system on which the disturbance has effect.

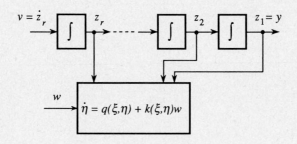

Fig. 4.8

We have thus found a *sufficient* condition for the existence of solutions to the problem of decoupling the output of a system from a disturbance, and explicitly constructed a decoupling feedback. It is not difficult to prove that the condition in question is also necessary, as we shall see in a moment. For convenience, we summarize all the results of interest in a formal statement in which, for more generality, we specify the decoupling feedback in terms of the functions $f(x)$, $g(x)$ and $h(x)$ characterizing the original description of the system.

Proposition 6.1. Suppose the system has relative degree r at x°. The problem of finding a feedback $u = \alpha(x) + \beta(x)v$, defined locally around x°, such that the output of the system is decoupled from the disturbance can be solved if and only if

$$L_p L_f^i h(x) = 0 \qquad \text{for all } 0 \leq i \leq r - 1 \qquad \text{and all } x \text{ near } x^\circ. \qquad (6.1)$$

If this is the case, then a solution is given by

$$u = -\frac{L_f^r h(x)}{L_g L_f^{r-1} h(x)} + \frac{v}{L_g L_f^{r-1} h(x)}.$$

Proof. We have only to prove the necessity. Let $u = \alpha(x) + \beta(x)v$ denote any feedback decoupling the output from the disturbance, and consider the corresponding closed loop

$$\dot{x} = f(x) + g(x)\alpha(x) + g(x)\beta(x)v + p(x)w$$
$$y = h(x).$$

By assumption, the output y has to be independent of w, and this has to be true also when $v(t) = 0$ for all t, i.e. for the system

$$x = f(x) + g(x)a(x) + p(x)w$$
$$y = h(x)$$

By repeating in the present case calculations similar to the ones done in Section 4.1, we obtain

$$y^{(1)}(t) = L_{f+g\alpha} h(x(t)) + L_p h(x(t))w(t).$$

from which we see that $y(t)$ can be independent from $w(t)$ only if $L_p h(x) = 0$ for all t such that $x(t)$ is near x°. Assuming this condition satisfied, we calculate $y^{(2)}(t)$ and get

$$y^{(2)}(t) = L_{f+g\alpha}^2 h(x(t)) + L_p L_{f+g\alpha} h(x(t))w(t)$$

Again, we conclude that $L_p L_{f+g\alpha} h(x)$ must be 0. The same arguments can be repeated for all the higher derivatives of $y(t)$, until we get

$$y^{(r)}(t) = L_{f+g\alpha}^r h(x(t)) + L_p L_{f+g\alpha}^{r-1} h(x(t))w(t)$$

and see that also $L_p L_{f+g\alpha}^{r-1} h(x)$ must be 0. We conclude that if the feedback decouples y from w, necessarily

$$L_p L_{f+g\alpha}^i h(x) = 0 \qquad \text{for all } 0 \leq i \leq r - 1 \qquad \text{and all } x \text{ near } x^\circ.$$

This is indeed the condition we wanted to prove because, as seen in the proof of Lemma 2.1, $L^i_{f+g\alpha}h(x) = L^i_f h(x)$ for all $0 \leq i \leq r-1$. •

Remark 6.2. Note that on the system thus decoupled one can further choose the new control v in order to achieve additional performances, like e.g. asymptotic stability. If the original system had an asymptotically stable zero dynamics, in view of the properties illustrated in Section 4.4, we see that this goal can be accomplished by means of a feedback of the form

$$v = -(c_0 h(x) + \ldots + c_{r-1}L^{r-1}_f h(x)) + \bar{v}.$$

Suppose, without loss of generality, $x^\circ = 0$, i. e. that $f(0) = 0$ and $h(0) = 0$. Let the coefficients

$$c_0, \ldots, c_{r-1}$$

be such that the vector field $f(x) + g(x)\alpha(x)$ has an asymptotically stable equilibrium at $x = 0$. Then, using the results illustrated in Section B.2, it is possible to conclude—as in Remark 4.4—that for each $\varepsilon > 0$ there exist $\delta > 0$ and $K > 0$ such that

$$\| x(0) \| < \delta \quad \text{and} \quad |w(t)| < K, |\bar{v}(t)| < K, \text{ for all } t \geq 0$$

imply $\| x(t) \| < \varepsilon$ for all $t \geq 0$. •

Sometimes, it is useful to write the condition of Proposition 4.1 in a slightly different manner. Recall that

$$L_p L^k_f h(x) = \langle dL^k_f(x), p(x)\rangle$$

and consider the codistribution

$$\Omega = \text{span}\{dh, dL_f h, \ldots, dL^{r-1}_f h\}.$$

Then, it is immediate to realize that the condition (6.1) is equivalent to the following one

$$p(x) \in \Omega^\perp(x) \qquad \text{for all } x \text{ near } x^\circ. \tag{6.2}$$

Sometimes it is possible to obtain "measurements" of the disturbance and use them in the design of the control law. If the disturbance w is available for measurements one can think to use a control

$$u = \alpha(x) + \beta(x)v + \gamma(x)w$$

which, besides to a feedback on x, includes a feedforward on the disturbance w. If this is the case, then decoupling the output from the disturbance is possible under conditions that are obviously weaker than those established before. Looking at the form of the closed loop

$$\dot{x} = f(x) + g(x)\alpha(x) + g(x)\beta(x)v + (g(x)\gamma(x) + p(x))w$$
$$y = h(x)$$

it is immediately understood that what is needed is the possibility of finding a function $\gamma(x)$ such that

$$(g(x)\gamma(x) + p(x)) \in \Omega^{\perp}(x) \qquad \text{for all } x \text{ near } x^{\circ}. \tag{6.3}$$

This condition is equivalent to

$$0 = L_{g\gamma+p}L_f^i h(x) = L_g L_f^i h(x)\gamma(x) + L_p L_f^i h(x)$$

for all $0 \leq i \leq r-1$ and all x near x°, and this, recalling the definition of relative degree, is in turn equivalent to

$$L_p L_f^i h(x) = 0 \qquad \text{for all } 0 \leq i \leq r-2$$
$$L_p L_f^{r-1} h(x) = -L_g L_f^{r-1} h(x)\gamma(x)$$

for all x near x°. The second one of these can always be satisfied, by choosing

$$\gamma(x) = -\frac{L_p L_f^{r-1} h(x)}{L_g L_f^{r-1} h(x)}$$

Thus, the necessary and sufficient condition for solving the problem of decoupling the disturbance from the output by means of a feedback that incorporates measurements of the disturbance is simply the first one. Note that this condition weakens the condition (6.1) of Proposition 6.1, in the sense that now $L_p L_f^i h(x) = 0$ must be zero for all the values of i up to $r-2$ included, but not necessarily for $i = r-1$.

If this is the case, a control law solving the problem of decoupling y from w is clearly given by

$$u = -\frac{L_f^r h(x)}{L_g L_f^{r-1} h(x)} + \frac{v}{L_g L_f^{r-1} h(x)} - \frac{L_p L_f^{r-1} h(x)}{L_g L_f^{r-1} h(x)}w.$$

Note that, geometrically, the condition (6.3) has the following interpretation: at each x, the vector $p(x)$ can be decomposed in the form

$$p(x) = c_1(x)g(x) + p_1(x)$$

where $c_1(x)$ is a real-valued function and $p_1(x)$ is a vector in $\Omega^\perp(x)$. This can be expressed in the form

$$p(x) \in \Omega^\perp(x) + \text{span}\{g(x)\} \qquad \text{for all } x \text{ near } x^\circ \tag{6.4}$$

thus showing to what extent the condition (6.2) can be weakened by allowing a feedback which incorporates measurements of the disturbance.

4.7 High Gain Feedback

In this Section we consider again the problem of the design of locally stabilizing feedback and we show that—under the stronger assumption that the zero dynamics are asymptotically stable *in the first approximation*—a nonlinear system can be locally stabilized by means of output feedback. First of all, we consider the case of a system having relative degree 1 at the point $x^\circ = 0$, and we show that asymptotic stabilization can be achieved by means of a memoryless linear feedback.

Proposition 7.1. Consider a system of the form (1.1), with $f(0) = 0$ and $h(0) = 0$. Suppose this system has relative degree 1 at $x = 0$, and suppose the zero dynamics are asymptotically stable in the first approximation, i.e. that the eigenvalues of the matrix

$$Q = \left[\frac{\partial q(\xi, \eta)}{\partial \eta} \right]_{(\xi, \eta) = (0,0)}$$

all have negative real part. Consider the closed loop system.

$$\dot{x} = f(x) + g(x)u \tag{7.1a}$$
$$u = -Kh(x) \tag{7.1b}$$

where
$$\begin{cases} K > 0 & \text{if } L_g h(0) > 0 \\ K < 0 & \text{if } L_g h(0) < 0. \end{cases}$$

Then, there exists a positive number K_o such that, for all K satisfying $|K| > K_o$ the equilibrium $x = 0$ of (7.1) is asymptotically stable.

Proof. An elegant proof of this result is provided by the Singular Perturbations Theory (see Appendix B). Suppose $L_g h(0) < 0$ (the other case can be dealt with in the same manner), and set

$$K = \frac{-1}{\varepsilon}.$$

Note that the closed loop system (7.1), rewritten in the form

$$\varepsilon \dot{x} = \varepsilon f(x) + g(x)h(x) = F(x, \varepsilon) \qquad (7.2)$$

can be interpreted as a system of the form (**B.3.5**). Since $F(x, 0) = g(x)h(x)$ and $g(0) \neq 0$ (because $L_g h(0) \neq 0$), in a neighborhood V of the point $x = 0$ the set of equilibrium points of $x' = F(x, 0)$ coincides with the set

$$E = \{x \in V : h(x) = 0\}$$

Moreover, since $dh(x)$ is nonzero at $x = 0$, one can always choose V so that the set E is a smooth $(n - 1)$-dimensional submanifold.

We apply the main Theorem of Section **B.3** to examine the stability properties of this system. To this end, we need to check the corresponding assumptions on the two "limit" subsystems (**B.3.6a**) and (**B.3.6b**). Note that, at each $x \in E$,

$$T_x E = \ker(dh(x)).$$

Moreover, it is easy to check that

$$V_x = \mathrm{span}\{g(x)\}.$$

In fact, at each $x \in E$,

$$J_x = \frac{\partial(g(x)h(x))}{\partial x} = g(x) \frac{\partial h(x)}{\partial x}$$

(because $h(x) = 0$) and therefore,

$$J_x g(x) = g(x) L_g h(x).$$

Thus, the vector $g(x)$ is an eigenvector of J_x, with eigenvalue $\lambda(x) = L_g h(x)$. At each $x \in E$, the system $x' = F(x, 0)$ has $(n - 1)$ trivial eigenvalues and one nontrivial eigenvalue. Since by assumption $\lambda(0) < 0$, we see that at each point x in a neighborhood of 0, the nontrivial eigenvalue $\lambda(x)$ of J_x is negative.

We will show now that the reduced vector field associated with the system (7.2) coincides with the zero dynamics vector field. The easiest way to see this is to express (7.1a) in normal form, that is

$$\dot{z} = b(z, \eta) + a(z, \eta)u$$
$$\dot{\eta} = q(z, \eta)$$

in which $z = h(x) \in \mathbf{R}$ and $\eta \in \mathbf{R}^{n-1}$. Accordingly, system (7.2) becomes

$$\varepsilon \begin{bmatrix} \dot{z} \\ \dot{\eta} \end{bmatrix} = \begin{bmatrix} \varepsilon b(z, \eta) + a(z, \eta) z \\ \varepsilon q(z, \eta) \end{bmatrix}.$$

In the coordinates of the normal form, the set E is the set of pairs (z, η) such that $z = 0$. Thus

$$f_R(x) = P_x \left[\frac{\partial F(x, \varepsilon)}{\partial \varepsilon} \right]_{\varepsilon = 0,\ x \in E} = \begin{bmatrix} 0 & I \end{bmatrix} \begin{bmatrix} b(0, \eta) \\ q(0, \eta) \end{bmatrix}$$

from which we deduce that the reduced system is given by

$$\dot{\eta} = q(0, \eta)$$

i.e. by the zero dynamics of (1.1). Since, by assumption, the latter is asymptotically stable in the first approximation at $\eta = 0$, it is possible to conclude that there exists $\varepsilon_0 > 0$ such that, for each $\varepsilon \in (0, \varepsilon_0)$ the system (7.2) hasan isolated equilibrium point x_ε near 0 which is asymptotically stable. Since $F(0, \varepsilon) = 0$ for all $\varepsilon \in (0, \varepsilon_0)$, we have necessarily $x_\varepsilon = 0$, and this completes the proof.

Remark 7.2. Note that this result is the nonlinear version of the well known fact that the root locus of a transfer function having relative degree 1 and all zeros in the left-half complex plane has all branches contained in the left-half complex plane for sufficiently large values of the loop gain. ●

We turn now our attention to the case of systems having higher relative degree, and we will show that the problem can be solved by reduction to the case of a system with relative degree 1. For, suppose we can replace the real output y by a "dummy" output of the form

$$w = k(x)$$

with $k(x)$ defined as

$$k(x) = L_f^{r-1} h(x) + c_{r-2} L_f^{r-2} h(x) + \ldots + c_1 L_f h(x) + c_0 h(x)$$

where $c_0, \ldots, c_{r-2}$ are real numbers.

We obtain in this way a new system

$$\dot{x} = f(x) + g(x) u$$
$$w = k(x)$$

having relative degree 1 at 0, because

$$L_g k(0) = L_g L_f^{r-1} h(0) \neq 0$$

In order to decide whether or not this new system can be stabilized by an output feedback of the form considered before, i.e. of the form

$$u = -Kw$$

in view of Proposition 7.1, we need to examine the asymptotic behavior of its zero dynamics. To this end, recall that the zero dynamics describes the internal behavior of a system constrained in such a way as to produce zero output. Observe also that, in the coordinates used to represent the normal form of the original system, the "dummy" output w is described by

$$w = z_r + c_{r-2} z_{r-1} + \ldots + c_1 z_2 + c_0 z_1.$$

The constraint $w = 0$ implies

$$z_r = -(c_{r-2} z_{r-1} + \ldots + c_1 z_2 + c_0 z_1).$$

Substituting this into the normal form of the original system and choosing the input $u(t)$ in order to impose $w(t) = 0$, one obtains the $(n-1)$-dimensional dynamics

$$\dot{z}_1 = z_2$$
$$\dot{z}_2 = z_3$$
$$\ldots$$
$$\dot{z}_{r-1} = -(c_{r-2} z_{r-1} + \ldots + c_1 z_2 + c_0 z_1)$$
$$\dot{\eta} = q(z_1, \ldots, z_{r-1}, -(c_{r-2} z_{r-1} + \ldots + c_1 z_2 + c_0 z_1), \eta)$$

which therefore describes the zero dynamics associated with the new output.

These equations have a "block triangular" form and from this it is easy to conclude, looking for instance at the corresponding jacobian matrix, that, if the zero dynamics of the original system is asymptotically stable in the first approximation, and if all the roots of the polynomial

$$n(s) = s^{r-1} + c_{r-2} s^{r-2} + \ldots + c_1 s^1 + c_0$$

have negative real part, the dynamics in question is also asymptotically stable in the first approximation. Thus, from Proposition 7.1, we can conclude that if the roots of the polynomial $n(s)$ all have negative real part, and K has the same sign as that of $L_g L_f^{r-1} h(0)$ the feedback

$$u = -K(L_f^{r-1} h(x) + c_{r-2} L_f^{r-2} h(x) + \ldots + c_1 L_f h(x) + c_0 h(x)) \qquad (7.3)$$

asymptotically stabilizes the equilibrium $x = 0$ of the system (7.1a)-(7.3).

From the feedback (7.3), which actually is a state feedback, it is possible to deduce an output feedback in the following way. Observe that the function $L_f^i h(x(t))$, for $0 \leq i \leq r-1$, coincides with the i-th derivative of the function $y(t)$ with respect to time. Thus the function $w(t)$ is related to $y(t)$ by

$$w(t) = y^{(r-1)}(t) + c_{r-2} y^{(r-2)}(t) + \ldots + c_1 y^{(1)}(t) + c_0 y(t)$$

and can therefore be interpreted as the output of a system obtained by cascade-connecting the original system with a linear filter having transfer function exactly equal to the polynomial $n(s)$. Clearly such a filter is not physically realizable, but it is not difficult to replace it by a suitable physically realizable approximation, without impairing the stability properties of the corresponding closed loop.

To this end, all we need is the following simple result.

Proposition 7.3. Suppose the system

$$\dot{x} = f(x) - g(x)k(x)K$$

is asymptotically stable in the first approximation (at the equilibrium $x = 0$). Then, if T is a sufficiently small positive number, also the system

$$\dot{x} = f(x) - g(x)\zeta$$
$$\dot{\zeta} = (1/T)(-\zeta + k(x)K)$$

is asymptotically stable in the first approximation (at $(x, \zeta) = (0, 0)$).

Proof. The proof of this result is another simple application of Singular Perturbation Theory. For, change the variable ζ into a new variable z defined by

$$z = -\zeta + k(x)K$$

and note that the system in question becomes

$$\dot{x} = f(x) - g(x)(-z + k(x)K)$$
$$T\dot{z} = -z + TK(\partial k / \partial x)[f(x) - g(x)(-z + k(x)K)] = -z + Tb(z, x).$$

This system has exactly the structure of the system (B.3.1), with $\varepsilon = T$. There is only one nontrivial eigenvalue, which is equal to -1, and the reduced system, which is given by

$$\dot{x} = f(x) - g(x)k(x)K$$

is by assumption asymptotically stable in the first approximation. Therefore, for sufficiently small positive T the equilibrium $(x,\zeta) = (0,0)$ is indeed asymptotically stable in the first approximation. ●

Note that the system discussed in this Proposition is nothing else than the system

$$\dot{x} = f(x) - g(x)u$$
$$y = k(x)$$

in closed loop with a linear system having transfer function (Fig. 4.9)

$$H(s) = \frac{-K}{1 + Ts}.$$

Thus we may interpret this result as the fact that the introduction of a "small time constant" in a stable control loop does not impair (at least locally) its asymptotic stability. Using this property $r-1$ times, we can immediately deduce the result indicated in the following statement.

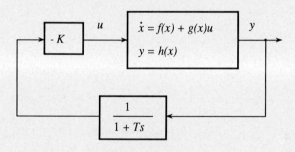

Fig. 4.9

Proposition 7.4. Suppose a system has relative degree r at $x^\circ = 0$ and its zero dynamics are asymptotically stable in the first approximation. Suppose also the roots of the polynomial

$$n(s) = s^{r-1} + c_{r-2}s^{r-2} + \ldots + c_1 s^1 + c_0$$

all have negative real part. A linear output feedback control with transfer function

$$H(s) = \frac{-Kn(s)}{(1 + Ts)^{r-1}}$$

stabilizes the system, provided K is a suitably large constant with the same sign as that of $L_g L_f^{r-1} h(0)$ and T is a sufficiently small positive constant.

4.8 Additional Results on Exact Linearization

We have illustrated in Section 4.2 a set of necessary and sufficient conditions for the existence of a (locally defined) state feedback and change of coordinates transforming the system (1.1a) into a linear and controllable one. Of course, if the conditions specified in Theorem 2.6 are not satisfied, there is no way to obtain a linear controllable system via feedback and change of coordinates. However, taking advantage of the construction indicated at the end of the Section, i.e. of the possibility of achieving always a decomposed system in which one of the two component subsystems is linear, one may wish at least to search for a feedback and a coordinates transformation which (if possible) *maximize* the dimension of the linear subsystem. In view of other results established in Section 4.2, the problem is clearly equivalent to that of finding a suitable "output" map $\lambda(x)$ for which the relative degree of the system at a point is the highest possible. As a matter of fact, the solution of such a problem is not much difficult, as the result hereafter discussed shows.

In the following statement, we make use of the notion of *involutive closure* of a distribution Δ, that has been introduced in Remark 1.3.9 and, in particular, of the following property.

Lemma 8.1. Consider a distribution Δ and suppose λ is a real-valued function such that $d\lambda(x^\circ) \neq 0$ and $d\lambda \in \Delta^\perp$. Then, in a neighborhood of x°, $d\lambda \in (\text{inv}(\Delta))^\perp$, where $\text{inv}(\Delta)$ denotes the involutive closure of Δ.

Proof. Consider the distribution

$$\Gamma = (\text{span}\{d\lambda\})^\perp.$$

This distribution is $(n-1)$-dimensional in a neighborhood of x°, and involutive, by Frobenius theorem. Moreover, by construction, $\Delta \subset \Gamma$. Since, by defintion, $\text{inv}(\Delta)$ is the smallest involutive distribution containing Δ, then $\text{inv}(\Delta) \subset \Gamma$, that is

$$\text{span}\{d\lambda\} \subset (\text{inv}(\Delta))^\perp. \quad \bullet$$

Theorem 8.2. Consider a pair of vector fields $f(x)$ and $g(x)$. Suppose, for some integer ν

$$\dim(\text{inv}(\text{span}\{g, ad_f g, \ldots, ad_f^{\nu-2} g\})) = k < n \tag{8.1}$$

for all x around x° and

$$\dim(\text{inv}(\text{span}\{g, ad_f g, \ldots, ad_f^{\nu-1} g\})) = n \tag{8.2}$$

at $x = x^\circ$. Then, there exists a function $\lambda(x)$ such that the system

$$\dot{x} = f(x) + g(x)u$$
$$y = \lambda(x)$$

has relative degree ν at x°. For any other possible choice of output map

$$y' = \lambda'(x)$$

the relative degree of the system is lower than or equal to ν.

Proof. The distribution

$$\text{inv}(\text{span}\{g, ad_f g, \ldots, ad_f^{\nu-2}g\}) \tag{8.3}$$

is involutive by construction and k-dimensional by assumption, with $k < n$. Thus, by Frobenius' Theorem, there exist $n - k$ functions $\lambda_1(x), \ldots, \lambda_{n-k}(x)$, whose differentials locally span the annihilator of (8.3). If we set, e.g., $\lambda(x) = \lambda_1(x)$ we have, by construction

$$L_g \lambda(x) = L_{ad_f g} \lambda(x) = \ldots = L_{ad_f^{\nu-2}g} \lambda(x) = 0$$

for all x near x°. Moreover, it is possible to show that

$$L_{ad_f^{\nu-1}g} \lambda(x^\circ) \neq 0$$

For, if this were not true, the nonzero covector $d\lambda(x^\circ)$ would be an element of $(\text{span}\{g, ad_f g, \ldots, ad_f^{\nu-1}g\})^\perp$ and, by Lemma 8.1, also an element of $(\text{inv}(\text{span}\{g, ad_f g, \ldots, ad_f^{\nu-1}g\}))^\perp$, that is a contradiction, because the latter has dimension 0 by assumption. Thus, by Lemma 1.3, we can conclude that the choice of $\lambda(x)$ as an output function for the system makes it having relative degree ν.

Now, consider any other output function $\lambda'(x)$ and suppose the corresponding relative degree is r. By Lemma 1.2 we have

$$d\lambda' \in (sp\{g, ad_f g, \ldots, ad_f^{r-2}g\})^\perp$$

and therefore, by Lemma 8.1, also

$$d\lambda' \in (\text{inv}(\text{span}\{g, ad_f g, \ldots, ad_f^{r-2}g\}))^\perp.$$

Since $d\lambda'(x^\circ) \neq 0$, we deduce that

$$\dim(\text{inv}(\text{span}\{g, ad_f g, \ldots, ad_f^{r-2}g\})) < n$$

for all x near x°. Using the assumptions (8.1)-(8.2), we conclude that $r \leq \nu$. •

Note that the result of the previous Theorem incorporates that of Theorem 2.6. As a matter of fact, if the conditions (i) and (ii) of Theorem 2.6 are satisfied, the integer ν defined in the previous statement is exactly equal to n. If the conditions in question are not satisfied, in order to find an output map $\lambda(x)$ which "maximizes" the relative degree of the system, one has to solve the partial differential equation

$$d\lambda(x)(\text{inv}(\text{span}\{g, ad_f g, \ldots, ad_f^{\nu-2}g\}(x))) = 0 \tag{8.4}$$

with ν defined as before. Once this solution has been constructed, then the feedback (2.18) will transform the system into one which, in suitable coordinates, contains a linear subsystem of maximal dimension.

Example 8.1. Consider the system

$$\dot{x} = \begin{bmatrix} x_2 - x_3^2 \\ x_3 + 2x_1^2 x_3 \\ x_1^2 \\ x_1 + x_3^2 \end{bmatrix} + \begin{bmatrix} 0 \\ 2x_3 \\ 1 \\ 0 \end{bmatrix} u.$$

In order to check whether or not this system can be transformed into a linear and controllable system via state feedback and coordinates transformation, we have to compute the vector fields $ad_f g, ad_f^2 g, ad_f^3 g$ and test the conditions of Theorem 2.6. Appropriate calculations show that

$$ad_f g(x) = \begin{bmatrix} 0 \\ -1 \\ 0 \\ -2x_3 \end{bmatrix} \qquad ad_f^2 g(x) = \begin{bmatrix} 1 \\ 0 \\ 0 \\ -2x_1^2 \end{bmatrix}.$$

Since

$$[g, ad_f g](x) = \begin{bmatrix} 0 \\ 0 \\ 0 \\ -2 \end{bmatrix}$$

one observes that

$$[g, ad_f g] \notin \text{span}\{g, ad_f g\}$$

and therefore the distribution $\text{span}\{g, ad_f g\}$ is not involutive. Thus, the conditions of Theorem 2.6 are violated (see Remark 2.9). However, in this case

$$\text{inv}(\text{span}\{g, ad_f g\}) = \text{span}\{g, ad_f g, [g, ad_f g]\} = \text{span}\left\{\begin{bmatrix} 0 \\ 1 \\ 0 \\ 0 \end{bmatrix}, \begin{bmatrix} 0 \\ 0 \\ 1 \\ 0 \end{bmatrix}, \begin{bmatrix} 0 \\ 0 \\ 0 \\ 1 \end{bmatrix}\right\}$$

and

$$\text{inv}(\text{span}\{g, ad_f g, ad_f^2 g\}) = \text{inv}(\text{span}\{g, ad_f g, [g, ad_f g], ad_f^2 g\})$$

$$= \text{span}\left\{\begin{bmatrix} 0 \\ 1 \\ 0 \\ 0 \end{bmatrix}, \begin{bmatrix} 0 \\ 0 \\ 1 \\ 0 \end{bmatrix}, \begin{bmatrix} 0 \\ 0 \\ 0 \\ 1 \end{bmatrix}, \begin{bmatrix} 1 \\ 0 \\ 0 \\ 0 \end{bmatrix}\right\}$$

so that the conditions of Theorem 8.2 are satisfied, with $\nu = 3$ and $k = 3$. Then, the maximal relative degree one can obtain for this system is $r = \nu = 3$. In order to find an output for which the relative degree is 3, one has to solve the differential equation (8.4), which yields, in this case

$$\lambda(x) = x_1.$$

From the previous discussion, it is clear that choosing a feedback

$$u = \frac{-L_f^3 \lambda(x) + v}{L_g L_f^2 \lambda(x)} = -x_1^2 + v$$

and new coordinates

$$z_1 = \lambda(x) = x_1$$
$$z_2 = L_f \lambda(x) = x_2 - x_3^2$$
$$z_3 = L_f^2 \lambda(x) = x_3$$

one obtains a system which contains a linear subsystem of dimension 3. Completing the choice of coordinates with

$$\eta = \eta(x) = x_4$$

yields

$$\dot{z}_1 = z_2$$
$$\dot{z}_2 = z_3$$
$$z_3 = v$$
$$\dot{\eta} = z_1 + z_3^2. \quad \bullet$$

We conclude the Section by discussing an additional problem. We have already observed in Remark 2.11 that if an Exact State Space Linearization Problem has been solved and the system has an output, the output map in the linearizing coordinates is not necessarily a linear map. Thus, one might pose the question of when there exist a feedback and a change of coordinates transforming the entire description of the system, output function included, into a linear and controllable one. An answer to this question is described in the following statement.

Theorem 8.3. Consider a system with relative degree r at $x = x^\circ$. Suppose also $f(x^\circ) = 0$ and $h(x^\circ) = 0$. There exists a feedback of the form (2.1) and a coordinates transformation $z = \Phi(x)$, defined locally around x°, changing the system (1.1) into a linear and controllable system

$$\dot{z} = Az + Bv$$
$$y = Cz$$

if and only if the following conditions are satisfied
  (i) the matrix $[g(x^\circ) \; ad_f g(x^\circ) \; \ldots \; ad_f^{n-2}g(x^\circ) \; ad_f^{n-1}g(x^\circ)]$ has rank n,
  (ii) the vector fields $\tilde{f}(x) = f(x) + g(x)\alpha(x)$ and $\tilde{g}(x) = g(x)\beta(x)$, with $\alpha(x)$ and $\beta(x)$ defined by

$$\alpha(x) = \frac{-L_f^r h(x)}{L_g L_f^{r-1} h(x)} \qquad \beta(x) = \frac{1}{L_g L_f^{r-1} h(x)}$$

are such that

$$[ad_{\tilde{f}}^i \tilde{g}, ad_{\tilde{f}}^j \tilde{g}](x) = 0 \tag{8.5}$$

for all $0 \leq i, j \leq n$, and all x near x°.

Remark 8.4. Note that the system

$$\dot{x} = f(x) + g(x)\alpha(x) + g(x)\beta(x)v = \tilde{f}(x) + \tilde{g}(x)v \tag{8.6a}$$
$$y = h(x) \tag{8.6b}$$

with $\alpha(x)$ and $\beta(x)$ chosen in the way indicated in (ii), has already a linear input-output response, by Proposition 2.12. Then, this Theorem shows that under the additional condition (8.5) it is possible to achieve linearity—via feedback and coordinatestransformation—also in the state space equations. On the other hand, since the conditions (i) and (ii) *imply* the conditions (i) and (ii) of Theorem 2.6 (as we shall see in a moment), this Theorem also describes to what extent the conditions of Theorem 2.6, necessary and sufficient to achieve linearity of the state space equations, must be strenghtened in order to achieve linearity also in the input-output response. In fact, as the reader may easily verify, since $\beta(x^\circ) \neq 0$, condition (i) implies

$$\text{rank}(\tilde{g}(x^\circ)\, ad_{\tilde{f}}\tilde{g}(x^\circ)\, \ldots\, ad_{\tilde{f}}^{n-1}\tilde{g}(x^\circ)) = n$$

and condition (ii) implies (see Remark **1**.3.7) that the distribution

$$\text{span}\{\tilde{g}, ad_{\tilde{f}}\tilde{g}, \ldots, ad_{\tilde{f}}^{n-1}\tilde{g}\}$$

is involutive. Thus system (8.6a), by Theorem 2.6, can be transformed into a linear and controllable system by means of state feedback and coordinates transformation. But since this system has been obtained from (1.1a) by means of a state feedback, namely $u = \alpha(x) + \beta(x)v$, then also (1.1a) can be transformed into a linear and controllable system by means of state feedback and coordinates transformation, i.e. must satisfy the conditions of Theorem 2.6. ●

Proof of Theorem 8.3. Sufficiency. For convenience, we separate the proof in several steps.

(i) Observe that, by construction, the system (8.6) satisfies

$$L_{\tilde{g}}L_{\tilde{f}}^{k}h(x) = 0 \qquad \text{for all } 0 \leq k \leq r - 2 \tag{8.7a}$$

(because the relative degree is invariant under feedback),

$$L_{\tilde{g}}L_{\tilde{f}}^{r-1}h(x) = (L_g L_f^{r-1}h(x))\beta(x) = 1 \tag{8.7b}$$

(because $L_{\tilde{f}}^{k}h(x) = L_f^{k}h(x)$ for all $k \leq r - 1$) and

$$L_{\tilde{f}}^{k}h(x) = 0 \qquad \text{for all } k \geq r \tag{8.7c}$$

(because $L_{\tilde{f}}^{r}h(x) = L_{f+g\alpha}L_f^{r-1}h(x) = L_f^{r}h(x) + L_g L_f^{r-1}h(x)\alpha(x) = 0$). Using the formula (1.2), we deduce from these

$$\langle dL_{\tilde{f}}^{s}h(x), ad_{\tilde{f}}^{k}\tilde{g}(x)\rangle \text{ is independent of } x \text{ for all } s, k \geq 0. \tag{8.7d}$$

(ii) As observed in the previous Remark, the system (8.6a) satisfies conditions (i) and (ii) of Theorem 2.6. Therefore, by Lemma 2.4, there exists a real-valued function $\lambda(x)$, defined in a neighborhood U of x°, satisfying

$$\langle d\lambda(x), ad_{\tilde{f}}^k \tilde{g}(x)\rangle = 0 \qquad \text{for all } x \text{ near } x^\circ,\ 0 \le k \le n-2$$

and the function

$$c(x) = \langle d\lambda(x), ad_{\tilde{f}}^{n-1} \tilde{g}(x)\rangle$$

is nonzero at the point x°. We show now that, because of the assumption (8.5), the function $\lambda(x)$ can always be chosen in such a way that $c(x) = 1$. For, recall that by construction the functions

$$z_i = L_{\tilde{f}}^{i-1}\lambda(x) \qquad 0 \le i \le n-1$$

have linearly independent differentials, so that they can be chosen as new coordinate functions near x°. As a consequence, there exists a function $\gamma(z_1, \ldots, z_n)$ such that

$$\gamma(\lambda(x), L_{\tilde{f}}\lambda(x), \ldots, L_{\tilde{f}}^{n-1}\lambda(x)) = c(x)$$

($\gamma(z)$ is exactly the function $c(x)$ expressed in the z coordinates).
    Observe also that, because of (8.5), for all $0 \le k \le n-2$

$$0 = \langle d\lambda(x), [ad_{\tilde{f}}^k \tilde{g}(x), ad_{\tilde{f}}^{n-1}\tilde{g}(x)]\rangle$$
$$= L_{ad_{\tilde{f}}^k \tilde{g}} L_{ad_{\tilde{f}}^{n-1}\tilde{g}}\lambda(x) - L_{ad_{\tilde{f}}^{n-1}\tilde{g}} L_{ad_{\tilde{f}}^k \tilde{g}}\lambda(x)$$
$$= L_{ad_{\tilde{f}}^k \tilde{g}} c(x).$$

Using this—with $k = 0$—in the previous expression for $c(x)$ we obtain

$$0 = L_{\tilde{g}}c(x) = \sum_{i=1}^n \frac{\partial \gamma}{\partial z_i} \frac{\partial L_{\tilde{f}}^{i-1}\lambda(x)}{\partial x} \tilde{g}(x) = (-1)^{n-1}\frac{\partial \gamma}{\partial z_n}c(x)$$

and we deduce

$$\frac{\partial \gamma}{\partial z_n} = 0.$$

Recursively, it is possible to show that

$$\frac{\partial \gamma}{\partial z_{n-1}} = \ldots = \frac{\partial \gamma}{\partial z_2} = 0$$

i.e. that $\gamma(z)$ depends only on z_1. In other words

$$c(x) = \gamma(\lambda(x))$$

where $\gamma(\zeta)$ is a real-valued function of the real variable ζ, defined in a neighborhood of $\lambda(x^\circ)$. Let $\psi(\zeta)$ be such that

$$\frac{\partial \psi}{\partial \zeta} = \frac{1}{\gamma(\zeta)}.$$

Then, the function $\tilde{\lambda}(x) = \psi(\lambda(x))$ satisfies

$$\langle d\tilde{\lambda}(x), ad_f^k \tilde{g}(x)\rangle = 0 \qquad \text{for all } 0 \le k \le n-2$$

$$\langle d\tilde{\lambda}(x), ad_f^{n-1} \tilde{g}(x)\rangle = \left[\frac{\partial \psi}{\partial \zeta}\right]_{\zeta = \lambda(x)} c(x) = \frac{1}{\gamma(\lambda(x))} c(x) = 1$$

for all x near x°.

(iii) The function $\lambda(x)$ considered in the previous step is such that

$$\langle dL_f^s \lambda(x), ad_f^k \tilde{g}(x)\rangle \text{ is independent of } x \tag{8.8}$$

for all s, k such that $0 \le s + k \le n-1$. We show now that, because of (8.5), this property holds for any value of s, k. For, observe that, if $v_1, \ldots, v_{n+1}$ is a collection of vector fields satisfying

$$\text{rank}[v_1(x) \ \ldots \ v_n(x)] = n$$
$$[v_i(x), v_j(x)] = 0 \qquad \text{for all } 1 \le i, j \le n+1$$

then

$$v_{n+1}(x) = \sum_{i=1}^{n} c_i v_i(x)$$

where $c_1, \ldots, c_n$ are real numbers. To see this, express $v_{n+1}(x)$ as

$$v_{n+1}(x) = \sum_{i=1}^{n} c_i(x) v_i(x)$$

and note that

$$0 = \sum_{i=1}^{n} [v_j(x), c_i(x) v_i(x)] = (L_{v_j} c_i(x)) v_i(x).$$

Thus

$$[L_{v_1} c_i(x) \ \ldots \ L_{v_n} c_i(x)] = dc_i(x)[v_1(x) \ \ldots \ v_n(x)] = 0.$$

i.e. $dc_i(x) = 0$, and $c_i(x)$ is independent of x for all $1 \le i \le n$. Using this property we deduce that

$$ad_{\tilde{f}}^n \tilde{g}(x) = \sum_{i=1}^{n} c_i^n \, ad_{\tilde{f}}^{i-1} \tilde{g}(x)$$

and, with a simple induction, also that

$$ad_{\tilde{f}}^k \tilde{g}(x) = \sum_{i=1}^{n} c_i^k \, ad_{\tilde{f}}^{i-1} \tilde{g}(x)$$

for all $k > n$, where the c_i^k's are real numbers. From this we have

$$\langle d\lambda(x), ad_{\tilde{f}}^k \tilde{g}(x) \rangle \text{ is independent of } x \text{ for all } k \geq 0$$

and, using again the formula (1.2), it is easy to conclude that (8.8) holds for any value of s, k.

(iv) Arrange (8.7d) and (8.8) in the matrix relation

$$\begin{bmatrix} d\lambda(x) \\ dL_{\tilde{f}}\lambda(x) \\ \cdots \\ dL_{\tilde{f}}^{n-1}\lambda(x) \\ dh(x) \end{bmatrix} [\tilde{g}(x) \; ad_{\tilde{f}}\tilde{g}(x) \; \dots \; ad_{\tilde{f}}^{n-1}\tilde{g}(x)] = \text{constant.}$$

The last row of the matrix on the left-hand side is linearly dependent from the first n ones through constant coefficients (because of the constancy of the right-hand side). Then, since the right factor of the left-hand side is nonsingular for all x near x°, we deduce that

$$dh(x) = \sum_{i=0}^{n-1} b_i dL_{\tilde{f}}^i \lambda(x)$$

where $b_0, \dots, b_{n-1}$ are real numbers. This implies

$$h(x) = \sum_{i=0}^{n-1} b_i L_{\tilde{f}}^i \lambda(x) + c \tag{8.9}$$

where c is a constant. Moreover, this constant is zero if $\lambda(x^\circ) = 0$, because of the assumptions $h(x^\circ) = 0$ and $f(x^\circ) = 0$.

(v) We know for the theory developed in Section 4.2 (see in particular (2.14)-

(2.15)) that the system (8.6a), after the feedback

$$v = \frac{-L_f^n \lambda(x)}{L_{\tilde{g}} L_f^{n-1} \lambda(x)} + \frac{1}{L_{\tilde{g}} L_f^{n-1} \lambda(x)} \tilde{v}$$

in the new coordinates

$$z_i = L_f^{i-1} \lambda(x) \qquad 0 \le i \le n-1$$

is a controllable linear system. But in these new coordinates the output map (8.6b) also is linear, as (8.9) shows. Thus, the proof of the sufficiency is completed.

The proof of the necessity requires only straightforward calculations, and is left to the reader. •

4.9 Observers With Linear Error Dynamics

We consider in this Section a problem which is in some sense dual of that considered in Section 4.2. We have seen that the solvabilty of the Exact State Space Linearization Problem enables us to design a feedback under which the system—in suitable local coordinates—becomes linear with prescribed eigenvalues. In the case of a linear system, the dual notion of assegnability via static state feedback is the existence of observers with prescribed eigenvalues. Moreover, it is known that the dynamics of an observer and that of the observation error (i.e. the difference between the unknown state and the estimated state) are the same. In view of this, if we wish to dualize the results developed so far, we are led to the problem of the synthesis of (nonlinear) observers yielding an error dynamics that, possibly after some suitable coordinates transformation, becomes linear and spectrally assignable.

For the sake of simplicity, we restrict ourselves to the consideration of systems without input and with scalar output, i.e. systems described by equations of the form

$$\dot{x} = f(x) \qquad\qquad (9.1a)$$
$$y = h(x) \qquad\qquad (9.1b)$$

with $y \in \mathbf{R}$.

Suppose there exists a coordinates transformation $z = \Phi(x)$ under which the vector field f and the output map h become respectively

$$\left[\frac{\partial \Phi}{\partial x}f(x)\right]_{x=\Phi^{-1}(z)} = Az + k(Cz)$$
$$h(\Phi^{-1}(z)) = Cz$$

where (A, C) is an observable pair and k is a n-vector valued function of a real variable.

If this is the case, then an observer of the form

$$\dot{\xi} = (A + GC)\xi - Gy + k(y)$$

yields an observation error (in the z coordinates)

$$e = \xi - z = \xi - \Phi(x)$$

governed by the differential equation

$$\dot{e} = (A + GC)e$$

which is linear and spectrally assignable (via the n-vector G of real numbers).

Motivated by these considerations, we examine the following problem, called the *Observer Linearization Problem*. Given a system without input (9.1), and an initial state x°, find (if possible) a neighborhood U° of x°, a coordinates transformation $z = \Phi(x)$ defined on U°, a mapping $k : h(U^\circ) \to \mathbf{R}^n$, such that

$$\left[\frac{\partial \Phi}{\partial x}f(x)\right]_{x=\Phi^{-1}(z)} = Az + k(Cz) \tag{9.2}$$
$$h(\Phi^{-1}(z)) = Cz \tag{9.3}$$

for all $z \in \Phi(U^\circ)$, for some suitable matrix A and row vector C, satisfying the condition

$$\operatorname{rank}\begin{bmatrix} C \\ CA \\ \cdots \\ CA^{n-1} \end{bmatrix} = n. \tag{9.4}$$

The conditions for the solvability of this problem can be described as follows.

Lemma 9.1. The Observer Linearization Problem is solvable only if

$$\dim(\operatorname{span}\{dh(x^\circ), dL_f h(x^\circ), \ldots, dL_f^{n-1}h(x^\circ)\}) = n. \tag{9.5}$$

Proof. The observability condition (9.4) implies the existence of a nonsingular $n \times n$ matrix T and a $n \times 1$ vector G such that

$$T(A+GC)T^{-1} = \begin{bmatrix} 0 & 0 & \cdots & 0 & 0 \\ 1 & 0 & \cdots & 0 & 0 \\ \cdots & \cdots & \cdots & \cdots & \cdots \\ 0 & 0 & \cdots & 1 & 0 \end{bmatrix} \qquad CT^{-1} = \begin{bmatrix} 0 & 0 & \cdots & 0 & 1 \end{bmatrix}.$$

$$(9.6)$$

Suppose (9.2) and (9.3) hold, and set

$$\tilde{z} = \tilde{\Phi}(x) = T\Phi(x)$$
$$\tilde{k}(y) = T(k(y) - Gy).$$

Then, it is easily seen that

$$h(\tilde{\Phi}^{-1}(\tilde{z})) = \begin{bmatrix} 0 & 0 & \cdots & 0 & 1 \end{bmatrix}\tilde{z}$$

$$\left[\frac{\partial \tilde{\Phi}}{\partial x} f(x)\right]_{x=\tilde{\Phi}^{-1}(\tilde{z})} = \begin{bmatrix} 0 & 0 & \cdots & 0 & 0 \\ 1 & 0 & \cdots & 0 & 0 \\ \cdots & \cdots & \cdots & \cdots & \cdots \\ 0 & 0 & \cdots & 1 & 0 \end{bmatrix} \tilde{z} + \tilde{k}(\begin{bmatrix} 0 & 0 & \cdots & 0 & 1 \end{bmatrix}\tilde{z})$$

From this, we deduce that there is no loss of generality in assuming that the pair (A, C) that makes (9.2) and (9.3) satisfied has directly the form specified by the rigth-hand sides of (9.6). Now, set

$$z = \Phi(x) = \mathrm{col}(z_1(x), \ldots, z_n(x)).$$

If (9.2) and (9.3) hold, we have, for all $x \in U^\circ$

$$h(x) = z_n(x)$$
$$\frac{\partial z_1}{\partial x} f(x) = k_1(z_n(x))$$
$$\frac{\partial z_2}{\partial x} f(x) = z_1(x) + k_2(z_n(x))$$
$$\cdots$$
$$\frac{\partial z_n}{\partial x} f(x) = z_{n-1}(x) + k_n(z_n(x))$$

where $k_1, \ldots, k_n$ denote the n components of the vector k.

Observe that

$$L_f h(x) = \frac{\partial z_n}{\partial x} f(x) = z_{n-1}(x) + k_n(z_n(x))$$

$$L_f^2 h(x) = \frac{\partial z_{n-1}}{\partial x} f(x) + \left[\frac{\partial k_n}{\partial y}\right]_{y=z_n} \frac{\partial z_n}{\partial x} f(x)$$

$$= z_{n-2}(x) + \left[\frac{\partial k_n}{\partial y}\right]_{y=z_n} \frac{\partial z_n}{\partial x} f(x) + k_{n-1}(z_n(x))$$

$$= z_{n-2}(x) + \tilde{k}_{n-1}(z_n(x), z_{n-1}(x))$$

where

$$\tilde{k}_{n-1}(z_n, z_{n-1}) = \frac{\partial k_n}{\partial z_n} z_{n-1} + \frac{\partial k_n}{\partial z_n} k_n(z_n) + k_{n-1}(z_n).$$

Proceeding in this way one obtains for each $L_f^i h(x)$, for $1 \leq i \leq n-1$, an expression of the form

$$L_f^i h(x) = z_{n-i}(x) + \tilde{k}_{n-i+1}(z_n(x), \ldots, z_{n-i+1}(x)).$$

Differentiating with respect to x and arranging all these expressions together, one obtains

$$
\begin{bmatrix} \dfrac{\partial h}{\partial x} \\ \dfrac{\partial L_f h}{\partial x} \\ \cdots \\ \dfrac{\partial L_f^{n-1} h}{\partial x} \end{bmatrix}
=
\begin{bmatrix} \dfrac{\partial h}{\partial z} \\ \dfrac{\partial L_f h}{\partial z} \\ \cdots \\ \dfrac{\partial L_f^{n-1} h}{\partial z} \end{bmatrix}
\frac{\partial z}{\partial x}
=
\begin{bmatrix} 0 & 0 & \cdots & 0 & 1 \\ 0 & 0 & \cdots & 1 & \star \\ \cdots & \cdots & \cdots & \cdots & \cdots \\ 1 & \star & \cdots & \star & \star \end{bmatrix}
\frac{\partial z}{\partial x}.
$$

This, because of the nonsingularity of the matrix on the right-hand side, proves the claim. ●

If the condition (9.5) is satisfied, then it is possible to define, in a neighborhood U° of x°, a unique vector field τ which satisfies the conditions

$$L_\tau h(x) = L_\tau L_f h(x) = \ldots = L_\tau L_f^{n-2} h(x) = 0$$
$$L_\tau L_f^{n-1} h(x) = 1$$

for all $x \in U^\circ$. As a matter of fact, one only needs to solve the set of equations

$$
\begin{bmatrix} dh(x) \\ dL_f h(x) \\ \vdots \\ dL_f^{n-2} h(x) \\ dL_f^{n-1} h(x) \end{bmatrix} \tau(x) = \begin{bmatrix} 0 \\ 0 \\ \vdots \\ 0 \\ 1 \end{bmatrix}. \tag{9.7}
$$

The construction of this vector field τ is useful in order to find necessary and sufficient conditions for the solution of our problem.

Lemma 9.2. The Observer Linearization Problem is solvable if and only if

(i) $\dim(\mathrm{span}\{dh(x^\circ), dL_f h(x^\circ), \ldots, dL_f^{n-1} h(x^\circ)\}) = n$
(ii) there exists a mapping F of some open set V of $\mathbf{R}^n$ onto a neighborhood U° of x° that satisfies the equation

$$
\frac{\partial F}{\partial z} = [\, \tau(x) \quad -ad_f \tau(x) \quad \cdots \quad (-1)^{n-1} ad_f^{n-1} \tau(x) \,]_{x=F(z)} \tag{9.8}
$$

for all $z \in V$, where τ is the unique vector field solution of (9.7).

Proof. Necessity. We already know that (i) is necessary. Suppose (9.2) and (9.3) are satisfied and set $F(z) = \Phi^{-1}(z)$ for all $z \in U^\circ$. Set also

$$
\theta(x) = \left(\frac{\partial F}{\partial z_1}\right)_{z=F^{-1}(x)}. \tag{9.9}
$$

We claim that

$$
ad_f^k \theta(x) = (-1)^k \left(\frac{\partial F}{\partial z_{k+1}}\right)_{z=F^{-1}(x)} \tag{9.10}
$$

for all $0 \le k \le n-1$. This equality, which is true by definition for $k = 0$, will be proved by induction, using the fact that (9.2) and (9.3) imply (see Proof of Lemma 9.1)

$$
f(x) = \left(\frac{\partial F}{\partial z} \tilde{f}(z)\right)_{z=F^{-1}(x)}
$$

with

$$
\tilde{f}(z) = \begin{bmatrix} k_1(z_n) \\ z_1 + k_2(z_n) \\ \cdots \\ z_{n-1} + k_n(z_n) \end{bmatrix}.
$$

In fact, suppose (9.10) is true for some $0 \le k < n-1$, and let e_i denote the i-th column of the $n \times n$ identity matrix. Then

$$ad_f^{k+1}\theta(x) = \left[f(x), (-1)^k \left(\frac{\partial F}{\partial z}e_{k+1}\right)_{z=F^{-1}(x)}\right]$$

$$= (-1)^k \left(\frac{\partial F}{\partial z}[\tilde{f}(z), e_{k+1}]\right)_{z=F^{-1}(x)} \cdot$$

$$= (-1)^{k+1}\left(\frac{\partial F}{\partial z}e_{k+2}\right)_{z=F^{-1}(x)}$$

Collecting all (9.10) together, we obtain

$$\frac{\partial F}{\partial z} = (\theta(x) - ad_f\theta(x) \ \cdots \ (-1)^{n-1}ad_f^{n-1}\theta(x))_{x=F(z)}.$$

If we show that θ necessarily coincides with the unique solution of (9.7), the proof is completed, because the p.d.e. (9.8) will coincide with the one just found.

To this end, observe that

$$(-1)^k L_{ad_f^k\theta}h(x) = \frac{\partial h}{\partial x}\left(\frac{\partial F}{\partial z_{k+1}}\right)_{z=F^{-1}(x)} = \left(\frac{\partial h(F(z))}{\partial z_{k+1}}\right)_{z=F^{-1}(x)}$$

but, since $h(\Phi^{-1}(z)) = z_n$, we have

$$L_{ad_f^k\theta}h(x) = 0$$

for all $0 \le k \le n-2$ and

$$(-1)^{n-1}L_{ad_f^{n-1}\theta}h(x) = 1.$$

Using Lemma 1.2, we deduce that

$$L_\theta L_f^k h(x) = 0$$

for all $0 \le k \le n-2$ and

$$L_\theta L_f^{n-1}h(x) = 1.$$

Thus, the vector field θ necessarily coincides with the unique solution of (9.7).

Sufficiency. Suppose (i) holds and let τ denote the solution of (9.7). Using

Lemma 1.2 one may immediately note (see (1.4)) that the matrix

$$
\begin{bmatrix}
dh(x) \\
dL_f h(x) \\
\cdots \\
dL_f^{n-1} h(x)
\end{bmatrix}
[\,\tau(x) \quad ad_f \tau(x) \quad \cdots \quad ad_f^{n-1} \tau(x)\,]
$$

has rank n for all x near x°. In particular, the vector fields $\{\tau, ad_f \tau, \ldots, ad_f^{n-1} \tau\}$ are linearly independent at all x near x°.

Let F denote a solution of the p.d.e. (9.8) and let z° be a point such that $F(z^\circ) = x^\circ$. From the linear independence of the vector fields on the right-hand side of (9.8) we deduce that the differential of F has rank n at z°, i.e. that F is a diffeormorphism of a neighborhood of z° onto a neighborhood of x°. Set $\Phi = F^{-1}$ and

$$
\tilde{f}(z) = \left(\frac{\partial \Phi}{\partial x} f(x) \right)_{x = \Phi^{-1}(z)}. \tag{9.11}
$$

By definition, the mapping F is such that

$$
\left(\frac{\partial F}{\partial z_{k+1}} \right)_{z = F^{-1}(x)} = (-1)^k ad_f^k \tau(x)
$$

so that

$$
\left(\frac{\partial \Phi}{\partial x} ad_f^k \tau(x) \right)_{x = \Phi^{-1}(z)} = (-1)^k e_{k+1} \tag{9.12}
$$

for all $0 \leq k \leq n-1$.

Using (9.11) and (9.12), one obtains, for all $0 \leq k \leq n-2$

$$
\begin{aligned}
(-1)^{k+1} e_{k+2} &= \left(\frac{\partial \Phi}{\partial x} ad_f^{k+1} \tau(x) \right)_{x = \Phi^{-1}(z)} \\
&= \left(\frac{\partial \Phi}{\partial x} [f, ad_f^k \tau](x) \right)_{x = \Phi^{-1}(z)} \\
&= [\tilde{f}(z), (-1)^k e_{k+1}] \\
&= (-1)^{k+1} \frac{\partial \tilde{f}}{\partial z_{k+1}}
\end{aligned}
$$

that is

$$
\frac{\partial \tilde{f}_{k+2}}{\partial z_{k+1}} = 1, \qquad \frac{\partial \tilde{f}_i}{\partial z_{k+1}} = 0 \qquad \text{for } i \neq k+2.
$$

From these, one deduces that $\tilde{f}_1$ depends only on z_n, and that $\tilde{f}_i$, for $2 \leq i \leq n$, is such that $\tilde{f}_i - z_{i-1}$ depends only on z_n. This proves that (9.2) holds. Moreover, since

$$L_{ad_f^k \tau} h(x) = 0 \qquad \text{for } 0 \leq k < n-1, \qquad L_{ad_f^{n-1} \tau} h(x) = (-1)^{n-1}$$

we deduce that

$$\frac{\partial h(F(z))}{\partial z_k} = 0 \qquad \text{for } 1 \leq k < n, \qquad \frac{\partial h(F(z))}{\partial z_n} = 1$$

thus proving (9.3). •

The integrability of the p.d.e. (9.8) may be expressed in terms of a property of the vector fields $\tau, ad_f \tau, \ldots, ad_f^{n-1} \tau$. To this end, one may use the following consequence of Frobenius Theorem.

Theorem 9.3. Let $\tau_1, \ldots, \tau_n$ be vector fields of $\mathbf{R}^n$. Consider the partial differential equations

$$\frac{\partial x}{\partial z_i} = \tau_i(x(z)) \tag{9.13}$$

where x denotes a mapping from an open set of $\mathbf{R}^n$ to an open set of $\mathbf{R}^n$. Let (z°, x°) be a point in $\mathbf{R}^n \times \mathbf{R}^n$ and suppose $\tau_1(x^\circ), \ldots, \tau_n(x^\circ)$ are linearly independent. There exists neighborhoods U° of x° and V° of z° and a diffeomorphism $x : V^\circ \to U^\circ$ solving the equation (9.13), and such that $x(z^\circ) = x^\circ$ if and only if

$$[\tau_i, \tau_j] = 0 \tag{9.14}$$

for all $1 \leq i, j \leq n$.

Proof. We limit ourselves to give a sketch of the proof of the sufficiency. To this end, set

$$\Delta_i = \text{span}\{\tau_1, \ldots, \tau_{i-1}, \tau_{i+1}, \ldots, \tau_n\}.$$

This distribution is involutive (because of (9.14) and has constant dimension $n-1$ in a neighborhood of x°. Therefore, by Frobenius Theorem, there exists a function ϕ_i whose differential spans $\Delta_i^\perp$, i.e. such that

$$\langle d\phi_i, \tau_j \rangle = 0$$

for all $j \neq i$. We claim that the differentials $d\phi_1, \ldots, d\phi_n$ are linearly independent for all x in a neighborhood of x°. For, let $c_i(x)$ denote the real-valued function

$$c_i(x) = \langle d\phi_i(x), \tau_i(x) \rangle$$

and note that $c_i(x^\circ) \neq 0$ because $d\phi_i(x^\circ) \neq 0$ and $\text{span}\{\tau_1, \ldots, \tau_n\}$ has dimension n at x°. If the differentials $d\phi_1(x^\circ), \ldots, d\phi_n(x^\circ)$ were linearly dependent, then for some nonzero row vector γ such that

$$\sum_{i=1}^{n} \gamma_i d\phi_i(x^\circ) = 0$$

we would have

$$0 = \sum_{i=1}^{n} \gamma_i \langle d\phi_i(x^\circ), \tau_j(x^\circ) \rangle = \gamma_j c_j(x^\circ)$$

and this would imply $c_j(x^\circ) = 0$ for some j, i.e. a contradiction.

As a consequence, the mapping $\xi = \Phi(x) = \text{col}(\phi_1(x), \ldots, \phi_n(x))$ is a local diffeomorphism at x°. By construction

$$\frac{\partial \Phi}{\partial x} [\tau_1(x) \quad \ldots \quad \tau_n(x)] = \text{diag}(c_1(x), \ldots, c_n(x)).$$

Moreover, using again (9.14), it is easy to see that $c_i(\Phi^{-1}(\xi))$ depends only on ξ_i. Thus, there exist functions $z_i = \mu_i(\xi_i)$ such that

$$\frac{\partial \mu_i}{\partial \xi_i} = \frac{1}{c_i(\Phi^{-1}(\xi))}$$

(recall that $c_i(x^\circ) \neq 0$). The composed function

$$z = T(x) = (\text{col}(\mu_1(\xi_1), \ldots, \mu_n(\xi_n)))_{\xi = \Phi(x)}$$

is such that

$$\left[\frac{\partial T}{\partial x} [\tau_1(x) \quad \ldots \quad \tau_n(x)] \right]_{x=T^{-1}(z)} = I$$

and therefore $x = T^{-1}(z)$ solves the p.d.e. (9.13). •

Merging Lemma 9.2 with Theorem 9.3 yields the desired result.

Theorem 9.4. The Observer Linearization Problem is solvable if and only if

(i) $\dim(\text{span}\{dh(x^\circ), dL_f h(x^\circ), \ldots, dL_f^{n-1} h(x^\circ)\}) = n$
(ii) the unique vector field τ solution of (9.7) is such that

$$[ad_f^i \tau, ad_f^j \tau] = 0 \tag{9.15}$$

for all $0 \leq i, j \leq n - 1$.

Remark 9.5. Using the Jacobi identity repeatedly, one can easily show that the condition (9.15) can be replaced by the condition

$$[\tau, ad_f^k \tau] = 0$$

for all $k = 1, 3, \ldots, 2n - 1$. •

In summary, one may proceed as follows in order to obtain an observer with linear (and spectrally assignable) error dynamics. If condition (i) holds, one finds first a vector field τ solving the equation (9.7). If also condition (ii) holds, one solves the p.d.e. (9.8) and finds a function F, defined in a neighborhood V^o of z^o, such that $F(z^o) = x^o$. Then one sets $\Phi = F^{-1}$. Eventually, one computes the mapping k as

$$k(z_n) = \begin{bmatrix} k_1(z_n) \\ k_2(z_n) \\ \cdots \\ k_n(z_n) \end{bmatrix} = \left[\frac{\partial \Phi}{\partial x} f(x) \right]_{x = \Phi^{-1}(z)} - \begin{bmatrix} 0 \\ z_1 \\ \cdots \\ z_{n-1} \end{bmatrix}.$$

At this point, the observer

$$\dot{\xi} = (A + GC)\xi - Gy + k(y)$$

with (A, C) in the form of the right-hand side of (9.6) yields the desired result.

4.10 Examples

We discuss in this Section two simple examples of physical control systems that can be modeled by equations of the form (1.1) and to which the design methodologies illustrated in the Chapter can be applied.

The first example is the one of a d.c. motor in which the rotor voltage is kept constant, while the stator voltage is used as a control variable (see fig. 4.10).

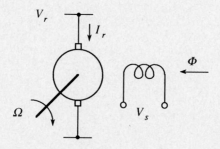

Fig. 4.10

The system in question is characterized by a set of three first order differential equations. The first one describes the electrical balance in the stator winding, and has the form

$$L_s \frac{dI_s}{dt} + R_s I_s = V_s$$

in which I_s represents the stator current, V_s the stator voltage, R_s and L_s the resistance and, respectively, inductance of the stator winding. The second equation describes the electrical balance in the rotor winding, and has the form

$$L_r \frac{dI_r}{dt} + R_r I_r = V_r - E$$

in which I_r represents the stator current, V_r the stator voltage (constant by assumption), R_r and L_r the resistance and, respectively, the inductance of the rotor winding, and E is the so-called "back e.m.f.". The third equation describes the mechanical balance of the load that, in the hypothesis of viscous friction only (namely friction torque purely proportional to the angular velocity) has the form

$$J \frac{d\Omega}{dt} + F\Omega = T$$

in which Ω denotes the angular velocity of the motor shaft, J the inertia of the load, F the viscous friction constant, and T the torque developed at the shaft. The coupling between the three equations is expressed by the relations

$$E = K_e \Phi \Omega$$
$$T = K_m \Phi I_r$$
$$\Phi = L_s I_s$$

in which Φ represents the flux associated with the stator winding, and K_e and K_m are constants. In the ideal hypothesis of 100% efficiency in the energy conversion, $K_e = K_m = K$.

Choosing, as state variables,

$$x_1 = I_s \qquad x_2 = I_r \qquad x_3 = \Omega$$

and considering the voltage V_s as the input variable, the equations in question can be rewritten in the form

$$\dot{x} = f(x) + g(x)u$$

with

$$f(x) = \begin{bmatrix} -\dfrac{R_s}{L_s} x_1 \\[2ex] -\dfrac{R_r}{L_r} x_2 + \dfrac{V_r}{L_r} - \dfrac{K L_s}{L_r} x_1 x_3 \\[2ex] \dfrac{-F}{J} x_3 + \dfrac{K L_s}{J} x_1 x_2 \end{bmatrix} \qquad g(x) = \begin{bmatrix} \dfrac{1}{L_s} \\[2ex] 0 \\[2ex] 0 \end{bmatrix} .$$

The first thing we want to check is whether or not this system is fully linearizable by state feedback and coordinates changes. To this end, we have to test the conditions (i) and (ii) of Theorem 2.6. Since

$$[f,g](x) = -\frac{\partial f}{\partial x_1}\frac{1}{L_s} = \begin{bmatrix} \dfrac{R_s}{L_s^2} \\ \dfrac{K}{L_r}x_3 \\ \dfrac{-K}{J}x_2 \end{bmatrix} \qquad [g,[f,g]](x) = \frac{\partial [f,g]}{\partial x_1}\frac{1}{L_s} = 0$$

we see that the distribution

$$D = \mathrm{span}\{g,[f,g]\}$$

has dimension 2 at each point of the dense set

$$U = \{x \in \mathbf{R}^3 : x_2 \neq 0 \text{ or } x_3 \neq 0\}$$

and is involutive on U. Thus, around any point of U the condition (ii) is fulfilled. In order to check the condition (i) we calculate also the vector field $[f,[f,g]](x)$ and we find that the condition in question is satisfied for all x° in an open and dense subset U° of U.

In order to transform the system into a linear and controllable one, it is necessary first to solve the partial differential equation

$$\frac{\partial \lambda}{\partial x}[g(x) \quad [f,g](x)] = 0$$

i.e.

$$[\frac{\partial \lambda}{\partial x_1} \quad \frac{\partial \lambda}{\partial x_2} \quad \frac{\partial \lambda}{\partial x_3}]\begin{bmatrix} \dfrac{1}{L_s} & \dfrac{R_s}{L_s^2} \\ 0 & \dfrac{K}{L_r}x_3 \\ 0 & -\dfrac{K}{J}x_2 \end{bmatrix} = [0 \quad 0].$$

An easy calculation shows that a possible solution is

$$\lambda(x) = L_r x_2^2 + J x_3^2$$

From this, the linearizing feedback and the linearizing coordinates are calculated by means of(2.14) and (2.15).

Next, we illustrate on this system the notion of zero dynamics and, to this end, an output map $h(x)$ has to be defined first. A natural output variable to look at in a motor is indeed the angular velocity of the shaft. More precisely, since the mode of control we are considering in this case (namely, holding V_r constant, and using V_s as an input) is particularly suited to the problem of controlling the velocity around a nominal *nonzero* value, we can choose as an output the quantity

$$y = h(x) = \Omega - \Omega^\circ = x_3 - \Omega^\circ$$

i.e. the deviation of the angular velocity Ω from a fixed reference value Ω°. The problem of zeroing the output, for this system, clearly corresponds to the search of all initial states and inputs which produce an angular velocity constantly equal to Ω°.

For the system thus defined we have

$$L_g h(x) = 0 \qquad L_g L_f h(x) = \frac{K}{J} x_2$$

and we see that the relative degree is $r = 2$ at each point in which $x_2 \neq 0$. Imposing zero output implies having the state evolving on the set

$$Z^\star = \{x \in \mathbf{R}^3 : h(x) = L_f h(x) = 0\}$$

that is on the manifold (see Fig. 4.11)

$$Z^\star = \{x \in \mathbf{R}^3 : x_3 = \Omega^\circ, x_1 x_2 = \frac{F\Omega^\circ}{L_s K}\}$$

and this can be accomplished, as shown in Section 4.3, by means of the input

$$u^\star(x) = \frac{-L_f^2 h(x)}{L_g L_f h(x)}.$$

The zero dynamics of a system describe its internal behavior when the input is set equal to $u^\star(x)$ and the initial conditions have been choosen on the manifold $Z^\star$. In the present example, the zero dynamics are 1-dimensional and can be easily obtained, for instance, by replacing the constraints

$$x_3 = \Omega^\circ \qquad x_1 = \frac{F\Omega^\circ}{KL_s x_2}$$

(which define the manifold $Z^\star$) in the system equations. Imposing these constraints one obtains

$$\dot{x}_2 = -\frac{R_r}{L_r} x_2 - \frac{F\Omega^{\circ 2}}{L_r x_2} + \frac{V_r}{L_r}$$

The differential equation thus found describes the projection, on the x_2-axis, of the motion of the system on the manifold $Z^\star$ on which the zero dynamics are defined. Note that $x_2 = 0$ is a singular value of the right-hand side, as expected from the shape of the manifold $Z^\star$ itself.

Suppose, for instance, $x_2 > 0$. The differential equation characterizing the zero dynamics has two equilibria, which correspond to the roots of the second order equation

$$R_r x_2^2 - V_r x_2 + F\Omega^{\circ 2} = 0.$$

These roots, on the (x_2, x_3) plane, span an ellipse (Fig. 4.11). This, in particular shows that only angular velocities satisfying the condition

$$4R_r F\Omega^{\circ 2} \leq V_r^2$$

can be imposed, and that, if $4R_r F\Omega^{\circ 2} < V_r^2$, the same fixed angular velocity Ω° can be obtained from two different steady-state values of the rotor current. Accordingly, on $Z^\star$ we find two equilibria x^a and x^b for the zero dynamics, with

$$x_2^a = \frac{V_r - \sqrt{V_r^2 - 4FR_r\Omega^{\circ 2}}}{2R_r} \qquad x_2^b = \frac{V_r + \sqrt{V_r^2 - 4FR_r\Omega^{\circ 2}}}{2R_r}.$$

The sign of the right-hand side of the differential equation defining the zero dynamics is negative for $0 < x_2 < x_2^a$, positive for $x_2^a < x_2 < x_2^b$ and again negative for $x_2^b < x_2 < \infty$. Thus it is easy to conclude that the point x^b is an *asymptotically stable* equilibrium for the zero dynamics, whereas the point x^a is an *unstable* equilibrium for the zero dynamics.

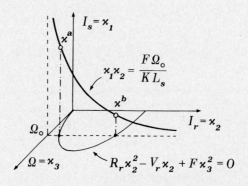

Fig. 4.11

The second example we want to consider is the one of a simple one-link robot arm, whose rotary motion about one end is controlled by means of an elastically coupled actuator. Elastic coupling between actuators and links is a phenomenon that cannot be neglected in many practical situations, and the experience has shown that robot arms in which the motion is transmitted by means of long shafts or transmission belts, or in which the actuator is an armonic drive, show a resonant behavior in the same range of frequencies used for control.

The effect of elastic coupling between actuators and links, that is commonly referred to as *joint elasticity*, can be modeled by inserting a linear torsional spring at each joint, between the shaft of the actuator and the end about which the link is rotating. In the case of a simple one-link arm, the model thus obtained is like the one illustrated in Fig 4.12. The system in question is described by means of two second order differential equations, one characterizing the mechanical balance of the actuator shaft, and the other one characterizing the mechanical balance of the link. Using q_1 and q_2 to denote the angular positions, with respect to a fixed reference frame, of the actuator shaft and—respectively—of the link, the actuator equation can be written in the form

$$J_1\ddot{q}_1 + F_1\dot{q}_1 + \frac{K}{N}(q_2 - \frac{q_1}{N}) = T$$

in which J_1 and F_1 represent inertia and viscous friction constants, K the elasticity constant of the spring which represents the elastic coupling with the joint, and N the transmission gear ratio. T is the torque produced at the actuator axis. On the other hand, the link equation can be written in the similar form

$$J_2\ddot{q}_2 + F_2\dot{q}_2 + K(q_2 - \frac{q_1}{N}) + mgd\cos q_2 = 0$$

in which m and d represent the mass and, respectively, the position of the center of gravity of the link.

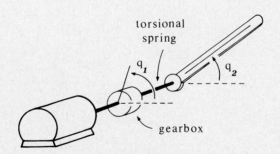

Fig. 4.12

Choosing the state vector

$$x = \text{col}(q_1, q_2, \dot{q}_1, \dot{q}_2)$$

the system can be represented in the form (1.1a), with input $u = T$, and

$$f(x) = \begin{bmatrix} x_3 \\ x_4 \\ \dfrac{-K}{J_1 N^2} x_1 + \dfrac{K}{J_1 N} x_2 - \dfrac{F_1}{J_1} x_3 \\ \dfrac{K}{J_2 N} x_1 - \dfrac{K}{J_2} x_2 - \dfrac{mgd}{J_2} \cos x_2 - \dfrac{F_2}{J_2} x_4 \end{bmatrix} \qquad g(x) = \begin{bmatrix} 0 \\ 0 \\ 1 \\ \dfrac{1}{J_1} \\ 0 \end{bmatrix}$$

As output of this system, it is natural to choose the angular position q_2 of the link with respect to the fixed reference frame, i.e.

$$y = h(x) = x_2.$$

An easy calculation shows that

$$L_f h(x) = f_2(x) = x_4$$
$$L_f^2 h(x) = f_4(x)$$
$$L_f^3 h(x) = \frac{\partial f_4}{\partial x_1} x_3 + \frac{\partial f_4}{\partial x_2} x_4 + \frac{\partial f_4}{\partial x_4} f_4(x)$$

and, therefore, since $f_4(x)$ does not depend on x_3,

$$L_g h(x) = L_g L_f h(x) = L_g L_f^2 h(x) = 0$$
$$L_g L_f^3 h(x) = \frac{\partial L_f^3 h}{\partial x_3} \frac{1}{J_1} = \frac{\partial f_4}{\partial x_1} \frac{1}{J_1} = \frac{K}{J_1 J_2 N}.$$

The system in question has relative degree $r = 4 = n$ at each point x° of the state space. Thus, on the basis of the results established at the beginning of Section 4.2, we conclude that this system can be exactly linearized via state feedback and coordinates transformation around any point x° of the state space. The linearizing feedback has the expression

$$u = \frac{-L_f^4 h(x) + v}{L_g L_f^3 h(x)}$$

and the linearizing coordinates are

$$z_1 = h(x), \qquad z_2 = L_f h(x)$$
$$z_3 = L_f^2 h(x), \qquad z_4 = L_f^3 h(x).$$

Note that, since by definition of relative degree,

$$h(x) = y, \qquad L_f h(x) = \frac{dy}{dt}$$
$$L_f^2 h(x) = \frac{d^2 y}{dt^2}, \qquad L_f^3 h(x) = \frac{d^3 y}{dt^3}$$

it is possible to identify the linearizing coordinates with the output and its first three derivatives with respect to time: these variables are in fact the angular position, velocity, acceleration and jerk of the link with respect to the fixed reference frame.

It may be of interest to linearize the system around a state x° having $x_2^\circ = 0$ (which corresponds to an horizontal position of the link). However, it is immediately seen that a state of this type cannot be an equilibrium for the vector field $f(x)$, because the constraints $f(x^\circ) = 0$ and $x_2^\circ = 0$ are not compatible, and therefore the corresponding linearized system will not be necessarily defined in a neighborhood of the point $z = 0$ (see Remark 2.10). In this case, one may try to render a point of this type an equilibrium by means of feedback, as described in Remark 2.10. The condition for this to be possible is that, at x°, the vector $f(x^\circ)$ is in the span of $g(x^\circ)$ or, in other words, that

$$f(x^\circ) + g(x^\circ)c = 0$$

for some real number c. In the present situation, this condition is satisfied for a state x° having $x_2^\circ = 0$, because it reduces to

$$x_3^\circ = 0$$
$$x_4^\circ = 0$$
$$\frac{-K}{N^2} x_1^\circ + c = 0$$
$$\frac{K}{N} x_1^\circ - mgd = 0$$

that can indeed be (uniquely) solved for c and x_1°. Thus, if instead of the former linearizing feedback one considers

$$u = \frac{-L_f^4 h(x) + v}{L_g L_f^3 h(x)} + c$$

with c satisfying the previous equation, the corresponding linearized system (in the same linearizing coordinates) will be defined around the point $z = 0$.

5 Elementary Theory of Nonlinear Feedback for Multi-Input Multi-Output Systems

5.1 Local Coordinates Transformations

In this Chapter we shall see how the theory developed for single-input single output systems can be extended to nonlinear systems having many inputs and many outputs. As a matter of fact, we shall see that a straightforward extension of most of the design procedures illustrated so far is possible, provided we restrict our consideration to a particular class of multivariable nonlinear systems, that will be specified in a moment. For multivariable nonlinear systems not belonging to this class, the solution of some important design problems like exact linearization, asymptotic stabilization, disturbance decoupling and noninteracting control requires more sophisticated techniques and will be treated separately in the next two Chapters.

In order to avoid unnecessary complications, we shall restrict our analysis to the consideration of systems having the same number m of input and output channels. Occasionally, we shall specify how the results should be adapted in order to include systems having a different number of inputs and outputs. The multivariable nonlinear systems we consider are described in state space form by equations of the following kind

$$\dot{x} = f(x) + \sum_{i=1}^{m} g_i(x)u_i \qquad (1.1a)$$

$$y_1 = h_1(x)$$
$$\cdots \qquad\qquad (1.1b)$$
$$y_m = h_m(x)$$

in which $f(x), g_1(x), \ldots, g_m(x)$ are smooth vector fields, and $h_1(x), \ldots, h_m(x)$ smooth functions, defined on an open set of $\mathbf{R}^n$. Whenever possible and convenient, these equations will be rewritten in the more condensed form

$$\dot{x} = f(x) + g(x)u$$
$$y = h(x)$$

having set

$$u = \mathrm{col}(u_1 \ldots u_m)$$
$$y = \mathrm{col}(y_1 \ldots y_m)$$

and where

$$g(x) = [\,g_1(x) \quad \ldots \quad g_m(x)\,]$$
$$h(x) = \mathrm{col}(h_1(x), \ldots h_m(x))$$

are respectively an $n \times m$-matrix and an m-vector.

The point of departure of the analysis is an appropriate multivariable version of the notion of relative degree, which, as a matter of fact, identifies the class of multivariable nonlinear systems we want to deal with in this Chapter. A multivariable nonlinear system of the form (1.1) has a (vector) *relative degree* $\{r_1, \ldots, r_m\}$ at a point x° if

(i)

$$L_{g_j} L_f^k h_i(x) = 0$$

for all $1 \leq j \leq m$, for all $1 \leq i \leq m$ for all $k < r_i - 1$, and for all x in a neighborhood of x°,

(ii) the $m \times m$ matrix

$$A(x) = \begin{bmatrix} L_{g_1} L_f^{r_1-1} h_1(x) & \cdots & L_{g_m} L_f^{r_1-1} h_1(x) \\ L_{g_1} L_f^{r_2-1} h_2(x) & \cdots & L_{g_m} L_f^{r_2-1} h_2(x) \\ \cdots & \cdots & \cdots \\ L_{g_1} L_f^{r_m-1} h_m(x) & \cdots & L_{g_m} L_f^{r_m-1} h_m(x) \end{bmatrix} \qquad (1.2)$$

is nonsingular at $x = x^\circ$.

Remark 1.1. It is immediately seen that this definition incorporates the one given at the beginning of the previous Chapter, for a single-input single-output nonlinear system. As far as the numbers $r_1, \ldots, r_m$ are concerned, note that each integer r_i is associated with the i-th output channel of the system. By definition, for all $k < r_i - 1$, the row vector

$$[\,L_{g_1} L_f^k h_i(x) \quad L_{g_2} L_f^k h_i(x) \quad \cdots \quad L_{g_m} L_f^k h_i(x)\,]$$

is zero for all x in a neighborhood of x° and, for $k = r_i - 1$, this row vector is nonzero (i.e. has at least a nonzero element at x°), because the matrix $A(x^\circ)$ is assumed to be nonsingular. As a consequence, in view also of the condition (i), we see that for each i there is at least one choice of j such that the (single-input single-output) system having output y_i and input u_j has exactly relative degree r_i at x° and, for any other possible choice of j (i.e. of input channel), the corresponding relative degree at x°—if any—is necessarily higher than or equal to r_i.

It is important to stress that the characterization of r_i as the integer such that

$$L_{g_j} L_f^k h_i(x) = 0 \qquad\qquad (1.3a)$$

for all $1 \leq j \leq m$, for all $1 \leq i \leq m$, for all $k < r_i - 1$, and for all x in a neighborhood of x°, and

$$L_{g_j} L_f^{r_i - 1} h_i(x^\circ) \neq 0 \qquad \text{for at least one } 1 \leq j \leq m \qquad (1.3b)$$

is only implied by, but not equivalent to, (i) and (ii). As a matter of fact, (ii) also incorporates the assumption that the matrix $A(x)$ is *nonsingular*. This assumption—although quite restrictive—plays a crucial role in making possible a straightforward extension of most of the results developed for single-input single-output systems.

Note, finally, that r_i is exactly the number of times one has to differentiate the i-th output $y_i(t)$ at $t = t^\circ$ in order to have at least one component of the input vector $u(t^\circ)$ explicitly appearing (see Section 4.1). •

The nonsingularity of $A(x^\circ)$ may be interpreted as the appropriate multivariable version of the assumption that the coefficient

$$a(x^\circ) = L_g L_f^{r-1} h(x^\circ)$$

is nonzero in a single-input single-output system. As we shall see, this greatly simplifies the problem of calculating normal forms and reduces it and several related issues to an essentially trivial extension of the theory illustrated so far. The deduction of normal forms is based on a proper choice of new local coordinates, specified in the following statements, which are multivariable versions of Lemma 4.1.2 and Proposition 4.1.4.

Lemma 1.2. Suppose the system has a (vector) relative degree $\{r_1, \ldots, r_m\}$ at x°. Then, the row vectors

$$dh_1(x^\circ), dL_f h_1(x^\circ), \ldots, dL_f^{r_1 - 1} h_1(x^\circ)$$
$$dh_2(x^\circ), dL_f h_2(x^\circ), \ldots, dL_f^{r_2 - 1} h_2(x^\circ)$$
$$\cdots$$
$$dh_m(x^\circ), dL_f h_m(x^\circ), \ldots, dL_f^{r_m - 1} h_m(x^\circ)$$

are linearly independent.

Proof. It is very similar of the proof of Lemma 4.1.4. Suppose, without loss of generality, that $r_1 \geq r_i$, $2 \leq i \leq m$. Consider the matrices

$$Q = \mathrm{col}(dh_1(x), \ldots, dL_f^{r_1 - 1} h_1(x), \ldots, dh_m(x), \ldots, dL_f^{r_m - 1} h_m(x))$$

and

$$P = \mathrm{col}(g_1(x), \ldots, g_m(x), \ldots, ad_f^{r_1-1}g_1(x), \ldots, ad_f^{r_1-1}g_m(x)).$$

Using Lemma 4.1.3 and the definition of relative degree, it is easy to see that the matrix QP, after possibly a reordering of the rows, exhibits a block triangular structure in which the diagonal blocks consist of rows of the matrix (1.2). This shows the linear independence of the rows of the matrix QP, i.e. that of the rows of the matrix Q and this completes the proof. •

Proposition 1.3. Suppose a system has a (vector) relative degree $\{r_1, \ldots, r_m\}$ at x°. Then

$$r_1 + \ldots + r_m \leq n.$$

Set, for $1 \leq i \leq m$

$$\phi_1^i(x) = h_i(x)$$
$$\phi_2^i(x) = L_f h_i(x)$$
$$\ldots$$
$$\phi_{r_i}^i(x) = L_f^{r_i-1} h_i(x).$$

If $r = r_1 + \ldots + r_m$ is strictly less than n, it is always possible to find $n - r$ more functions $\phi_{r+1}(x), \ldots, \phi_n(x)$ such that the mapping

$$\Phi(x) = \mathrm{col}[\phi_1^1(x), \ldots, \phi_{r_1}^1(x), \ldots, \phi_1^m(x), \ldots, \phi_{r_m}^m(x), \phi_{r+1}(x), \ldots, \phi_n(x)]$$

has a jacobian matrix which is nonsingular at x° and therefore qualifies as a local coordinates transformation in a neighborhood of x°. The value at x° of these additional functions can be chosen arbitrarily. Moreover, if the distribution

$$G = \mathrm{span}\{g_1, \ldots, g_m\}$$

is involutive near x°, it is always possible to choose $\phi_{r+1}(x), \ldots, \phi_n(x)$ in such a way that

$$L_{g_j} \phi_i(x) = 0 \tag{1.4}$$

for all $r + 1 \leq i \leq n$, for all $1 \leq j \leq m$, and all x around x°.

Proof. We only have to prove the second part of the statement, namely the possibility of choosing the remaining $n - r$ new coordinates in such a way that

(1.4) holds. Since the matrix $A(x^o)$ can be written as

$$
A(x^o) = \begin{bmatrix} dL_f^{r_1-1}h_1(x^o) \\ dL_f^{r_2-1}h_2(x^o) \\ \cdots \\ dL_f^{r_m-1}h_m(x^o) \end{bmatrix} [g_1(x^o) \quad g_2(x^o) \quad \cdots \quad g_m(x^o)]
$$

from the nonsingularity of this matrix we deduce that the m vectors $g_1(x^o), \ldots, g_m(x^o)$ are linearly independent. Thus, the distribution G is m-dimensional near x^o. Since the distribution is also involutive by assumption, by Frobenius' Theorem we deduce the existence of $n - m$ real-valued functions $\lambda_1(x), \ldots, \lambda_{n-m}(x)$, defined in a neighborhood of x^o, such that

$$
\text{span}\{d\lambda_1, \ldots, d\lambda_{n-m}\} = G^\perp.
$$

Consider now the codistribution

$$
\Omega = \text{span}\{dL_f^k h_i : 0 \le k \le r_i - 1, 1 \le i \le m\}
$$

which has dimension r, and note that

$$
G(x^o) \cap \Omega^\perp(x^o) = 0. \tag{1.5}
$$

For, if this were not true, there would exist a nonzero vector in $G(x^o)$, i.e. a vector of the form

$$
g = \sum_{i=1}^m c_i g_i(x^o)
$$

that would annihilate all vectors of $\Omega(x^o)$, but this is a contradiction, because

$$
\begin{bmatrix} dL_f^{r_1-1}h_1(x^o) \\ dL_f^{r_2-1}h_2(x^o) \\ \cdots \\ dL_f^{r_m-1}h_m(x^o) \end{bmatrix} g = A(x^o) \begin{bmatrix} c_1 \\ c_2 \\ \cdots \\ c_m \end{bmatrix} = 0
$$

implies $c_1 = c_2 = \ldots = c_m = 0$, by the nonsingularity of $A(x^o)$.

Since (1.5) implies

$$
\dim(G^\perp(x^o) + \Omega(x^o)) = n
$$

the proof can be continued exactly as in the proof of the corresponding Proposition 4.1.4. •

Remark 1.4. Note that this result incorporates that of Proposition 4.1.4. An important point to be stressed is that the choice of the additional functions $\phi_i(x)$'s satisfying the condition (1.4) is possible if and only if the distribution G spanned by the vector fields $g_1(x), \ldots, g_m(x)$ is *involutive*. Such a condition is always satisfied when the system has only one input, because this set consists of only one vector, and this is why a similar assumption was not mentioned earlier. •

Remark 1.5. The reader will have no difficulties in proving that most of the results established so far can be extended to a system having a different number of inputs and outputs, provided that the condition (ii), namely the nonsingularity of the matrix $A(x)$, is replaced by the assumption that this matrix has rank equal to the number of its rows (i.e. to the number of output channels). Note that this implies dealing with a system having a number of inputs larger than or equal to the number of outputs. As a matter of fact, under this assumption Lemma 1.2 is still true, and from this one deduces that the same type of coordinates transformation introduced in the Proposition 1.3 can be considered. •

The calculations leading to the description of the system in the new coordinates are exactly the same done earlier for single-input single-output nonlinear systems. As a matter of fact, differentiating with respect to time, one obtains, e.g., for the first set of new coordinates

$$\frac{d\phi_1^1}{dt} = \phi_2^1(t)$$

$$\ldots$$

$$\frac{d\phi_{r_1-1}^1}{dt} = \phi_{r_1}^1(t)$$

$$\frac{d\phi_{r_1}^1}{dt} = L_f^{r_1} h_1(x(t)) + \sum_{j=1}^{m} L_{g_j} L_f^{r_1-1} h_1(x(t)) u_j(x).$$

Note that the coefficient that multiply $u_j(t)$ in the latter equation is exactly equal to the $(1, j)$ entry of the matrix $A(x)$.

Consistently with the notations already introduced in the previous Chapter, set now

$$\xi^i = \begin{bmatrix} \xi_1^i \\ \xi_2^i \\ \ldots \\ \xi_{r_i}^i \end{bmatrix} = \begin{bmatrix} \phi_1^i(x) \\ \phi_2^i(x) \\ \ldots \\ \phi_{r_i}^i(x) \end{bmatrix}$$

for $1 \leq i \leq m$,

$$\xi = (\xi^1, \ldots, \xi^m)$$

$$\eta = \begin{bmatrix} \eta_1 \\ \eta_2 \\ \ldots \\ \eta_{n-r} \end{bmatrix} = \begin{bmatrix} \phi_{r+1}(x) \\ \phi_{r+2}(x) \\ \ldots \\ \phi_n(x) \end{bmatrix}$$

and

$$a_{ij}(\xi, \eta) = L_{g_j} L_f^{r_i - 1} h_i(\Phi^{-1}(\xi, \eta)) \qquad \text{for } 1 \le i, j \le m$$

$$b_i(\xi, \eta) = L_f^{r_i} h_i(\Phi^{-1}(\xi, \eta)) \qquad \text{for } 1 \le i \le m$$

Then, the equations in question can be rewritten as

$$\dot{\xi}_1^i = \xi_2^i$$

$$\ldots$$

$$\dot{\xi}_{r_i-1}^i = \xi_{r_i}^i$$

$$\dot{\xi}_{r_i}^i = b_i(\xi, \eta) + \sum_{j=1}^m a_{ij}(\xi, \eta) u_j \tag{1.6}$$

$$y_i = \xi_1^i$$

for $1 \le i \le m$. As far as the remaining set of new coordinates is concerned, we cannot expect any special form for the corresponding equations. If the distribution spanned by the vector fields $g_1(x), \ldots, g_m(x)$ is not involutive (which is likely to be the most general case), we can only write generically, with a vector notation

$$\dot{\eta} = q(\xi, \eta) + \sum_{i=1}^m p_i(\xi, \eta) u_i = q(\xi, \eta) + p(\xi, \eta) u. \tag{1.7}$$

Otherwise, if the distribution in question is involutive, it is always possible to choose the remaining set of coordinates $\phi_{r+1}, \ldots, \phi_n$ in such a way as to obtain an equation of the type

$$\dot{\eta} = q(\xi, \eta).$$

However, as we observed earlier, this is not always very easy because it involves, in general, solving partial differential equations for $\phi_{r+1}, \ldots, \phi_n$.

The equations (1.6) and (1.7) characterize the *normal form* of the equations describing (locally around a point x°) a nonlinear system, with m inputs and m outputs, having a (vector) relative degree $\{r_1, \ldots, r_m\}$ at x°. Observe, in particular, that if x° is an equilibrium point of $f(x)$, if $h_1(x^\circ) = \ldots = h_m(x^\circ) = 0$, and if $\phi_{r+1}(x^\circ) = \ldots = \phi_n(x^\circ) = 0$, the normal form thus found is defined in a neighborhood of the point $(\xi, \eta) = (0, 0)$. Note also that—in the equation

(1.6)—the coefficients $a_{ij}(\xi, \eta)$ are exactly the entries of the matrix (1.2), with x replaced by $\Phi^{-1}(\xi, \eta)$, and the coefficients $b_i(\xi, \eta)$, are the entries of a vector

$$b(x) = \begin{bmatrix} L_f^{r_1} h_1(x) \\ L_f^{r_2} h_2(x) \\ \ldots \\ L_f^{r_m} h_m(x) \end{bmatrix} \tag{1.8}$$

again with x replaced by $\Phi^{-1}(\xi, \eta)$.

We conclude the Section by discussing the interpretation of the equations (1.7), thus illustrating the multivariable version of the analysis developed in Section 4.3. This will provide an appropriate extension of the notion of *zero dynamics* for a system having relative degree $\{r_1, \ldots, r_m\}$. The idea is always that of solving first the *Problem of Zeroing the Output*, i.e. to find initial conditions and inputs consistent with the constraint that the output function $y(t)$ is identically zero for all times in a neighborhood of $t = 0$, and then to analyze the corresponding internal dynamics. Calculations similar to those carried out at the beginning of Section 4.3 show that, if $y(t) = 0$ for all t, then

$$h_1(x(t)) = L_f h_1(x(t)) = \ldots = L_f^{r_1 - 1} h_1(x(t)) = 0$$
$$h_2(x(t)) = L_f h_2(x(t)) = \ldots = L_f^{r_2 - 1} h_2(x(t)) = 0$$
$$\ldots$$

i.e. $\xi(t) = 0$ for all t near 0.

Imposing the derivative of order r_i of $y_i(t)$ being zero, for all $1 \leq i \leq m$, constrains the inputs $u_1(t), \ldots, u_m(t)$ to be solutions of the system of equations

$$0 = y_i^{(r_i)}(t) = b_i(0, \eta(t)) + \sum_{j=1}^{m} a_{ij}(0, \eta(t)) u_j(t) \qquad \text{for } 1 \leq i \leq m$$

which, using a vector notation, can be rewritten as

$$b(0, \eta(t)) + A(0, \eta(t)) u(t) = 0.$$

Recall now that the matrix (1.2) is nonsingular at $x = x^\circ$ by definition. Thus the matrix $A(\xi, \eta)$ is nonsingular at $(\xi, \eta) = (0, 0)$, and the above equation can be solved for $u(t)$ if $\eta(t)$ is close to 0.

From these considerations we deduce, in close analogy with the results established in Section 4.3, that if the output $y(t)$ has to be 0 for all t, then necessarily the initial state of the system must be set to a value such that $\xi(0) = 0$, whereas $\eta(0) = \eta^\circ$ can be chosen arbitrarily. According to the value of $\eta°$, the input must be set as

$$u(t) = -[A(0, \eta(t))]^{-1} b(0, \eta(t))$$

with $\eta(t)$ solution of a differential equation of the form

$$\dot{\eta}(t) = q_0(0, \eta(t)) \tag{1.9}$$

where $q_0(\xi, \eta)$ is defined as

$$q_0(\xi, \eta) = q(\xi, \eta) - p(\xi, \eta)[A(\xi, \eta)]^{-1} b(\xi, \eta)$$

with initial condition $\eta(0) = \eta°$. Note also that for each set of initial data $\xi = 0$ and $\eta = \eta°$ the input thus defined is the *unique* input capable to keep $y(t)$ identically zero for all times.

The dynamics (1.9) characterize the internal dynamics consistent with the constraint $y(t) = 0$, and are called the *zero dynamics* of the system.

Moving from these calculations to a coordinate-free setting, the reader will have no difficulties in realizing that, in order to yield $y(t) = 0$ for all times, the system must evolve on the subset

$$Z^\star = \{ x \in \mathbf{R}^n : L_f^k h_i(x) = 0, 0 \leq k \leq r_i - 1, 1 \leq i \leq m \}$$

under the effect of an input $u(t)$ solution of the equation

$$b(x(t)) + A(x(t))u(t) = 0.$$

Moreover, an easy calculation (similar to the corresponding one presented towards the end of Section 4.3), shows that the state feedback thus obtained, namely

$$u^\star(x) = -A^{-1}(x)b(x)$$

is such that the vector field

$$f^\star(x) = f(x) + g(x)u^\star(x)$$

is tangent to $Z^\star$. As a consequence, any trajectory of the closed loop system

$$\dot{x} = f^\star(x)$$

starting at a point of $Z^\star$ remains in $Z^\star$ (for small values of t) and the restriction $f^\star(x)|_{Z^\star}$ of $f^\star(x)$ to $Z^\star$, which is a well-defined vector field of $Z^\star$, describes—in a coordinate-free setting—the zero dynamics of the system.

The *Problem of Reproducing a Reference Output* function

$$y_R(t) = \text{col}(y_{1R}(t), y_{2R}(t), \ldots, y_{mR}(t))$$

is dealt with in a similar manner. Setting

$$\xi_R(t) = \begin{bmatrix} \xi_R^1(t) \\ \xi_R^2(t) \\ \cdots \\ \xi_R^m(x) \end{bmatrix} \qquad \text{with} \qquad \xi_R^i(t) = \begin{bmatrix} y_{iR}(t) \\ y_{iR}^{(1)}(t) \\ \cdots \\ y_{iR}^{(r_i-1)}(t) \end{bmatrix} \qquad \text{for } 1 \leq i \leq m$$

we find that the problem is solved if and only if
(i) the initial state of the system is such that $\xi(0) = \xi_R(0)$, whereas $\eta(0) = \eta^\circ$ can be chosen arbitrarily,
(ii) the input $u(t)$ is set as

$$u(t) = A^{-1}(\xi_R(t), \eta(t))(-b(\xi_R(t), \eta(t)) + \begin{bmatrix} y_{1R}^{(r_1)}(t) \\ \cdots \\ y_{mR}^{(r_m)}(t) \end{bmatrix}) \qquad (1.10a)$$

where $\eta(t)$ denotes the solution of the differential equation

$$\dot{\eta} = q(\xi_R(t), \eta) + p(\xi_R(t), \eta)A^{-1}(\xi_R(t), \eta)(-b(\xi_R(t), \eta) + \begin{bmatrix} y_{1R}^{(r_1)}(t) \\ \cdots \\ y_{mR}^{(r_m)}(t) \end{bmatrix})$$
$$(1.10b)$$

with initial condition $\eta(0) = \eta^\circ$.

For each set of initial data $\xi(0) = \xi_R(0)$ and $\eta(0) = \eta^\circ$ the input thus defined is the *unique* input capable of keeping $y(t) = y_R(t)$ for all times. The (forced) dynamics (1.10b) correspond to the dynamics describing the internal behavior of the system when input and initial conditions have been chosen in such a way as to constrain the output to track exactly $y_R(t)$. Thus, the relations (1.10) describe a system with input $y_R(t)$, output $u(t)$ and state $\eta(t)$ that can be interpreted as a realization of the *inverse* of the original system.

5.2 Exact Linearization Via Feedback

The purpose of this Section is to illustrate how a system having m inputs can be transformed into a linear and controllable system by means of feedback and change of coordinates in the state space, thus extending to multi-input systems the results already discussed in Section 4.2.

The appropriate multivariable version of the state feedback considered in the corresponding single-input single-output problem is the one in which each input u_i depends on the state x of the system and on the new reference inputs $v_1, \ldots, v_m$ as

$$u_i = \alpha_i(x) + \sum_{j=1}^{m} \beta_{ij}(x)v_j \tag{2.1}$$

where $\alpha_i(x)$ and $\beta_{ij}(x)$, for $1 \leq i, j \leq m$, are smooth functions defined on an open subset of $\mathbf{R}^n$. Note that the number of components of the new reference input

$$v = \mathrm{col}(v_1, \ldots, v_m)$$

has been chosen—for simplicity—exactly equal to the number of components of the original input u.

The composition of (2.1) with the system (1.1) yields a closed-loop system having the same structure and described by equations of the form

$$\dot{x} = f(x) + \sum_{i=1}^{m} g_i(x)\alpha_i(x) + \sum_{i=1}^{m}\left(\sum_{j=1}^{m} g_j(x)\beta_{ji}(x)\right)v_i \tag{2.2a}$$

$$y_1 = h_1(x)$$

$$\cdots \tag{2.2b}$$

$$y_m = h_m(x).$$

Using for (2.1) the more condensed expression

$$u = \alpha(x) + \beta(x)v \tag{2.3}$$

in which

$$\alpha(x) = \begin{bmatrix} \alpha_1(x) \\ \vdots \\ \alpha_m(x) \end{bmatrix} \qquad \beta(x) = \begin{bmatrix} \beta_{11}(x) & \cdots & \beta_{1m}(x) \\ \cdots & \cdots & \cdots \\ \beta_{m1}(x) & \cdots & \beta_{mm}(x) \end{bmatrix}$$

are an m-vector and, respectively, an $m \times m$ matrix, the closed loop (2.2) can be rewritten in more convenient way as

$$\dot{x} = f(x) + g(x)\alpha(x) + g(x)\beta(x)v \tag{2.4a}$$

$$y = h(x). \tag{2.4b}$$

We also systematically assume that the matrix $\beta(x)$ is *nonsingular* for all x. Accordingly, the feedback (2.2) is called a *regular static state feedback*.

As anticipated, the main problem dealt with in this Section is that of using feedback and coordinates transformation to the purpose of changing a nonlinear system into a linear and controllable one. Formally, the problem in question can be stated in the following way.

State-Space Exact Linearization Problem. Given a set of vector fields $f(x)$ and $g_1(x), \ldots, g_m(x)$ and an initial state x°, find (if possible), a neighborhood U of x°, a pair of feedback functions $\alpha(x)$ and $\beta(x)$ defined on U, a coordinates transformation $z = \Phi(x)$ also defined on U, a matrix $A \in \mathbf{R}^{n \times n}$ and a matrix $B \in \mathbf{R}^{n \times m}$, such that

$$\left[\frac{\partial \Phi}{\partial x} (f(x) + g(x)\alpha(x)) \right]_{x = \Phi^{-1}(z)} = Az \qquad (2.5)$$

$$\left[\frac{\partial \Phi}{\partial x} (g(x)\beta(x)) \right]_{x = \Phi^{-1}(z)} = B \qquad (2.6)$$

and

$$\operatorname{rank}(B \quad AB \quad \cdots \quad A^{n-1}B) = n.$$

The point of departure of our discussion will be the normal form developed and illustrated in the previous Section. Consider a nonlinear system having (vector) relative degree $\{r_1, \ldots, r_m\}$ at x° and suppose that the sum $r = r_1 + r_2 + \ldots + r_m$ is exactly equal to the dimension n of the state space. If this is the case, the set of functions

$$\phi_k^i(x) = L_f^{k-1} h_i(x) \qquad \text{for } 1 \le k \le r_i, \ 1 \le i \le m$$

defines completely a local coordinates transformation at x°. In the new coordinates the system is described by m sets of equations of the form

$$\dot{\xi}_1^i = \xi_2^i$$

$$\cdots$$

$$\dot{\xi}_{r_i-1}^i = \xi_{r_i}^i$$

$$\dot{\xi}_{r_i}^i = b_i(\xi) + \sum_{j=1}^m a_{ij}(\xi) u_j(t)$$

for $1 \le i \le m$, and no extra equations are involved.

Now, recall that in a neighborhood of the point $\xi^\circ = \Phi^{-1}(x^\circ)$ the matrix $A(\xi)$

is nonsingular and therefore the equations

$$v = \begin{bmatrix} v_1 \\ v_2 \\ \vdots \\ v_m \end{bmatrix} = b(\xi) + A(\xi)u$$

can be solved for u. As a matter of fact, the input u solving these equations has the form of a state feedback

$$u = A^{-1}(\xi)[-b(\xi) + v].$$

Imposing this feedback yields a system characterized by the m sets of equations

$$\dot{\xi}_1^i = \xi_2^i$$

$$\cdots$$

$$\dot{\xi}_{r_i-1}^i = \xi_{r_i}^i$$
$$\dot{\xi}_{r_i}^i = v_i$$

for $1 \leq i \leq m$, which is clearly linear and controllable.

From these calculations, which extend in a trivial way the ones performed at the beginning of Section 4.2, we see that the conditions that the system—for some choice of output functions $h_1(x), \ldots, h_m(x)$—has a (vector) relative degree $\{r_1, \ldots, r_m\}$ at x°, and that $r_1 + r_2 + \ldots + r_m = n$, imply the existence of a coordinates transformation and a state feedback, defined locally around x° which solve the State Space Exact Linearization Problem. Note that, in terms of the original description of the system, the *linearizing feedback* has the form

$$u = \alpha(x) + \beta(x)v$$

with $\alpha(x)$ and $\beta(x)$ given by

$$\alpha(x) = -A^{-1}(x)b(x) \qquad \beta(x) = A^{-1}(x)$$

with $A(x)$ and $b(x)$ as in (1.2) and (1.8), whereas the *linearizing coordinates* are defined as

$$\xi_k^i(x) = L_f^{k-1}h_i(x) \qquad \text{for } 1 \leq k \leq r_i, \ 1 \leq i \leq m.$$

We show now that the conditions in question are also necessary.

Lemma 2.1. Suppose the matrix $g(x^\circ)$ has rank m. Then, the State Space Exact Linearization Problem is solvable if and only if there exist a neighborhood U of x° and m real-valued functions $h_1(x), \ldots, h_m(x)$ defined on U, such that the system

$$\dot{x} = f(x) + g(x)u$$
$$y = h(x)$$

has some (vector) relative degree $\{r_1, \ldots, r_m\}$ at x° and $r_1 + r_2 + \ldots + r_m = n$.

Proof. We need only to show the necessity. We follow very closely the proof of Lemma 4.2.4. First of all, it is shown that the integers r_i, $1 \leq i \leq m$, are invariant under a regular feedback.

Recall that, for any $\alpha(x)$

$$L_{f+g\alpha}^k h_i(x) = L_f^k h_i(x) \qquad \text{for all } 0 \leq k \leq r_i - 1, \ 1 \leq i \leq m.$$

From this, one deduces that

$$L_{(g\beta)_j} L_{f+g\alpha}^k h_i(x) = L_{(g\beta)_j} L_f^k h_i(x) = \sum_{s=1}^{m} L_{g_s} L_f^k h_i(x) \beta_{sj}(x) = 0$$

for all $0 \leq k < r_i - 1$, for all $1 \leq i, j \leq m$, and all x near x°. Moreover

$$[L_{(g\beta)_1} L_{f+g\alpha}^{r_i-1} h_i(x^\circ) \quad \cdots \quad L_{(g\beta)_m} L_{f+g\alpha}^{r_i-1} h_i(x^\circ)] =$$

$$[L_{g_1} L_f^{r_i-1} h_i(x^\circ) \quad \cdots \quad L_{g_m} L_f^{r_i-1} h_i(x^\circ)] \beta(x^\circ)$$

and thus, if the matrix $\beta(x^\circ)$ is nonsingular

$$[L_{(g\beta)_1} L_{f+g\alpha}^{r_i-1} h_i(x^\circ) \quad \cdots \quad L_{(g\beta)_m} L_{f+g\alpha}^{r_i-1} h_i(x^\circ)] \neq [0 \quad \cdots \quad 0].$$

This completes the proof of the fact that the integers r_i, $1 \leq i \leq m$, are invariant under regular feedback.

We return now to the proof of the necessity. Since, by assumption, the matrix $g(x^\circ)$ has rank m, from (2.6) we deduce that any B satisfying this relation has also rank m. Therefore, without loss of generality, as in the proof of Lemma 4.2.4, we may assume that the matrices A and B considered in the statement of the Problem have the form (Brunowsky canonical form)

$$A = \text{diag}(A_1, \ldots, A_m) \qquad B = \text{diag}(b_1, \ldots, b_m)$$

where A_i is the $\kappa_i \times \kappa_i$ matrix

$$A_i = \begin{bmatrix} 0 & 1 & 0 & \cdots & 0 \\ 0 & 0 & 1 & \cdots & 0 \\ \vdots & \vdots & \vdots & \ddots & \vdots \\ 0 & 0 & 0 & \cdots & 1 \\ 0 & 0 & 0 & \cdots & 0 \end{bmatrix}$$

and b_i is the $\kappa_i \times 1$ vector

$$b_i = \text{col}(0, \ldots, 0, 1).$$

Now, decompose the vector $z = \Phi(x)$ in the form

$$z = \text{col}(z^1, \ldots, z^m)$$

and set

$$y_i = [1 \quad 0 \quad \ldots 0] z^i \qquad \text{for } 1 \le i \le m \tag{2.7}$$

with $\dim(z^i) = \kappa_i$, for $1 \le i \le m$. A straigthforward calculation shows that the linear system

$$\dot{z} = Az + Bv$$

with output functions defined as in (2.7) has (vector) relative degree $\{\kappa_1, \ldots, \kappa_m\}$ and $\kappa_1 + \kappa_2 + \ldots + \kappa_m = n$. Thus, since a vector relative degree is invariant under regular feedback and coordinates transformation, the proof is completed. •

Remark 2.2. Note that the condition that the matrix $g(x^\circ)$ has rank m is indeed necessary for the existence of any set of m "output" functions such that the system has some relative degree at x° because, as already observed in the proof of Proposition 1.3, this matrix is a factor of the matrix (1.2). If the matrix $g(x)$ has a rank $\rho < m$, but this rank is *constant* for all x near x°, then the State Space Exact Linearization Problem is solvable if and only if there exist ρ functions $h_1(x), \ldots, h_\rho(x)$ defined in a neighborhood U of x°, such that the system has some (vector) relative degree $\{r_1, \ldots, r_\rho\}$ at x° (see Remark 1.5) and $r_1 + r_2 + \ldots + r_\rho = n$. As a matter of fact, if the matrix $g(x)$ has constant rank $\rho < m$, it is possible to find a nonsingular matrix $\beta(x)$ such that

$$g(x)\beta(x) = [g'(x) \quad 0]$$

with $g'(x)$ consisting of ρ columns and having rank ρ. Thus, a preliminary feedback of the form

$$u = \beta(x) \begin{bmatrix} u' \\ u'' \end{bmatrix}$$

changes the original system into the system

$$\dot{x} = f(x) + g'(x)u'$$

which satisfies the assumptions of Lemma 2.1, for $m = \rho$. •

On the basis of this result, we proceed now to describe how, under suitable conditions on the vector fields $f(x), g_1(x), \ldots, g_m(x)$, it is possible to find m functions $h_1(x), h_2(x), \ldots, h_m(x)$ satisfying the requirements of the previous Lemma. Extending the solution of the corresponding problem dealt with in the previous Chapter (Lemma 4.2.5), the conditions in question will be stated in terms of properties of suitable distributions spanned by vector fields of the form

$$g_1(x), \ldots, g_m(x), ad_f g_1(x), \ldots, ad_f g_m(x), \ldots, ad_f^{n-1} g_1(x), \ldots ad_f^{n-1} g_m(x).$$

More precisely, having set

$$G_0 = \text{span}\{g_1, \ldots, g_m\}$$
$$G_1 = \text{span}\{g_1, \ldots, g_m, ad_f g_1, \ldots, ad_f g_m\}$$
$$\ldots$$
$$G_i = \text{span}\{ad_f^k g_j : 0 \le k \le i, 1 \le j \le m\}$$

for $i = 0, 1, \ldots, n - 1$, the following result will be proved.

Lemma 2.3. Suppose the matrix $g(x^\circ)$ has rank m. Then, there exist a neighborhood U of x° and m real-valued functions $\lambda_1(x), \lambda_2(x), \ldots, \lambda_m(x)$ defined on U, such that the system

$$\dot{x} = f(x) + g(x)u$$
$$y = \lambda(x)$$

has some (vector) relative degree $\{r_1, \ldots, r_m\}$ at x°, with

$$r_1 + r_2 + \ldots + r_m = n$$

if and only if:
  (i) for each $0 \le i \le n - 1$, the distribution G_i has constant dimension near x°;
  (ii) the distribution G_{n-1} has dimension n;
  (iii) for each $0 \le i \le n - 2$, the distribution G_i is involutive.

Note that, in view of this result and of the previous discussion, we can state the conditions for the solvability of the State Space Exact Linearization Problem in the following way.

Theorem 2.4. Suppose the matrix $g(x^\circ)$ has rank m. Then, the State Space Exact Linearization Problem is solvable if and only if
(i) for each $0 \leq i \leq n - 1$, the distribution G_i has constant dimension near x°;
(ii) the distribution G_{n-1} has dimension n;
(iii) for each $0 \leq i \leq n - 2$, the distribution G_i is involutive.

We provide now a proof of this result, and—in doing this—we also indicate how the functions $\lambda_1(x), \ldots, \lambda_m(x)$ can be constructed.

Proof of Lemma 2.3. The proof that the conditions in question are sufficient is conceptually similar to that of the corresponding result of Chapter 4 (Lemma 4.2.5), but—unfortunately—not as straightforward as that one. The main issue is to find solutions $\lambda_1(x), \ldots, \lambda_m(x)$ of equations of the form

$$L_{g_j} L_f^k \lambda_i(x) = 0 \qquad \text{for all } 0 \leq k \leq r_i - 2, 1 \leq j \leq m \tag{2.8}$$

and to impose, as a nontriviality condition, the nonsingularity of the matrix (1.2). In addition, one has also to make sure that $r_1 + r_2 + \ldots + r_m = n$.

The equations (2.8) are clearly equivalent (by Lemma 4.1.3) to equations of the form

$$\langle d\lambda_i(x), ad_f^k g_j(x) \rangle = 0 \qquad \text{for all } x \text{ near } x^\circ, \text{ all } 0 \leq k \leq r_i - 2, \text{ all } 1 \leq j \leq m$$

and this suggests that, for each value of i, the differential $d\lambda_i(x)$ must be a covector belonging to the codistribution

$$(\text{span}\{ad_f^k g_j : 0 \leq k \leq r_i - 2, \ 1 \leq j \leq m\})^\perp = G_{r_i-2}^\perp$$

On the basis of this observation, it is convenient to proceed in this way. Recall that the distributions $G_0, \ldots, G_{n-1}$ all have constant dimension near x° (assumption (i)) and that in particular G_{n-1} has dimension n (assumption (ii)). Thus, there exists an integer, that we shall denote by κ and which is less than or equal to n, such that

$$\dim(G_{\kappa-2}) < n$$
$$\dim(G_{\kappa-1}) = n.$$

Set

$$m_1 = n - \dim(G_{\kappa-2})$$

and note that, since $G_{\kappa-2}$ is involutive (assumption (iii)), by Frobenius' Theorem there exist m_1 functions, that we shall denote by $\lambda_i(x)$, $1 \leq i \leq m_1$, such that

$$\text{span}\{d\lambda_i : 1 \leq i \leq m_1\} = G^\perp_{\kappa-2}.$$

By construction, these functions are such that

$$\langle d\lambda_i(x), ad^k_f g_j(x)\rangle = 0 \qquad \text{for all } x \text{ around } x^\circ$$
$$0 \leq k \leq \kappa - 2, \; 1 \leq j \leq m, \; 1 \leq i \leq m_1$$

and this, by Lemma 4.1.3 implies

$$L_{g_j} L^k_f \lambda_i(x) = 0 \tag{2.9}$$

for all x around x°, $0 \leq k \leq \kappa - 2$, $1 \leq j \leq m$, $1 \leq i \leq m_1$. Moreover, we claim that the $m_1 \times m$ matrix

$$A_1(x) = \{a^1_{ij}(x)\} = \{L_{g_j} L^{\kappa-1}_f \lambda_i(x)\}$$

has rank m_1 at x°. For, suppose this is false. Then, using (2.9) and again Lemma 4.1.3, we would have that

$$\sum_{i=1}^{m_1} c_i L_{g_j} L^{\kappa-1}_f \lambda_i(x^\circ) = \sum_{i=1}^{m_1} (-1)^{\kappa-1} c_i \langle d\lambda_i(x^\circ), ad^{\kappa-1}_f g_j(x^\circ)\rangle = 0$$

$$\text{for all } 1 \leq j \leq m$$

for some set of real numbers c_i, $1 \leq i \leq m_1$. But this, together with (2.9) implies

$$\sum_{i=1}^{m_1} c_i \langle d\lambda_i(x^\circ), ad^k_f g_j(x^\circ)\rangle = 0 \qquad \text{for all } 0 \leq k \leq \kappa - 1, \; 1 \leq j \leq m.$$

This shows that

$$\sum_{i=1}^{m_1} c_i d\lambda_i(x^\circ) \in G^\perp_{\kappa-1}(x^\circ).$$

Since $G_{\kappa-1}$ has dimension n, the vector on the left-hand side must be zero, and this in turn implies all c_i's are zero, because the row vectors $d\lambda_1(x^\circ), \ldots, d\lambda_{m_1}(x^\circ)$ are by construction linearly independent.

The properties thus established, namely the equalities (2.9) and the fact that $A^1(x^\circ)$ has full rank, show that the functions $\lambda_i(x)$, $1 \leq i \leq m_1$, are good candidates to the solution of the problem. As a matter of fact, if the integer m_1 is exactly equal to m (note that $m_1 \leq m$, always, because $A^1(x^\circ)$ has m columns and rank m_1) then these functions indeed solve the problem. For, if this is the case, (2.9) imply that the matrix $A^1(x)$ is exactly equal to the matrix (1.2) with

$$r_1 = r_2 = \ldots = r_m = \kappa.$$

Thus, the system with outputs $\lambda_i(x)$, $1 \leq i \leq m$, has (vector) relative degree $\{\kappa, \kappa, \ldots, \kappa\}$. Moreover, $r_1 + r_2 + \ldots + r_m = n$, because

$$m\kappa \leq n$$

(see Proposition 1.3), and

$$n = \dim(G_{\kappa-1}) \leq m\kappa$$

by construction. This shows that the functions thus found satisfy the required conditions.

If the integer m_1 is strictly less than m, the set $\{\lambda_i(x) : 1 \leq i \leq m_1\}$ provides only a partial solution of the problem, and then one has to continue the search for an additional set of $m - m_1$ new functions. The idea is to move one step backward, look at $G_{\kappa-3}$, and try to find the new functions among those whose differentials span $G_{\kappa-3}^{\perp}$. In order to show how these new functions must be constructed, we need first to show a preliminary property. We claim that
(a) the codistribution

$$\Omega_1 = \mathrm{span}\{d\lambda_1, \ldots, d\lambda_{m_1}, dL_f\lambda_1, \ldots, dL_f\lambda_{m_1}\}$$

has dimension $2m_1$ around x°,
(b) $\Omega_1 \subset G_{\kappa-3}^{\perp}$
The proof of (b) is immediate. As a matter of fact, the differentials $d\lambda_i(x)$, $1 \leq i \leq m_1$, which are in $G_{\kappa-2}^{\perp}$ by construction, are also in $G_{\kappa-3}^{\perp}$ because $G_{\kappa-3} \subset G_{\kappa-2}$. The differentials $dL_f\lambda_i(x)$, $1 \leq i \leq m_1$, by (2.9) and Lemma 4.1.3, are such that

$$\langle dL_f\lambda_i(x), ad_f^k g_j(x)\rangle = 0$$

for all x around x°, all $0 \leq k \leq \kappa - 3$, $1 \leq j \leq m$, $1 \leq i \leq m_1$. Therefore these differentials are in $G_{\kappa-3}^{\perp}$. To prove (a), suppose is false at $x = x^\circ$ and there exist numbers c_i and d_i, $1 \leq i \leq m_1$, such that

$$\sum_{i=1}^{m_1}(c_i d\lambda_i(x^\circ) + d_i dL_f\lambda_i(x^\circ)) = 0.$$

This would imply

$$\langle\sum_{i=1}^{m_1}(c_i d\lambda_i(x^\circ) + d_i dL_f\lambda_i(x^\circ)), ad_f^{\kappa-2}g_j(x^\circ)\rangle = 0$$

for all $1 \le j \le m$. This, in turn, implies

$$\sum_{i=1}^{m_1} d_i \langle d\lambda_i(x^\circ), ad_f^{\kappa-1} g_j(x^\circ) \rangle = 0.$$

By the linear independence of the $d\lambda_i(x)$'s and the linear independence of the rows of the matrix $A^1(x)$, we conclude that all d_i's and c_i's must be zero.

From (a) and (b), we deduce that the dimension of $G_{\kappa-3}^\perp$ is larger than or equal to $2m_1$. Suppose is larger, and set

$$m_2 = \dim(G_{\kappa-3}^\perp) - 2m_1.$$

Since $G_{\kappa-3}$ is involutive (assumption (iii)), invoking Frobenius' Theorem we know that $G_{\kappa-3}^\perp$ is spanned by $2m_1 + m_2$ exact differentials. Properties (a) and (b) already identify $2m_1$ of such differentials (namely, those spanning Ω_1). Thus, we can conclude that there exist m_2 additional functions, that we shall denote by $\lambda_i(x)$, $m_1 + 1 \le i \le m_1 + m_2$, such that

$$G_{\kappa-3}^\perp = \Omega_1 + \text{span}\{d\lambda_i(x), m_1 + 1 \le i \le m_1 + m_2\}. \tag{2.10}$$

Observe that, by construction, these new functions are such that

$$L_{g_j} L_f^k \lambda_i(x) = 0 \qquad \text{for all } x \text{ near } x^\circ,$$
$$0 \le k \le \kappa - 3, \ 1 \le j \le m, \ m_1 + 1 \le i \le m_1 + m_2. \tag{2.11}$$

Moreover, we claim that
(c) the $(m_1 + m_2) \times m$ matrix

$$A^2(x) = \{a_{ij}^2(x)\}$$

with

$$\begin{cases} a_{ij}^2(x) = \langle d\lambda_i(x), ad_f^{\kappa-1} g_j(x) \rangle & \text{if } 1 \le i \le m_1 \\ a_{ij}^2(x) = \langle d\lambda_i(x), ad_f^{\kappa-2} g_j(x) \rangle & \text{if } m_1 + 1 \le i \le m_1 + m_2 \end{cases}$$

has rank equal to $m_1 + m_2$ at $x = x^\circ$.

For, suppose there exist real numbers c_i, $1 \le i \le m_1$ and d_i, $m_1 + 1 \le i \le m_1 + m_2$, such that

$$-\sum_{i=1}^{m_1} c_i \langle d\lambda_i(x^\circ), ad_f^{\kappa-1} g_j(x^\circ) \rangle + \sum_{i=m_1+1}^{m_1+m_2} d_i \langle d\lambda_i(x^\circ), ad_f^{\kappa-2} g_j(x^\circ) \rangle = 0.$$

Then, using Lemma 4.1.3, it follows that

$$\langle \sum_{i=1}^{m_1} c_i dL_f \lambda_i(x^\circ) + \sum_{i=m_1+1}^{m_1+m_2} d_i d\lambda_i(x^\circ), ad_f^{\kappa-2} g_j(x^\circ) \rangle = 0$$

i.e.

$$\langle \sum_{i=1}^{m_1} c_i dL_f \lambda_i(x^\circ) + \sum_{i=m_1+1}^{m_1+m_2} d_i d\lambda_i(x^\circ) \rangle \in (\text{span}\{ad_f^{\kappa-2} g_j(x^\circ) : 1 \le j \le m\})^\perp.$$

By construction, the vector on the left-hand side of this relation belongs also to $G_{\kappa-3}^\perp$, and therefore we have

$$\langle \sum_{i=1}^{m_1} c_i dL_f \lambda_i(x^\circ) + \sum_{i=m_1+1}^{m_1+m_2} d_i d\lambda_i(x^\circ) \rangle \in G_{\kappa-2}^\perp.$$

Since the codistribution $G_{\kappa-2}^\perp$ is spanned by $d\lambda_1, \ldots, d\lambda_{m_1}$, the previous relation shows that

$$\langle \sum_{i=1}^{m_1} c_i dL_f \lambda_i(x^\circ) + \sum_{i=m_1+1}^{m_1+m_2} d_i d\lambda_i(x^\circ) \rangle \in \text{span}\{d\lambda_i(x^\circ) : 1 \le i \le m_1\}$$

but this is in contradiction with the property expressed by (2.10), unless all c_i's and d_i's are zero.

The properties thus illustrated enable us to prove that, if $m_1 + m_2$ is equal to m (note that $m_1 + m_2 \le m$, always, because $A^2(x^\circ)$ has m columns and rank $m_1 + m_2$), the set of functions $\{\lambda_i : 1 \le i \le m\}$ is a solution of the problem. As a matter of fact, from (2.9), (2.10) and the nonsingularity of the matrix $A^2(x^\circ)$, we immediately deduce that the system has a (vector) relative degree $\{r_1, \ldots, r_m\}$, with

$$r_1 = r_2 = \ldots = r_{m_1} = \kappa$$
$$r_{m_1+1} = r_{m_1+2} = \ldots = r_m = \kappa - 1.$$

Moreover, $r_1 + r_2 + \ldots + r_m = n$, because

$$n = \dim(G_{\kappa-2}) + m_1 \le m(\kappa - 1) + m_1 = m_1\kappa + m_2(\kappa - 1) \le n.$$

If $m_1 + m_2$ is strictly less than m (and note that this includes also the case $m_2 = 0$), one has to continue, searching for an additional set of functions among those whose differentials span $G_{\kappa-4}^\perp$.

After $\kappa - 1$ iterations of this procedure, one has found $m_{\kappa-1}$ functions with the property that the differentials

$$\begin{cases} d\lambda_i(x), dL_f\lambda_i(x), \ldots, dL_f^{\kappa-2}\lambda_i(x) & \text{for } 1 \leq i \leq m_1 \\ d\lambda_i(x), dL_f\lambda_i(x), \ldots, dL_f^{\kappa-3}\lambda_i(x) & \text{for } m_1+1 \leq i \leq m_1+m_2 \\ \ldots \\ d\lambda_i(x), dL_f\lambda_i(x) & \text{for } m_1+\ldots+m_{\kappa-3}+1 \leq i \leq m_1+\ldots+m_{\kappa-2} \\ d\lambda_i(x) & \text{for } m_1+\ldots+m_{\kappa-2}+1 \leq i \leq m_1+\ldots+m_{\kappa-1} \end{cases}$$

are a basis of $G_0^\perp$. Since G_0 has dimension m by assumption, the total number of differentials in this table is equal to

$$n - m = \dim(G_0^\perp) = (\kappa - 1)m_1 + (\kappa - 2)m_2 + \ldots + m_{\kappa-1}. \qquad (2.12)$$

With arguments similar to those used to prove the property (a) above, it is possible to prove that also the $\kappa m_1 + (\kappa - 1)m_2 + \ldots + 2m_{\kappa-1}$ differentials

$$\begin{cases} d\lambda_i(x), dL_f\lambda_i(x), \ldots, dL_f^{\kappa-1}\lambda_i(x) & \text{for } 1 \leq i \leq m_1 \\ d\lambda_i(x), dL_f\lambda_i(x), \ldots, dL_f^{\kappa-2}\lambda_i(x) & \text{for } m_1+1 \leq i \leq m_1+m_2 \\ \ldots \\ d\lambda_i(x), dL_f\lambda_i(x), dL_f^2\lambda_i(x) & \text{for } m_1+\ldots+m_{\kappa-3}+1 \leq i \leq m_1+\ldots+m_{\kappa-2} \\ d\lambda_i(x), dL_f\lambda_i(x) & \text{for } m_1+\ldots+m_{\kappa-2}+1 \leq i \leq m_1+\ldots+m_{\kappa-1} \end{cases}$$

are independent in a neighborhood of x°. Thus, we may deduce that $n - (\kappa m_1 + (\kappa - 1)m_2 + \ldots + 2m_{\kappa-1}) \geq 0$. If this inequality is strict, set

$$m_\kappa = n - (\kappa m_1 + (\kappa - 1)m_2 + \ldots + 2m_{\kappa-1})$$

and note that, by (2.12),

$$m_1 + \ldots + m_\kappa = m.$$

Clearly, there exist m_κ functions $\lambda_i(x)$, $m_1 + \ldots + m_{\kappa-1} + 1 \leq i \leq m$, such that the differentials

$$d\lambda_i(x) \qquad \text{for } m_1 + \ldots + m_{\kappa-1} + 1 \leq i \leq m$$

together with those of the previous table form a set of exactly n independent differentials (in a neighborhood of x°). With arguments similar to those used to prove property (c) above, it is possible to prove that the system, with outputs

$\lambda_i(x)$, $1 \leq i \leq m$, has relative degree $\{r_1, \ldots, r_m\}$ at x^0, with

$$\begin{cases} r_i = \kappa & \text{for } 1 \leq i \leq m_1 \\ r_i = \kappa - 1 & \text{for } m_1 + 1 \leq i \leq m_1 + m_2 \\ \ldots \\ r_i = 2 & \text{for } m_1 + \ldots + m_{\kappa-2} + 1 \leq i \leq m_1 + \ldots + m_{\kappa-1} \\ r_i = 1 & \text{for } m_1 + \ldots + m_{\kappa-1} + 1 \leq i \leq m. \end{cases} \qquad (2.13)$$

Moreover, $r_1 + r_2 + \ldots + r_m = n$, and thus the proof of the sufficiency is complete.

The proof of the necessity is quite straightforward and is left, as an exercise, to the reader. •

Remark 2.5. It may be interesting to observe that the conditions stated in Theorem 2.4, in the case of a single-input system, reduce exactly to those described in Theorem 4.2.6. For, if this is the case, i.e. if $m = 1$, the distribution G_i reduces to

$$G_i = \text{span}\{g, ad_f g, \ldots, ad_f^i g\}.$$

The condition (ii) above, i.e. $\dim(G_{n-1}) = n$, implies that $\dim(G_i) = i + 1$. i.e. the condition (i). This being the case, the involutivity of G_{n-2} implies (see Remark 4.2.9) also that of $G_0, \ldots, G_{n-3}$. •

Remark 2.6. Note that if $m_2 = 0$, no useful function can be found at the second iteration of the procedure and one has to proceed directly with the third iteration. If this is the case, then it is clear that the condition "$G_{\kappa-3}$ is involutive" (which is part of the conditions (iii)) is superfluous because, as shown in the proof, it is in fact implied by the involutivity of $G_{\kappa-2}$. The same consideration is of course true for any $G_{\kappa-i}$ such that $m_{i-1} = 0$.

Thus, strictly speaking, the requirement (iii) is in some sense redundand, because the involutivity of some distributions of the sequence $G_0, \ldots, G_{n-2}$ might imply that of the others. On the other hand, the way the condition was presented is much simpler, in that does not require identifying what distributions must necessarily be involutive in order to let the procedure go. •

Remark 2.7. The arguments illustrated in the proof enable us to identify the numbers $r_1, \ldots, r_m$ directly in terms of the dimensions of the distributions $G_0, \ldots, G_{n-2}$ (well-defined by assumption). For, it suffices to use (2.13) and to keep in mind that

$$m_{i-1} = n - \dim(G_{\kappa-i}) - (i-1)m_1 - (i-2)m_2 - \ldots - 2m_{i-2}.$$

Remark 2.8. Note that if the system were linear, conditions (i) and (iii) of Lemma 2.3 would be automatically satisfied and condition (ii) would reduce to the condition that the system is controllable. In this case the previous construction will end up with a set of *linear* functions $\lambda_i(x)$, $1 \leq i \leq m$. Using these functions in the expressions of the linearizing feedback and of the linearizing coordinates would produce a linear feedback and a linear coordinates change that brings the system to its Brunowsky canonical form. •

We illustrate now in a simple example how a system satisfying the conditions of Theorem 2.4 can be transformed into a linear and controllable system via feedback and coordinates changes.

Example 2.1. Consider the system

$$\dot{x} = \begin{bmatrix} x_2 + x_2^2 \\ x_3 - x_1 x_4 + x_4 x_5 \\ x_2 x_4 + x_1 x_5 - x_5^2 \\ x_5 \\ x_2^2 \end{bmatrix} + \begin{bmatrix} 0 \\ 0 \\ \cos(x_1 - x_5) \\ 0 \\ 0 \end{bmatrix} u_1 + \begin{bmatrix} 1 \\ 0 \\ 1 \\ 0 \\ 1 \end{bmatrix} u_2.$$

In this system the distribution $G_0 = \text{span}\{g_1, g_2\}$ has dimension $2 = m$ in a neighborhood of $x^\circ = 0$. Moreover, since

$$[g_1, g_2](x) = 0$$

using Remark 1.3.7 we see that the distribution in question is also involutive. Consider now

$$G_1 = \text{span}\{g_1, g_2, ad_f g_1, ad_f g_2\}$$

where

$$ad_f g_1(x) = \begin{bmatrix} 0 \\ -\cos(x_1 - x_5) \\ -x_2 \sin(x_1 - x_5) \\ 0 \\ 0 \end{bmatrix} \quad ad_f g_2(x) = \begin{bmatrix} 0 \\ -1 \\ -(x_1 - x_5) \\ -1 \\ 0 \end{bmatrix}.$$

This distribution has maximal dimension 4 at $x^\circ = 0$. Therefore its dimension is constant near x°. Moreover, since

$$[g_1, ad_f g_1](x) = [g_1, ad_f g_2](x) = [g_2, ad_f g_1](x) = [g_2, ad_f g_2](x) = 0$$

and

$$[ad_f g_1, ad_f g_2](x) = \sin(x_1 - x_5)g_1(x)$$

this distribution is also involutive.

Finally, similar calculations show that the distribution

$$G_2 = \mathrm{span}\{g_1, g_2, ad_f g_1, ad_f g_2, ad_f^2 g_1, ad_f^2 g_2\}$$

has maximal dimension 5 at $x^\circ = 0$, and therefore at each x in a neighborhood of $x^\circ = 0$.

Since by definition $G_{i-1} \subset G_i$ for any $i \geq 1$, and G_2 has a dimension which is equal to the dimension n of the state space, we see that $G_2 = G_3 = G_4$ and G_2, G_3 are (trivially) involutive. The system satisfies the hypotheses of Theorem 2.4.

In order to solve the State Space Exact Linearization Problem, we have to construct two functions $\lambda_1(x)$ and $\lambda_2(x)$ according to the procedure indicated in the proof of Lemma 2.3. Since in this case $\kappa = 3$, one has to consider first the codistribution $G_1^\perp$. This codistribution has dimension 1. Therefore, there exists a real-valued function $\lambda_1(x)$ such that

$$\mathrm{span}\{d\lambda_1\} = G_1^\perp.$$

As a matter of fact, it is not difficult to check that the function

$$\lambda_1(x) = x_1 - x_5$$

does the job. Then, we know from the proof of Lemma 2.3 that the function $L_f \lambda_1(x)$ has a differential which is linearly independent from that of $\lambda_1(x)$ and that

$$\mathrm{span}\{d\lambda_1(x), dL_f \lambda_1(x)\} \subset G_0^\perp.$$

The left-hand side of this relation has dimension 2, whereas the right-hand side has dimension 3. Therefore, there exists another real-valued function $\lambda_2(x)$ such that

$$\mathrm{span}\{d\lambda_1(x), dL_f \lambda_1(x), d\lambda_2(x)\} = G_0^\perp.$$

Since
$$d\lambda_1(x) = [1 \quad 0 \quad 0 \quad 0 \quad -1]$$
$$dL_f \lambda_1(x) = dx_2 = [0 \quad 1 \quad 0 \quad 0 \quad 0]$$

a function $\lambda_2(x)$ whose differential is linearly independent of $d\lambda_1(x)$ and $dL_f \lambda_1(x)$ and is annihilated by the vectors of G_0 is indeed the function

$$\lambda_2(x) = x_4.$$

At this point, the procedure illustrated in the proof of Lemma 2.3 is terminated. By construction, the two functions $\lambda_1(x)$ and $\lambda_2(x)$ are such that

$$L_{g_1}\lambda_1(x) = L_{g_2}\lambda_1(x) = L_{g_1}L_f\lambda_1(x) = L_{g_2}L_f\lambda_1(x) = 0$$

$$L_{g_1}\lambda_2(x) = L_{g_2}\lambda_2(x) = 0$$

and, moreover, the matrix

$$\begin{bmatrix} L_{g_1}L_f^2\lambda_1(x) & L_{g_2}L_f^2\lambda_1(x) \\ L_{g_1}L_f\lambda_2(x) & L_{g_1}L_f\lambda_2(x) \end{bmatrix}$$

is nonsingular at $x = 0$. Thus, the system in question, with dummy outputs $y_1 = \lambda_1(x)$ and $y_2 = \lambda_2(x)$ will have relative degree $\{r_1, r_2\} = \{3, 2\}$, with $r_1 + r_2 = 5 = n$. $\bullet$

5.3 Noninteracting Control

In a multivariable system, in addition to standard synthesis problems like exact linearization (already examined in the previous Section), asymptotic stabilization, disturbance decoupling, output tracking (that will be briefly discussed in Section 5.5), one may wish to use feedback in order to reduce the system, at least from an input-output point of view, to an aggregate of independent single-input single-output channels. This problem, known as the problem of *noninteracting control*, will be discussed in the present Section. For convenience, we start from a formal definition.

Noninteracting Control Problem. Given a nonlinear system of the form

$$\dot{x} = f(x) + \sum_{i=1}^{m} g_i(x)u_i$$

$$y_1 = h_1(x)$$

$$\dots$$

$$y_m = h_m(x)$$

and an initial point x°, find a regular static state feedback control law

$$u_i = \alpha_i(x) + \sum_{j=1}^{m} \beta_{ij}(x)v_j$$

defined in a neighborhood U of x°, with the property that in the corresponding closed loop system

$$\dot{x} = f(x) + \sum_{i=1}^{m} g_i(x)\alpha_i(x) + \sum_{j=1}^{m} \left(\sum_{i=1}^{m} g_i(x)\beta_{ij}(x)\right) v_j$$

$$y_1 = h_1(x)$$

$$\dots$$

$$y_m = h_m(x)$$

each output channel y_i, $1 \leq i \leq m$, is affected only by the corresponding input channel v_i and not by v_j, if $j \neq i$.

Remark 3.1. The property that the output channel y_i, is affected only by the corresponding input channel v_i can be formally expressed in the following terms: for any initial state x° and for any pair of input functions $v^a(t)$ and $v^b(t)$, defined for $t \geq 0$, which are equal in the i-th component but possibly different in the other ones, i.e. satisfy the condition

$$(v^a(t))_i = (v^b(t))_i \qquad \text{for all } t \geq 0$$

the corresponding outputs $y^a(t)$ and $y^b(t)$ are also equal in the i-th component, i.e. satisfy

$$(y^a(t))_i = (y^b(t))_i \qquad \text{for all } t \geq 0$$

Note also that the nonsingularity of $\beta(x)$ takes care of avoiding trivial solutions, namely solutions in which—in the closed loop—some output is not affected at all by any input. As we shall see in a moment, if the output y_i is affected by the input vector u of the open loop then y_i is also affected by the input vector v of the closed loop, whenever a regular feedback is implemented. •

The main result about the Noninteracting Control Problem is that this problem is solvable if and only if the system has some vector relative degree, i.e. belongs to the special class of multivariable systems considered throughout this Chapter. The sufficiency is discussed first.

Suppose that the system has been given the normal form illustrated in Section 5.1 and suppose the following feedback law is imposed

$$\begin{bmatrix} u_1 \\ u_2 \\ \vdots \\ u_m \end{bmatrix} = -A^{-1}(\xi, \eta)b(\xi, \eta) + A^{-1}(\xi, \eta) \begin{bmatrix} v_1 \\ v_2 \\ \vdots \\ v_m \end{bmatrix}. \tag{3.1}$$

An immediate calculation shows that the imposition of this feedback yields a system characterized by the m sets of equations

$$\dot{\xi}_1^i = \xi_2^i$$

$$\cdots$$

$$\dot{\xi}_{r_i-1}^i = \xi_{r_i}^i$$

$$\dot{\xi}_{r_i}^i = v_i$$

$$y_i = \xi_1^i$$

for $1 \leq i \leq m$, together with an additional set of the form

$$\dot{\eta} = q(\xi, \eta) - p(\xi, \eta) A^{-1}(\xi, \eta) b(\xi, \eta) + P(\xi, \eta) A^{-1}(\xi, \eta) v.$$

The structure of these equations (which correspond to the block diagram of Fig. 5.1), shows that the noninteraction requirement has been achieved. As a matter of fact, the input v_1 controls only the output y_1, throughout a chain of r_1 integrators, the input v_2 controls only the output y_2, throughout a chain of r_2 integrators, etc. If $r = r_1 + r_2 + \ldots + r_m$ is not equal to n, in the closed loop system an unobservable part is present, which behaves like a "sink", namely is affected by all inputs and all the states, but has no effect on the outputs. If, on the other hand, $r = n$, no "sink" is present and the closed loop system consists— as already shown in the previous Section—only of m chains of r_i integrators each. We observe also that in either cases the input-output behavior of the closed-loop thus obtained is that of a *linear* system, characterized by a transfer function matrix of the form

$$H(s) = \begin{bmatrix} \frac{1}{s^{r_1}} & 0 & \cdots & 0 \\ 0 & \frac{1}{s^{r_2}} & \cdots & 0 \\ \vdots & \vdots & \ddots & \vdots \\ 0 & 0 & \cdots & \frac{1}{s^{r_m}} \end{bmatrix}.$$

Although the use of the normal form is very helpful in understanding how the noninteracting control problem can be solved, it is clear that the achievement of an input-output noninteractive behavior is independent of the coordinates used in the state space description. Thus, we deduce that a feedback of the form

$$u = \alpha(x) + \beta(x)v \qquad (3.2a)$$

with $\alpha(x)$ and $\beta(x)$ given by

$$\alpha(x) = -A^{-1}(x)b(x) \qquad \beta(x) = A^{-1}(x) \qquad (3.2b)$$

with $A(x)$ and $b(x)$ as in (1.2) and (1.8) (which is the expression of (3.1) in the original state space coordinates) solves the Noninteracting Control Problem. We shall refer to this as to the *Standard Noninteractive Feedback*.

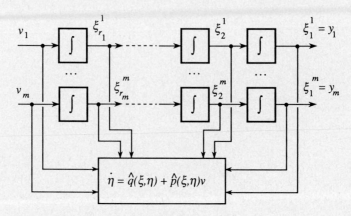

Fig. 5.1

It is clear from the previous discussion that for any system in which the matrix $A(x)$ is *nonsingular*, at a point $x = x^\circ$, the noninteracting control problem can be solved by means of a static state feedback, which is defined for all x in a neighborhood of the point x°. In general, the solution thus described is defined only *locally* in the state space, namely for all x near x° at which the matrix $A(x)$ is nonsingular. It is also possible to show that the nonsingularity of $A(x)$ is a *necessary* condition for the existence of solutions to this problem, at least as far as a regular feedback (i.e. with nonsingular $\beta(x)$) is sought.

For, recall that the integers $\{r_1, \ldots, r_m\}$ are not modified by static state feedback if the matrix $\beta(x)$ is nonsingular (see proof of Lemma 2.1). Suppose the system has been rendered noninteractive and observe that, in the corresponding closed loop

$$L_{g\beta_j} L^k_{f+g\alpha} h_i(x) = 0$$

for all $0 \le k < r_i - 1$, for all $1 \le i, j \le m$, all x near x°, and

$$[L_{g\beta_1} L^{r_i-1}_{f+g\alpha} h_i(x^\circ) \quad \cdots \quad L_{g\beta_m} L^{r_i-1}_{f+g\alpha} h_i(x^\circ)] \ne [0 \quad \cdots \quad 0].$$

Thus, the r_i-th derivative with respect to time of the i-th output $y_i(t)$ of the closed-loop system has the following expression

$$y_i^{(r_i)}(t) = L^{r_i}_{f+g\alpha} h_i(x(t)) + \sum_{j=1}^m L_{g\beta_j} L^{r_i-1}_{f+g\alpha} h_i(x(t)) v_i(t)$$

for each $1 \le i \le m$. If the system in question is noninteractive, $y_i(t)$ must depend only on $v_i(t)$ and this implies that

$$L_{g\beta_j} L_{f+g\alpha}^{r_i-1} h_i(x^\circ) = 0$$

for all $j \neq i$. This shows that the matrix

$$\bar{A}(x^\circ) = \{\bar{a}_{ij}(x^\circ)\} = \{L_{g\beta_j} L_{f+g\alpha}^{r_i-1} h_i(x^\circ)\}$$

is a diagonal matrix.

Recall now (see again proof of Lemma 2.1) that this matrix is related to the matrices $A(x^\circ)$ and $\beta(x^\circ)$ by the expression

$$\bar{A}(x^\circ) = A(x^\circ)\beta(x^\circ)$$

and that each of the rows of $A(x^\circ)$ is nonzero. Therefore, the matrix in question is necessarily nonsingular (because it is diagonal and none of its rows is zero) and we can conclude that also the matrix $A(x^\circ)$ is nonsingular.

We synthesize all the previous discussion in a formal statement.

Proposition 3.2. Consider a multivariable nonlinear system with m inputs and m outputs

$$\dot{x} = f(x) + \sum_{i=1}^{m} g_i(x) u_i$$

$$y_1 = h_1(x)$$

$$\cdots$$

$$y_m = h_m(x).$$

Suppose

$$L_{g_j} L_f^k h_i(x) = 0$$

for all $1 \leq j \leq m$, for all $1 \leq i \leq m$, for all $k < r_i - 1$, and for all x in a neighborhood of x°, and

$$[L_{g_1} L_f^{r_i-1} h_i(x^\circ) \quad \cdots \quad L_{g_m} L_f^{r_i-1} h_i(x^\circ)] \neq [0 \quad \cdots \quad 0]$$

for all $1 \leq i \leq m$. Then, the Noninteracting Control Problem is solvable if and only if the matrix $A(x^\circ)$ is nonsingular, i.e. if the system has a vector relative degree $\{r_1, \ldots, r_m\}$ at x°.

In view of its importance in connection with the solution of the noninteracting control problem, the matrix $A(x)$ is sometimes called the *decoupling matrix* of the system (in this case "decoupling" means "separation of the individual input-output channels"). From the previous Proposition we see that the class of systems having a vector relative degree at some point x° and the class of systems in which the noninteracting control problem can be solved, locally around x°, by means of static state feedback actually coincide. In other words, we may say that the special class of multivariable nonlinear system considered in this Chapter is exactly the class of those systems that can be made noninteractive via static state feedback.

Remark 3.3. The previous analysis can easily be extended to deal with systems having a number m of inputs which is larger than the number p of outputs. In this case, the Noninteracting Control Problem is the one of finding a regular static state feedback and a partition of the input vector v into p disjoint sets

$$v = \mathrm{col}(v_1, v_2, \ldots, v_p)$$

such that, in the corresponding closed loop, each output channel $y_i, 1 \leq i \leq p$, is affected only by the corresponding set of inputs v_i (and not by v_j, if $j \neq i$). A rather straightforward extension of Proposition 3.2 shows that the problem in question is solvable if and only if the matrix $A(x^\circ)$ has rank p (i.e. equal to the number of output channels).

The proof of the necessity is almost identical to that of Proposition 3.2. As far as the sufficiency is concerned, the proof is based on an appropriate extension of the normal form. The reader will have no difficulty in understanding that, in the case of systems with $m > p$ inputs and p outputs, a normal form similar to the one utilized so far can be developed under the assumption that the matrix $A(x)$ has rank p, because under this assumption the choice of local coordinates indicated in the Proposition 1.3 is still valid (see Remark 1.5). The normal form thus deduced has a structure which is identical to that of the one discussed in Section 5.1, with the only formal difference that $m > p$ input components are present in the appropriate places. If $A(x)$ has rank p, the equations

$$\begin{bmatrix} L_f^{r_1} h_1(x) - v_1^T \\ L_f^{r_2} h_2(x) - v_2^T \\ \vdots \\ L_f^{r_p} h_p(x) - v_p^T \end{bmatrix} + A(x)u = 0$$

can be solved for u, for any partition $v_1, \ldots, v_p$ of the vector v. The imposition of the corresponding feedback yields a closed loop system in which, for $1 \leq i \leq p$, v_i affects only y_i. •

We conclude this Section with some considerations about the stability of a system which has been made noninteractive by means of static state-feedback. From the block diagram of Fig. 5.1, we see that the internal structure of the noninteractive closed-loop obtained using the feedback (3.2) consists of m chains of r_i integrators each, all feeding the (unobservable) subsystem

$$\dot{\eta} = q(\xi, \eta) - p(\xi, \eta) A^{-1}(\xi, \eta) b(\xi, \eta) + p(\xi, \eta) A^{-1}(\xi, \eta) v.$$

Imposing on this system an additional feedback of the form

$$v_i = -c_0^i \xi_1^i - c_1^i \xi_2^i - \ldots - c_{r_i-1}^i \xi_{r_i}^i + \bar{v}_i$$

that, in the original coordinates, reads as

$$v_i = -c_0^i h_i(x) - c_1^i L_f h_i(x) - \ldots - c_{r_i-1}^i L_f^{r_i-1} h_i(x) + \bar{v}_i \qquad (3.3)$$

for $1 \leq i \leq m$, yields an overall closed-loop which is still noninteractive, and characterized by equations of the form

$$\dot{\xi}^i = \begin{bmatrix} 0 & 1 & 0 & \cdots & 0 \\ 0 & 0 & 1 & \cdots & 0 \\ \vdots & \vdots & \vdots & \ddots & \vdots \\ 0 & 0 & 0 & \cdots & 1 \\ -c_0^i & -c_1^i & -c_2^i & \cdots & -c_{r_i-1}^i \end{bmatrix} \xi^i + \begin{bmatrix} 0 \\ 0 \\ \vdots \\ 0 \\ 1 \end{bmatrix} \bar{v}_i$$

$$y_i = \begin{bmatrix} 1 & 0 & 0 & \cdots & 0 \end{bmatrix} \xi^i$$

for $1 \leq i \leq m$, and

$$\dot{\eta} = q(\xi, \eta) - p(\xi, \eta) A^{-1}(\xi, \eta) b(\xi, \eta) + p(\xi, \eta) A^{-1}(\xi, \eta) v.$$

In particular, the system thus obtained has a *linear input-output behavior*, characterized by the diagonal transfer functions matrix

$$H(s) = \begin{bmatrix} \dfrac{1}{d_1(s)} & 0 & \cdots & 0 \\ 0 & \dfrac{1}{d_2(s)} & \cdots & 0 \\ \vdots & \vdots & \ddots & \vdots \\ 0 & 0 & \cdots & \dfrac{1}{d_m(s)} \end{bmatrix}$$

with

$$d_i(s) = c_0^i + c_1^i s + \ldots + c_{r_i-1}^i s^{r_i-1} + s^{r_i} \qquad (3.4)$$

As far as the internal asymptotic stability is concerned, we see from the previous equations that the system has essentially the same structure as the one we obtained, via a similar feedback, in Section 4.4. Thus, using the results of Section B.1 we can conclude that if the zero dynamics of the system are asymptotically stable, and the polynomials (3.4) all have roots in the left-half complex plane, the system in question is locally asymptotically stable at $(\xi, \eta) = (0, 0)$.

Remark 3.4. It is apparent from the previous discussion that the asymptotic stability of the zero dynamics is a *sufficient* condition to achieve noninteracting control with an internally asymptotically stable closed-loop. However, it must be stressed that such a condition is *not* in general a necessary one. As a matter of fact, there may exist systems whose zero dynamics are not asymptotically stable (or even unstable) in which the achievement of noninteractive control by means of an internally asymptotically stable closed-loop is still possible. The precise characterization of those nonlinear systems that can be rendered noninteractive and simultaneously internally stable by means of static state feedback requires a rather more sophisticated analysis, that will be pursued in the next Chapters. ●

Remark 3.5. The previous analysis considers only the property of (internal) asymptotic stability, i.e. the asymptotic behavior (of the closed loop system) in the particular situation in which all the reference inputs $v_1, \ldots, v_m$ are set to zero. In general, the equations describing the closed loop system have the form

$$\dot{x}(t) = \tilde{f}(x) + \sum_{i=1}^{m} \tilde{g}_i(x)\bar{v}_i$$

Suppose, without loss of generality, that $x^\circ = 0$ is an equilibrium of the system, i.e. that $f(0) = 0$, and that $h(0) = 0$. If the zero dynamics of the system are asymptotically stable, and the feedback as been chosen as the composition of the standard noninteractive feedback (3.2) with the stabilizing feedback (3.3) (of course, with the polynomials (3.4) having all roots in the left-half complex plane), the vector field $\tilde{f}(x)$ has an asymptotically stable equilibrium at x°. Thus, using the results illustrated in Section B.2 it is possible, as in Section 4.4, to conclude that for each ε there exist δ and K such that

$$\| x^\circ \| < \delta, |v_i(t)| < K \qquad \text{for all } t \geq 0, 1 \leq i \leq m$$

imply $\| x(t) \| < \varepsilon$ for all $t \geq 0$. ●

5.4 Exact Linearization of the Input-Output Response

In Section 5.2, we have shown that if a system has relative degree $\{r_1, \ldots, r_m\}$ at a point x° and

$$r_1 + r_2 + \ldots + r_m = n$$

then this system can be rendered linear by means of feedback and change of coordinates. If the latter condition is not satisfied (but the system continues to have relative degree $\{r_1, \ldots, r_m\}$ at a certain point), one can at least obtain a system whose *input-output behavior* is linear. As a matter of fact, we have already shown in Section 5.3 that a result of this kind can always be achieved by means of the so-called Standard Noninteractive Feedback

$$u(x) = -A^{-1}(x)b(x) + A^{-1}(x)v.$$

The possibility of using feedback in order to achieve linearity in the input-output response is not restricted to systems having a certain (vector) relative degree at a point of interest, but holds for a broader class of systems: we shall see in this Section how this broader class can be characterized and how a feedback producing a linear input-output behavior can be designed. To this end, we need to start with a precise formulation of what we mean by achieving "linear input-output behavior" via feedback. Looking again at a nonlinear system having relative degree $\{r_1, \ldots, r_m\}$ on which the Standard Noninteractive Feedback had been imposed, we find that its outputs $y_i(t)$, for $1 \leq i \leq m$, are related to the input by expressions of the form

$$y_i(t) = \psi_i(t)\xi^i(0) + \int_0^t k_i(t-s)v_i(s)\,ds$$

where

$$\psi_i(t) = [\,1 \quad t \quad \frac{t^2}{2} \quad \cdots \quad \frac{t^{r_i-1}}{(r_i-1)!}\,] \qquad k_i(t) = \frac{t^{r_i-1}}{(r_i-1)!}$$

and $\xi^i(0)$ represents the value at time $t = 0$ of certain components of the state vector, in the coordinates associated with the normal form.

The latter is (obviously) linear in the input and in the initial state. However, the linearity in the initial state is due only to the fact that special coordinates have been chosen, and does not hold anymore if $\xi^i(0)$ is expressed as a (generally nonlinear) function of the initial value x° of the state in the original coordinates. Nevertheless, in any case the response is always given by the sum of the response under zero input, which is a function of the time and of the initial condition only, and of a response depending on the input and not on the initial state, which is linear in the input itself. In other words, the response has a structure of the following kind

$$y(t) = Q(t, x^\circ) + \sum_{i=1}^{m} \int_0^t w_i(t - \tau_1)v_i(\tau_1)\,d\tau_1. \tag{4.1}$$

Comparing this with the general expression of the Volterra series expansion of the input-output response of a nonlinear system (see (3.2.3)), we may thus conclude that a nonlinear system having relative degree $\{r_1, \ldots, r_m\}$ subject to the Standard Noninteractive Feedback is characterized by an output response in which the first order kernels $w_i(t, \tau_1)$ depend only on the difference $t - \tau_1$ and not on x°, and all kernels of order higher than one are vanishing.

Note also that if the first order kernels of a Volterra series expansion depend only on the difference $t - \tau_1$ and not on x°, then necessarily all the kernels of higher order are vanishing, and therefore the condition that the $w_i(t, \tau_1)$'s depend only on the difference $t - \tau_1$ and not on x° is *necessary and sufficient* for the response of a nonlinear system to be of the form (4.1).

On the basis of these observations, our goal is now to try to use feedback in order to achieve (on a class of systems possibly broader that the one of systems having some vector relative degree) a response in which all the first order kernels of a Volterra series expansion depend only on the difference $t - \tau_1$ and not on x°. In order to simplify the formulation of the problem, note that if one considers the Taylor series expansion (3.2.9b) of $w_i(t, \tau_1)$, it is easily found that a necessary and sufficient condition for this kernel to be independent of x° and dependent only on $t - \tau_1$, or—in other words—for a response of the form (4.1) to hold, is that

$$L_{g_i} L_f^k h_j(x) = \text{independent of } x \qquad (4.2)$$

for all $k \geq 0$ and all $1 \leq i, j \leq m$.

In general, the conditions (4.2) will not be satisfied for a specific nonlinear system. If this is the case, we may wish to have them satisfied via feedback, as expressed in the following statement.

Input-Output Exact Linearization Problem. Given a set of $m+1$ vector fields $f(x), g_1(x), \ldots, g_m(x)$, a set of m real-valued functions $h_1(x), \ldots, h_m(x)$ and an initial state x°, find (if possible), a neighborhood U of x° and a pair of feedback functions $\alpha(x)$ and $\beta(x)$ defined on U, such that for all $k \geq 0$ and all $1 \leq i, j \leq m$

$$L_{(g\beta)_i} L_{(f+g\alpha)}^k h_j(x) = \text{independent of } x \text{ on } U. \qquad (4.3)$$

First of all, we show that—because of finite dimensionality of the underlying system—the apparently infinite set of conditions (4.3) is actually completely determined by a finite subset of them. It is possible to prove, in fact, the following result.

Lemma 4.1. Suppose (4.3) holds for all $0 \leq k \leq 2n - 1$ and all $1 \leq i, j \leq m$. Then (4.3) holds for all $k \geq 0$ and all $1 \leq i, j \leq m$.

Proof. We can indeed prove the result for a (infinite) set of functions of the form

$$L_{g_i} L_f^k h_j(x) \qquad (4.4)$$

($k \geq 0$ and $1 \leq i, j \leq m$) thus simplifying the notation. First of all, recall (see Lemma 1.9.5) that, given any neighborhood U of x°, on an open and dense subset U' of U the largest codistribution Q invariant under the vector fields $f, g_1, \ldots, g_m$

which contains span$\{dh_1, \ldots, dh_m\}$ is locally spanned by exact differentials of the type

$$\omega = dL_{g_{i_1}} \cdots L_{g_{i_r}} h_j$$

with $r \leq n - 1$, $0 \leq i_r \leq m$, and $g_0 = f$. Since, by assumption, the functions (4.4) are constant on U' for all $0 \leq k \leq 2n - 1$ and all $1 \leq i, j \leq m$, we deduce that

$$dL_{g_{i_1}} \cdots L_{g_{i_r}} h_j = 0$$

whenever $i_k \neq 0$, $1 \leq k \leq r$, so that Q is necessarily spanned by differentials of the type $dL_f^k h_j$, with $1 \leq j \leq m$ and $0 \leq k \leq n - 1$. Let q denote the dimension of Q at a point of U' and define, in a neighborhood V of this point, new local coordinates $(\zeta_1, \zeta_2) = \Phi(x)$, where the q elements of ζ_2 are chosen in the set $\{L_f^k h_j(x) : 1 \leq j \leq m, 0 \leq k \leq n - 1\}$. Then, by Proposition 1.7.2, the vector fields $f, g_1, \ldots, g_m$ and the functions $h_1, \ldots, h_m$ are transformed into

$$f(\zeta_1, \zeta_2) = \begin{bmatrix} f_1(\zeta_1, \zeta_2) \\ f_2(\zeta_2) \end{bmatrix} \qquad g_i(\zeta_1, \zeta_2) = \begin{bmatrix} g_{1i}(\zeta_1, \zeta_2) \\ g_{2i}(\zeta_2) \end{bmatrix} \qquad h_i(\zeta_1, \zeta_2) = h_i(\zeta_2).$$

Replacing the expressions thus found into (4.4), we obtain

$$L_{g_i} L_f^k h_i(\Phi^{-1}(\zeta_1, \zeta_2)) = L_{g_{2i}} L_{f_2}^k h_i(\zeta_2)$$

so that the constancy of the (4.4) with respect to x (on the neighborhood V) is equivalent to the constancy of the functions on the right-hand side with respect to ζ_2.

We use now again the assumption that the functions in question are constant for all $0 \leq k \leq 2n - 1$ and all $1 \leq i, j \leq m$, and we note that this implies (see the formula (4.1.2))

$$\langle dL_{f_2}^s h_i(\zeta_2), ad_{f_2}^r g_{2i}(\zeta_2) \rangle = (-1)^r L_{g_{2i}} L_{f_2}^{s+r} h_i(\zeta_2) = \text{independent of } \zeta_2$$

for all r, s such that $0 \leq r + s \leq 2n - 1$. Recall that, by construction, for each value of $1 \leq k \leq q$, there exist some $1 \leq j \leq m, 0 \leq s \leq n - 1$, such that

$$(\zeta_2)_k = L_f^s h_j(\Phi^{-1}(\zeta_1, \zeta_2)) = L_{f_2}^s h_j(\zeta_2)$$

where $(\zeta_2)_k$ denotes the k-th component of ζ_2. Replacing this into the previous expression yields

$$\langle d(\zeta_2)_k, ad_{f_2}^r g_{2i}(\zeta_2) \rangle = k\text{-th component of } ad_{f_2}^r g_{2i}(\zeta_2) = \text{independent of } \zeta_2$$

for all $1 \leq i \leq m$ and all $0 \leq r \leq n$. In other words, the vector fields $ad_{f_2}^r g_{2i}(\zeta_2)$ are constant vector fields for all $1 \leq i \leq m$ and all $0 \leq r \leq n$.

Let R denote the smallest distribution invariant under the vector fields $f_2, g_{21}, \ldots, g_{2m}$ and containing the vector fields $g_{21}, \ldots, g_{2m}$. Recalling the algorithm described in Section 1.8, it is easy to realize that this distribution can be expressed (because of the constancy of the vector fields $ad_{f_2}^k g_{2i}(\zeta_2)$) as

$$R = \text{span}\{ad_{f_2}^k g_{2i} : 1 \leq i \leq m, 0 \leq k \leq n-1\}$$

and that, for any $1 \leq i \leq m$,

$$ad_{f_2}^n g_{2i} \in R.$$

Since $ad_{f_2}^n g_{2i}$ also is a constant vector field, we conclude that the latter can be expressed as a linear combination, with constant coefficients, of vector fields of the set $\{ad_{f_2}^k g_{2i} : 1 \leq i \leq m, 0 \leq k \leq n-1\}$, and the same property holds for any vector field of the form $ad_{f_2}^{n+s} g_{2i}$, with $s > 0$ (as a simple induction argument shows).

Exactly as in the step (iii) of Theorem 4.8.3, this fact can be used to show that

$$L_{g_{2i}} L_{f_2}^k h_j(\zeta_2) = \text{independent of } \zeta_2$$

for all $k \geq 0$ and all $1 \leq i, j \leq m$. Thus, the functions (4.4) are constant on a neighborhood V of every point x of a dense subset U' of U. Being smooth, they are constant on all U and this complets the proof. •

We come back now to the Input-Output Exact Linearization Problem. Our goal is to find necessary and sufficient conditions under which this problem is solvable, and to show how a pair of feedback functions $\alpha(x)$ and $\beta(x)$ which actually solves the problem can be constructed. First of all, from the data $f(x), g_j(x), h_i(x), 1 \leq i, j \leq m$, we construct the set of real-valued functions $L_{g_j} L_f^k h_i(x)$, $0 \leq k \leq 2n-1$, and we arrange all these functions into a set of $m \times m$ matrices of the form

$$T_k(x) = \begin{bmatrix} L_{g_1} L_f^k h_1(x) & \cdots & L_{g_m} L_f^k h_1(x) \\ \vdots & \cdots & \vdots \\ L_{g_1} L_f^k h_m(x) & \cdots & L_{g_m} L_f^k h_m(x) \end{bmatrix} \qquad 0 \leq k \leq 2n-1.$$

As a matter of fact, the possibility of solving the problem in question depends on a property of the set of matrices thus constructed. This property can be expressed in different forms, depending on how the data $T_k(x)$, $0 \leq k \leq 2n-1$, are arranged.

One way of arranging these data is to consider a formal power series $T(s, x)$ in the indeterminate s, defined as

$$T(s, x) = \sum_{k=0}^{\infty} T_k(x) s^{-k-1}. \tag{4.5}$$

We will see below that the problem in question may be solved if and only if $T(s, x)$ satisfies a suitable separation condition.

Another equivalent condition for the existence of solutions is based on the construction of a sequence of Toeplitz matrices, denoted $M_k(x)$, $0 \leq k \leq 2n - 1$, and defined as

$$M_k(x) = \begin{bmatrix} T_0(x) & T_1(x) & \cdots & T_k(x) \\ 0 & T_0(x) & \cdots & T_{k-1}(x) \\ \vdots & \vdots & \ddots & \vdots \\ 0 & 0 & \cdots & T_0(x) \end{bmatrix}. \tag{4.6}$$

In this case one is interested in the special situation in which linear dependence between rows may be tested by taking linear combinations with constant coefficients only.

In view of the relevance of this particular property throughout all the subsequent analysis, we discuss the point with a little more detail. Let $M(x)$ be a $p \times m$ matrix whose entries are smooth real-valued functions. We say that x° is a *regular* point of M if there exists a neighborhood U of x° with the property that

$$\text{rank}(M(x)) = \text{rank}(M(x^\circ)) \tag{4.7}$$

for all $x \in U$. In this case, the integer $\text{rank}(M(x^\circ))$ is denoted $r_K(M)$; clearly $r_K(M)$ depends on the point x°, because on a neighborhood V of another point x^1, $\text{rank}(M(x^1))$ may be different.

With the matrix M we will associate another notion of "rank", in the following way. Let x° be a regular point of M, U an open set on which (4.7) holds, and $\bar{M}$ a matrix whose entries are the restrictions to U of the corresponding entries of M. We consider the vector space defined by taking *linear combinations of rows of $\bar{M}$ over the field* $\mathbf{R}$, the set of real numbers, and denote $r_{\mathbf{R}}(M)$ its dimension (note that again $r_{\mathbf{R}}(M)$ may depend on x°). Clearly, the two integers $r_{\mathbf{R}}(M)$ and $r_K(M)$ are such that

$$r_{\mathbf{R}}(M) \geq r_K(M). \tag{4.8}$$

The equality of these two integers may easily be tested in the following way. Note that both remain unchanged if M is multiplied on the left by a nonsingular matrix of real numbers. Let us call a *row-reduction* of M the process of multiplying M on the left by a nonsingular matrix V of real numbers with the purpose of annihilating the maximum number of rows in VM (here also the row-reduction process may depend on the point x°). Then, it is trivially seen that the two sides of (4.8) are equal if and only if any process of row-reduction of M leaves a number of nonzero rows in VM which is equal to $r_K(M)$.

We may now return to the original synthesis problem and prove the main result.

Theorem 4.2. There exists a solution at x° to the Input-Output Exact Linearization Problem if and only if either one of the following equivalent conditions is satisfied

(a) there exists a formal power series

$$K(s) = \sum_{k=0}^{\infty} K_k s^{-k-1}$$

whose coefficients are $m \times m$ matrices of real numbers, and a formal power series

$$R(s, x) = R_{-1}(x) + \sum_{k=0}^{\infty} R_k(x) s^{-k-1}$$

whose coefficients are $m \times m$ matrices of smooth functions defined on a neighborhood U of x°, with invertible $R_{-1}(x)$, which factorize the formal power series $T(s, x)$ as follows

$$T(s, x) = K(s) \cdot R(s, x) \tag{4.9}$$

(b) for all $0 \le i \le 2n - 1$, the point x° is a regular point of the Toeplitz matrix M_i and

$$r_{\mathbf{R}}(M_i) = r_K(M_i). \tag{4.10}$$

The proof of this Theorem consists in the following steps. First of all we introduce a recursive algorithm, known as the Structure Algorithm, which operates on the sequence of matrices $T_k(x)$. Then, we prove the sufficiency of (b), essentially by showing that this assumption makes it possible to continue the Structure Algorithm at each stage and that from the data thus extracted one may construct a feedback solving the problem. Then, we complete the proof that (a) is necessary and that (a) implies (b).

Remark 4.3. For the sake of notational compactness, from this point on we make systematic use of the following notation. Let γ be a $s \times 1$ vector of smooth functions and $\{g_1, \ldots, g_m\}$ a set of vector fields. We let $L_g \gamma$ denote the $s \times m$ matrix whose i-th column is the vector $L_{g_i} \gamma$, i.e.

$$L_g \gamma = [\, L_{g_1} \gamma \quad \cdots \quad L_{g_m} \gamma \,]. \quad \bullet$$

Structure Algorithm. *Step 1.* Let x° be a regular point of T_0 and suppose $r_{\mathbf{R}}(T_0) = r_K(T_0)$. Then, there exists a nonsingular matrix of real numbers, denoted by

$$V_1 = \begin{bmatrix} P_1 \\ K_1^1 \end{bmatrix}$$

where P_1 performs row permutations, such that

$$V_1 T_0(x) = \begin{bmatrix} S_1(x) \\ 0 \end{bmatrix}$$

where $S_1(x)$ is an $r_0 \times m$ matrix and rank $(S_1(x^\circ)) = r_0$. Set

$$\delta_1 = r_0$$
$$\gamma_1(x) = P_1 h(x)$$
$$\bar{\gamma}_1(x) = K_1^1 h(x)$$

and note that

$$L_g \gamma_1(x) = S_1(x)$$
$$L_g \bar{\gamma}_1(x) = 0$$

If $T_0(x) = 0$, then P_1 must be considered as a matrix with no rows and K_1^1 is the identity matrix.

Step i. Consider the matrix

$$\begin{bmatrix} L_g \gamma_1(x) \\ \vdots \\ L_g \gamma_{i-1}(x) \\ L_g L_f \bar{\gamma}_{i-1}(x) \end{bmatrix} = \begin{bmatrix} S_{i-1}(x) \\ L_g L_f \bar{\gamma}_{i-1}(x) \end{bmatrix}$$

and let x° be a regular point of this matrix. Suppose

$$r_R \begin{bmatrix} S_{i-1} \\ L_g L_f \bar{\gamma}_{i-1} \end{bmatrix} = r_K \begin{bmatrix} S_{i-1} \\ L_g L_f \bar{\gamma}_{i-1} \end{bmatrix}. \tag{4.11}$$

Then, there exists a nonsingular matrix of real numbers, denoted by

$$V_i = \begin{bmatrix} I_{\delta_1} & \cdots & 0 & 0 \\ \vdots & \ddots & \vdots & \vdots \\ 0 & \cdots & I_{\delta_{i-1}} & 0 \\ 0 & \cdots & 0 & P_i \\ K_1^i & \cdots & K_{i-1}^i & K_i^i \end{bmatrix}$$

where P_i performs row permutations, such that

$$V_i \begin{bmatrix} L_g\gamma_1(x) \\ \vdots \\ L_g\gamma_{i-1}(x) \\ L_gL_f\bar{\gamma}_{i-1}(x) \end{bmatrix} = \begin{bmatrix} S_i(x) \\ 0 \end{bmatrix}$$

where $S_i(x)$ is an $r_{i-1} \times m$ matrix and $\mathrm{rank}(S_i(x^\circ)) = r_{i-1}$. Set

$$\delta_i = r_{i-1} - r_{i-2}$$
$$\gamma_i(x) = P_i L_f \bar{\gamma}_{i-1}(x)$$
$$\bar{\gamma}_i(x) = K_1^i \gamma_1(x) + \cdots + K_{i-1}^i \gamma_{i-1}(x) + K_i^i L_f \bar{\gamma}_{i-1}(x)$$

and note that

$$\begin{bmatrix} L_g\gamma_1(x) \\ \vdots \\ L_g\gamma_i(x) \end{bmatrix} = S_i(x)$$
$$L_g\bar{\gamma}_i(x) = 0.$$

If the condition (4.11) is satisfied but the last $m - r_{i-2}$ rows of the matrix depend on the first r_{i-2}, then the step degenerates, P_i must be considered as a matrix with no rows, K_i^i is the identity matrix, $\delta_i = 0$ and $S_i(x) = S_{i-1}(x)$. •

As we said before, this algorithm may be continued at each stage if and only if the assumption (b) is satisfied, because of the following fact.

Lemma 4.4. Let x° be a regular point of T_0 and suppose $r_R(T_0) = r_K(T_0)$. Then x° is a regular point of

$$\begin{bmatrix} S_{i-1} \\ L_g L_f \bar{\gamma}_{i-1} \end{bmatrix}$$

and the condition (4.11) holds for all $2 \leq i \leq k$ if and only if x° is a regular point of T_i and the condition (4.10) holds for all $1 \leq i \leq k - 1$.

Proof. We sketch the proof for the case $k = 2$. Recall that

$$M_1 = \begin{bmatrix} T_0 & T_1 \\ 0 & T_0 \end{bmatrix} = \begin{bmatrix} L_g h & L_g L_f h \\ 0 & L_g h \end{bmatrix}.$$

Moreover, let V_1, γ_1 and $\bar{\gamma}_1$ be defined as in the first step of the algorithm.

Multiply M_1 on the left by

$$V = \begin{bmatrix} V_1 & 0 \\ 0 & V_1 \end{bmatrix}.$$

As a result, one obtains

$$VM_1 = \begin{bmatrix} V_1 L_g h & L_g V_1 L_f h \\ 0 & V_1 L_g h \end{bmatrix} = \begin{bmatrix} L_g P_1 h & L_g L_f P_1 h \\ 0 & L_g L_f K_1^1 h \\ 0 & L_g P_1 h \\ 0 & 0 \end{bmatrix} =$$

$$= \begin{bmatrix} S_1 & L_g L_f \gamma_1 \\ 0 & L_g L_f \bar{\gamma}_1 \\ 0 & S_1 \\ 0 & 0 \end{bmatrix}$$

note that $r_{\mathbf{R}}(S_1) = r_K(S_1)$. Thus, because of the special structure of VM_1, x° is a regular point of M_1 and the condition $r_{\mathbf{R}}(M_1) = r_K(M_1)$ is satisfied if and only if x° is a regular point of

$$\begin{bmatrix} L_g L_f \bar{\gamma}_1 \\ S_1 \end{bmatrix}$$

and

$$r_{\mathbf{R}} \begin{bmatrix} L_g L_f \bar{\gamma}_1 \\ S_1 \end{bmatrix} = r_K \begin{bmatrix} L_g L_f \bar{\gamma}_1 \\ S_1 \end{bmatrix}$$

i.e. the condition (4.11) holds for $i = 2$. For higher values of k one may proceed by induction. ●

From this, we see that the Structure Algorithm may be continued up to the k-th step if and only if the condition (4.10) holds for all i up to $k - 1$. The Structure Algorithm may be continued up to the $2n$-th step if and only if the assumption (b) is satisfied.

Proof of Theorem 4.2. Sufficiency of (b): construction of the linearizing feedback. If the Structure Algorithm can be continued up to the $2n$-th iteration, two possibilities may occur. Either there is a step $q \leq 2n$ such that the matrix

$$\begin{bmatrix} L_g\gamma_1(x) \\ \vdots \\ L_g\gamma_{q-1}(x) \\ L_gL_f\bar{\gamma}_{q-1}(x) \end{bmatrix}$$

has rank m at x°. Then the algorithm terminates. Formally, one can still set $P_q =$ identity, $V_q =$ identity

$$\gamma_q(x) = P_q L_f \bar{\gamma}_{q-1}(x)$$

and

$$\begin{bmatrix} S_{q-1}(x) \\ L_g\gamma_q(x) \end{bmatrix} = S_q(x)$$

and consider $K_1^q, \ldots, K_q^q$ as matrices with no rows. Or, else, from a certain step on all further steps are degenerate. In this case, let q denote the index of the last nondegenerate step. Then, for all $q < j \leq 2n$, P_j will be a matrix with no rows, K_j^j the identity and $\delta_j = 0$.

From the functions $\gamma_1, \ldots, \gamma_q$ generated by the Structure Algorithm, one may construct a linearizing feedback in the following way. Set

$$\Gamma(x) = \begin{bmatrix} \gamma_1(x) \\ \vdots \\ \gamma_q(x) \end{bmatrix}$$

and recall that $S_q = L_g\Gamma$ is an $r_{q-1} \times m$ matrix, of rank r_{q-1} at x°. Then the equations

$$(L_g\Gamma(x))\alpha(x) = -L_f\Gamma(x) \qquad (4.12a)$$

$$(L_g\Gamma(x))\beta(x) = -[I_{r_{q-1}} \quad 0] \qquad (4.12b)$$

on a suitable neighborhood U of x° are solved by a pair of smooth functions α and β.

Necessity of (b): proof that the above feedback solves the problem. Set $\tilde{f}(x) = f(x) + g(x)\alpha(x)$ and $\tilde{g}(x) = g(x)\beta(x)$. We show first that, for all $0 \leq k \leq 2n-1$

$$P_1 L_{\tilde{g}} L_{\tilde{f}}^k(x) = \text{ independent of } x \qquad (4.13a)$$

$$P_i K_{i-1}^{i-1} \cdots K_1^1 L_{\tilde{g}} L_f^k(x) = \text{ independent of } x \qquad (4.13b)$$

for all $2 \leq i \leq q$ and that

$$K_q^q K_{q-1}^{q-1} \cdots K_1^1 L_{\tilde{g}} L_f^k(x) = \text{ independent of } x \qquad (4.13c)$$

To this end, note that (4.12) imply

$$L_{\bar{f}} \gamma_i = 0 \qquad (4.14a)$$
$$L_{\tilde{g}} \gamma_i = \text{ independent of } x \qquad (4.14b)$$

for all $1 \leq i \leq q$. Moreover, since $L_g \bar{\gamma}_i = 0$ for all $i \geq 1$, also

$$L_{\bar{f}} \bar{\gamma}_i = L_f \bar{\gamma}_i \qquad (4.14c)$$
$$L_{\tilde{g}} \bar{\gamma}_i = 0 \qquad (4.14d)$$

for all $1 \leq i$. Using (4.14) repeatedly, it is easy to see that, if $k \geq i$

$$
\begin{aligned}
K_i^i \cdots K_1^1 L_f^k h &= K_i^i \cdots K_2^2 L_f^k \bar{\gamma}_1 \\
&= K_i^i \cdots K_3^3 L_f^{k-1} \bar{\gamma}_2 \\
&\cdots \\
&= K_i^i L_f^{k-i+2} \bar{\gamma}_{i-1} \\
&= L_f^{k-i+1} \bar{\gamma}_i
\end{aligned}
\qquad (4.15)
$$

and, if $k < i$

$$K_i^i \cdots K_1^1 L_f^k h = K_i^i \cdots K_{k+1}^{k+1} L_{\bar{f}} \bar{\gamma}_k. \qquad (4.16)$$

These expressions hold for every $i \geq 1$ (recall that, if $i > q$, K_i^i is an identity matrix).

Thus, if $i \leq q$ and $k \geq i-1$ we get from (4.15)

$$P_i K_{i-1}^{i-1} \cdots K_1^1 L_{\tilde{g}} L_f^k h = L_{\tilde{g}} P_i L_f^{k-i+2} \bar{\gamma}_{i-1} = L_{\tilde{g}} L_f^{k-i+1} \gamma_i$$

which is either independent of x (if $k = i - 1$) or zero, while for $i \leq q$ and $k < i-1$ we get from (4.16)

$$P_i K_{i-1}^{i-1} \cdots K_1^1 L_{\tilde{g}} L_f^k h = P_i \cdots K_{k+2}^{k+2} L_{\tilde{g}} \Big(\bar{\gamma}_{k+1} - \sum_{j=1}^{k} K_j^{k+1} \gamma_j \Big).$$

The right-hand side of this expression is again independent of x and this completes the proof of (4.13b).

Moreover, if $k \geq q$, (4.15) yields

$$
\begin{aligned}
K_q^q \cdots K_1^1 L_{\tilde{g}} L_f^k h &= K_k^k \cdots K_1^1 L_{\tilde{g}} L_f^k h \\
&= L_{\tilde{g}} L_f \bar{\gamma}_k \\
&= L_{\tilde{g}} K_{k+1}^{k+1} L_f \bar{\gamma}_k \\
&= L_{\tilde{g}} \Big(\bar{\gamma}_{k+1} - \sum_{j=1}^{q} K_j^{k+1} \gamma_j \Big)
\end{aligned}
$$

and this, together with (4.16) written for $i = q$, which holds for $k < q$, shows that also (4.13c) is true. Finally, (4.13a) is also true because $P_1 L_{\tilde{g}} L_f^k h = L_{\tilde{g}} L_f^k \gamma_1$ and the latter is either independent of x (if $k = 0$) or zero.

Suppose now that the matrix

$$
H = \begin{bmatrix}
P_1 & & \\
P_2 K_1^1 & & \\
\vdots & & \\
P_q K_{q-1}^{q-1} \cdots K_1^1 & & \\
K_q^q K_{q-1}^{q-1} \cdots K_1^1 & &
\end{bmatrix}
\tag{4.17}
$$

is square and nonsingular. This, together with the (4.13) already proved, shows in fact that

$$
L_{\tilde{g}} L_f^k h(x) = \text{independent of } x
$$

for all $0 \leq k \leq 2n - 1$ and, in view of Lemma 4.1, proves the sufficiency of (b). But the nonsingularity of (4.17) is a straightforward consequence of the fact that this matrix may be deduced from the matrix $(V_q \ldots V_1)$ by means of elementary row operations.

Necessity of (a). Let

$$
\hat{\beta}(x) = \beta^{-1}(x)
$$
$$
\hat{\alpha}(x) = -\beta^{-1}(x)\alpha(x)
$$

and let

$$
\tilde{T}_k(x) = L_{\tilde{g}} L_f^k h(x).
$$

If the feedback pair α and β is such as to make $\tilde{T}_k(x)$ independent of x for all

k (i.e. to solve the problem), then

$$L_f^k h = L_f^k h + \tilde{T}_{k-1}\hat{\alpha} + \tilde{T}_{k-2}L_f\hat{\alpha} + \ldots + \tilde{T}_0 L_f^{k-1}\hat{\alpha}. \tag{4.18}$$

This expression may be easily proved by induction. In fact, one has

$$L_f^{k+1} h = L_{(\tilde{f}+\tilde{g}\hat{\alpha})}L_f^k h + L_f(\tilde{T}_{k-1}\hat{\alpha} + \ldots + \tilde{T}_0 L_f^{k-1}\hat{\alpha})$$
$$= L_f^{k+1} h + L_{\tilde{g}}L_f^k h\hat{\alpha} + \tilde{T}_{k-1}L_f\hat{\alpha} + \ldots + \tilde{T}_0 L_f^k \hat{\alpha}.$$

From (4.18) one then deduces

$$L_g L_f^k h = (L_{\tilde{g}}L_f^k h)\hat{\beta} + \tilde{T}_{k-1}L_g\hat{\alpha} + \tilde{T}_{k-2}L_g L_f\hat{\alpha}(x) + \ldots + \tilde{T}_0 L_g L_f^{k-1}\hat{\alpha}$$

or

$$T_k(x) = \tilde{T}_k\hat{\beta}(x) + \tilde{T}_{k-1}L_g\hat{\alpha}(x) + \ldots + \tilde{T}_0 L_g L_f^{k-1}\hat{\alpha}(x). \tag{4.19}$$

Now, consider the formal power series

$$K(s) = \sum_{k=0}^{\infty} \tilde{T}_k s^{-k-1}$$

$$R(s, x) = \hat{\beta}(x) + \sum_{k=0}^{\infty} (L_g L_f^k \hat{\alpha}(x)) s^{-k-1}$$

and note that the latter is invertible (i.e. the coefficient of the 0-th power of s is an invertible matrix). At this point, the expression (4.19) tells us exactly that the Cauchy product of the two series thus defined is equal to the series (4.5), thus proving the necessity of (a).

(a)$\Rightarrow$(b). If (4.9) is true, we may write

$$M_k(x) = \begin{bmatrix} K_0 & K_1 & \cdots & K_k \\ 0 & K_0 & \cdots & K_{k-1} \\ \vdots & \vdots & \ddots & \vdots \\ 0 & 0 & \cdots & K_0 \end{bmatrix} \begin{bmatrix} R_{-1}(x) & R_0(x) & \cdots & R_{k-1}(x) \\ 0 & R_{-1}(x) & \cdots & R_{k-2}(x) \\ \vdots & \vdots & \ddots & \vdots \\ 0 & 0 & \cdots & R_{-1}(x) \end{bmatrix}.$$

The factor on the left of this matrix is a matrix of real numbers, whereas the factor on the right is nonsingular at x°, as a consequence of the nonsingularity of $R_{-1}(x)$. Thus x° is a regular point of M_k and the condition (4.10) holds. ●

Remark 4.5. We stress again the importance of the Structure Algorithm as a test for the fulfillment of the conditions (a) (or (b)) as well as a procedure for the construction of the linearizing feedback. ●

Remark 4.6. An obviuos sufficient condition for the solvability of the Input-Output Exact Linearization Problem is that the system has relative degree $\{r_1, \ldots, r_m\}$ at x°. The reader may easily verify that, if this is the case, the Structure Algorithm can be continued up to a step $q = \max\{r_1, \ldots, r_m\}$, yielding $S_q(x) = A(x)$. •

Example 4.1. Consider the system

$$\dot{x} = \begin{bmatrix} x_1^2 + x_2 \\ x_1 x_3 \\ -x_1 + x_3 \\ 0 \\ x_5 + x_3^2 \end{bmatrix} + \begin{bmatrix} 0 \\ 0 \\ 1 \\ 1 \\ x_2 \end{bmatrix} u_1 + \begin{bmatrix} 0 \\ 1 \\ 0 \\ 0 \\ 0 \end{bmatrix} u_2$$

$$y = \begin{bmatrix} x_3 \\ x_4 \end{bmatrix}.$$

On this system, the Structure Algorithm proceeds as follows. Construct the matrix

$$T_0(x) = L_g h(x) = \begin{bmatrix} 1 & 0 \\ 1 & 0 \end{bmatrix}.$$

This matrix satisfies the condition (4.10), and one can set

$$V_1 = \begin{bmatrix} 1 & 0 \\ 1 & -1 \end{bmatrix}$$

that yields

$$S_1(x) = \begin{bmatrix} 1 & 0 \end{bmatrix}$$
$$\gamma_1(x) = x_3$$
$$\bar{\gamma}_1(x) = x_3 - x_4.$$

Consider now the matrix

$$\begin{bmatrix} S_1(x) \\ L_g L_f \bar{\gamma}_1(x) \end{bmatrix} = \begin{bmatrix} 1 & 0 \\ 1 & 0 \end{bmatrix}$$

which still satisfies the condition (4.11). Thus we can proceed with the algorithm, and set

$$V_2 = \begin{bmatrix} 1 & 0 \\ 1 & -1 \end{bmatrix}$$

that yields

$$S_2(x) = [1 \quad 0]$$

$$\bar{\gamma}_2(x) = \gamma_1(x) - L_f\bar{\gamma}_1(x) = x_1$$

(no $\gamma_2(x)$ exists, because $r_1 = r_0 = 1$).

At the third step, we consider the matrix

$$\begin{bmatrix} S_2(x) \\ L_g L_f \bar{\gamma}_2(x) \end{bmatrix} = \begin{bmatrix} 1 & 0 \\ 0 & 1 \end{bmatrix}$$

which now has $r_2 = 2$. Thus, the algorithm terminates, with $q = 3$, and

$$\gamma_3(x) = L_f\bar{\gamma}_2(x) = x_2 + x_1^2.$$

The system can be rendered linear from an input-output point of view, by means of the feedback $u = \alpha(x) + \beta(x)v$, with $\alpha(x)$ and $\beta(x)$ solutions of (see (4.12))

$$\begin{bmatrix} L_g\gamma_1(x) \\ L_g\gamma_3(x) \end{bmatrix} \alpha(x) = - \begin{bmatrix} L_f\gamma_1(x) \\ L_f\gamma_3(x) \end{bmatrix}$$

$$\begin{bmatrix} L_g\gamma_1(x) \\ L_g\gamma_3(x) \end{bmatrix} \beta(x) = \begin{bmatrix} 1 & 0 \\ 0 & 1 \end{bmatrix}$$

i.e. with

$$\alpha(x) = - \begin{bmatrix} -x_1 + x_3 \\ 2x_1^3 + 2x_1x_2 + x_1x_3 \end{bmatrix}$$

and $\beta(x)$ the identity matrix.

Note that this system does not have any relative degree, because the matrix (1.2), which in this case coincides with $T_0(x)$, is singular, nor the state-input equations can be exactly linearized by means of feedback and coordinates changes, because the distribution $G = \text{span}\{g_1, g_2\}$ is not involutive.

5.5 Exercises and Examples

For a multivariable system having a (vector) relative degree at a point x° of interest, it is possible to solve design problems like asymptotic output reproduction, disturbance decoupling, and model matching, exactly in the same way as done in Chapter 4 for single-input single-output systems. The corresponding procedures are straightforward extensions of those already illustrated, and their derivation is left, as an exercise, to the reader.

Exercise 5.1. Show how the problem of asymptotically tracking a prescribed reference output $y_R(t)$ can be solved. Discuss the asymptotic properties of the internal response of the corresponding closed loop system.

Exercise 5.2. Consider a system of the form

$$\dot{x} = f(x) + \sum_{i=1}^{m} g_i(x)u_i + p(x)w \qquad (5.1a)$$

$$y = h(x). \qquad (5.1b)$$

Prove the following result.

Proposition 5.1. There exists a feedback of the form $u = \alpha(x) + \beta(x)v$ which renders the output y of (5.1) independent of the disturbance w if and only if

$$L_p L_f^k h_i(x) = 0 \qquad \text{for all } 0 \le k \le r_i - 1,\ 1 \le i \le m.$$

There exists a feedback of the form $u = \alpha(x) + \beta(x)v + g(x)w$ which renders the output y of (5.1) independent of the disturbance w if and only if

$$L_p L_f^k h_i(x) = 0 \qquad \text{for all } 0 \le k \le r_i - 2,\ 1 \le i \le m.$$

In the next problem we do not assume the system having a relative degree, but we consider the weaker hypothesis that the system can be rendered linear from an input-output point of view.

Exercise 5.3. Consider a system of the form

$$\dot{x} = f(x) + \sum_{i=1}^{m} g_i(x)u_i \qquad (5.2a)$$

$$y = h(x) \qquad (5.2b)$$

and a *linear* reference model

$$\dot{\zeta} = A\zeta + Bw \qquad (5.3a)$$

$$y_R = C\zeta. \qquad (5.3b)$$

Suppose that the system (5.2) satisfies the conditions of Theorem 4.2 for solvability of the Input-Output Exact Linearization Problem. Let $P_i, K_1^i, \ldots, K_i^i$ be the set of matrices determined at the i-th step of the Structure Algorithm (performed on the set of data $f(x), g_1(x), \ldots, g_m(x), h(x)$). Set

$$C_1 = P_1 C$$

and, for $i \ge 2$

$$C_i = P_i \bar{C}_{i-1} A$$

$$\bar{C}_i = K_1^i C_1 + \ldots + K_{i-1}^i C_{i-1} + K_i^i \bar{C}_{i-1} A.$$

Set also

$$D = \text{col}(C_1, \ldots, C_q).$$

Prove the following result.

Proposition 5.2. If and only if

$$\bar{C}_i B = 0 \qquad \text{for all } i \geq 1 \tag{5.4}$$

there exists a feedback of the form

$$\dot{\zeta} = \gamma(\zeta, x) + \delta(\zeta, x)w$$
$$u = \alpha(\zeta, x) + \beta(\zeta, x)w$$

yielding, for the corresponding closed loop, an input-output response of the form

$$y(t) = Q(t, \zeta^\circ, x^\circ) + \int_0^t Ce^{A(t-\sigma)}Bw(\sigma)\,d\sigma. \tag{5.5}$$

In particular, if (5.4) holds, then one can obtain a response of the form (5.5) by choosing

$$\gamma(\zeta, x) = A, \qquad\qquad \delta(\zeta, x) = B$$
$$\alpha(\zeta, x) = \alpha(x) - \beta(x)DA\zeta, \qquad \beta(\zeta, x) = -\beta(x)DB$$

where $\alpha(x)$ and $\beta(x)$ are solutions of (4.12).

Hint. Construct an "error" system

$$\dot{x} = f(x) + \sum_{i=1}^m g_i(x)u_i$$
$$\dot{\zeta} = A\zeta + Bw$$
$$e = h(x) - C\zeta$$

and solve for this one the Input-Output Exact Linearization Problem. •

We discuss now an elementary application of the design methodologies illustrated in this Chapter to the system which describes the control of the rotation of a rigid spacecraft around its center of mass. Recall (see Section 1.5) that the system in question can be modeled by equations of the form

$$\dot{\omega} = J^{-1}S(\omega)J\omega + J^{-1}T$$
$$\dot{R} = S(\omega)R$$

in which R is a 3×3 orthogonal matrix (with $\det(R) = 1$), which describes the rotation of the spacecraft with respect to an inertially fixed reference frame, and ω is a 3-dimensional vector which expresses its angular velocity (with respect to a reference frame fixed to the spacecraft). In what follows, we assume—as in Section 2.5—that the external control force is exerted by a set of gas jets; accordingly, we set

$$T = Bu$$

where u is a vector which represents the magnitudes of the control torques, and B is a constant matrix. In particular, we assume that 3 independent control torques are available, so that the matrix B is nonsingular.

Our purpose is to obtain, by means of a feedback of the form (2.1), a system in which each component of the new reference input controls, independently of the other ones, the rotation of the spacecraft around one of its reference axis. As customary in aircraft and space mechanics, the manoeuvre needed to rotate the spacecraft—from an initial position in which its reference axes are aligned with the ones of the fixed reference frame—to a generic attitude R, can be executed in the following way. A rotation (yaw) of an angle ψ around the axis a_3, followed by a rotation ($pitch$) of an angle θ around the resulting axis a_2, followed by a rotation ($roll$) of an angle ϕ around the resulting axis a_1 (see Fig.5.2).

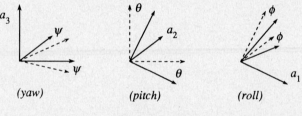

(yaw)   (pitch)   (roll)

Fig. 5.2

The three elementary rotations thus described can be represented, as any rotation, by means of an orthogonal matrix whose entries are direction cosines. An immediate calculation shows that the matrices corresponding to the three elementary rotations in question are, respectively

$$R(\psi) = \begin{bmatrix} \cos\psi & \sin\psi & 0 \\ -\sin\psi & \cos\psi & 0 \\ 0 & 0 & 1 \end{bmatrix}$$

$$R(\theta) = \begin{bmatrix} \cos\theta & 0 & -\sin\theta \\ 0 & 1 & 0 \\ \sin\theta & 0 & \cos\theta \end{bmatrix}$$

$$R(\phi) = \begin{bmatrix} 1 & 0 & 0 \\ 0 & \cos\phi & \sin\phi \\ 0 & -\sin\phi & \cos\phi \end{bmatrix}.$$

Note that

$$R(\psi) = e^{(A_1\psi)}, \qquad R(\theta) = e^{-(A_2\theta)}, \qquad R(\phi) = e^{(A_3\phi)}$$

where the matrices A_1, A_2, A_3 are the three matrices, already introduced in Section 2.5

$$A_1 = \begin{bmatrix} 0 & 1 & 0 \\ -1 & 0 & 0 \\ 0 & 0 & 0 \end{bmatrix} \qquad A_2 = \begin{bmatrix} 0 & 0 & 1 \\ 0 & 0 & 0 \\ -1 & 0 & 0 \end{bmatrix} \qquad A_3 = \begin{bmatrix} 0 & 0 & 0 \\ 0 & 0 & 1 \\ 0 & -1 & 0 \end{bmatrix}.$$

Thus, the manoeuvre previously described brings the attitude of the spacecraft—from an initial value $R = I$ in which its reference axes are aligned with the ones of the fixed reference frame—to a final value R given by

$$R = e^{(A_3\phi)}e^{-(A_2\theta)}e^{(A_1\psi)}$$

This expression, can be interpreted as a smooth mapping

$$F : \mathbf{R}^3 \to SO(3)$$

which assigns to each triplet (ψ, θ, ϕ) an element

$$R = F(\psi, \theta, \phi) = e^{(A_3\phi)}e^{-(A_2\theta)}e^{(A_1\psi)} \tag{5.6}$$

of the set $SO(3)$ of orthogonal 3×3 matrices (whose determinant is equal to 1). It is easy to show that the mapping F is locally *invertible*, in a neighborhood of the value $R = I$ (this is, in fact, a consequence of the property that the mapping in question has rank 3 at the point $(\psi, \theta, \phi) = 0$, because

$$\left[\frac{\partial F}{\partial \phi}\right]_{(\psi,\theta,\phi)=0} = A_3 \qquad \left[\frac{\partial F}{\partial \theta}\right]_{(\psi,\theta,\phi)=0} = -A_2 \qquad \left[\frac{\partial F}{\partial \psi}\right]_{(\psi,\theta,\phi)=0} = A_1$$

and the three matrices A_1, A_2, A_3 are linearly independent). In other words, there exists a neighborhood U of the point $R = I$ in $SO(3)$ with the property that, for each $R \in U$, the relation (5.6) can be satisfied by one and only one triplet (ψ, θ, ϕ). Moreover, the mapping

$$F^{-1} : U \to \mathbf{R}^3$$

which assigns to each $R \in U$ the (unique) triplet $(\psi, \theta, \phi) = F^{-1}(R)$ which satisfies (5.6), is a smooth mapping.

We see from these arguments that the three angles (ψ, θ, ϕ) can be used to *parametrize*, locally around the point $R = I$, the set of rotation matrices which define the attitude of the spacecraft. Considering these three quantities as *outputs* of the control system, one can pose the problem of finding, if there exists, a static state-feedback of the form (2.1), namely

$$u = \alpha(R, \omega) + \sum_{i=1}^{3} \beta_i(R, \omega) v_i \qquad (5.7)$$

which renders the angle ψ dependent only on the input v_1, the angle θ dependent only on the input v_2, and the angle ϕ dependent only on the input v_3, that is to solve—for the system in question—the noninteracting control problem.

Note that the functions which characterize the feedback (5.7) are formally expressed as functions of the state (R, ω) of the system. However, if the value of the attitude R belongs to the set U in which the mapping (5.6) is invertible, we can replace R by $F(\psi, \theta, \phi)$, and therefore rewrite the right-hand side of (5.7) as a function of the six variables $(\psi, \theta, \phi, \omega_1, \omega_2, \omega_3)$.

In order to check whether or not the noninteracting control problem is solvable, one has to calculate the integers r_1, r_2, r_3 and check whether or not the matrix (1.2) is invertible. However, the calculation of quantities of the form

$$L_{g_j} L_f^k h_i(x)$$

cannot be directly pursued in this case, because an explicit expression of the function $h_i(x)$, which is the i-th component of mapping

$$\begin{bmatrix} y_1 \\ y_2 \\ y_3 \end{bmatrix} = \begin{bmatrix} \psi \\ \theta \\ \phi \end{bmatrix} = F^{-1}(R)$$

is not available. Instead, we calculate r_1, r_2, r_3 and the matrix (1.2) indirectly, by appealing to the interpretation of r_i as the least integer for which the r_i-th derivative of y_i with respect to time depends explicitly on the input.

The problem is to differentiate with respect to t the functions $\psi(t)$, $\theta(t)$, $\phi(t)$.

To this end, it is to convenient compare the expression of

$$\dot{R} = \frac{d}{dt}e^{(A_3\phi(t))}e^{-(A_2\theta(t))}e^{(A_1\psi(t))}$$

with

$$\dot{R} = S(\omega(t))R(t).$$

Since

$$\frac{d}{dt}(e^{(A_3\phi(t))}e^{-(A_2\theta(t))}e^{(A_1\psi(t))}) = (\dot{\phi}A_3 - \dot{\theta}e^{(A_3\phi)}A_2e^{-(A_3\phi)}$$
$$+ \dot{\psi}e^{(A_3\phi)}e^{-(A_2\theta)}A_1e^{(A_2\theta)}e^{-(A_3\phi)})R$$

and R is an invertible matrix, we obtain from these expression a relation

$$\dot{\phi}A_3 - \dot{\theta}e^{(A_3\phi)}A_2e^{-(A_3\phi)} + \dot{\psi}e^{(A_3\phi)}e^{-(A_2\theta)}A_1e^{(A_2\theta)}e^{-(A_3\phi)} = S(\omega)$$

which must be solved for $\dot{\psi}, \dot{\theta}, \dot{\phi}$.

Observe that all matrices in this expression are skew-symmetric and that, in particular

$$e^{(A_3\phi)}A_2e^{-(A_3\phi)} = \begin{bmatrix} 0 & \sin\phi & \cos\phi \\ -\sin\phi & 0 & 0 \\ -\cos\phi & 0 & 0 \end{bmatrix}$$

$$e^{(A_3\phi)}e^{-(A_2\theta)}A_1e^{(A_2\theta)}e^{-(A_3\phi)} = \begin{bmatrix} 0 & \cos\theta\cos\phi & -\cos\theta\sin\phi \\ -\cos\theta\cos\phi & 0 & -\sin\theta \\ \cos\theta\sin\phi & \sin\theta & 0 \end{bmatrix}.$$

Solving the previous relation for $\dot{\psi}, \dot{\theta}, \dot{\phi}$ yields, after some simple calculations

$$\mathrm{col}(\dot{\psi}, \dot{\theta}, \dot{\phi}) = M(\psi, \theta, \phi)\omega$$

where

$$M(\psi, \theta, \phi) = \begin{bmatrix} 0 & \sin\phi\sec\theta & \cos\phi\sec\theta \\ 0 & \cos\phi & -\sin\phi \\ 1 & \sin\phi\tan\theta & \cos\phi\tan\theta \end{bmatrix}$$

is a matrix which, as shown, depends only on (ψ, θ, ϕ), which is invertible for all (ψ, θ, ϕ) in a neighborhood of the origin.

Since no component of the first derivative of $y(t)$ depends explicitly on the input u, we go to the second derivative. Clearly,

$$y^{(2)} = \frac{dM}{dy}\frac{dy}{dt} + M\frac{d\omega}{dt} = \frac{dM}{dy}M\omega + MJ^{-1}S(\omega)J\omega + MJ^{-1}Bu.$$

The second derivative of $y(t)$ has a form of the type

$$y^{(2)} = b(\psi,\theta,\phi,\omega_1,\omega_2,\omega_3) + A(\psi,\theta,\phi)u.$$

From this we deduce that $r_1 = r_2 = r_3 = 2$. Moreover, since the matrix

$$A(\psi,\theta,\phi) = M(\psi,\theta,\phi)J^{-1}B$$

is invertible at $(\psi,\theta,\phi) = (0,0,0)$, we conclude that the system has relative degree $\{2,2,2\}$ at this point and the noninteracting control problem is solvable. A static state feedback which solves this problem is given by

$$u = A^{-1}(\psi,\theta,\phi)(v - b(\psi,\theta,\phi,\omega_1,\omega_2,\omega_3)). \tag{5.8}$$

Note also that, since the state space of the system has dimension 6 (see Section 1.5), the condition

$$n = r_1 + r_2 + r_3$$

is also satisfied and the system is exactly linearizable. In fact, in the coordinates

$$x_1 = \mathrm{col}(\psi,\theta,\phi)$$
$$x_2 = M(\psi,\theta,\phi)\omega$$

the closed loop system obtained by means of the feedback (5.8) becomes

$$\dot{x}_1 = x_2$$
$$\dot{x}_2 = v$$

6 Geometric Theory of State Feedback: Tools

6.1 The Zero Dynamics

In the last two Chapters of these notes, we discuss problems similar to the ones analyzed in Chapters **4** and **5**, for more general classes of multivariable nonlinear systems. We will not assume anymore that the systems we deal with have some vector relative degree at the point of interest (i.e., in particular, that the matrix (5.1.2) is nonsingular at x°), but we will replace this hypothesis by some milder regularity assumptions, namely the constancy of the dimensions of certain distributions (and/or of the ranks of certain mappings) around a given point. For every nonlinear system of the class we consider in these notes, these assumptions are satisfied at each point of an open and dense set in the state space. In addition to these more "technical" hypotheses, we will assume sometimes that the system—viewed as a mapping between inputs and outputs—is "invertibile", in a sense that will be precised later on.

For convenience, we describe in the present Chapter a number of basic tools that are of fundamental importance in the analysis of this broader class of systems, and we defer to the last Chapter the illustration of how these tools are used in the solution of specific control problems. We begin by discussing—in a rather general setting—the problem of how the output of a nonlinear system of the form

$$\dot{x} = f(x) + g(x)u \tag{1.1a}$$

$$y = h(x) \tag{1.1b}$$

with the *same* number m of input and output components, and state x defined on an open subset U of $\mathbf{R}^n$, can be set to zero by means of a proper choice of initial state and input.

Consider a point x° in the state space of (1.1) and suppose that $f(x^\circ) = 0$ and $h(x^\circ) = 0$. Thus, if the initial state of (1.1) at time $t = 0$ is equal to x° and the input $u(t)$ is zero for all $t \geq 0$, then also the output $y(t)$ is zero for all $t \geq 0$. Our purpose is to identify, if possible, the set of all pairs consisting of an initial state and an input function which produce an identically zero output. To this end, it is convenient to introduce first some terminology.

Let M be a smooth connected submanifold of U which contains the point x°. The manifold M is said to be *locally controlled invariant* at x° if there exists a smooth mapping $u : M \to \mathbf{R}^m$, and a neighborhood U° of x°, such that the vector field $\tilde{f}(x) = f(x) + g(x)u(x)$ is *tangent* to M for all $x \in M \cap U^\circ$ or, what is the same, M is *locally invariant* under the vector field $\tilde{f}(x)$.

An *output zeroing submanifold* of (1.1) is a smooth connected submanifold M of U which contains the point x° and satisfies
(i) for each $x \in M$, $h(x) = 0$;
(ii) M is locally controlled invariant at x°.

In other words, an output zeroing submanifold is a submanifold M of the state space with the property that—for some choice of feedback control $u(x)$—the trajectories of the closed loop system

$$\dot{x} = f(x) + g(x)u(x)$$
$$y = h(x)$$

which start in M stay in M for all times in a neighborhood of the time $t = 0$, and the corresponding output is identically zero in the meanwhile.

If M and M' are two connected smooth submanifolds of U which both contain the point x°, it is said that M *locally contains* M' (or, M *coincides with* M') if, for some neighborhood U° of x°, $M \cap U^\circ \supset M' \cap U^\circ$ (or $M \cap U^\circ = M' \cap U^\circ$). An output zeroing submanifold M is *locally maximal* if, for some neighborhood U° of x°, any other output zeroing submanifold M' satisfies $M \cap U^\circ \supset M' \cap U^\circ$.

In general, it is not clear whether or not a locally maximal output zeroing submanifold might exist at all. However, under some mild regularity assumptions, in a neighborhood of x° a manifold $Z^\star$ satisfying the said requirements can be found rather easily, as the recursive construction that we will describe in a moment shows. Note that requirement (i) implies $h(x^\circ) = 0$, i.e. that the point x° belongs to the inverse image, noted $h^{-1}(0)$, of the point $y = 0$ with respect to the output mapping h. This motivates the consideration of the sequence of nested submanifolds $M_0 \supset M_1 \supset \ldots \supset M_k \supset \ldots$ defined in the following way.

Zero Dynamics Algorithm. Step 0: set $M_0 = h^{-1}(0)$. Step k: suppose that, for some neighborhood U_{k-1} of x°, $M_{k-1} \cap U_{k-1}$ is a smooth submanifold, let M_{k-1}^c denote the connected component of $M_{k-1} \cap U_{k-1}$ which contains the point x° (M_{k-1}^c is nonempty because $f(x^\circ) = 0$) and define M_k as

$$M_k = \{x \in M_{k-1}^c : f(x) \in \mathrm{span}\{g_1(x), \ldots, g_m(x)\} + T_x M_{k-1}^c\}. \quad \bullet \qquad (1.2)$$

The following statements describe conditions under which the sequence thus defined converges to a locally maximal output zeroing submanifold.

Proposition 1.1. Suppose that, for each $k \geq 0$, there exists a neighborhood U_k of x° such that $M_k \cap U_k$ is a smooth submanifold. Then, for some $k^\star < n$

and some neighborhood U_{k^*} of x°, $M_{k^*+1} = M_{k^*}^c$. Suppose also that

$$\dim(\text{span}\{g_1(x^\circ), \ldots, g_m(x^\circ)\}) = m, \qquad (1.3a)$$

and that the subspace $\text{span}\{g_1(x), \ldots, g_m(x)\} \cap T_x M_{k^*}^c$ has constant dimension for all $x \in M_{k^*}^c$. Then, the manifold $Z^* = M_{k^*}^c$ is a locally maximal output zeroing submanifold for (1.1).

Proposition 1.2. If, in addition

$$\text{span}\{g_1(x^\circ), \ldots, g_m(x^\circ)\} \cap T_{x^\circ} Z^* = 0, \qquad (1.3b)$$

then there exists a *unique* smooth mapping $u^* : Z^* \to \mathbf{R}^m$ such that the vector field

$$f^*(x) = f(x) + g(x)u^*(x)$$

is tangent to Z^*. •

Proof of Proposition 1.1. Since all M_k's are locally smooth submanifolds and $M_k \supset M_{k+1}$, a dimensionality argument shows that for some integer $k^* < n$ and some neighborhood U_{k^*} of x°, $M_{k^*+1} = M_{k^*}^c$. Set $Z^* = M_{k^*}^c$. Since $M_{k^*+1} = Z^*$, then by construction, at each point x of Z^*, there exists a vector $u \in \mathbf{R}^m$ such that

$$f(x) + g(x)u \in T_x Z^*. \qquad (1.4)$$

Since Z^* is a smooth submanifold, in a neighborhood U' of x° it is possible to define a submersion $H : U' \to \mathbf{R}^q$ (where $q = n - \dim(Z^*)$), such that

$$Z^* \cap U' = \{x \in U' : H(x) = 0\}.$$

As a consequence, since $T_x Z^* = \ker(dH(x))$ at each $x \in (Z^* \cap U')$, the condition (1.4) can be reexpressed in the equivalent form

$$\langle dH(x), f(x) + g(x)u \rangle = 0.$$

From the fact that this equation can be solved for u we deduce that

$$\langle dH(x), f(x) \rangle \in \text{Im}(\langle dH(x), g(x) \rangle) \qquad (1.5)$$

at each point x of $Z^* \cap U^\circ$.

If (1.3a) holds and the subspace $\text{span}\{g_1(x), \ldots, g_m(x)\} \cap T_x Z^*$ has constant dimension for all $x \in Z^*$ near x°, the matrix $\langle dH(x), g(x) \rangle$ has constant rank at each $x \in Z^*$ near x°. Therefore from (1.5) we deduce the existence—on some neighborhood $U'' \subset U'$ of x°—of a smooth mapping $u^* : U'' \to \mathbf{R}^m$ such that

$$f(x) + g(x)u^*(x) \in T_x Z^*$$

for all $x \in Z^\star \cap U''$. Thus, $Z^\star$ is locally controlled invariant at x°.

$Z^\star$ is also such that $h(x) = 0$ for all $x \in Z^\star$, by construction. Observe now that any other output zeroing submanifold Z' is necessarily such that $Z' \subset M_k$ near x°, for all $k \geq 0$. This is proved, by induction, showing that $Z' \subset M_{k-1}^c$ implies $Z' \subset M_k$. In fact

$$x \in Z' \Rightarrow f(x) \in \mathrm{span}\{g_1(x), \ldots, g_m(x)\} + T_x Z'$$
$$\Rightarrow f(x) \in \mathrm{span}\{g_1(x), \ldots, g_m(x)\} + T_x M_{k-1}^c$$
$$\Rightarrow x \in M_k.$$

From this we deduce that Z' is locally contained into $Z^\star$, i.e. that $Z^\star$ is locally maximal. ●

Proof of Proposition 1.2. Note that, if (1.3b) holds, the matrix $\langle dH(x), g(x)\rangle$ has rank m for all x near x°. For, the identity $\langle dH(x), g(x)\rangle \gamma = 0$ would imply either $g(x)\gamma = 0$, which is contradicted by (1.3a) or $g(x)\gamma \in \ker(dH)$ which is contradicted by (1.3b). As a consequence, in this case the mapping $u^\star(x)$ found in the proof of Proposition 1.1 is unique. ●

Suppose the hypotheses listed in the previous Propositions are satisfied. Since the vector field $f^\star(x)$ is tangent to $Z^\star$, the *restriction* $f^\star(x)|_{Z^\star}$ of $f^\star(x)$ to $Z^\star$ is a well defined vector field of $Z^\star$ (in what follows, whenever there is no danger of confusion, in order to simplify the notation we will often use $f^\star(x)$ instead of $f^\star(x)|_{Z^\star}$). The submanifold $Z^\star$ is called the (local) *zero dynamics submanifold* and the vector field $f^\star(x)$ of $Z^\star$ is called the *zero dynamics vector field*. The pair $(Z^\star, f^\star)$ is called the *zero dynamics* of the system (1.1).

By construction, the dynamical system

$$\dot{x} = f^\star(x), \qquad x \in Z^\star \tag{1.6}$$

identifies the internal dynamical behavior induced on the system when the output has been forced, by proper choice of initial state and input, to remain zero for some interval of time. In fact, setting $x(0) = x^\circ \in Z^\star$ and

$$u(t) = u^\star(x(t))$$

where $x(t)$ is the solution of (1.6) initialized at $x^\circ \in Z^\star$, an output $y(t)$ is obtained which is identically zero as long as $x(t)$ remains in $Z^\star$ (i.e. for some open interval of the time axis).

Remark 1.3. The approach followed here arrives at a *local* characterization of the notion of zero dynamics. Of course, a global characterization would also—in principle—be possible, identifying $Z^\star$ as the largest (with respect to inclusion) smooth submanifold of $h^{-1}(0)$ having the property that, for some smooth input

$u^\star : Z^\star \to \mathbf{R}^m$, the vector field $f^\star = f + gu^\star$ is tangent to $Z^\star$. •

Remark 1.4. The hypoteses (1.3a) and (1.3b) of Propositions 1.1 and 1.2 can be interpreted as a special property of *invertibility* for the system. As a matter of fact , from the proofs of these Propositions, it is easy to see that if (and only if) these two conditions are satisfied then, for each initial state x' (in a neighborhood of x°), any two pairs (x', u^a) and (x', u^b) producing zero output are necessarily equal, i.e. have $u^a = u^b$. •

Remark 1.5. It is useful to show what the arguments illustrated reduces to in the case of a linear system. As far as the Zero Dynamics Algorithm is concerned, the reader should not have much difficulty in realizing that

$$M_0 = \ker(C)$$

and

$$M_k = \{x \in M_{k-1} : Ax \in \mathrm{Im}(B) + M_{k-1}\}.$$

Thus, all M_k's are *subspaces* of the state space. The smoothness assumptions are indeed satisfied, and there is an integer $k^\star < n$ such that $M_{k^\star+1} = M_{k^\star}$. The subspace $V^\star = M_{k^\star}$ is by construction the largest subspace of $\ker(C)$ satisfying

$$AV^\star \subset V^\star + \mathrm{Im}(B).$$

The hypotheses (1.3) reduce to the following ones

$$\dim(\mathrm{Im}(B)) = m, \qquad V^\star \cap \mathrm{Im}(B) = 0$$

and these, as is known from the theory of linear systems, are exactly conditions under which the transfer function matrix $C(sI - A)^{-1}B$ (which is square because by assumption the system has the same number m of input and output components) is *invertible* (see also Remark 1.4).

These two conditions imply the existence of a unique state feedback $u^\star(x)$, which now is a linear function of x, namely

$$u^\star(x) = Fx$$

such that $f^\star(x) = Ax + Bu^\star(x)$ is tangent to $V^\star$, i. e. such that $(A + BF)x$ is in $V^\star$ for all $x \in V^\star$, namely,

$$(A + BF)V^\star \subset V^\star.$$

The subspace $V^\star$ is *invariant* under the linear mapping $(A + BF)$ and the restriction $(A + BF)|_{V^\star}$ identifies a linear dynamical system, defined on $V^\star$, whose dynamics are by definition the zero dynamics of the original system. We will show now that the eigenvalues of $(A + BF)|_{V^\star}$ (more precisely, its *invariant polynomials*) coincide with the so-called *transmission zeros* (more precisely, with the *transmission polynomials*) of the transfer function matrix of the system. This property enables us to extend to the case of multivariable systems the interpretation, already given in the Remark 4.3.1, of the zero dynamics as a nonlinear analogue of the notion of "zeros" of a linear system.

For, recall that the transmission polynomials of a multivariable minimal linear system are defined as the invariant factors of the so-called *system matrix*:

$$\begin{bmatrix} sI - A & B \\ C & 0 \end{bmatrix}.$$

(1.7)

We will prove that the invariant factors of this matrix coincide with those of the linear mapping $(A + BF)|_{V^\star}$. To this end, choose suitable new coordinates

$$\bar{x} = \mathrm{col}(x_1, x_2)$$

such that the subspace $V^\star$ assumes the form

$$V^\star = \{(x_1, x_2) \in \mathbf{R}^n : x_1 = 0\}$$

and such that

$$\mathrm{Im}(B) \subset \{(x_1, x_2) \in \mathbf{R}^n : x_2 = 0\}.$$

This is indeed possible because $V^\star \cap \mathrm{Im}(B) = 0$. Accordingly, the matrices A, B, C will be represented in a form of the type

$$\begin{bmatrix} A_{11} & A_{12} \\ A_{21} & A_{22} \end{bmatrix} \qquad \begin{bmatrix} B_1 \\ 0 \end{bmatrix} \qquad [C_1 \quad 0]$$

(the special structure of C is due to the fact that $V^\star \subset \ker(C)$).

Observe that, since $AV^\star \subset V^\star + \mathrm{Im}(B)$, the matrices A_{12} and B_1 satisfy the condition

$$\mathrm{Im}(A_{12}) \subset \mathrm{Im}(B_1).$$

Therefore there exists a (unique, because B_1 has rank m) matrix F_2 such that

$$B_1 F_2 = -A_{12}.$$

Setting

$$F = [0 \quad F_2]$$

it follows that

$$A + BF = \begin{bmatrix} A_{11} & 0 \\ A_{21} & A_{22} \end{bmatrix}.$$

We see from this that $V^\star$ is invariant under $(A + BF)$ and, in particular, that A_{22} is a representation of the linear mapping $(A + BF)|_{V^\star}$. Moreover, it is easy to see that the maximality of $V^\star$ implies the nonsingularity of the matrix

$$\begin{bmatrix} sI - A_{11} & B_1 \\ C_1 & 0 \end{bmatrix} \tag{1.8}$$

for all $s \in \mathbf{C}$. In fact, suppose that for some s° the matrix in question is singular. Then, there exists vectors x_1° and u such that

$$s^\circ x_1^\circ - A_{11} x_1^\circ + B_1 u = 0, \qquad C_1 x_1^\circ = 0$$

If this is the case, then the subspace

$$V = V^\star + \mathrm{span}\{\mathrm{col}(x_1^\circ, 0)\}$$

satisfies

$$AV \subset V + \mathrm{Im}(B), V \subset \ker(C)$$

i.e. a contradiction, because V properly contains $V^\star$ and $V^\star$ is the largest subspace satisfying these conditions.

Observe now that

$$\begin{bmatrix} sI - A - BF & B \\ C & 0 \end{bmatrix} = \begin{bmatrix} sI - A & B \\ C & 0 \end{bmatrix} \begin{bmatrix} I & 0 \\ -F & I \end{bmatrix} \tag{1.9}$$

and therefore that the invariant polynomials of the matrix (1.7) and those of the left-hand side of (1.9) coincide. Replacing in the latter the expressions previously established for $A + BF$, B, C and taking a permutation of rows and columns, one obtains a matrix—whose invariant polynomials still coincide with those of (1.7)- –of the form

$$\begin{bmatrix} sI - A_{11} & B_1 & 0 \\ C_1 & 0 & 0 \\ -A_{21} & 0 & sI - A_{22} \end{bmatrix} \tag{1.10}$$

Since the submatrix (1.8) is nonsingular for all $s \in \mathbf{C}$, the invariant polynomials of (1.10) coincide with those of A_{22}, and this shows that the invariant polynomials of (1.7) are exactly those of $(A + BF)|_{V^\star}$.

Before closing this Remark, note also that the nonsingularity of the matrix (1.8) for all $s \in \mathbf{C}$ implies, in view of a well known controllability condition, that the pair (A_{11}, B_1) is a *controllable* pair. From this, it can be immediately deduced that, if all the eigenvalues of the matrix A_{22} have negative real part, the pair

$$\begin{bmatrix} A_{11} & 0 \\ A_{21} & A_{22} \end{bmatrix} \qquad \begin{bmatrix} B_1 \\ 0 \end{bmatrix}$$

is stabilizable. Since the latter has been deduced from the original pair (A, B) by means of a regular feedback, which preserves stabilizability, we can conclude that a suffcient condition for a linear system to be stabilizable by means of static state feedback is that its zero dynamics are asymptotically stable. We will find in Section 7.1 a nonlinear version of this condition. ●

We proceed now to illustrate how the zero dynamics algorithm can be implemented in practice on a given system of the form (1.1) and, in doing so, we also show how the various regularity assumptions indicated in the description of the algorithm (namely, the smoothness of $M_k \cap U_k$ for all $k \geq 1$) as well as the hypothesis (1.3b) can be tested.

At the beginning, M_0 is defined as the set of points where the mapping h is zero; if the differential dh of this mapping has constant rank, say s_0, near x°, for some neighborhood U_0, then the set $M_0 \cap U_0$ is a smooth $(n - s_0)$-dimensionalmanifold. If s_0 is strictly less than the number m of components of h, without loss of generality it is assumed that exactly the first s_0 rows of dh are independent (otherwise, the order of the rows of h is changed). Thus, if S_0 denotes the matrix which selects the first s_0 rows of an s-dimensional vector, namely the $(s_0 \times s)$ matrix

$$S_0 = [I \quad 0]$$

the mapping

$$H_0(x) = S_0 h(x)$$

is clearly such that

$$M_0 \cap U_0 = \{x \in U_0 : H_0(x) = 0\}.$$

Let M_0^c denote the connected component of $M_0 \cap U_0$ which contains the point x°. At the first step of the Zero Dynamics Algorithm, M_1 can be calculated (see also the proof of Proposition 1.1) as the set of all $x \in M_0^c$ such that

$$\langle dH_0(x), f(x) + g(x)u \rangle = L_f H_0(x) + L_g H_0(x)u = 0 \qquad (1.11)$$

is solvable in u. Suppose the matrix $L_g H_0(x)$ has constant rank, say r_0, for all $x \in M_0^c$ (note that the constancy of the rank is required only on this submanifold and not on the whole U_0). Then, the space of solutions γ of the linear equation

$$\gamma L_g H_0(x) = 0$$

has constant dimension (namely, $s_0 - r_0$) for all $x \in M_0^c$. Since $L_g H_0(x)$ is smooth, for some neighborhood $U_0' \subset U_0$ of x° it is possible to define an $(s_0 - r_0) \times s_0$ matrix $R_0(x)$ of smooth functions of x, such that at each $x \in (M_0^c \cap U_0')$ the rows of $R_0(x)$ span the space of solutions of this equation. In particular

$$R_0(x)L_g H_0(x) = 0$$

for all $x \in (M_0^c \cap U_0')$, and it is immediately seen that at each point $x \in (M_0^c \cap U_0')$ the equation (1.11) is solvable in u if and only if x is such that

$$R_0(x)L_f H_0(x) = 0.$$

Setting

$$\Phi_0(x) = R_0(x)L_f H_0(x)$$

then for some neighborhood U_1 of x° the set $M_1 \cap U_1$ can be expressed in the form

$$M_1 \cap U_1 = \{x \in U_1 : H_0(x) = 0 \text{ and } \Phi_0(x) = 0\}.$$

If the smooth mapping $\mathrm{col}(H_0(x), \Phi_0(x))$ has constant rank, say $s_0 + s_1$, near x°, then the previous construction can be iterated once more. Note that since the vector $\Phi_0(x)$ has $(s_0 - r_0)$ rows, then

$$s_1 \leq s_0 - r_0. \tag{1.12a}$$

At step $k+1 \geq 2$, the iteration is started with mappings $H_{k-1}(x)$ and $\Phi_{k-1}(x)$, where H_{k-1} is such that the rank of dH_{k-1} is exactly equal to the number $s_0 + \ldots + s_{k-1}$ of its rows, and, for some neighborhood U_k of x°,

$$M_k \cap U_k = \{x \in U_k : H_{k-1}(x) = 0 \text{ and } \Phi_{k-1}(x) = 0\}$$

Suppose the mapping $\mathrm{col}(H_{k-1}(x), \Phi_{k-1}(x))$ has constant rank $(s_0 + \ldots + s_k)$ near x° (thus, for a suitable choice of U_k, the set $M_k \cap U_k$ is a smooth manifold). Without loss of generality, assume that the first $(s_0 + \ldots + s_k)$ rows of the differential of this mapping are linearly independent (otherwise, change the order in the last set of rows, namely in Φ_{k-1}), let S_{k-1} denote the matrix which selects the first s_k rows of $\Phi_{k-1}(x)$, and set

$$H_k(x) = \mathrm{col}(H_{k-1}(x), S_{k-1}\Phi_{k-1}(x)).$$

Obviously

$$M_k \cap U_k = \{x \in U_k : H_k(x) = 0\}.$$

One has to look now at the equation

$$L_f H_k(x) + L_g H_k(x)u = 0.$$

If the matrix $L_g H_k(x)$ has constant rank r_k for all $x \in M_k^c$ (the connected component of $M_k \cap U_k$ through x^o) it is possible to find a matrix $R_k(x)$ of smooth functions whose $(s_0 + \ldots + s_k - r_k)$ rows span (locally around x^o) the space of solutions of

$$\gamma L_g H_k(x) = 0.$$

Clearly it is possible to choose

$$R_k(x) = \begin{bmatrix} R_{k-1}(x) & 0 \\ P_{k-1}(x) & Q_{k-1}(x) \end{bmatrix}$$

where $P_{k-1}(x)$ and $Q_{k-1}(x)$ are suitable matrices, because by construction $R_{k-1}(x)L_g H_{k-1}(x) = 0$ for all x near x^o in M_k^c. Moreover, since $R_{k-1}(x)$ has $(s_0 + \ldots + s_{k-1} - r_{k-1})$ rows, $P_{k-1}(x)$ and $Q_{k-1}(x)$ have $(s_k - r_k + r_{k-1})$ rows each. From this choice of $R_k(x)$, one can define a new mapping $\Phi_k(x)$ as

$$\Phi_k(x) = P_{k-1}(x)L_f H_{k-1}(x) + Q_{k-1}(x)L_f S_{k-1}\Phi_{k-1}(x)$$

and continue the construction. Note that, since Φ_k has $(s_k - r_k + r_{k-1})$ rows, then

$$s_{k+1} \le s_k - r_k + r_{k-1}. \tag{1.12b}$$

Remark 1.6. Note that the integers s_k and r_k introduced in the previous calculations can be characterized also as

$$s_k = \dim(M_{k-1}) - \dim(M_k)$$
$$r_k = \dim(\text{span}\{g_1(x), \ldots, g_m(x)\}) - \dim(\text{span}\{g_1(x), \ldots, g_m(x)\} \cap T_x M_k)$$

Thus, in particular, the invertibility hypothesis (1.3) can be expressed in the form

$$r_{k^*} = \text{rank}(L_g H_{k^*}(x^o)) = m. \quad \bullet$$

Before proceeding further with the analysis of some properties of the construction thus described, we illustrate it with the aid of two simple examples.

Example 1.1. Consider a system of the form (1.1), with 2 inputs and 2 outputs, defined on $\mathbf{R}^4$, with

$$f(x) = \begin{bmatrix} 0 \\ x_4 \\ \lambda x_3 + x_4 \\ 0 \end{bmatrix} \qquad g_1(x) = \begin{bmatrix} 1 \\ x_3 \\ 0 \\ 0 \end{bmatrix} \qquad g_2(x) = \begin{bmatrix} 0 \\ 0 \\ 0 \\ 1 \end{bmatrix}$$

$$h_1(x) = x_1$$
$$h_2(x) = x_2.$$

First of all, note that this system has no relative degree, because the matrix (5.1.2), which in this case has the form

$$L_g h(x) = \begin{bmatrix} 1 & 0 \\ x_3 & 0 \end{bmatrix}$$

has rank 1 for all x. Proceeding with the zero dynamics algorithm, we see that dh has rank 2 for all x. Thus $s_0 = 2$, $H_0 = h$ and

$$M_0 = \{x \in \mathbf{R}^4 : x_1 = x_2 = 0\}.$$

We construct the matrix $L_g H_0(x)$, whose rank r_0—as already observed—is 1 for all x and we set

$$R_0(x) = [-x_3 \quad 1].$$

Thus

$$\Phi_0(x) = x_4.$$

Since the rank of the mapping $\mathrm{col}(H_0(x), \Phi_0(x))$ is 3 for all x, we have $s_1 = 1$, $H_1(x) = \mathrm{col}(x_1, x_2, x_4)$ and

$$M_1 = \{x \in \mathbf{R}^4 : x_1 = x_2 = x_4 = 0\}.$$

The matrix

$$L_g H_1(x) = \begin{bmatrix} 1 & 0 \\ x_3 & 0 \\ 0 & 1 \end{bmatrix}$$

has rank $r_1 = 2$ at all $x \in M_1$, and the algorithm terminates. We have $Z^\star = M_1$, and the unique $u^\star(x)$ that keeps the state of the system evolving on $Z^\star$ must solve, at each $x \in Z^\star$, the equation

$$L_f H_1(x) + L_g H_1(x) u^\star(x) = \begin{bmatrix} 0 \\ x_4 \\ 0 \end{bmatrix} + \begin{bmatrix} 1 & 0 \\ x_3 & 0 \\ 0 & 1 \end{bmatrix} u^\star(x).$$

Thus

$$u^\star(x) = 0.$$

Accordingly, the zero dynamics of the system—those of $f^\star(x)|_{Z^\star}$—are described by

$$\dot{x}_3 = \lambda x_3. \quad \bullet$$

Example 1.2. Consider a system of the form (1.1), with 2 inputs and 2 outputs, defined on $\mathbf{R}^5$, with

$$f(x) = \begin{bmatrix} x_2 \\ x_4 \\ x_1 x_4 \\ x_5 \\ x_3 \end{bmatrix} \qquad g_1(x) = \begin{bmatrix} 1 \\ x_3 \\ 0 \\ x_5 \\ 1 \end{bmatrix} \qquad g_2(x) = \begin{bmatrix} 0 \\ x_2 \\ 1 \\ x_2 \\ 1 \end{bmatrix}$$

$$h_1(x) = x_1$$
$$h_2(x) = x_2.$$

This system has no relative degree at $x^\circ = 0$, because the matrix (5.1.2), which in this case has the form

$$L_g h(x) = \begin{bmatrix} 1 & 0 \\ x_3 & x_2 \end{bmatrix}$$

is singular at $x^\circ = 0$. Proceeding with the zero dynamics algorithm, we find $s_0 = 2$, $H_0 = h$ and

$$M_0 = \{x \in \mathbf{R}^5 : x_1 = x_2 = 0\}.$$

The matrix $L_g H_0(x)$ has rank $r_0 = 1$ for all $x \in M_0$ and the algorithm can be continued. Choosing

$$R_0(x) = [-x_3 \quad 1]$$

the product

$$R_0(x) L_g H_0(x) = [0 \quad x_2]$$

vanishes at each $x \in M_0$. Then

$$\Phi_0(x) = x_4 - x_2 x_3.$$

The rank of the mapping $\mathrm{col}(H_0(x), \Phi_0(x))$ is 3 for all x, we have $s_1 = 1$, $H_1(x) = \mathrm{col}(x_1, x_2, x_4 - x_2 x_3)$, and

$$M_1 = \{x \in \mathbf{R}^5 : x_1 = x_2 = x_4 = 0\}.$$

The matrix

$$L_g H_1(x) = \begin{bmatrix} 1 & 0 \\ x_3 & x_2 \\ x_5 - x_3^2 & -x_2 x_3 \end{bmatrix}$$

has still rank $r_1 = 1$ at all $x \in M_1$. We choose now

$$R_1(x) = \begin{bmatrix} -x_3 & 1 & 0 \\ x_3^2 - x_5 & 0 & 1 \end{bmatrix}$$

thus obtaining

$$\Phi_1(x) = -x_2 x_5 + x_2 x_3^2 - x_3 x_4 - x_1 x_2 x_4 + x_5.$$

The mapping $\mathrm{col}(H_1(x), \Phi_1(x))$ has rank 4, and we may set

$$H_2(x) = \mathrm{col}(x_1, x_2, x_4 - x_2 x_3, -x_2 x_5 + x_2 x_3^2 - x_3 x_4 - x_1 x_2 x_4 + x_5)$$

and

$$M_2 = \{x \in \mathbf{R}^5 : x_1 = x_2 = x_4 = x_5 = 0\}.$$

The matrix $L_g H_2(x)$ has rank 2 at $x^\circ = 0$, and therefore the algorithm terminates. We have $Z^\star = M_2$, and the unique $u^\star(x)$ that keeps the state of the system evolving on $Z^\star$ is a solution of

$$(L_f H_2(x) + L_g H_2(x) u^\star(x))|_{x \in Z^\star} = 0$$

that is

$$u^\star(x) = \mathrm{col}(0, -x_3).$$

Accordingly, the zero dynamics of the system—those of $f^\star(x)|_{Z^\star}$—are described by

$$\dot{x}_3 = -x_3. \quad \bullet$$

The constructions indicated above, and consequently the various regularity assumptions about the ranks of the mappings $\mathrm{col}(H_k(x), \Phi_k(x))$ and about the ranks of the matrices $L_g H_k(x)$, are apparently depending on the choice that, at each iteration, is made of the matrices $R_k(x)$ which annihilate $L_g H_k(x)$. However, we shall see in a moment that this is not the case if the invertibility hypotheses (1.3) are assumed. To this end, the following result is helpful.

Proposition 1.7. Assume the following
(i) $dh(x)$ has constant rank for all x around x^o and, for some choice of matrices $R_0(x), \dots, R_{k^\star - 1}(x)$, the differentials of the mappings $\mathrm{col}(H_k(x), \Phi_k(x))$, i.e. the matrices $\mathrm{col}(dH_k(x), d\Phi_k(x))$, have constant rank for all x around x^o, $0 \leq k \leq k^\star - 1$,
(ii) the matrices $L_g H_k(x)$ have constant rank for all $x \in M_k$ around x^o, $0 \leq k \leq k^\star - 1$,
(iii) the matrix $L_g H_{k^\star}(x^o)$ has rank m.
Then $s_0 = m$, $s_1 = s_0 - r_0$, and for all $k > 1$, $s_{k+1} = s_k - r_k + r_{k-1}$. As a consequence

$$H_0(x) = h(x) \tag{1.13a}$$

and

$$H_{k+1}(x) = \begin{bmatrix} H_k(x) \\ \Phi_k(x) \end{bmatrix} \tag{1.13b}$$

(i.e. there is no need to discard rows of $\Phi_k(x)$ in order to define $H_{k+1}(x)$).

Moreover, any other choice of matrices $R_0(x), \dots, R_{k^\star - 1}(x)$ is such that the conditions (i), (ii), (iii) are still satisfied.

Proof. Recall that

$$s_0 \leq m \tag{1.12c}$$

By definition, $s_{k^\star + 1} = 0$ and, by assumption, $r_{k^\star} = m$. Using this and all (1.12) together, we obtain

$$m = r_{k^\star} \leq s_{k^\star} + r_{k^\star - 1} \leq s_{k^\star - 1} + r_{k^\star - 2} \leq \dots \leq s_1 + r_0 \leq s_0 \leq m$$

thus concluding that the sign of equality must hold in all of (1.12). This proves the first part of the statement because, since $s_{k+1} = s_k - r_k + r_{k-1}$, all the rows of $\mathrm{col}(dH_k(x), d\Phi_k(x))$ are linearly independent, and the selection matrix S_k reduces to the identity matrix.

In order to prove the second part, we will show first, by induction, that a different choice of $R_0(x), \dots, R_k(x)$ (i.e. a different choice of $R_0(x)$ and $P_k(x)$, $Q_k(x)$, for $k \geq 0$) yields a sequence of mappings $\tilde{H}_0(x), \tilde{\Phi}_0(x), \dots, \tilde{\Phi}_k(x)$ related to the former by

$$\tilde{H}_0(x) = H_0(x)$$
$$\tilde{\Phi}_0(x) = T_0(x)\Phi_0(x) + V_0(x)$$

$$\cdots$$

$$\tilde{\Phi}_k(x) = F_k(x)H_k(x) + T_k(x)\Phi_k(x) + V_k(x)$$

(1.14)

where $T_i(x)$ is a matrix which is nonsingular at each $x \in M_i$ and $V_i(x)$ vanishes on M_i, for all $0 \leq i \leq k$.

To this end, we show—again by induction—that these relations imply, for each $x \in M_k$,

$$L_{g_i}\tilde{H}_k(x) = S_k(x)L_{g_i}H_k(x)$$

where $S_k(x)$ is a nonsingular matrix, for all $0 \leq i \leq m$. In fact, since

$$L_{g_i}\tilde{H}_{k+1}(x) = \begin{bmatrix} L_{g_i}\tilde{H}_k(x) \\ L_{g_i}\tilde{\Phi}_k(x) \end{bmatrix} = \begin{bmatrix} L_{g_i}\tilde{H}_k(x) \\ L_{g_i}(F_k(x)H_k(x) + T_k(x)\Phi_k(x) + V_k(x)) \end{bmatrix}$$

and since, at each $x \in M_{k+1}$

$$(L_{g_i}F_k(x))H_k(x) + (L_{g_i}T_k(x))\Phi_k(x) = 0$$
$$L_{g_i}V_k(x) = G_k(x)L_{g_i}H_k(x)$$

for some suitable matrix $G_k(x)$ (the latter because the differentials of the entries of $V_k(x)$ are linear combinations of the differentials of the entries of $H_k(x)$ at each $x \in M_{k+1}$), we have

$$L_{g_i}\tilde{H}_{k+1}(x) = \begin{bmatrix} S_k(x) & 0 \\ F_k(x) + G_k(x) & T_k(x) \end{bmatrix} L_{g_i}H_{k+1}(x) = S_{k+1}(x)L_{g_i}H_{k+1}(x).$$

Recall now that $\tilde{R}_{k+1}(x)$ is a matrix whose rows, at each $x \in M_{k+1}$ near x°, are a basis of the space of solutions of the homogeneous linear equation $\gamma L_g\tilde{H}_{k+1}(x) = 0$. Thus, in view of the expression established for $L_{g_i}\tilde{H}_{k+1}(x)$, it is concluded that $\tilde{R}_{k+1}(x)$ has necessarily a form of the type

$$\tilde{R}_{k+1}(x) = M(x)R_{k+1}(x)S_k^{-1}(x) + L_k(x)$$

in which $M(x)$ is nonsingular for all $x \in M_{k+1}$ and $L_k(x)$ vanishes on M_{k+1}. Moreover, since it is requested that the upper-right block of $\tilde{R}_{k+1}(x)$ be zero (and so is the corresponding block of $R_{k+1}(x)$), the $M(x)$ must have, for each

$x \in M_{k+1}$, the form

$$M(x) = \begin{bmatrix} M_{11}(x) & 0 \\ M_{21}(x) & M_{22}(x) \end{bmatrix}$$

Using these expressions in the construction of $\tilde{\Phi}_{k+1}(x)$, a simple calculation shows that the latter can be given the form

$$\tilde{\Phi}_{k+1}(x) = F_{k+1}(x)H_{k+1}(x) + T_{k+1}(x)\Phi_{k+1}(x) + V_{k+1}(x)$$

thus proving the correctness of (1.14).

From the expressions thus proved, using again the fact that $H_0(x)$ and the $\Phi_i(x)$'s, $0 \le i \le k-1$, are vanishing for all $x \in M_k$ near x°, it follows that

$$d\tilde{H}_k(x) = S_k(x)dH_k(x)$$
$$L_g\tilde{H}_k(x) = S_k(x)L_gH_k(x)$$

for each $x \in M_k$ near x°, where $S_k(x)$ is a nonsingular matrix, and this completes the proof of the second part of the statement. •

This result essentially shows that the regularity assumptions (i) and (ii), if the invertibility hypothesis (iii) is satisfied, do not depend on the particular choice of matrices introduced at each iteration of the algorithm. In view of this property, we will say that a point x° is a *regular point* of the zero dynamics algorithm if the conditions (i), (ii), (iii) of Proposition 1.7 are satisfied.

We show now that $h(x)$ and the mappings $\Phi_k(x)$ constructed at each step of the zero dynamics algorithm are heplful in defining a new set of local coordinates around x°, which induces on the equations describing the system a structure of special interest (although not as simple as the normal form analyzed in the previous Chapter). The point of departure is the following result.

Proposition 1.8. If x° is a regular point of the zero dynamics algorithm, the differentials of the entries of

$$\Phi(x) = \text{col}(h(x), \Phi_0(x), \dots, \Phi_{k^*-1}(x)) \tag{1.15}$$

are linearly independent at x°.

Proof. It is immediate from Proposition 1.7. •

The next step of our program is that of using the components of the mapping (1.15) in order to define a new set of local coordinates in the state space. However, before proceeding with this, it is convenient to explain the forthcoming constructions with the aid of a simple example.

Example 1.3. Consider a system with $m = 3$ and suppose the zero dynamics algorithm proceeds in the following way.

Step 1. Let $\langle dh, g \rangle = L_g h$ have the form

$$\langle dh, g \rangle = \begin{bmatrix} L_g h_1 \\ L_g h_2 \\ 0 \end{bmatrix}$$

and rank 1 at each $x \in M_0 = \{x : h_1(x) = h_2(x) = h_3(x) = 0\}$, locally around x°. Then, there exists a smooth function γ, defined locally around x°, such that

$$L_g h_2(x) = -\gamma(x) L_g h_1(x) + \sigma_2(x)$$

with $\sigma_2(x) = 0$ for all $x \in M_0$ (note that $\sigma_2(x)$ is not necessarily zero if x is not in M_0, because the rank of $L_g h(x)$ is not necessarily 1 at one of these points). We can set

$$R_0(x) = \begin{bmatrix} \gamma(x) & 1 & 0 \\ 0 & 0 & 1 \end{bmatrix}$$

and therefore

$$\Phi_0(x) = \begin{bmatrix} \gamma L_f h_1 + L_f h_2 \\ L_f h_3 \end{bmatrix} = \begin{bmatrix} \phi_2 \\ \phi_3 \end{bmatrix}.$$

Step 2. Consider the matrix

$$L_g H_1 = \begin{bmatrix} L_g h \\ L_g \Phi_0 \end{bmatrix}$$

which is 5×3, and suppose it has rank 2 at each $x \in M_1 = \{x \in M_0 : \phi_2(x) = \phi_3(x) = 0\}$, locally around x°, with the first and fourth rows being linearly independent (note that the second one is already dependent on the first one and the third one is vanishing). If δ_1 and δ_2 are smooth functions, defined locally around x° such that

$$L_g \phi_3(x) = -\delta_1(x) L_g h_1(x) - \delta_2(x) L_g \phi_2(x) + \sigma_3(x)$$

with $\sigma_3(x) = 0$ for all $x \in M_1$, we can use

$$R_1(x) = \begin{bmatrix} R_0(x) & [0 \ \ 0] \\ [\delta_1(x) \ \ 0 \ \ 0] & [\delta_2(x) \ \ 1] \end{bmatrix}$$

and then set

$$\Phi_1(x) = \delta_1(x) L_f h_1(x) + \delta_2(x) L_f \phi_2(x) + L_f \phi_3(x) = \psi_3(x).$$

Step 3. Suppose now that the matrix

$$L_g H_2 = \begin{bmatrix} L_g H_1 \\ L_g \Phi_1 \end{bmatrix}$$

which is 6×3 has rank 3 at x^o (thus, in particular, its first, fourth and sixth rows will be linearly independent). If this is the case, then the algorithm terminates, and $Z^\star$ can be locally described, in a neighborhood U of x^o, as

$$Z^\star = \{x \in U : h_1(x) = h_2(x) = h_3(x) = \phi_2(x) = \phi_3(x) = \psi_3(x) = 0\}.$$

The input $u^\star(x)$ which renders the vector field $f^\star(x) = f(x) + g(x)u^\star(x)$ tangent to $Z^\star$ must solve, at each $z \in Z^\star$, the equation

$$L_f H_2(x) + L_g H_2(x)u^\star(x) = 0$$

and is therefore given by

$$u^\star = - \begin{bmatrix} L_g h_1 \\ L_g \phi_2 \\ L_g \psi_3 \end{bmatrix}^{-1} \begin{bmatrix} L_f h_1 \\ L_f \phi_2 \\ L_f \psi_3 \end{bmatrix}$$

(note that the equation for $u^\star(x)$ apparently consists of 6 scalar equation, in which—however—the second, third and fifth one are automatically solved at each $x \in Z^\star$).

By the Proposition 1.8, the functions h_1, h_2, h_3, ϕ_2, ϕ_3, ψ_3 have linearly independent differentials at x^o, so that they can be chosen as a partial set of new local coordinates. Denoting by η the set of complementary coordinates (with $\eta(x^o) = 0$), it is easy to check that in the new coordinates the system is described by

$$\begin{aligned}
\dot{y}_1 &= L_f h_1 + L_g h_1 u \\
\dot{y}_2 &= L_f h_2 + L_g h_2 u = L_f h_2 - \gamma L_g h_1 u + \sigma_2 u \\
&= \phi_2 - \gamma(L_f h_1 + L_g h_1 u) + \sigma_2 u \\
\dot{y}_3 &= L_f h_3 + L_g h_3 u = L_f h_3 = \phi_3 \\
\dot{\phi}_2 &= L_f \phi_2 + L_g \phi_2 u \\
\dot{\phi}_3 &= L_f \phi_3 + L_g \phi_3 u = L_f \phi_3 - (\delta_1 L_g h_1 + \delta_2 L_g \phi_2)u + \sigma_3 u \\
&= \psi_3 - \delta_1(L_f h_1 + L_g h_1 u) - \delta_2(L_f \phi_2 + L_g \phi_2 u) + \sigma_3 u \\
\dot{\psi}_3 &= L_f \psi_3 + L_g \psi_3 u \\
\dot{\eta} &= f_0(\eta, y_1, y_2, y_3, \phi_2, \phi_3, \psi_3) + g_0(\eta, y_1, y_2, y_3, \phi_2, \phi_3, \psi_3)u.
\end{aligned}$$

Note that, setting $u = u^*$, one obtains

$$\dot{y}_1 = 0$$
$$\dot{y}_2 = \phi_2 + \sigma_2 u^*$$
$$\dot{y}_3 = \phi_3$$
$$\dot{\phi}_2 = 0$$
$$\dot{\phi}_3 = \psi_3 - \sigma_3 u^*$$
$$\dot{\psi}_3 = 0$$
$$\dot{\eta} = f_0(\eta, y_1, y_2, y_3, \phi_2, \phi_3, \psi_3) + g_0(\eta, y_1, y_2, y_3, \phi_2, \phi_3, \psi_3)u^*$$

and from this, since both σ_2 and σ_3 are 0 on Z^*, we see that the zero dynamics—in the new coordinates—is described by

$$\dot{\eta} = f^*(\eta) = f_0(\eta, 0, \ldots, 0) + g_0(\eta, 0, \ldots, 0)u^*(\eta, 0, \ldots, 0). \quad \bullet$$

To extend the constructions described in this example is not too much difficult. What is needed is essentially to give an appropriate notation to the entries of $h(x)$ and of $\Phi_0(x), \ldots, \Phi_{k^*-1}(x)$. To this end, we suppose first of all—without loss of generality—that the outputs of the system have been rearranged in such a way that in the matrix

$$\begin{bmatrix} dH_{k-1}(x) \\ d\Phi_{k-1}(x) \end{bmatrix} g(x)$$

the last $(s_k - r_k + r_{k-1})$ rows are dependent on the previous ones (at each x on M_k, near x^o). If this is the case, then the matrix $Q_{k-1}(x)$ may be chosen of the form

$$Q_{k-1}(x) = [\bar{Q}_{k-1}(x) \quad I].$$

Now, set $T_0(x) = h(x)$ and, recalling that $\Phi_{k-1}(x)$ has by construction s_k entries, let $T_k(x)$ be a vector consisting of exactly m elements in which the first $(m - s_k)$ ones are zero, while the last ones coincide with those of $\Phi_{k-1}(x)$, $1 \leq k \leq k^*$. Then, in the rectangular $m \times (k^* + 1)$ matrix

$$T(x) = [T_0(x) \quad T_1(x) \quad \cdots \quad T_{k^*}(x)]$$

each row consists of a sequence of nonzero entries, say in number of n_i (where i is the index of the row), followed by a sequence of zero entries. Moreover, by construction,

$$n_1 \leq n_2 \leq \ldots \leq n_m.$$

Now set, for $1 \leq k \leq n_i$, $1 \leq i \leq m$, $\xi_k^i(x)$ equal to the entry of $T(x)$ on the i-th row and k-th column, and

$$\xi^i = \text{col}(\xi_1^i, \xi_2^i, \ldots, \xi_{n_i}^i) \tag{1.16}$$

Using these functions as new coordinates, together with an extra set η consisting of $(n - s_0 - \ldots - s_{k^*})$ components, the equations of the system can be put into a form that, to some extent, generalizes the normal form introduced in the previous Chapter. Note that, by construction, the new coordinates $\xi_k^i(x)$ are such that $\xi_k^i(x^\circ) = 0$, for all for $1 \leq k \leq n_i$, $1 \leq i \leq m$, and one can always choose the complementary set of coordinates $\eta(x)$ such that $\eta(x^\circ) = 0$.

Proposition 1.9. Suppose x° is a regular point of the zero dynamics algorithm. In the local coordinates $z = \Phi(x) = (\xi^1, \ldots, \xi^m, \eta)$ defined by (1.16), the system (1.1) assumes the form (in which x stands for $\Phi^{-1}(z)$)

$$\dot{\xi}_1^1 = \xi_2^1$$
$$\cdots$$
$$\dot{\xi}_{n_1-1}^1 = \xi_{n_1}^1$$
$$\dot{\xi}_{n_1}^1 = b^1(x) + a^1(x)u$$
$$\cdots$$
$$\dot{\xi}_1^2 = \xi_2^2 + \delta_{11}^2(x)(b^1(x) + a^1(x)u) + \sigma_1^2(x)u$$
$$\cdots$$
$$\dot{\xi}_{n_2-1}^2 = \xi_{n_2}^2 + \delta_{n_2-1,1}^2(x)(b^1(x) + a^1(x)u) + \sigma_{n_2-1}^2(x)u$$
$$\dot{\xi}_{n_2}^2 = b^2(x) + a^2(x)u$$
$$\cdots$$
$$\dot{\xi}_1^i = \xi_2^i + \sum_{j=1}^{i-1}\delta_{1j}^i(b^j(x) + a^j(x)u) + \sigma_1^i(x)u$$
$$\cdots$$
$$\dot{\xi}_{n_i-1}^i = \xi_{n_i}^i + \sum_{j=1}^{i-1}\delta_{n_i-1,j}^i(b^j(x) + a^j(x)u) + \sigma_{n_i-1}^j(x)u$$
$$\dot{\xi}_{n_i}^i = b^i(x) + a^i(x)u$$
$$\dot{\eta} = f_0(\xi^1, \ldots, \xi^m, \eta) + g_0(\xi^1, \ldots, \xi^m, \eta)u$$

and

$$y_i = \xi_1^i$$

for $i = 1, \ldots, m$.

In particular

$$a^i(x) = L_g\xi_{n_i}^i(x) \tag{1.17a}$$
$$b^i(x) = L_f\xi_{n_i}^i(x). \tag{1.17b}$$

The coordinate functions $\xi_k^i(x)$, the coefficients $\delta_{k,j}^i(x)$ and $\sigma_k^i(x)$ are such that

$$\xi_{k+1}^i(x) = -\sum_{j=1}^{i-1} \delta_{k,j}^i(x)b^j(x) + L_f\xi_k^i(x) \quad 1 \le k \le n_i-1, 2 \le i \le m \ (1.18a)$$

$$L_g\xi_k^i(x) = \sum_{j=1}^{i-1} \delta_{k,j}^i(x)a^j(x) + \sigma_k^i(x) \quad 1 \le k \le n_i-1, 2 \le i \le m. \quad (1.18b)$$

In the new coordinates, the submanifold $Z^\star$ is described as

$$Z^\star = \{x \in U : \xi^i(x) = 0, 1 \le i \le m\}$$

and the functions $\sigma_k^i(x)$ vanish on $Z^\star$.

The matrix

$$A(x) = \mathrm{col}(a^1(x), \ldots, a^m(x)) \tag{1.19}$$

is nonsingular at x°, and the unique $u^\star(x)$ which solves the equation

$$b^i(x) + a^i(x)u^\star(x) = 0 \qquad 1 \le i \le m \tag{1.20}$$

is such that $f^\star(x) = f(x) + g(x)u^\star(x)$ is tangent to $Z^\star$. Thus, the zero dynamics of the system, in the new coordinates, are described by

$$\dot{\eta} = f^\star(\eta) = f_0(0, \ldots, 0, \eta) + g_0(0, \ldots, 0, \eta)u^\star(0, \ldots, 0, \eta). \tag{1.21}$$

Proof. It is quite simple, although a little tedious, and is left as an exercise to the reader. We suggest to check first the (1.18) that, on the basis of the definitions (1.17), descend directly from the properties of the zero dynamics algorithm; then the special form of the system equations follows trivially. •

Remark 1.10. It is important to note that the results illustrated in this Section, in particular the generalized normal form and the corresponding characterization of the zero dynamics, incorporate completely the results discussed in Section 5.1.

In fact, suppose the integers $r_1, \ldots, r_m$ are such that the vector

$$[\, L_{g_1}L_f^k h_i(x) \quad L_{g_2}L_f^k h_i(x) \quad \ldots \quad L_{g_m}L_f^k h_i(x)\,]$$

is zero for all x near x° and for all $k < r_i - 1$, and nonzero for $k = r_i - 1$ at $x = x^\circ$. Without much effort, it is possible to realize that, after possibly a reordering of the outputs, the integers $r_1, \ldots, r_m$ thus defined are related to the integers $n_1, \ldots, n_m$ associated with the generalized normal form in the following way

$$r_1 = n_1, \qquad r_i \le n_i \qquad \text{for } 2 \le i \le m,$$

and also that

$$\delta^i_{kj}(x) = 0 \qquad \text{for all } 1 \le k \le r_i - 1, 1 \le j \le i - 1, 2 \le i \le m$$
$$\sigma^i_k(x) = 0 \qquad \text{for all } 1 \le k \le r_i - 1, 2 \le i \le m.$$

If, *in addition*, the matrix (5.1.2) is nonsingular, i.e. the system has vector relative degree $\{r_1, \ldots, r_m\}$ at x°, then $r_i = n_i$ for all $1 \le i \le m$, and the previous normal form reduces exactly to the one introduced in Section 5.1.

6.2 Controlled Invariant Distributions

In the next Sections of this Chapter we develop a series of results that are very helpful in studying the effect of a static state feedback on a nonlinear system of the form (1.1). In accordance with the set-up already established in Chapter 5, we consider a feedback control law of the form (5.2.3), namely

$$u = \alpha(x) + \beta(x)v \tag{2.1}$$

with α and β defined on a neighborhood U° of the point of interest (which sometimes could be the state space U on which the system (1.1) is defined), and $\beta(x)$ nonsingular for all x.

The effect of this feedback is that of changing the original system (1.1) into one of the same structure, noted

$$\dot{x} = \tilde{f}(x) + \sum_{i=1}^m \tilde{g}_i(x)v_i$$

in which we have set

$$\tilde{f}(x) = f(x) + \sum_{i=1}^m g_i(x)\alpha_i(x)$$
$$\tilde{g}_i(x) = f(x) + \sum_{j=1}^m g_j(x)\beta_{ji}(x).$$

In more condensed form, the latter will be almost always rewritten as

$$\tilde{f}(x) = f(x) + g(x)\alpha(x)$$
$$\tilde{g}(x) = g(x)\beta(x).$$

The purpose for which feedback is introduced is to obtain a dynamics with some nice properties that the original dynamics does not have. As we shall see later on, a typical situation is the one in which a modification is required in order to obtain the invariance of a given distribution Δ under the vector fields which characterize the new dynamics. This kind of problem is usually dealt with in the following way.

A distribution Δ is said to be *controlled invariant* on U if there exists a feedback pair (α, β) defined on U with the property that Δ is invariant under the vector fields $\tilde{f}, \tilde{g}_1, \ldots, \tilde{g}_m$, i.e. if

$$[\tilde{f}, \Delta](x) \subset \Delta(x) \tag{2.2a}$$

$$[\tilde{g}_i, \Delta](x) \subset \Delta(x) \qquad \text{for } 1 \leq i \leq m \tag{2.2a}$$

for all $x \in U$.

A distribution Δ is said to be *locally controlled invariant* if for each $x \in U$ there exists a neighborhood U° of x with the property that Δ is controlled invariant on U°. In view of the previous definition, this requires the existence of a feedback pair (α, β) defined on U° such that (2.2) are true for all $x \in U^\circ$.

The notion of local controlled invariance lends itself to a simple geometric test. If we set

$$G = \text{span}\{g_1, \ldots, g_m\}$$

we may express the test in question in the following terms.

Lemma 2.1. Let Δ be an involutive distribution. Suppose Δ, G and $\Delta + G$ are nonsingular on U. Then Δ is locally controlled invariant if and only if

$$[f, \Delta] \subset \Delta + G \tag{2.3a}$$

$$[g_i, \Delta] \subset \Delta + G \qquad \text{for } 1 \leq i \leq m. \tag{2.3b}$$

Proof of Lemma 2.1 (Necessity). Suppose Δ is locally controlled invariant. Let U° be a neighborhood of x° and (α, β) a feedback pair defined on U° which makes (2.2) satisfied on U°. Let τ be a vector field of Δ. Then we have

$$[\tilde{f}, \tau] = [f + g\alpha, \tau] = [f, \tau] + \sum_{j=1}^m [g_j, \tau]\alpha_j - \sum_{j=1}^m (L_\tau \alpha_j) g_j$$

$$[\tilde{g}_i, \tau] = \sum_{j=1}^m [g_j \beta_{ji}, \tau] = \sum_{j=1}^m [g_j, \tau]\beta_{ji} - \sum_{j=1}^m (L_\tau \beta_{ji}) g_j$$

for $1 \leq i \leq m$.

Since β is invertible, one may solve the last m equalities for $[g_j, \tau]$, obtaining

$$[g_j, \tau] \in \sum_{j=1}^m [\tilde{g}_j, \Delta] + G$$

for $1 \leq j \leq m$. Therefore, from (2.2b) we deduce (2.3b). Moreover, since

$$[f, \tau] \in [\tilde{f}, \Delta] + \sum_{j=1}^{m} [g_j, \Delta] + G$$

again from (2.2) and (2.3b) we deduce (2.3a). •

Remark 2.2. Note that in proving the necessity of conditions (2.2) we have not yet used the hypothesis that Δ, G and $\Delta + G$ are nonsingular. •

In order to prove the sufficiency, we first need the following interesting result which is a consequence of the Frobenius Theorem.

Theorem 2.3. Let U and V be open sets in $\mathbf{R}^m$ and $\mathbf{R}^n$ respectively. Let $x_1, \ldots, x_m$ denote coordinates of a point x in $\mathbf{R}^m$ and $y_1, \ldots, y_n$ coordinates of a point y in $\mathbf{R}^n$. Let $\Gamma^1, \ldots, \Gamma^m$ be smooth functions

$$\Gamma^i : U \to \mathbf{R}^{n \times n}.$$

Consider the set of partial differential equations

$$\frac{\partial y(x)}{\partial x_i} = \Gamma^i(x) y(x) \qquad 1 \leq i \leq m \tag{2.4}$$

where y denotes a function

$$y : U \to V.$$

Given a point $(x^\circ, y^\circ) \in U \times V$ there exist a neighborhood U° of x° in U and a unique smooth function

$$y : U^\circ \to V$$

which satisfies the equations (2.4) and is such that $y(x^\circ) = y^\circ$ if and only if the functions $\Gamma^1, \ldots, \Gamma^m$ satisfy the conditions

$$\frac{\partial \Gamma^i}{\partial x_k} - \frac{\partial \Gamma^k}{\partial x_i} + \Gamma^i \Gamma^k - \Gamma^k \Gamma^i = 0 \qquad 1 \leq i, k \leq m \tag{2.5}$$

for all $x \in U$.

Proof. Necessity. Suppose that for all (x°, y°) there is a function y which satisfies (2.4). Then from the property

$$\frac{\partial^2 y}{\partial x_i \partial x_k} = \frac{\partial^2 y}{\partial x_k \partial x_i}$$

one has

$$\frac{\partial}{\partial x_i} \left(\Gamma^k(x) y(x) \right) = \frac{\partial}{\partial x_k} \left(\Gamma^i(x) y(x) \right)$$

Expanding the derivatives on both sides and evaluating them at $x = x^\circ$ one obtains

$$\left[\left[\frac{\partial \Gamma^k}{\partial x_i}\right]_{x^\circ} + \Gamma^k(x^\circ)\Gamma^i(x^\circ)\right]y^\circ = \left[\left[\frac{\partial \Gamma^i}{\partial x_k}\right]_{x^\circ} + \Gamma^i(x^\circ)\Gamma^k(x^\circ)\right]y^\circ$$

which, due to arbitrariness of x°, y°, yields the condition (2.5).

Sufficiency. The proof of this part consists of the following steps.

(i) It is shown that the fulfillment of (2.5) enables us to define on $U \times V$ a certain involutive distribution Δ, of dimension m.

(ii) Using Frobenius Theorem, one can find a neighborhood $U' \times V'$ of (x°, y°) and a local coordinates transformation

$$F : (x, y) \mapsto \xi$$

defined on $U' \times V'$, with the property that

$$\Delta(x, y) = \text{span}\left\{\left[\frac{\partial}{\partial \xi_1}\right]_{(x,y)}, \dots, \left[\frac{\partial}{\partial \xi_m}\right]_{(x,y)}\right\}$$

for all $(x, y) \in U' \times V'$

(iii) From the transformation F one constructs a solution of (2.4).

As for the step (i), the distribution Δ is defined, at each $(x, y) \in U \times V$, by

$$\Delta(x, y) = \text{span}\left\{\left[\frac{\partial}{\partial x_i}\right] + \sum_{h=0}^{n}\sum_{k=0}^{n} \Gamma_{hk}^i(x)y_k\left[\frac{\partial}{\partial y_h}\right] : 1 \le i \le m\right\}.$$

In other words, $\Delta(x, y)$ is spanned by m tangent vectors whose coordinates with respect to the canonical basis $\{[\frac{\partial}{\partial x_1}], \dots, [\frac{\partial}{\partial x_n}], [\frac{\partial}{\partial y_1}], \dots, [\frac{\partial}{\partial y_n}]\}$ of the tangent space to $U \times V$ at (x, y) have the form

$$\begin{bmatrix} 1 \\ 0 \\ 0 \\ \vdots \\ 0 \\ \Gamma^1(x)y \end{bmatrix}, \begin{bmatrix} 0 \\ 1 \\ 0 \\ \vdots \\ 0 \\ \Gamma^2(x)y \end{bmatrix}, \dots, \begin{bmatrix} 0 \\ 0 \\ 0 \\ \vdots \\ 1 \\ \Gamma^m(x)y \end{bmatrix}.$$

These m vectors are linearly independent at all (x, y) and so the distribution Δ is nonsingular and of dimension m. Moreover, it is an easy computation to check that if the "integrability" condition (1.14) is satisfied, then Δ is involutive.

The possibility of constructing the coordinates transformation described in (ii) is a straightforward consequence of Frobenius theorem. The function F thus defined is such that if v is a vector in Δ, the last n components of F_*v are vanishing. Since, moreover, the tangent vectors $\left[\frac{\partial}{\partial y_1}\right], \ldots, \left[\frac{\partial}{\partial y_n}\right]$ span a subspace which is complementary to $\Delta(x, y)$ at all (x, y) and F is nonsingular, one may easily conclude that the function

$$\xi = F(x, y)$$

is such that jacobian matrix

$$\begin{bmatrix} \dfrac{\partial \xi_{m+1}}{\partial y_1} & \cdots & \dfrac{\partial \xi_{m+1}}{\partial y_m} \\ \cdots & \cdots & \cdots \\ \dfrac{\partial \xi_{m+n}}{\partial y_1} & \cdots & \dfrac{\partial \xi_{m+n}}{\partial y_n} \end{bmatrix} \tag{2.6}$$

is nonsingular at all $(x, y) \in U' \times V'$.

Without loss of generality we may assume that

$$\xi_i(x^\circ, y^\circ) = 0$$

for all $m + 1 \leq i \leq m + n$. As a consequence, the integral submanifold of Δ passing through (x°, y°) is defined by the set of equations

$$\xi_{m+i}(x, y) = 0 \qquad 1 \leq i \leq n.$$

Since the matrix (2.6) is nonsingular , thanks to the implicit function theorem, the above equations may be solved for y, yielding a set of functions

$$y_i = \eta_i(x) \qquad 1 \leq i \leq n \tag{2.7}$$

defined in a neighborhood $U_0 \subset U'$ of x°. Moreover

$$\eta_i(x^\circ) = y_i^\circ \qquad 1 \leq i \leq n.$$

The functions (2.7) satisfy the differential equations (2.4) and therefore, are the required solutions. As a matter of fact, the functions

$$\phi_i(x, y) = y_i - \eta_i(x) \qquad 1 \leq i \leq n$$

are constant on the integral submanifold of Δ passing through (x°, y°) and, therefore, if v is a vector in Δ,

$$d\phi_i v = 0 \qquad 1 \leq i \leq n$$

at all pairs $(x, \eta(x))$. These equations, taking for v each one of the m vectors used to define Δ, yield exactly

$$\frac{\partial \eta_i}{\partial x_j} = (\Gamma^j(x)\eta(x))_i \qquad 1 \le i \le n, \ 1 \le j \le m. \quad \bullet$$

Proof of Lemma 2.1 (Sufficiency). Recall that, by assumption, Δ, G, and $\Delta + G$ are nonsingular; let d denote the dimension of Δ and let

$$p = \dim G - \dim \Delta \cap G.$$

Given any $x^\circ \in U$ it is possible to find a neighborhood U° of x° and an $m \times n$ nonsingular matrix B, whose (i,j)-th element b_{ij} is a smooth real-valued function defined on U, such that, for

$$\hat{g}_i = \sum_{j=1}^{m} g_j b_{ji} \qquad 1 \le i \le m$$

the following is true

$$\begin{aligned} \mathrm{span}\{\hat{g}_{p+1}, \dots, \hat{g}_m\} &\subset \Delta \\ (\Delta + G) &= \Delta \oplus \mathrm{span}\{\hat{g}_1, \dots, \hat{g}_p\}. \end{aligned} \tag{2.8}$$

The tangent vectors $\hat{g}_1(x), \dots, \hat{g}_p(x)$ are clearly linearly independent at all $x \in U^\circ$.

Now, observe that if the assumption (2.3b) is satisfied, then also

$$[\hat{g}_i, \Delta] \subset \Delta + G \tag{2.9}$$

and let $\tau_1, \dots, \tau_d$ be a set of vector fields which locally span Δ around x°. From (2.8) and (2.9) we deduce the existence of a unique set of smooth real-valued functions c_{ji}^k, defined locally around x°, and a vector field $\delta_i^k \in \Delta$ defined locally around x° such that

$$[\hat{g}_i, \tau_k] = \sum_{j=1}^{p} c_{ji}^k \hat{g}_j + \delta_i^k \tag{2.3b'}$$

for all $1 \le i \le m$ and $1 \le k \le d$. Using the same arguments and setting

$$\hat{g}_0 = f$$

from (2.3a) and (2.9) we deduce the existence of a unique set of real-valued smooth functions c_{j0}^k and a vector field $\delta_0^k \in \Delta$, defined locally around x°, such that

$$[\hat{g}_0, \tau_k] = \sum_{j=1}^{p} c_{j0}^{k} \hat{g}_j + \delta_0^k. \qquad (2.3a')$$

Now, suppose there exists a nonsingular $m \times m$ matrix $\hat{B}$, whose (i,j)-th element $\hat{b}_{ij}$ is a smooth real-valued function defined locally around x, such that

$$-L_{\tau_k}\hat{b}_{hi} + \sum_{j=1}^{m} c_{hj}^{k} \hat{b}_{ji} = 0 \qquad (2.10)$$

for $1 \leq k \leq d$, $1 \leq h \leq p$, $1 \leq i \leq m$. Then, it is easy to see that

$$\left[\sum_{h=1}^{m} \hat{g}_h \hat{b}_{hi}, \tau_k\right] \in \Delta \qquad (2.11)$$

for $1 \leq i \leq m$, $1 \leq k \leq d$. For,

$$\left[\sum_{h=1}^{m} \hat{g}_h \hat{b}_{hi}, \tau_k\right] = -\sum_{h=1}^{m} (L_{\tau_k}\hat{b}_{hi})\hat{g}_h + \sum_{j=1}^{m} \hat{b}_{ji}[\hat{g}_j, \tau_k]$$

$$= -\sum_{h=1}^{m} (L_{\tau_k}\hat{b}_{hi})\hat{g}_h + \sum_{j=1}^{m} \hat{b}_{ji} \sum_{h=1}^{p} c_{hj}^{k} \hat{g}_h + \bar{\delta}_i^k = \bar{\delta}_i^k$$

where $\bar{\delta}_i^k$ is a vector field in Δ. Since $\tau_1, \ldots, \tau_k$ locally span Δ, (2.11) implies that

$$\left[\sum_{h=1}^{m} \hat{g}_h \hat{b}_{hi}, \Delta\right] \subset \Delta.$$

Therefore, the matrix

$$\beta = B\hat{B}$$

is such that (2.2b) is satisfied.

Using similar arguments, one can see that if there exists an $m \times 1$ vector $\hat{a}$, whose i-th element $\hat{a}_i$ is a smooth real-valued function defined locally around x°, such that

$$-L_{\tau_k}\hat{a}_h + \sum_{j=1}^{m} c_{hj}^{k} \hat{a}_j + c_{h0}^{k} = 0 \qquad (2.12)$$

for $1 \leq k \leq d$, $1 \leq h \leq p$, then

$$\left[\hat{g}_0 + \sum_{h=1}^{m} \hat{g}_h \hat{a}_h, \tau_k\right] \in \Delta \qquad (2.13)$$

for $1 \leq k \leq d$. For,

$$[\hat{g}_0 + \sum_{h=1}^{m} \hat{g}_h \hat{a}_h, \tau_k] = -\sum_{h=1}^{p} (L_{\tau_k} \hat{a}_h) \hat{g}_h + \sum_{j=1}^{m} \hat{a}_j \sum_{h=1}^{p} c_{hj}^k \hat{g}_h + \sum_{h=1}^{p} c_{h0}^k \hat{g}_h + \bar{\delta}^k = \bar{\delta}^k$$

where $\bar{\delta}^k$ is a vector field in Δ. From this one deduces that the vector

$$\alpha = B\hat{a}$$

is such that (2.2a) is satisfied.

Thus, we have seen that the possibility of finding $\hat{B}$ and $\hat{a}$ which satisfy (2.10) and (2.11) enables us to construct a pair of feedback functions that makes (2.2) satisfied. In order to complete the proof, we have to show now that (2.10) and (2.12) can be solved for $\hat{B}$ and $\hat{a}$.

Since Δ is nonsingular and involutive, we may assume, without loss of generality, that our choice of local coordinates is such that

$$\tau_k = \frac{\partial}{\partial x_k} \qquad 1 \leq k \leq d.$$

The equations (2.10) and (2.12) may be rewritten as a set of partial differential equations of the form (2.4) by simply setting

$$\Gamma_k = \begin{bmatrix} c_{11}^k & \cdots & c_{1m}^k & c_{10}^k \\ \cdots & \cdots & \cdots & \cdots \\ c_{p1}^k & \cdots & c_{pm}^k & c_{p0}^k \\ 0 & \cdots & 0 & 0 \\ \cdots & \cdots & \cdots & \cdots \\ 0 & \cdots & 0 & 0 \end{bmatrix} \qquad 1 \leq k \leq d.$$

As a matter of fact, for each fixed i, the equations (2.10) correspond to an equation for the i-th column of $\hat{B}$, of the form

$$\frac{\partial}{\partial x_k} \begin{bmatrix} \hat{b}_i \\ 0 \end{bmatrix} = \Gamma^k \begin{bmatrix} \hat{b}_i \\ 0 \end{bmatrix} \qquad 1 \leq k \leq d \qquad (2.14)$$

(where $\hat{b}_i$ stands for the i-th column of $\hat{B}$) and the equations (2.12) correspond to

$$\frac{\partial}{\partial x_k} \begin{bmatrix} \hat{a} \\ 1 \end{bmatrix} = \Gamma^k \begin{bmatrix} \hat{a} \\ 1 \end{bmatrix} \qquad 1 \leq k \leq d. \qquad (2.15)$$

Both these equations have exactly the form

$$\frac{\partial y}{\partial x_k} = \Gamma^k y \qquad 1 \le k \le d \tag{2.16}$$

the unknown vector y being $m+1$ dimensional. Since now the functions Γ^k depend also on the coordinates $x_{d+1}, \ldots, x_n$ (with respect to which no derivative of y is considered), in order to achieve uniqueness, the value of y must be specified, for a given $x_1^o, \ldots, x_d^o$, at each $x_{d+1}, \ldots, x_n$. For consistency, the last component of the initial value of the solution sought for the equations (2.14) must be set equal to zero, whereas the last component of the initial value of the solution sought for the equation (2.15) must be set equal to 1. In addition, the first m components of the initial values of the solutions sought for each of the equations (2.14) must be columns of a nonsingular $m \times m$ matrix, in order to let $\hat{B}$ be nonsingular.

The solvability of an equation of the form (2.14) depends, as we have seen, on the fulfillment of the integrability conditions (2.5). This, in turn, is implied by (2.3). Consider the Jacobi identity

$$-[[\hat{g}_i, \tau_k], \tau_h] + [[\hat{g}_i, \tau_h], \tau_k] = [\hat{g}_i, [\tau_h, \tau_k]]$$

for any $0 \le i \le m$. Using for $[\hat{g}_i, \tau_k]$ and $[\hat{g}_i, \tau_k]$ the expressions given by (2.3a') or (2.3b') and taking $\tau_k = \frac{\partial}{\partial x_k}$, $\tau_h = \frac{\partial}{\partial x_h}$ one easily obtains

$$\left[\sum_{j=1}^p c_{ji}^k \hat{g}_j + \delta_i^k, \frac{\partial}{\partial x_h}\right] - \left[\sum_{j=1}^p c_{ji}^h \hat{g}_j + \delta_i^h, \frac{\partial}{\partial x_k}\right] = 0.$$

This yields

$$-\sum_{j=1}^p \frac{\partial c_{ji}^k}{\partial x_h} \hat{g}_j + \sum_{j=1}^p c_{ji}^k \left(\sum_{s=1}^p c_{sj}^h \hat{g}_s + \delta_j^h\right) + \left[\delta_i^k, \frac{\partial}{\partial x_h}\right]$$

$$+\sum_{j=1}^p \frac{\partial c_{ji}^h}{\partial x_k} \hat{g}_j + \sum_{j=1}^p c_{ji}^h \left(\sum_{s=1}^p c_{sj}^k \hat{g}_s + \delta_j^k\right) - \left[\delta_i^h, \frac{\partial}{\partial x_k}\right] = 0.$$

Now, recall that $\frac{\partial}{\partial x_h}$ and $\frac{\partial}{\partial x_k}$ are both vector fields of Δ, which is involutive. Therefore, also $[\delta_i^k, \frac{\partial}{\partial x_h}]$ and $[\delta_i^h, \frac{\partial}{\partial x_k}]$ are in Δ. Since Δ and span$\{\hat{g}_1, \ldots, \hat{g}_p\}$ are direct summands and $\hat{g}_1, \ldots, \hat{g}_p$ are linearly independent, the previous equality implies

$$-\frac{\partial c_{ji}^k}{\partial x_h} + \frac{\partial c_{ji}^h}{\partial x_k} + \sum_{s=1}^p c_{js}^h c_{si}^k - \sum_{s=1}^p c_{js}^k c_{si}^h = 0$$

for $1 \leq j \leq p$, $0 \leq i \leq m$, $1 \leq h, k \leq d$, which is easily seen to be identical to the condition (2.5). ●

We see from this Lemma that, under reasonable assumptions (namely, the nonsingularity of Δ, G and $\Delta + G$) an involutive distribution is locally controlled invariant if and only if the conditions (2.3) are satisfied. These conditions are of special interest because they don't invoke the existence of feedback functions α and β, as the definition does, but are expressed only in terms of the vector fields $f, g_1, \ldots, g_m$ which characterize the given control system and of the distribution itself. The fulfillment of conditions (2.3) implies the existence of a pair of feedback functions which make Δ invariant under the new dynamics but the actual construction of such a feedback pair generally involves the solution of a set of partial differential equations, as we have seen in the proof of Lemma (2.1). There are cases, however, in which the solution of partial differential equations may be avoided and these, luckily enough, include some situations of great importance in control theory. These will be examined in the next Section.

We conclude this Section with an interesting result which describes a uniqueness property of any feedback which renders invariant a given distribution.

Lemma 2.4. Let x° be an equilibrium point of the vector field $f(x)$. Suppose Δ is a nonsingular and involutive controlled invariant distribution and suppose also

$$\dim(G) = m$$
$$\Delta \cap G = \{0\}.$$

Let α^1 and α^2 be any two feedback functions such that $[f + g\alpha^i, \Delta] \subset \Delta$ for $i = 1, 2$ and $\alpha^1(x^\circ) = \alpha^2(x^\circ) = 0$. Let M_{x° be the maximal integral submanifold of Δ which contains the point x°. Then

$$\alpha^1(x) = \alpha^2(x) \tag{2.17}$$

for each $x \in M_{x^\circ}$.

Proof. Let $\beta(x)$ be a nonsingular matrix such that $[g\beta, \Delta] \subset \Delta$. Proving (2.17) is equivalent to prove that $\beta(x)\alpha^1(x) = \beta(x)\alpha^2(x)$. Using the fact that $[f + g\alpha^i, \Delta] \subset \Delta$, one deduces that

$$[f + (g\beta)\beta^{-1}\alpha^1 - f - (g\beta)\beta^{-1}\alpha^2, \Delta] \subset \Delta$$

that is

$$[(g\beta)\beta^{-1}(\alpha^1 - \alpha^2), \tau] \subset \Delta$$

for all vector fields τ of Δ, which yields

$$[(g\beta)\beta^{-1}(\alpha^1 - \alpha^2), \tau]$$
$$= \sum_{i=1}^{m} ((\beta^{-1}(\alpha^1 - \alpha^2))_i [(g\beta)_i, \tau] - L_\tau(\beta^{-1}(\alpha^1 - \alpha^2))_i(g\beta)_i) \in \Delta.$$

Using the fact that $[(g\beta)_j, \tau] \subset \Delta$ for all $1 \leq j \leq m$, $\Delta \cap G = \{0\}$ and the fact that the $(g\beta)_i$'s are linearly independent for all x, one deduces that

$$L_\tau \beta^{-1}(\alpha^1 - \alpha^2) = 0 \text{ for all } \tau \in \Delta.$$

This implies that $\beta^{-1}(\alpha^1 - \alpha^2)(x)$ is constant on M_{x° and therefore, since $\alpha^1(x^\circ) = \alpha^2(x^\circ)$, (2.17) must follow. •

6.3 The Maximal Controlled Invariant Distribution in $\ker(dh)$

The notion of controlled invariant distribution is of particular interest in the problem of using feedback for the purpose of rendering some outputs of the system independent of certain inputs. In fact, suppose a control system of the form

$$\dot{x} = f(x) + \sum_{i=1}^{m} g_i(x)u_i + p(x)w$$
$$y = h(x)$$

is given, in which the additional input w represents an undesired perturbation that affects the behavior of the system through the vector field p, and consider the following problem: find, if possible, a static state feedback (of the form (2.1)), with the property that, in the corresponding closed-loop system

$$\dot{x} = f(x) + g(x)\alpha(x) + \sum_{i=1}^{m}(g(x)\beta(x))_i v_i + p(x)w$$
$$y = h(x)$$

the perturbation w has no influence on the output y.

In view of some results established in Chapter 3 (see Theorem 3.3.12 and Remark 3.3.13), this problem has a solution if and only if there is a distribution Δ which is

(i) invariant under the vector fields $\tilde{f} = f + g\alpha$, $\tilde{g}_i = (g\beta)_i$, $1 \leq i \leq m$, and p which characterize the closed-loop system,

(ii) contains the vector field p,

(iii) is contained in the distribution

$$\ker(dh) = \bigcap_{j=1}^{m} \ker(dh_j) = (\text{span}\{dh_1, \ldots, dh_m\})^{\perp}.$$

According to the terminology established in the previous Section, we see from (i) that Δ is a *controlled invariant* distribution, invariant under the vector field p, that—as (ii) and (iii) specify—satisfies

$$p \in \Delta \subset \ker(dh). \tag{3.1}$$

On the basis of this simple observation, we can conclude that the problem of using feedback in order to make the output of a given system independent of a certain input implies the problem of finding a distribution Δ which is controlled invariant for the system (1.1) and satisfies the constraint (3.1). Among the conditions which this distribution must satisfy there is also the invariance under the vector field p but, as it is immediate to check, this is not really an additional constraint if the distribution itself is *involutive*. In fact, if (3.1) is satisfied, p is a vector field of Δ and, if the latter is involutive, the invariance under p is achieved by definition.

Note also that, if the distribution in question is involutive and *nonsingular*, then in a neighborhood of each point x in the state space it is possible to change the coordinates (see e.g. Remark **3.3.11**) in such a way that the closed loop system

$$\dot{x} = \tilde{f}(x) + \tilde{g}(x)v + p(x)w$$
$$y = h(x)$$

is locally represented by equations of the form

$$\dot{x}_1 = \tilde{f}_1(x_1, x_2) + \tilde{g}_1(x_1, x_2)v + p(x_1, x_2)w$$
$$\dot{x}_2 = \tilde{f}_2(x_2) + \tilde{g}_2(x_2)v$$
$$y = h(x_2).$$

We see from this that the disturbance w has no effect on the output y, just because the feedback has rendered *unobservable* the closed-loop system. In fact, all pairs of states whose x_2 components are equal, produce identical outputs under any input. We observe then that seeking a pair of functions α and β which makes (i) and (iii) satisfied for some distribution Δ essentially corresponds to search for a feedback that induces a certain amount of unobservability into the system.

The problem of finding, for the system (1.1), a (possibly involutive) controlled invariant distribution which satisfies the constraint (3.1) can be dealt with in the following way. First of all, it is examined whether or not the family of all controlled invariant distributions of (1.1) which are contained in $\ker(dh)$ has a *maximal* element (in the sense of distributions inclusion, i.e. an element which contains all other members of the family). Then, it is checked whether or not the maximal element thus defined is involutive and contains the vector field p. We shall see in this Section how a program of this kind can be accomplished.

As explained in the previous Section, a necessary condition for a distribution to be controlled invariant is that the conditions (2.3) are satisfied, and these conditions—which turn out to be also sufficient, at least for local controlled invariance, under some mild regularity assumptions—are particularly interesting because they do not involve explicitly the feedback functions α and β. Motivated by this, we are naturally led to consider the family, noted $\mathcal{J}(f, g, \ker(dh))$, of all smooth distributions which satisfy the conditions (2.3) and are contained in $\ker(dh)$. Since this family is closed under distribution addition (in fact, a trivial calculation shows that if Δ_1 and Δ_2 satisfy (2.3) then also $\Delta_1 + \Delta_2$ satisfies this condition), then this family has a well defined maximal element, namely the sum of all the members of the family. In view of Lemma 2.1, the maximal element of $\mathcal{J}(f, g, \ker(dh))$ is the natural candidate in the search for the maximal locally controlled invariant distribution contained in $\ker(dh)$.

The calculation of the maximal element of $\mathcal{J}(f, g, \ker(dh))$ is made possible by the following recursive construction.

Controlled invariant distribution algorithm. Step 0: set $\Omega_0 = \text{span}\{dh\}$. Step k: set

$$\Omega_k = \Omega_{k-1} + L_f(\Omega_{k-1} \cap G^\perp) + \sum_{i=1}^{m} L_{g_i}(\Omega_{k-1} \cap G^\perp). \tag{3.2}$$

Remark 3.1. Note that the codistribution $\Omega_{k-1} \cap G^\perp$, being defined as an intersection of codistributions, may fail to be smooth. However, it is still possible to define the codistribution $L_f(\Omega_{k-1} \cap G^\perp)$, as the one which is spanned by all covector fields of the form $L_f\omega$, with ω smooth covector field in $\Omega_{k-1} \cap G^\perp$.

Lemma 3.2. Suppose there exists an integer $k^\star$ such that $\Omega_{k^\star+1} = \Omega_{k^\star}$. Then $\Omega_k = \Omega_{k^\star}$ for all $k > k^\star$. If $\Omega_{k^\star} \cap G^\perp$ and $\Omega_{k^\star}^\perp$ are smooth, then $\Omega_{k^\star}^\perp$ is the maximal element of $\mathcal{J}(f, g, \ker(dh))$.

Proof. The first part of the statement is a trivial consequence of the definitions. As for the other, note first that from the equality $\Omega_{k^\star+1} = \Omega_{k^\star}$ we deduce

$$L_{g_i}(\Omega_{k^\star} \cap G^\perp)) \subset \Omega_{k^\star}$$

for $1 \leq i \leq m$ and also for $i = 0$ if we set $f = g_0$, as sometimes we did before. Let

ω be a covector field in $\Omega_{k^\star} \cap G^\perp$, and τ a vector field in $\Omega_{k^\star}^\perp$. In the expression

$$\langle L_{g_i}\omega, \tau \rangle = L_{g_i}\langle \omega, \tau \rangle - \langle \omega, [g_i, \tau] \rangle$$

we have

$$\langle L_{g_i}\omega, \tau \rangle = 0$$

because $L_{g_i}\omega \in \Omega_{k^\star}$ and

$$\langle \omega, \tau \rangle = 0$$

because $\tau \in \Omega_{k^\star}^\perp + G$. Thus

$$\langle \omega, [g_i, \tau] \rangle = 0.$$

Since $\Omega_{k^\star} \cap G^\perp$ is smooth by assumption, $[g_i, \tau]$ annihilates every covector in $\Omega_{k^\star} \cap G^\perp$, i.e.

$$[g_i, \tau] \in \Omega_{k^\star}^\perp + G$$

for $0 \leq 1 \leq m$. Thus, $\Omega_{k^\star}^\perp$ is a member of $\mathcal{J}(f, g, \ker(dh))$. Let $\bar{\Delta}$ be any other element of this collection. We will prove that $\bar{\Delta} \subset \Omega_{k^\star}^\perp$. First of all, note that if ω is a covector field in $\bar{\Delta}^\perp \cap G^\perp$ and τ a vector field in $\bar{\Delta}$ we have

$$\langle L_{g_i}\omega, \tau \rangle = 0$$

so that (recall that $\bar{\Delta}$ is a smooth distribution)

$$L_{g_i}(\bar{\Delta}^\perp \cap G^\perp) \subset \bar{\Delta}^\perp.$$

Suppose

$$\bar{\Delta}^\perp \supset \Omega_k$$

for some $k \geq 0$. Then

$$\Omega_{k+1} \subset \Omega_k + L_f(\bar{\Delta}^\perp \cap G^\perp) + \sum_{i=1}^{m} L_{g_i}(\bar{\Delta}^\perp \cap G^\perp) \subset \bar{\Delta}^\perp.$$

Thus, since $\Omega_0 = \text{span}\{dh\} \subset \bar{\Delta}^\perp$, we deduce that

$$\bar{\Delta} \subset \Omega_{k^\star}^\perp$$

and $\Omega_{k^\star}^\perp$ is the maximal element of $\mathcal{J}(f, g, \ker(dh))$. ●

It is important to observe that the algorithm (3.2) is invariant under feedback transformation.

Lemma 3.3. Let $\tilde{f}, \tilde{g}_1, \ldots, \tilde{g}_m$ be any set of vector fields deduced from $f_1, g_1, \ldots, g_m$ by setting $\tilde{f} = f + g\alpha$, $\tilde{g}_i = (g\beta)_i$, $1 \leq i \leq m$; then each codistribution Ω_k of the sequence (3.2) is such that

$$\Omega_k = \Omega_{k-1} + L_{\tilde{f}}(G^\perp \cap \Omega_{k-1}) + \sum_{i=1}^{m} L_{\tilde{g}_i}(G^\perp \cap \Omega_{k-1})$$

Proof. Recall that, given a covector field ω, a vector field τ and a scalar function γ,

$$L_{(\tau\gamma)}\omega = (L_\tau\omega)\gamma + \langle\omega, \tau\rangle d\gamma.$$

If ω is a covector field in $G^\perp \cap \Omega_{k-1}$, then

$$L_{\tilde{f}}\omega = L_f\omega + \sum_{i=1}^{m}(L_{g_i}\omega)\alpha_i + \sum_{i=1}^{m}\langle\omega, g_i\rangle d\alpha_i$$

$$L_{\tilde{g}_i}\omega = \sum_{j=1}^{m}(L_{g_j}\omega)\beta_{ji} + \sum_{j=1}^{m}\langle\omega, g_j\rangle d\beta_{ji}$$

But $\langle\omega, g_j\rangle = 0$ because $\omega \in G^\perp$ and therefore

$$L_{\tilde{f}}(G^\perp \cap \Omega_{k-1}) + \sum_{i=1}^{m} L_{\tilde{g}_i}(G^\perp \cap \Omega_{k-1}) \subset L_f(G^\perp \cap \Omega_{k-1}) + \sum_{i=1}^{m} L_{g_i}(G^\perp \cap \Omega_{k-1}).$$

Since β is invertible, one may also write $f = \tilde{f} - \tilde{g}\beta^{-1}\alpha$ and $g_i = (\tilde{g}\beta^{-1})_i$ and, using the same arguments, prove the reverse inclusion. The two sides of the inclusion are thus equal and the Lemma is proved. •

For convenience, we introduce a terminology which is useful to indicate the convergence of the sequence (3.2) in a finite number of stages. We set

$$\Delta^\star = (\Omega_0 + \Omega_1 + \ldots + \Omega_k + \ldots)^\perp \tag{3.3}$$

and we say that $\Delta^\star$ is *finitely computable* if there exists an integer $k^\star$ such that, in the sequence (3.2), $\Omega_{k^\star} = \Omega_{k^\star+1}$. If this is the case, then obviously $\Delta^\star = \Omega_{k^\star}^\perp$.

In the Lemma 3.2 we have seen that if $\Delta^\star$ is finitely computable and if $(\Delta^\star)^\perp \cap G^\perp$ and $\Delta^\star$ are smooth, then $\Delta^\star$ is the maximal element of $\mathcal{J}(f, g, \ker(dh))$. In order to let this distribution be locally controlled invariant all we need are the assumptions of Lemma 2.1, as stated below.

Lemma 3.4. Suppose $\Delta^\star$ is finitely computable. Suppose G, $\Delta^\star$, $\Delta^\star + G$ are nonsingular. Then $\Delta^\star$ is involutive and is the largest locally controlled invariant distribution contained in $\ker(dh)$.

Proof. First, observe that the assumption of nonsingularity on G, $\Delta^\star$, $\Delta^\star + G$ indeed implies the smoothness of $(\Delta^\star)^\perp \cap G^\perp$ and $\Delta^\star$. So, we need only to show that $\Delta^\star$ is involutive.

For, let d denote the dimension of $\Delta^\star$. At any point x° one may find a neighborhood U° of x° and vector fields $\tau_1, \ldots, \tau_d$ such that

$$\Delta^\star = \text{span}\{\tau_1, \ldots, \tau_d\}$$

on U°. Consider the distribution

$$D = \text{span}\{\tau_i : 1 \leq i \leq d\} + \text{span}\{[\tau_i, \tau_j] : 1 \leq i, j \leq d\}$$

and suppose, for the moment, that D is nonsingular on U°. Then, every vector field τ in D can be expressed as the sum of a vector field τ' in $\Delta^\star$ and a vector field τ'' of the form

$$\tau'' = \sum_{i=1}^{d} \sum_{j=1}^{d} c_{ij} [\tau_i, \tau_j]$$

where c_{ij}, $1 \leq i, j \leq d$, are smooth real-valued functions defined on U°.

We want to show that

$$[g_k, D] \subset D + G$$

for all $0 \leq k \leq m$. In view of the above decomposition of any vector field τ in D, this amounts to show that

$$[g_k, [\tau_i, \tau_j]] \subset D + G.$$

The expression of the vector field on the left-hand side via Jacobi identity yields

$$[g_k, [\tau_i, \tau_j]] = [\tau_i, [g_k, \tau_j]] - [\tau_j, [g_k, \tau_i]].$$

The vector field $[g_k, \tau_j]$ is in $\Delta^\star + G$ and therefore, because of the nonsingularity of $\Delta^\star$ and $\Delta^\star + G$, it can be written as the sum of a vector field τ in $\Delta^\star + G$ and a vector field g in G. Since $[\tau_i, g] \in \Delta^\star + G$ for any $g \in G$, we have

$$[\tau_i, [g_k, \tau_j]] = [\tau_i, \tau + g] \in D + \Delta^\star + G = D + G$$

and we conclude that D is such that

$$[g_k, D] \subset D + G$$

for all $0 \leq k \leq m$.

Now, recall that ker(dh) is involutive by assumption, and therefore that

$$D \subset \text{ker}(dh).$$

From this and from the previous inclusions we deduce that D is an element of $\mathcal{J}(f, g, \text{ker}(dh))$. Since $D \supset \Delta^\star$ by construction and $\Delta^\star$ is the maximal element of $\mathcal{J}(f, g, \ker(dh))$ we see that

$$D = \Delta^\star.$$

Thus, any Lie bracket of vector fields of $\Delta^\star$, which is in D by construction, is still in $\Delta^\star$ and the latter is an involutive distribution.

If we drop the assumption that D has constant dimension on U°, we can still conclude that D coincides with $\Delta^\star$ on the subset $\bar{U} \subset U^\circ$ consisting of all regular points of D. Then, using Lemma 1.3.4, we can as well prove that $D = \Delta^\star$ on the whole of U°. ●

In practice, the largest locally controlled invariant distribution contained in $\ker(dh)$ can be calculated, in a neighborhood of a fixed point x°, in the following manner. Suppose Ω_{k-1} has constant dimension, say σ_{k-1}, near x° and let this codistribution be spanned by the rows of a $(\sigma_{k-1} \times n)$ matrix W_{k-1} of smooth functions. In order to calculate a basis of Ω_k, using (3.2), one has to determine first the intersection $\Omega_{k-1} \cap G^\perp$. A covector ω in $\Omega_{k-1} \cap G^\perp(x)$, being a linear combination of the rows of $W_{k-1}(x)$ which annihilates the vectors of $G(x)$, has the form $\omega = \gamma W_{k-1}(x)$, with γ solution of

$$\gamma W_{k-1}(x)g(x) = 0. \tag{3.4}$$

If the matrix

$$A_{k-1}(x) = W_{k-1}(x)g(x)$$

has constant rank, say ρ_{k-1}, near x°, the space of solutions of (3.4) has constant dimension $(\sigma_{k-1} - \rho_{k-1})$ near x° and there exists a $(\sigma_{k-1} - \rho_{k-1}) \times \sigma_{k-1}$ matrix of smooth functions, noted $S_{k-1}(x)$, whose rows span, at each x, this space. As a consequence, $\Omega_{k-1} \cap G^\perp(x)$ is spanned by the rows of the matrix $S_{k-1}(x)W_{k-1}(x)$. From this, using (3.2) and also recalling Remark 1.6.9, it is concluded that Ω_k can be described in the form

$$\Omega_k = \Omega_{k-1} + \text{span}\{L_f(S_{k-1}W_{k-1})_i : 1 \leq i \leq \sigma_{k-1} - \rho_{k-1}\} +$$
$$+ \text{span}\{L_{g_j}(S_{k-1}W_{k-1})_i : 1 \leq i \leq \sigma_{k-1} - \rho_{k-1}, 1 \leq j \leq m\}$$

(where $(S_{k-1}W_{k-1})_i$ denotes the i-th row of $S_{k-1}W_{k-1}$). From the covector fields indicated on the right-hand side, one can easily find a basis for Ω_k, if the latter has constant dimension σ_k near x°.

Of course, the recursive construction is initialized by setting $W_0(x) = dh(x)$. If $\sigma_{k-1} = \sigma_k$ for some k, then by definition

$$L_f(\Omega_{k-1} \cap G^\perp) \subset \Omega_{k-1}$$
$$L_{g_j}(\Omega_{k-1} \cap G^\perp) \subset \Omega_{k-1}, 1 \leq j \leq m$$

and the construction terminates. In other words, if appropriate regularity conditions are satisfied (namely, the constancy of the dimensions of Ω_k and of $\Omega_k \cap G^\perp$), after a finite number $k^\star$ of iterations the condition $\Omega_{k^\star+1} = \Omega_{k^\star}$ of Lemma 3.2 is achieved.

Remark 3.5. Note that the integer ρ_k, the rank of A_k, can be characterized as

$$\rho_k = \dim(\Omega_k) - \dim(\Omega_k \cap G^\perp). \quad \bullet$$

For convenience, we incorporate into a suitable definition all the regularity conditions introduced in the previous discussion and we say that the point x° is a *regular point* of the controlled invariant distribution algorithm if, in a neighborhood of x°, the distribution G and the codistributions Ω_k and $\Omega_k \cap G^\perp$, for all $k \geq 0$, are nonsingular.

Proposition 3.6. Suppose x° is a regular point of the controlled invariant distribution algorithm. Then the hypotheses of Lemma 3.4 are locally satisfied, i.e., in a neighborhood U° of x°, $\Delta^\star$ is finitely computable and G, $\Delta^\star$ and $\Delta^\star + G$ are nonsingular.

Proof. It is an immediate consequence of the previous discussion. $\bullet$

Example 3.1. Consider again the system already discussed in the Example 1.1. In this case,

$$W_0(x) = \begin{bmatrix} 1 & 0 & 0 & 0 \\ 0 & 1 & 0 & 0 \end{bmatrix}$$

and

$$A_0(x) = \begin{bmatrix} 1 & 0 \\ x_3 & 0 \end{bmatrix}.$$

Thus, $\sigma_0 = 2$ and $\rho_0 = 1$. We can choose

$$S_0(x) = \begin{bmatrix} -x_3 & 1 \end{bmatrix}$$

and we find

$$\Omega_0 \cap G^\perp(x) = \text{span}\{S_0(x)W_0(x)\} = \text{span}\{\omega\}$$

with $\omega = \begin{bmatrix} -x_3 & 1 & 0 & 0 \end{bmatrix}$. Now, observe that

$$L_f\omega = f^T(x)\left[\frac{\partial \omega^T}{\partial x}\right]^T + \omega(x)\left[\frac{\partial f}{\partial x}\right] = \begin{bmatrix} (\lambda x_3 + x_4) & 0 & 0 & 1 \end{bmatrix}$$

and

$$L_{g_1}\omega = \begin{bmatrix} 0 & 0 & 1 & 0 \end{bmatrix}$$
$$L_{g_2}\omega = \begin{bmatrix} 0 & 0 & 0 & 0 \end{bmatrix}.$$

From these, since

$$\Omega_1 = \Omega_0 + \mathrm{span}\{L_f\omega, L_{g_1}\omega, L_{g_2}\omega\}$$

we conclude that $k^* = 1$, $\Omega_{k^*}(x) = (\mathbf{R}^4)^*$ for all x. As a consequence, $\Delta^* = 0$ for all x.

Example 3.2. Consider a system of the form (1.1), with 2 inputs and 2 outputs, defined on $\mathbf{R}^5$, with

$$f(x) = \begin{bmatrix} x_2 \\ 0 \\ x_1 x_4 \\ x_3^2 \\ x_1 \end{bmatrix} \qquad g_1(x) = \begin{bmatrix} 1 \\ x_3 \\ 0 \\ x_5 \\ 1 \end{bmatrix} \qquad g_2(x) = \begin{bmatrix} 0 \\ 0 \\ 1 \\ x_1 \\ x_2 x_3 \end{bmatrix}$$

$$h_1(x) = x_1$$
$$h_2(x) = x_2.$$

In this case, again

$$W_0(x) = \begin{bmatrix} 1 & 0 & 0 & 0 & 0 \\ 0 & 1 & 0 & 0 & 0 \end{bmatrix}$$

and

$$A_0 = \begin{bmatrix} 1 & 0 \\ x_3 & 0 \end{bmatrix}$$

Thus, $\sigma_0 = 2$ and $\rho_0 = 1$. We can use the same $S_0(x)$ as in the previous example, having

$$\Omega_0 \cap G^\perp(x) = \mathrm{span}\{S_0(x)W_0(x)\} = \mathrm{span}\{\omega\}$$

with $\omega = [-x_3 \quad 1 \quad 0 \quad 0 \quad 0]$. Since

$$L_f\omega = [-x_1 x_4 \quad -x_3 \quad 0 \quad 0 \quad 0]$$
$$L_{g_1}\omega = [0 \quad 0 \quad 1 \quad 0 \quad 0]$$
$$L_{g_2}\omega = [-1 \quad 0 \quad 0 \quad 0 \quad 0]$$

we can choose, as a basis of Ω_1, the rows of the matrix

$$W_1(x) = \begin{bmatrix} 1 & 0 & 0 & 0 & 0 \\ 0 & 1 & 0 & 0 & 0 \\ 0 & 0 & 1 & 0 & 0 \end{bmatrix}.$$

We calculate now

$$A_1(x) = W_1(x)g(x) = \begin{bmatrix} 1 & 0 \\ x_3 & 0 \\ 0 & 1 \end{bmatrix}$$

whose rank ρ_2 is 2 for all x. Therefore, the construction terminates. In fact, we can set

$$S_1(x) = [-x_3 \quad 1 \quad 0]$$

and find that $S_1(x)W_1(x)$ has its (single) row coincident with the one already found for $S_0(x)W_0(x)$. This clearly implies

$$L_f(\Omega_1 \cap G^\perp) \subset \Omega_1$$
$$L_{g_j}(\Omega_1 \cap G^\perp) \subset \Omega_1 \qquad 1 \le j \le m$$

i.e., $k^\star = 1$. Thus, $\Omega_{k^\star}$ is spanned by the rows of $W_1(x)$ and

$$\Delta^\star = \ker(W_1) = \operatorname{span}\{ \begin{bmatrix} 0 \\ 0 \\ 0 \\ 1 \\ 0 \end{bmatrix}, \begin{bmatrix} 0 \\ 0 \\ 0 \\ 0 \\ 1 \end{bmatrix} \}. \quad \bullet$$

We have seen before that, if the hypotheses of Lemma 3.4 hold, the distribution $\Delta^\star$ is the largest locally controlled invariant distribution contained in ker(dh). This means that there exists feedback functions α and β, defined locally in a neighborhood of each given point, such that this distribution is left invariant by the vector fields $\tilde{f} = f + g\alpha$, and $\tilde{g}_i = (g\beta)_i$, $1 \le i \le m$. However, for the actual construction of these feedback functions the only result available so far is the one described in the proof of Lemma 2.1, that is a solution of a special set of partial differential equations. If a slightly stronger set of hypotheses is assumed, it is possible to avoid the solution of partial differential equations, and to find α and β at the end of a recursive procedure that involves only solving linear (x-dependent) algebraic equations. This result is summarized, for convenience, in the next statement.

Proposition 3.7. Suppose x° is a regular point for the controlled invariant distribution algorithm. Then, in a neighborhood U° of x°, the following properties hold. For each $k \ge 0$, there exists a σ_k-dimensional vector of smooth functions

$$\Lambda_k = \operatorname{col}(\lambda_1, \ldots, \lambda_{\rho_k}, \lambda_{\rho_k+1}, \ldots, \lambda_{\sigma_k})$$

such that

$$\Omega_k = \text{span}\{d\lambda_i : 1 \le i \le \sigma_k\} \tag{3.5}$$

and

$$\text{span}\{d\lambda_i : 1 \le i \le \rho_k\} \cap G^\perp = 0. \tag{3.6}$$

Moreover, Ω_{k+1} can be expressed in the form

$$\begin{aligned}
\Omega_{k+1} &= \Omega_k + \text{span}\{dL_{f+ga}\lambda_i : \rho_k + 1 \le i \le \sigma_k\} \\
&\quad + \text{span}\{dL_{(g\beta)_j}\lambda_i : \rho_k + 1 \le i \le \sigma_k, 1 \le j \le m\}
\end{aligned} \tag{3.7}$$

where α and β are solutions of

$$\langle d\lambda_i(x), f(x) + g(x)\alpha(x)\rangle = 0 \qquad 1 \le i \le \rho_k \tag{3.8a}$$

$$\langle d\lambda_i(x), g(x)\beta_j(x)\rangle = \delta_{ij} \qquad 1 \le i \le \rho_k \tag{3.8b}$$

and $\beta_j(x)$ denotes the j-th column of $\beta(x)$.

As a consequence $\Delta^\star$, the largest locally controlled invariant distribution contained in $\ker(dh)$, can be expressed in the form

$$\Delta^\star = \bigcap_{i=1}^{\sigma_{k^\star}} \ker(d\lambda_i).$$

A pair of feedback functions that solve (3.8) for $k = k^\star$ is such that

$$[f + g\alpha, \Delta^\star] \subset \Delta^\star$$

$$[(g\beta)_i, \Delta^\star] \subset \Delta^\star \qquad 1 \le i \le m.$$

Remark 3.8. Note that, obviously,

$$\Lambda_0 = \text{col}(h_1, \ldots, h_m).$$

Because of (3.6), the row vectors $\langle d\lambda_i, g(x)\rangle$, $1 \le i \le \rho_k$, are linearly independent for all x near x°. Thus, the equations (3.8) can always be solved. In particular, because of the special form of the right-hand side of (3.8b), the latter can always be solved by a matrix $\beta(x)$ which is nonsingular in a neighborhood of x°. ●

Remark 3.9. Note also that the involutivity of the distribution $\Delta^\star$, that was proved in Proposition 3.4 under some weaker hypotheses, now follows trivially from the fact that $(\Delta^\star)^\perp$ is spanned by exact differentials. ●

Proof of Proposition 3.7. We proceed by induction, since an expression of the form (3.5) certainly holds for $k = 0$, because $\Omega_0 = \text{span}\{dh_1, \ldots, dh_m\}$. Suppose (3.5) holds. Since by assumption the intersection $\Omega_k \cap G^\perp$ has constant dimension $\sigma_k - \rho_k$, then it is always possible to reorder the entries of Λ_k in such a way that also (3.6) holds. Because of (3.6) no linear combination of $d\lambda_1, \ldots, d\lambda_{\rho_k}$ can be in $G^\perp$, and we deduce that $\Omega_k \cap G^\perp$ is spanned by vectors of the form

$$\omega = d\lambda_i + c_{i1}d\lambda_1 + \ldots + c_{i\rho_k}d\lambda_{\rho_k}$$

where $c_{i1}, \ldots, c_{i\rho_k}$ are suitable functions and $(\rho_k + 1) \leq i \leq \sigma_k$. Recall now that the controlled invariant distribution algorithm is invariant under feedback (Lemma 3.3), so that we can calculate Ω_{k+1} as

$$\Omega_{k+1} = \Omega_k + \sum_{i=0}^{m} L_{\tilde{g}_i}(\Omega_k \cap G^\perp)$$

assuming that the feedback functions α and β are exactly those given by (3.8). The derivative of ω along $\tilde{g}_j$, $0 \leq j \leq m$, has the form

$$L_{\tilde{g}_j}\omega = dL_{\tilde{g}_j}\lambda_i + \sum_{s=1}^{\rho_k}(L_{\tilde{g}_j}c_{is})d\lambda_s + \sum_{s=1}^{\rho_k} c_{is}dL_{\tilde{g}_j}\lambda_s$$

Since, by construction $\langle d\lambda_s, \tilde{g}_j \rangle$ is either 0 or 1, the third term of this sum is zero. On the other hand, the second term is already in Ω_k, because it is a combination of covectors of Ω_k. We see from this that

$$L_{\tilde{g}_j}\omega = dL_{\tilde{g}_j}\lambda_i + \omega \qquad \text{with} \qquad \omega \in \Omega_k \tag{3.9}$$

and therefore

$$\Omega_{k+1} = \Omega_k + \text{span}\{dL_{\tilde{g}_j}\lambda_s : \rho_k + 1 \leq s \leq \sigma_k, 0 \leq j \leq m\}.$$

This proves (3.7). At this point it is clearly possible to choose, in the set of functions whose differentials span the second term of this sum, an additional set of $(\sigma_{k+1} - \sigma_k)$ new functions, that will be denoted by $\lambda_{\sigma_k+1}, \ldots, \lambda_{\sigma_{k+1}}$, such that

$$\Omega_{k+1} = \text{span}\{d\lambda_i : 1 \leq i \leq \sigma_{k+1}\}$$

thus proving the validity of (3.5) for $k + 1$.

The last part of the statement is a trivial consequence of the previous construction. As a matter of fact, consider again the expression (3.9) of the derivative along $\tilde{g}_j$, $0 \leq j \leq m$, of a covector field in $\Omega_k \cap G^\perp$. If the algorithm terminates at $k = k^\star$, then

$$L_{\tilde{g}_j}(\Omega_k \cap G^\perp) \subset \Omega_k$$

and therefore we see from (3.9) that

$$L_{\tilde{g}_j} d\lambda_i \in \Omega_{k^*}$$

for $(\rho_{k^*} + 1) \leq i \leq \sigma_{k^*}$. On the other hand, this relation is valid also for $1 \leq i \leq \rho_{k^*}$, because, in this case, by construction

$$L_{\tilde{g}_j} d\lambda_i = dL_{\tilde{g}_j} \lambda_i = 0.$$

Therefore, since the $d\lambda_i$'s, $1 \leq i \leq \sigma_{k^*}$, span Ω_{k^*}, we obtain that

$$L_{\tilde{g}_j} \Omega_{k^*} \subset \Omega_{k^*}$$

i.e. that Ω_{k^*} is invariant under $\tilde{g}_j$, $0 \leq j \leq m$. Since Ω_{k^*} is nonsingular, and therefore smooth, in view of Lemma 1.6.7 we conclude that $\Delta^* = \Omega_{k^*}^\perp$ is invariant under the new dynamics. •

It is quite interesting to establish a relationship between the controlled invariant distribution algorithm and some concepts introduced earlier, like the zero dynamics algorithm. To this end, observe that if x° is a regular point of the controlled invariant distribution algorithm, the distribution Δ^* is nonsingular and involutive in a neighborhood U° of x°. Thus, by Corollary 2.1.8, Δ^* has the maximal integral manifolds property on U°, i.e. U° is partitioned into maximal integral submanifolds of Δ^*. Let L_{x° denote the integral submanifold of Δ^* which contains the point x°. In what follows, we will characterize the relation existing between L_{x° and the zero dynamics manifold Z^*.

Proposition 3.10. Suppose x° is a regular point for the controlled invariant distribution algorithm, and $\dim(G(x^\circ)) = m$. Suppose also that

$$\sum_{i=1}^{m} L_{g_i}(\Omega_k \cap G^\perp) \subset \Omega_k \tag{3.10}$$

for all $k \geq 0$. Then the assumptions of Proposition 1.1 hold, and for all $x \in Z^*$ in a neighborhood of x°

$$\Delta^*(x) = T_x Z^*.$$

As a consequence, Z^* locally coincides with the integral submanifold L_{x° of Δ^*.

Proof. We prove, by induction, that if the assumption (3.10) holds, then in a neighborhood U° of x°,

$$M_k \cap U^\circ = \{x \in U^\circ : \Lambda_k(x) = 0\} \tag{3.11}$$

This is true, by definition, for $k = 0$. Suppose is true for some $k \geq 0$. Since the differentials $d\lambda_i(x)$ are by assumption linearly independent at x°, and the matrix

$$\mathrm{col}(\langle d\lambda_1(x), g(x)\rangle, \ldots, \langle d\lambda_{\sigma_k}(x), g(x)\rangle) = L_g\Lambda_k(x)$$

has constant rank ρ_k near x°, then, according to the zero dynamics algorithm, M_{k+1} is obtained in the following way. Let $R_k(x)$ be a matrix whose rows—at each x—form a basis in the space of all vectors γ such that $\gamma L_g\Lambda_k(x) = 0$. Then

$$M_{k+1} \cap U^\circ = \{x \in U^\circ : \Lambda_k(x) = 0, R_k(x)L_f\Lambda_k(x) = 0\}.$$

On the other hand, if the assumption (3.10) is satisfied, Ω_{k+1} is given (see the proof of Proposition 3.7) by

$$\Omega_{k+1} = \mathrm{span}\{d\lambda_i : 1 \le i \le \sigma_k\} + \mathrm{span}\{dL_{f+g\alpha}\lambda_i : \rho_k + 1 \le i \le \sigma_k\}.$$

Observe that, by definition of $\alpha(x)$ and of $R_k(x)$

$$
\begin{aligned}
L_{f+g\alpha}\Lambda_k(x) = 0 &\Leftrightarrow \langle d\lambda_i, f + g\alpha\rangle = 0, \ 1 \le i \le \rho_k, \text{ and } R_k(x)L_{f+g\alpha}\Lambda_k(x) = 0 \\
&\Leftrightarrow 0 = R_k(x)L_{f+g\alpha}\Lambda_k(x) = R_k(x)L_f\Lambda_k(x) + R_k(x)L_g\Lambda_k(x)\alpha(x) \\
&\Leftrightarrow 0 = R_k(x)L_f\Lambda_k(x)
\end{aligned}
$$

Thus,

$$x \in M_{k+1} \cap U^\circ \Leftrightarrow \Lambda_k(x) = 0 \text{ and } L_{f+g\alpha}\Lambda_k(x) = 0 \Leftrightarrow \Lambda_{k+1}(x) = 0$$

and this proves the assertion (3.11). •

Remark 3.11. Note that, in case the condition (3.10) holds, then

$$s_0 + \ldots + s_k = \sigma_k \text{ and } r_k = \rho_k. \quad \bullet$$

There are two special classes of systems which satisfy the assumption (3.10): the linear systems, and the nonlinear systems having a relative degree at the point x°. We discuss first the case of a linear system.

Corollary 3.12. In a linear system, the zero dynamics algorithm and the controlled invariant distribution algorithm produce the same result. More precisely, let $V^\star$ denote the largest subspace of $\ker(C)$ satisfying

$$AV^\star \subset V^\star + \mathrm{Im}(B).$$

Then,

$$Z^\star = V^\star$$
$$\Delta^\star(x) = V^\star \text{ at each } x \in \mathbf{R}^n.$$

Proof. In this case, the controlled invariant distribution algorithm proceeds as follows. Note that the codistribution $\Omega_0 = \mathrm{span}\{dh\}$ is spanned by constant covector fields, namely the rows $c_1, \ldots, c_m$ of the matrix C. Suppose also Ω_k is spanned by constant covector fields, the σ_k rows of a matrix W_k. Then the intersection $\Omega_k \cap G^\perp$ is also spanned by constant vector fields, the $(\sigma_k - \rho_k)$ rows of the matrix $S_k W_k$, in which S_k is a matrix whose rows span the space of solutions γ of the equation

$$\gamma A_k = \gamma W_k B = 0.$$

Since g_j is a constant vector field, the j-th column of the matrix B, it is immediately deduced that

$$L_{g_j}(S_k W_k)_i = 0$$

and this implies (see also Remark 1.6.9) that

$$\sum_{i=1}^{m} L_{g_i}(\Omega_k \cap G^\perp) \subset \Omega_k$$

i.e. that the condition (3.10) holds. Moreover

$$L_f(S_k W_k)_i = (S_k W_k)_i A$$

and this shows that also Ω_{k+1} is spanned by constant covector fields. For each x, the codistributions Ω_k and $\Omega_k \cap G^\perp$ have indeed constant dimension and any point x° is a regular point for the controlled invariant distribution algorithm. The hypoheses of Proposition 3.10 are satisfied, and the result follows. ●

We consider now the case of nonlinear systems having a vector relative degree at a point x°. In order to prove that for these systems the hypotheses of Proposition 3.10 are satisfied, we prove first a property related to the notion of controlled invariant distribution.

Lemma 3.13. Suppose the integers $r_1, \ldots, r_m$ are such that the vector

$$[\, L_{g_1} L_f^k h_i(x) \quad L_{g_2} L_f^k h_i(x) \quad \cdots \quad L_{g_m} L_f^k h_i(x) \,]$$

is zero for all x near x° and for all $k < r_i - 1$, and nonzero for $k = r_i - 1$ at $x = x^\circ$. Then, in a neighborhood U° of the point x°, every controlled invariant distribution contained in $\ker(dh)$ is also contained in the distribution D defined by

$$D = \bigcap_{i=1}^{m} \bigcap_{k=1}^{r_i} \ker(dL_f^{k-1} h_i). \tag{3.13}$$

Suppose D is a smooth distribution. A pair of feedback functions (α, β) defined on U is such that

$$[f + g\alpha, D] \subset D \qquad\qquad (3.14a)$$
$$[(g\beta)_i, D] \subset D \qquad 1 \le i \le m \qquad (3.14b)$$

if and only if

$$d(\langle dL_f^{r_i-1}h_i, f(x) + g(x)\alpha(x)\rangle) \in D^\perp \text{ for all } 1 \le i \le m \qquad (3.15a)$$
$$d(\langle dL_f^{r_i-1}h_i, g(x)\beta_j(x)\rangle) \in D^\perp \text{ for all } 1 \le i, j \le m. \qquad (3.15b)$$

In particular, if the system has relative degree $\{r_1, \ldots, r_m\}$ at x°, i.e. if the matrix $A(x)$ defined by (5.1.2) is nonsingular, then D satisfies (3.14), with $\alpha(x)$ and $\beta(x)$ solutions of

$$A(x)\alpha(x) + b(x) = 0 \qquad\qquad (3.16a)$$
$$A(x)\beta(x) = I \qquad\qquad (3.16b)$$

(where $b(x)$ is the vector defined by (5.1.8)).

Proof. Let Δ be a locally controlled invariant distribution contained in ker(dh). Then, by definition, $\Delta \subset (\text{span}\{dh_i\})^\perp$ for all $1 \le i \le m$. Moreover, for some locally defined feedback α, $[\tilde{f}, \Delta] \subset \Delta$. Suppose $\Delta \subset (\text{span}\{dL_f^k h_i\})^\perp$ for some $k < r_i - 1$; then, using the property

$$L_{f+g\alpha}L_f^k h_i = L_f^{k+1}h_i$$

we have, for any vector field τ of Δ

$$0 = \langle dL_f^k h_i, [\tilde{f}, \tau]\rangle = L_{\tilde{f}}\langle dL_f^k h_i, \tau\rangle - \langle dL_{\tilde{f}}L_f^k h_i, \tau\rangle = \langle dL_f^{k+1}h_i, \tau\rangle$$

i.e. $\Delta \subset (\text{span}\{dL_f^{k+1}h_i\})^\perp$. This proves that $\Delta \subset D$.

Now, suppose there exists a pair of feedback functions that makes (3.14) satisfied. Let τ be a vector field in D. Then

$$\langle dL_f^k h_i, \tau\rangle = 0$$
$$\langle dL_f^k h_i, [\tilde{f}, \tau]\rangle = 0$$
$$\langle dL_f^k h_i, [\tilde{g}_j, \tau]\rangle = 0$$

for all $1 \le i, j \le m$, $0 \le k \le r_i - 1$. From the second one, written for $k = r_i - 1$, we deduce

$$0 = L_{\tilde{f}}\langle dL_f^{r_i-1}h_i, \tau\rangle - \langle dL_{\tilde{f}}L_f^{r_i-1}h_i, \tau\rangle = \langle d(\langle dL_f^{r_i-1}h_i, f + g\alpha\rangle), \tau\rangle$$

i.e. the condition (3.15a). Similarly, from the third one we obtain (3.15b). Conversely, if the conditions (3.15) hold, then the previous equalities are true for $k = r_i - 1$. For other values of $k < r_i - 1$, these equalities hold for any feedback (α, β) by definition of r_i. Thus, we deduce that D is invariant under $\tilde{f}$ and $\tilde{g}_i$, $1 \leq i \leq m$, if and only if (α, β) are solutions of (3.15).

The third part of the statement is a trivial consequence of the second one. In fact, if the matrix $A(x)$ is nonsingular for all x in a neighborhood of x°, the equations (3.16) have a (unique) solution, and this solution trivially satisfies (3.15), because in this case

$$\langle dL_f^{r_i-1} h_i(x), f(x) + g(x)\alpha(x) \rangle, \qquad \langle dL_f^{r_i-1} h_i(x), (g(x)\beta(x))_j \rangle$$

are constant. ●

Using this Lemma, it is not difficult to see that, in the case of a system having relative degree $\{r_1, \ldots, r_m\}$ at a point x°, the condition (3.10) is satisfied. This and other properties of interest are collected in the following statement.

Corollary 3.14. Suppose the system has relative degree $\{r_1, \ldots, r_m\}$ at a point x°. Then this point is a regular point of the controlled invariant distribution algorithm and the condition (3.10) holds. In particular $\Delta^\star$, the largest locally controlled invariant distribution contained in $\ker(dh)$, can be expressed, in a neighborhood U° of x°, as

$$\Delta^\star = \bigcap_{i=1}^{m} \bigcap_{k=1}^{r_i} \ker(dL_f^{k-1} h_i)$$

and is rendered invariant by the standard noninteractive feedback (5.3.2). The result of Proposition 3.10 apply and,

$$Z^\star = \{x \in U^\circ : L_f^{k-1} h_i(x) = 0, \ 1 \leq k \leq r_i, \ 1 \leq 1 \leq m\} \qquad (3.17)$$

Proof. It is left, as an exercise, to the reader. ●

Remark 3.15. It is immediate to check that the results stated in the last part of Lemma 3.12 are also valid in case the system has a number p of outputs which is less than the number m of inputs, provided that the matrix (5.1.2) has rank p at the point x° (see also Remark 5.1.5). Thus, in particular, they are valid for a system with only one output $y_i = h_i(x)$, because in this case, by definition, the matrix in question reduces to a single nonzero row. As a byproduct, it is found that the largest locally controlled invariant distribution contained in $\ker(dh_i)$, noted $\Delta_i^\star$, has the form

$$\Delta_i^\star = \bigcap_{k=1}^{r_i} \ker(dL_f^{k-1}h_i). \quad \bullet \tag{3.18}$$

In general, if the condition (3.10) is not satisfied, it is not possible to identify the zero dynamics manifold $Z^\star$ with an integral submanifold of $\Delta^\star$. In other words, the problems of using feedback in order to constrain the output of a system to be zero for a certain time and the problem of using feedback in order to induce a certain amount of unobservability, in a general nonlinear setting, are not equivalent (although in a linear system they *are*, as the statement of Corollary 3.12 shows). There is however always a relation between $Z^\star$ and the integral submanifolds of $\Delta^\star$, which is expressed in the following statement.

Proposition 3.16. Suppose x° is both a regular point for the controlled invariant distribution algorithm, and for the zero dynamics algorithm. Let L_{x° denote the integral submanifold of $\Delta^\star$ which contains the point x°. Then L_{x° is a locally controlled invariant submanifold and $h(x) = 0$ at each point $x \in L_{x^\circ}$, i.e. L_{x° is an output zeroing manifold for (1.1). As a consequence, L_{x° is locally contained in $Z^\star$.

Proof. Recall that, if the assumptions are satisfied, $(\Delta^\star)^\perp$ is spanned by the differentials of certain functions λ_i, $1 \leq i \leq \sigma_{k^\star}$. Thus, for some neighborhood U of x°,

$$L_{x^\circ} = \{x \in U : \lambda_i(x) = \lambda_i(x^\circ), \ 1 \leq i \leq \sigma_{k^\star}\}.$$

Suppose $\alpha(x)$ is a function which solves the equation (3.8a) for $k = k^\star$. Since $f(x^\circ) = 0$ by assumption, we can always suppose that $\alpha(x^\circ) = 0$, and therefore that the point x° is an equilibrium of the vector field $\tilde{f}(x) = f(x) + g(x)\alpha(x)$. The statement of Proposition 3.7 says that the distribution $\Delta^\star$ is invariant under the vector field $\tilde{f}(x)$, and therefore, according to the interpretation of invariance given in Section 1.6, the flow of $\tilde{f}(x)$ locally carries L_{x° into another integral submanifold of $\Delta^\star$. But the point x° is fixed under the flow of $\tilde{f}(x)$ and we conclude that the flow of $\tilde{f}(x)$ carries L_{x° into itself; in other words, $\tilde{f}(x)$ is tangent to L_{x°.

We have in this way that at each point of L_{x° near x° there exists a smooth mapping, namely $\alpha(x)$, with the property that $f(x)+g(x)\alpha(x)$ is tangent to L_{x°. Thus L_{x° is locally controlled invariant. The other statements are immediate consequences. $\bullet$

Note that the previous analysis also clarifies, to some extent, the difference between the notion of a controlled invariant submanifold and that of a controlled invariant distribution.

6.4 Controllability Distributions

A distribution Δ is said to be a *controllability distribution* on U if it is involutive and there exist a feedback pair (α, β) defined on U and a subset I of the index set $\{1, \ldots, m\}$ with the property that $\Delta \cap G = \text{span}\{\tilde{g}_i : i \in I\}$, and Δ is the smallest distribution which is invariant under the vector fields $\tilde{f}, \tilde{g}_1, \ldots, \tilde{g}_m$ and contains $\tilde{g}_i$ for all $i \in I$.

A distribution Δ is said to be a *local controllability distribution* if for each $x^\circ \in U$ there exists a neighborhood U° of x° with the property that Δ is a controllability distribution on U°.

It is clear that, by definition, a (local) controllability distribution is (locally) controlled invariant. Therefore, according to the result of Lemma 2.1, such a distribution must satisfy (2.3) (recall that the necessity of (2.3) is not dependent on the assumptions made in Lemma 2.1 but only on the controlled invariance and the nonsingularity of β). Therefore, it is interesting to look for the extra condition to be added to (2.3) in order to let a given controlled invariant distribution become a local controllability distribution. To this purpose, it is useful to introduce the following algorithm.

Controllability Distribution Algorithm. Let Δ be a fixed distribution. Step 0: set $S_0 = \Delta \cap G$. Step k: set

$$S_k = \Delta \cap \left([f, S_{k-1}] + \sum_{j=1}^{m} [g_j, S_{k-1}] + G\right). \tag{4.1}$$

Lemma 4.1. The sequence (4.1) is nondecreasing. If there exists an integer $k^\star$ such that $S_{k^\star} = S_{k^\star+1}$, then $S_k = S_{k^\star}$ for all $k > k^\star$.

Proof. We need only to prove that $S_k \supset S_{k-1}$. This is clearly true for $k = 1$. If true for some k, then

$$\left([f, S_k] + \sum_{j=1}^{m} [g_i, S_k]\right) \supset \left([f, S_{k-1}] + \sum_{j=1}^{m} [g_i, S_{k-1}]\right)$$

and, therefore,

$$S_{k+1} \supset S_k \quad \bullet$$

Remark 4.2. Note that we may as well represent S_k as

$$S_k = \Delta \cap \left([f, S_{k-1}] + \sum_{j=1}^{m} [g_j, S_{k-1}] + G\right) + S_{k-1}$$

or as

$$S_k = \Delta \cap ([f, S_{k-1}] + \sum_{j=1}^{m} [g_j, S_{k-1}] + S_{k-1} + G)$$

The last one comes from the first and from the modular distributive rule, which holds because $S_{k-1} \subset \Delta$. •

As we did for the algorithm (3.2) we introduce now a terminology which will be used in order to remind both the convergence of the sequence (4.1) in a finite number of stages and the dependence of its final element on the distribution Δ. We set

$$\mathcal{S}(\Delta) = (S_0 + S_1 + \ldots + S_k + \ldots) \qquad (4.2)$$

and we say that $\mathcal{S}(\Delta)$ is *finitely computable* if there exists an integer $k^\star$ such that, in the sequence (4.1), $S_{k^\star} = S_{k^\star+1}$. If this is the case, then obviously $\mathcal{S}(\Delta) = S_{k^\star}$.

An interesting property of the algorithm (4.1) is the following one.

Lemma 4.3. Let $\tilde{f}, \tilde{g}_1, \ldots, \tilde{g}_m$ be any set of vector fields deduced from $f, g_1, \ldots, g_m$ by setting $\tilde{f} = f + g\alpha$ and $\tilde{g}_i = (g\beta)_i$, $1 \leq i \leq m$, with β invertible. Then each distribution S_k of the sequence (4.1) is such that

$$S_k = \Delta \cap ([\tilde{f}, S_{k-1}] + \sum_{j=1}^{m} [\tilde{g}_j, S_{k-1}] + G).$$

Proof. Let τ be a vector field of S_{k-1}. Then, we have

$$[\tilde{f}, \tau] = [f + g\alpha, \tau] = [f, \tau] + \sum_{j=1}^{m} ([g_j, \tau]\alpha_j - (L_\tau \alpha_j)g_j)$$

$$[\tilde{g}_i, \tau] = [(g\beta)_i, \tau] = \sum_{j=1}^{m} ([g_j, \tau]\beta_{ji} - (L_\tau \beta_{ji})g_j).$$

Therefore

$$[\tilde{f}, S_{k-1}] + \sum_{j=1}^{m} ([\tilde{g}_j, S_{k-1}] + G \subset [f, S_{k-1}] + \sum_{j=1}^{m} ([g_j, S_{k-1}] + G$$

But, since β is invertible, then $f = \tilde{f} - g\beta^{-1}\alpha$ and $g_i = (\tilde{g}\beta^{-1})_i$ so that, by doing the same computations, it is found that the reverse inclusion holds. The two sides are thus equal and the Lemma is proved. •

From this it is now possible to deduce the desired "intrinsic" characterization of a local controllability distribution.

Lemma 4.4. Let Δ be an involutive distribution. Suppose Δ, G, $\Delta + G$ are nonsingular and that $\mathcal{S}(\Delta)$ is finitely computable. Then Δ is a local controllability distribution if and only if

$$[f, \Delta] \subset \Delta + G \tag{4.3a}$$

$$[g_i, \Delta] \subset \Delta + G \qquad 1 \leq i \leq m \tag{4.3b}$$

$$\mathcal{S}(\Delta) = \Delta. \tag{4.3c}$$

Proof (Necessity). Suppose Δ is a local controllability distribution. Then it is locally controlled invariant and (2.3) are satisfied. Moreover, locally around each x there exists a feedback pair (α, β) with the property that $\Delta \cap G = \text{span}\{\tilde{g}_i, i \in I\}$, where I is a subset of $\{1, \dots, m\}$, and Δ is the smallest distribution which is invariant under $\tilde{f}, \tilde{g}_1, \dots, \tilde{g}_m$ and contains $\tilde{g}_i$ for all $i \in I$. Consider the sequence of distributions defined by setting

$$\Delta_0 = \Delta \cap G \tag{4.4a}$$

$$\Delta_k = [\tilde{f}, \Delta_{k-1}] + \sum_{i=1}^{m} [\tilde{g}_i, \Delta_{k-1}] + \Delta_{k-1}. \tag{4.4b}$$

It is easily seen, by induction, that

$$\Delta_k \subset \Delta$$

for all k. This is true for $k = 0$ and, if true for some $k > 0$, the invariance of Δ under $\tilde{f}, \tilde{g}_1, \dots, \tilde{g}_m$ shows that $\Delta_{k+1} \subset \Delta$. Therefore, one has

$$\Delta_k = \Delta \cap ([\tilde{f}, \Delta_{k-1}] + \sum_{i=1}^{m} [\tilde{g}_i, \Delta_{k-1}] + \Delta_{k-1} + G)$$

i.e., from Lemma 4.3 (see also Remark 4.2)

$$\Delta_k = S_k. \tag{4.5}$$

Note also that, by definition, $\Delta_0 = \text{span}\{\tilde{g}_i : i \in I\}$. Thus, the sequence of distributions generated by the algorithm (4.4) is exactly the same as the one yielding $\langle \tilde{f}, \tilde{g}_1, \dots, \tilde{g}_m | \text{span}\{\tilde{g}_i : i \in I\} \rangle$, the smallest distribution invariant under $\tilde{f}, \tilde{g}_1, \dots, \tilde{g}_m$ and containing $\text{span}\{\tilde{g}_i : i \in I\}$. From (4.5) and from the assumption that $\mathcal{S}(\Delta)$ is finitely computable we know that there is an integer k^* such that $\Delta_{k^*} = \Delta_{k^*+1}$. Therefore, in view of Lemma 1.8.2, the largest distribution in the sequence (4.4) is exactly $\langle \tilde{f}, \tilde{g}_1, \dots, \tilde{g}_m | \text{span}\{\tilde{g}_i : i \in I\} \rangle$. From this, one concludes that the largest distribution in the sequence (4.4) must coincide with Δ, i.e., again from (4.5), that the condition (4.3c) is satisfied.

Sufficiency. We know from Lemma 2.1 that if Δ is involutive, if G, Δ and $G + \Delta$ are nonsingular and if the conditions (2.3) are satisfied, then locally around each x there exists a pair of feedback functions (α, β) with the property that Δ is invariant under $\tilde{f}, \tilde{g}_1, \ldots, \tilde{g}_m$. From this fact one may deduce that

$$\Delta \cap ([\tilde{f}, S_{k-1}] + \sum_{i=1}^{m} [\tilde{g}_i, S_{k-1}] + G) + S_{k-1}$$

$$= [\tilde{f}, S_{k-1}] + \sum_{i=1}^{m} [\tilde{g}_i, S_{k-1}] + \Delta \cap G + S_{k-1}$$

$$= [\tilde{f}, S_{k-1}] + \sum_{i=1}^{m} [\tilde{g}_i, S_{k-1}] + S_{k-1}.$$

In view of Lemma 4.3 and Remark 4.2, this shows that

$$S_k = [\tilde{f}, S_{k-1}] + \sum_{i=1}^{m} [\tilde{g}_i, S_{k-1}] + S_{k-1}.$$

Without loss of generality, we may assume that $\tilde{g}_1, \ldots, \tilde{g}_m$ are such that $\Delta \cap G = \text{span}\{\tilde{g}_i : i \in I\}$ for some index set I. In fact, $\Delta \cap G$ is nonzero because otherwise $\mathcal{S}(\Delta)$ would be zero, thus contradicting (4.3c). Since $\Delta \cap G$ is nonsingular, one may find a new feedback function $\bar{\beta}$ and construct new vector fields $\bar{g}_i = (\tilde{g}\bar{\beta})_i$, $1 \leq i \leq m$, such that, for some index set I, $\text{span}\{\bar{g}_i : i \in I\} = \Delta \cap G$ and $\bar{g}_i = \tilde{g}_i$ for $i \notin I$. This new set of vector fields still keeps Δ invariant because $\bar{g}_i \in \Delta$ for $i \in I$ and Δ is involutive.

So, $S_0 = G \cap \Delta = \text{span}\{\tilde{g}_i : i \in I\}$, and the sequence of distributions S_k coincides with the sequence of distributions yielding $\langle \tilde{f}, \tilde{g}_1, \ldots, \tilde{g}_m | \text{span}\{\tilde{g}_i : i \in I\}\rangle$. Since, by assumption, for some k^*, $S_{k^*} = S_{k^*+1}$ we deduce from Lemma 1.8.2 that S_{k^*} is the smallest distribution which is invariant under $\tilde{f}, \tilde{g}_1, \ldots, \tilde{g}_m$ and contains $\text{span}\{\tilde{g}_i : i \in I\}$. But (4.3c) says that S_{k^*} coincides with Δ and this completes the proof. $\bullet$

In view of the use of the notion of local controllability distribution in problems of decoupling or noninteracting control, it is useful to be able to construct a "maximal" local controllability distribution contained in a given distribution. To this end one may use the following result.

Lemma 4.5. Let Δ be an involutive distribution. Suppose G, Δ, $G + \Delta$ are nonsingular and

$$[f, \Delta] \subset \Delta + G$$

$$[g_i, \Delta] \subset \Delta + G \qquad 1 \leq i \leq m.$$

Moreover, suppose $\mathcal{S}(\Delta)$ is finitely computable and nonsingular. Then $\mathcal{S}(\Delta)$ is the largest local controllability distribution contained in Δ.

Proof. As in the proof of Lemma 4.4 (sufficiency) it is easily seen that the assumptions imply that locally around each x there exists a pair of feedback functions with the property that $\Delta \cap G = \text{span}\{\tilde{g}_i : i \in I\}$ and $\mathcal{S}(\Delta)$ is the smallest distribution which is invariant under $\tilde{f}, \tilde{g}_1, \ldots, \tilde{g}_m$ and contains $\text{span}\{\tilde{g}_i : i \in I\}$. Moreover, since

$$\text{span}\{\tilde{g}_i : i \in I\} \subset \mathcal{S}(\Delta) \subset \Delta$$

and $\Delta \cap G = \text{span}\{\tilde{g}_i : i \in I\}$, it is seen that

$$\mathcal{S}(\Delta) \cap G = \text{span}\{\tilde{g}_i : i \in I\}.$$

Thus $\mathcal{S}(\Delta)$ is a local controllability distribution.

Let $\bar{\Delta}$ be another local controllability distribution contained in Δ. Then, by definition, in a neighborhood U° of each x there exists a feedback pair $(\bar{\alpha}, \bar{\beta})$ with the property that $\bar{\Delta} \cap G = \text{span}\{\bar{g}_i : i \in \bar{I}\}$ for some subset $\bar{I}$ of $\{1, \ldots, m\}$, and $\bar{\Delta}$ is invariant under $\bar{f}, \bar{g}_1, \ldots, \bar{g}_m$, where $\bar{f} = f + g\bar{\alpha}$ and $\bar{g}_i = (g\bar{\beta})_i$ for $1 \leq i \leq m$. Consider the sequence of distributions

$$\bar{\Delta}_0 = \text{span}\{\bar{g}_i : i \in \bar{I}\}$$
$$\bar{\Delta}_k = [\bar{f}, \bar{\Delta}_{k-1}] + \sum_{i=1}^{m} [\bar{g}_i, \bar{\Delta}_{k-1}] + \bar{\Delta}_{k-1}.$$

Note that $\bar{\Delta}_k \subset \bar{\Delta} \subset \Delta$. Thus

$$\bar{\Delta}_k \subset \Delta \cap ([\bar{f}, \bar{\Delta}_{k-1}] + \sum_{i=1}^{m} [\bar{g}_i, \bar{\Delta}_{k-1}] + \bar{\Delta}_{k-1} + G).$$

Since $\bar{\Delta}_0 = \bar{\Delta} \cap G \subset \Delta \cap G = S_0$, it is easy to show, by induction, by means of Lemma 4.3 and Remark 4.2, that $\bar{\Delta}_k \subset S_k$ for all $k \geq 0$, i.e.

$$\bar{\Delta}_k \subset \mathcal{S}(\Delta).$$

Now recall (see Lemma 1.8.5) that there exists a dense subset of U° with the property that at each x, $\bar{\Delta}(x) = \bar{\Delta}_{n-1}(x)$. Thus, we have that

$$\bar{\Delta}(x) \subset \mathcal{S}(\Delta)(x)$$

for all x in a dense subset. Since $\bar{\Delta}$ is smooth and $\mathcal{S}(\Delta)$ is nonsingular, this implies $\bar{\Delta} \subset \mathcal{S}(\Delta)$. ●

Using the same arguments it is also possible to prove the following characterization of the maximal controllability distribution contained in a given distribution Δ.

Lemma 4.6. Let Δ be an involutive controlled invariant distribution. Let (α, β) be a pair of feedback functions such that

$$[\tilde{f}, \Delta] \subset \Delta$$
$$[\tilde{g}_i, \Delta] \subset \Delta \qquad \text{for } 1 \le i \le m$$
$$\Delta \cap G = \text{span}\{\tilde{g}_i : i \in I\}$$

for some suitable subset I of $\{1, \ldots, m\}$. Consider the sequence

$$\Delta_0 = \text{span}\{\tilde{g}_i : i \in I\}$$
$$\Delta_k = \Delta_{k-1} + [\tilde{f}, \Delta_{k-1}] + \sum_{i=1}^{m} [\tilde{g}_i, \Delta_{k-1}].$$

Then $\mathcal{S}(\Delta)$ is finitely computable if and only if $\Delta_{k^\star} = \Delta_{k^\star+1}$ for some $k^\star$. If this is the case, $\mathcal{S}(\Delta) = \Delta_{k^\star}$. Moreover, if $\Delta_{k^\star}$ is nonsingular, then $\Delta_{k^\star}$ is the largest controllability distribution contained in Δ.

The previous results can be used, for instance, in order to find the maximal controllability distribution in $\ker(dh)$, if so is requested. To this end, using the results of Section 6.3 one first finds—provided the assumptions of Lemma 3.4 are satisfied—the distribution $\Delta^\star$, which is the largest locally controlled invariant distribution contained in $\ker(dh)$. Then, using Lemma 4.5, it is possible to conclude that if $\mathcal{S}(\Delta^\star)$ satisfies the assumptions of this Lemma, then $\mathcal{S}(\Delta^\star)$ is exactly the largest local controllability distribution contained in $\ker(dh)$. In fact, $\mathcal{S}(\Delta^\star)$ is not only the largest controllability distribution in $\Delta^\star$ but also the largest controllability distribution in $\ker(dh)$ because any controllability distribution contained in $\ker(dh)$, being locally controlled invariant, must be contained in $\Delta^\star$. Alternatively, using Lemma 4.6, one can compute a feedback (α, β) which renders $\Delta^\star$ invariant, and then find $\mathcal{S}(\Delta^\star)$ by means of the algorithm (1.8.1). In this case, the finite computability of $\mathcal{S}(\Delta^\star)$ is implied by the existence of an integer $k^\star$ such that $\Delta_{k^\star} = \Delta_{k^\star+1}$.

7 Geometric Theory of State Feedback: Applications

7.1 Asymptotic Stabilization via State Feedback

In this last Chapter we show how, on the basis of the concepts introduced and developed in Chapter 6, a number of relevant synthesis problems, like asymptotic stabilization, output regulation, disturbance decoupling, noninteracting control with internal stability, can be solved for a multivariable nonlinear system, under assumptions that are considerably weaker than those considered in Chapters 4 and 5 (namely the existence of a relative degree at a certain point of interest).

We begin by considering the problem of local asymptotic stabilization at a certain equilibrium point. Our purpose is to extend the results developed in Section 4.4, by showing that if the zero dynamics of a system are asymptotically stable at this point, the system itself can be locally asymptotically stabilized via state feedback. Of course, as stressed at the beginning of that Section, our results are of special relevance only in case the linear approximation of the system is not stabilizable.

To this end, suppose that the system satisfies the regularity assumptions described in Section 6.1, so that the functions (6.1.15) can be taken as a (partial) set of local coordinates around the point x°, and the generalized normal form illustrated in Proposition 6.1.9 can be defined. Suppose, without loss of generality, that $x^\circ = 0$ and choose the input u which satisfies the equations

$$b^i(x) + a^i(x)u = v_i, \qquad 1 \le i \le m \tag{1.1}$$

where the $b^i(x)$'s and $a^i(x)$'s are defined by (6.1.17). Note (see (6.1.20)) that the input thus defined is related to the (unique) input $u^\star(x)$, which imposes the vector field $f^\star(x) = f(x) + g(x)u^\star(x)$ to be tangent to the zero dynamics manifold $Z^\star$, by the following relation

$$u = u^\star(x) + A^{-1}(x)v \tag{1.2}$$

where $A(x)$ is the matrix (6.1.19). The effect of this feedback is to modify the normal form of the equations describing the system into one having a structure of the following type (recall that, on the right-hand sides, x stands for $\Phi^{-1}(z)$, and $z = (\xi^1, \ldots, \xi^m, \eta))$

$$\dot\xi^1 = A_{11}\xi^1 + b_1 v_1$$
$$\dot\xi^2 = A_{22}\xi^2 + b_2 v_2 + D_{21}(x)v_1 + S_2(x)(u^\star(x) + A^{-1}(x)v)$$

$$\cdots$$

$$\dot\xi^m = A_{mm}\xi^m + b_m v_m + \sum_{j=1}^{m-1} D_{mj}(x)v_j + S_m(x)(u^\star(x) + A^{-1}(x)v)$$
$$\dot\eta = f_0(x) + g_0(x)(u^\star(x) + A^{-1}(x)v)$$

in which

$$A_{ii} = \begin{bmatrix} 0 & 1 & 0 & \cdots & 0 \\ 0 & 0 & 1 & \cdots & 0 \\ \cdots & \cdots & \cdots & \cdots & \cdots \\ 0 & 0 & 0 & \cdots & 1 \\ 0 & 0 & 0 & \cdots & 0 \end{bmatrix} \qquad b_i = \begin{bmatrix} 0 \\ 0 \\ \vdots \\ 0 \\ 1 \end{bmatrix} \qquad 1 \le i \le m$$

and

$$D_{ij}(x) = \begin{bmatrix} \delta^i_{1j}(x) \\ \vdots \\ \delta^i_{n_i-1,j}(x) \\ 0 \end{bmatrix} \qquad S_i(x) = \begin{bmatrix} \sigma^i_1(x) \\ \vdots \\ \sigma^i_{n_i-1}(x) \\ 0 \end{bmatrix}$$

for $2 \le i \le m$, $1 \le j \le m-1$. Since the coefficients $\sigma^i_k(x)$ are vanishing at each $x \in Z^\star$, so are the matrices $S_i(x)$.

In view of the fact that, by construction, $u^\star(0) = 0$ and $S_i(0) = 0$ for $2 \le i \le m$, it is immediate to observe that the *linear approximation* at $z = 0$ of first m sets of the equations thus found is *controllable*. As a matter of fact, the equations in question have the form

$$\dot\xi^1 = A_{11}\xi^1 + b_1 v_1$$
$$\dot\xi^2 = A_{22}\xi^2 + b_2 v_2 + D_{21}(0)v_1 + \tilde f_2(z) + \tilde g_2(z)v$$

$$\cdots$$

$$\dot\xi^m = A_{mm}\xi^m + b_m v_m + D_{m1}(0)v_1 + \ldots + D_{m,m-1}(0)v_{m-1} + \tilde f_m(z) + \tilde g_m(z)v$$

with $\tilde g_i(z)$ vanishing at $z = 0$, and $\tilde f_i(z)$ vanishing at $z = 0$ together with its first order derivatives, and the pair

$$A = \begin{bmatrix} A_{11} & 0 & \cdots & 0 \\ 0 & A_{22} & \cdots & 0 \\ \cdots & \cdots & \ddots & \cdots \\ 0 & 0 & \cdots & A_{mm} \end{bmatrix} \qquad B = \begin{bmatrix} b_1 & 0 & \cdots & 0 \\ D_{21}(0) & b_2 & \cdots & 0 \\ \cdots & \cdots & \ddots & \cdots \\ D_{m1}(0) & D_{m2}(0) & \cdots & b_m \end{bmatrix}$$

is indeed a controllable pair.

Set now

$$\xi = \mathrm{col}(\xi^1, \ldots, \xi^m),$$

rewrite the equations in question in the more condensed form

$$\dot{\xi} = A\xi + Bv + \tilde{f}(\xi, \eta) + \tilde{g}(\xi, \eta)v \tag{1.3a}$$
$$\dot{\eta} = q(\xi, \eta) + p(\xi, \eta)v \tag{1.3b}$$

and note that, by construction,

$$\dot{\eta} = q(0, \eta)$$

characterizes the *zero dynamics* of the system. Moreover $\tilde{f}(0, \eta) = 0$.

We easily deduce from this that, if the zero dynamics of the system are asymptotically stable, any linear feedback

$$v = F\xi$$

which stabilizes $(A + BF)$ will also asymptotically stabilize the equilibrium $(\xi, \eta) = (0, 0)$ of (1.3). In fact, the corresponding closed loop system will have the form of the equations (**B**.2.1), and the hypotheses of the corresponding Lemma are satisfied.

For convenience, the result thus established is summarized in a formal statement.

Proposition 1.1. Consider a nonlinear system of the form (**6**.1.1). Suppose x° is a regular point for the zero dynamics algorithm. Suppose x° is an asymptotically stable equilibrium of the zero dynamics. Then, there exists a matrix F such that the feedback

$$u = u^\star(x) + A^{-1}(x)F\xi(x) \tag{1.2}$$

asymptotically stabilizes the corresponding closed loop system at the equilibrium point $x = x^\circ$. $\bullet$

We stress again that this result—as the corresponding result presented in

Section 4.2—does not require asymptotic stability in the first approximation for the zero dynamics, so that it may be useful in order to solve critical problems of local asymptotic stabilization.

Remark 1.2. The result we have established, namely the fact that the linear approximation at $x = 0$ of the equation (1.3a) is controllable, can be interpreted as a nonlinear version of a property of linear systems that was already observed at the end of Remark 6.1.5, namely the controllability of the pair (A_{11}, B_1) in (6.1.8). •

Remark 1.3. Observe that, by means of essentially the same arguments as the ones used above, it is possible to prove the following result. Let x° be a regular point for the zero dynamics algorithm for the system (6.1.1), and suppose this system has an asymptotically stable zero dynamics (at the equilibrium point $x = x^\circ$). Then, there exists a smooth mapping $k : U^\circ \to \mathbf{R}^m$, where U° is a neighborhood of x°, such that the system (6.1.1a), with output

$$y = k(x)$$

has relative degree $\{1, \dots, 1\}$ at x° and a zero dynamics which is still asymptotically stable (at the equilibrium point $x = x^\circ$). The proof of this is left as an exercise to the reader. •

7.2 Tracking and Regulation

We illustrate in this Section a more general approach to the design problem already considered in Section 4.5, i.e. the problem of asymptotically tracking a prescribed reference output . Given a system of the form

$$\dot{x} = f(x) + g(x)u \qquad (2.1a)$$
$$y = h(x) \qquad (2.1b)$$

that will be henceforth referred to as the controlled *plant*, and a reference output $y_R(t)$, the problem is to find a feedback control law which is able to impose on the error

$$e(t) = y(t) - y_R(t)$$

a behavior which asymptotically decays to zero as time tends to infinity.

For convenience, it is assumed that the desired reference output—as often occurs in practice—is not just a fixed function of time, but rather coincides with the output of some autonomous (nonlinear) dynamical system. In particular, it is assumed that

$$y_R(t) = -q(w(t)) \qquad (2.2a)$$

where $q(w)$ is a fixed mapping, and $w(t)$ is a function of time which satisfies a differential equation of the form

$$\dot{w} = s(w). \tag{2.2b}$$

The dynamical system thus defined, which generates the set of all the reference functions of time the output of the plant is required to track, is called the *exosystem*.

It is also possible, since this does not complicate much more the setup, to consider the more general situation in which the behavior of the plant is affected by some undesired *disturbance*. As in Section 4.6, a phenomenon of this type can be modeled by means of an additional term of the form

$$p(x)d(t)$$

on the rigth-hand side of (2.1a). Assuming that the function of time $d(t)$ belongs to the set of solutions of some nonlinear differential equation, there is no loss of generality in considering the latter as part of the exosystem, and therefore it is possible to set (after having redefined, if needed, the matrix $p(x)$ in a obvious way) $d(t) = w(t)$.

Incorporating into a single set of equations the plant (2.1), the exosystem (2.2), the error equation and the perturbation term, one is led then to consider a composite system having the form

$$\dot{x} = f(x) + g(x)u + p(x)w \tag{2.3a}$$
$$\dot{w} = s(w) \tag{2.3b}$$
$$e = h(x) + q(w) \tag{2.3c}$$

for which a feedback control u is sought , yielding

$$\lim_{t \to \infty} e(t) = 0$$

for any possible initial condition. If this property is satisfied, it is said that *regulation* of the output y of plant (2.1) has been achieved. Of course, since the control u is implemented by means of a feedback structure, it is also required that the corresponding closed loop has some *stability* property.

As usual, it is assumed that the plant (2.1), with m inputs and m outputs, has a state x defined in a neighborhood U of the origin in $\mathbf{R}^n$; the state w of the exosystem is defined on a neighborhood W of the origin of $\mathbf{R}^s$. In addition to the standing smoothness hypotheses on $f(x)$, $g(x)$ and $h(x)$, it is assumed that $s(x)$ is a smooth vector field, and $q(x)$ is a smooth mapping, both defined on W. In particular, it is supposed that $f(0) = 0$, $s(0) = 0$, $h(0) = 0$, $q(0) = 0$. Thus, for $u = 0$, the composite system (2.3) has an equilibrium state $(x, w) = (0, 0)$ yielding zero error.

The control action to (2.1) can be provided either by state feedback or by error feedback. In the first case (see Fig. 7.1a), the input u is a function of the states x and w of the plant and, respectively, of the exosystem, namely

$$u = \alpha(x, w) \tag{2.4}$$

and this, together with (2.3), yields a closed loop system having the form

$$\dot{x} = f(x) + g(x)\alpha(x, w) + p(x)w \tag{2.5a}$$
$$\dot{w} = s(w). \tag{2.5b}$$

In particular, it is assumed that $\alpha(0, 0) = 0$, so that the closed loop (2.5) has an equilibrium at $(x, w) = (0, 0)$. In the case of error feedback (see Fig. 7.1b), the input u is a function of the state z of an auxiliary dynamical system driven by the error e. In other words, u is the output of a dynamical system of the form

$$u = \theta(z) \tag{2.6a}$$
$$\dot{z} = \eta(z, e) \tag{2.6b}$$

with state variable z defined on a neighborhood Z of the origin of $\mathbf{R}^\nu$. Composing (2.3) with (2.6) yields in this case a closed loop system of the form

$$\dot{x} = f(x) + g(x)\theta(z) + p(x)w \tag{2.7a}$$
$$\dot{z} = \eta(z, h(x) + q(w)) \tag{2.7b}$$
$$\dot{w} = s(w). \tag{2.7c}$$

Again, it is assumed that $\eta(0, 0) = 0$ and $\theta(0) = 0$, so that the triplet $(x, z, w) = (0, 0, 0)$ is an equilibrium of the closed loop (2.7).

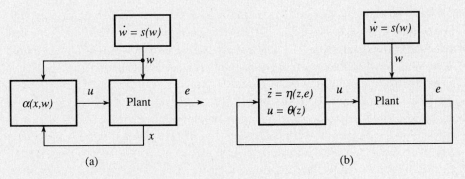

(a)   (b)

Fig. 7.1

In the following two statements we give a precise formulation of two types of control problems, which formalize the requirements of achieving output regulation with stability, by means of state feedback and—respectively—error feedback.

State Feedback Regulator Problem. Given a nonlinear system of the form (2.3), find, if possible, a mapping $\alpha(x, w)$ such that
(i) the equilibrium $x = 0$ of

$$\dot{x} = f(x) + g(x)\alpha(x, 0) \tag{2.8}$$

is asymptotically stable in the first approximation,
(ii) there exists a neighborhood $V \subset U \times W$ of $(0,0)$ such that, for each initial condition $(x(0), w(0)) \in V$, the solution of (2.5) satisfies

$$\lim_{t \to \infty} (h(x(t)) + q(w(t)) = 0.$$

Error Feedback Regulator Problem. Given a nonlinear system of the form (2.3), find, if possible, an integer ν and two mappings $\theta(z)$ and $\eta(z, e)$, such that
(iii) the equilibrium $(x, z) = (0, 0)$ of

$$\dot{x} = f(x) + g(x)\theta(z) \tag{2.9a}$$
$$\dot{z} = \eta(z, h(x)) \tag{2.9b}$$

is asymptotically stable in the first approximation,
(iv) there exists a neighborhood $V \subset U \times Z \times W$ of $(0,0,0)$ such that, for each initial condition $(x(0), z(0), w(0)) \in V$, the solution of (2.7) satisfies

$$\lim_{t \to \infty} (h(x(t)) + q(w(t)) = 0.$$

Remark 2.1. Note that the requirements (i) and (iii) are rather strong, in that they provide stability in the first approximation for the closed loop system. A characterization of this kind is clearly more convenient than simple asymptotic stability, because less sensitive to possible parameter variations, but more demanding, in that requires (see Section 4.4) asymptotic stabilizability of the linear approximation of the plant. The possibility of fulfilling (i) and (iii) depends entirely on the properties of the *linear approximation* of the plant at $x = 0$, and the design of a feedback providing either one of these two properties is a problem whose solution requires only standard results from linear system theory. However, as we shall see in a moment, the simultaneous fulfillment of (i) and (ii) (respectively (iii) and (iv)) is a problem whose solution requires a specific nonlinear analysis. •

Since, as we have just remarked, the linear approximation of the systems in question has a determinant role in the solution of either regulator problems, we first set up an appropriate notation in which these approximations are explicitly shown. For, note that the closed loop system (2.5) can be written in the form

$$\dot{x} = (A + BK)x + (P + BL)w + \phi(x, w)$$
$$\dot{w} = Sw + \psi(w)$$

where $\phi(x, w)$ and $\psi(w)$ vanish at the origin with their first order derivatives, and A, B, P, K, L, S are matrices defined by

$$A = \left[\frac{\partial f}{\partial x}\right]_{x=0} \qquad S = \left[\frac{\partial s}{\partial w}\right]_{w=0} \qquad K = \left[\frac{\partial \alpha}{\partial x}\right]_{x=0, w=0}$$

$$B = g(0) \qquad P = p(0) \qquad L = \left[\frac{\partial \alpha}{\partial w}\right]_{x=0, w=0}.$$

On the other hand, the closed loop system (2.7) can be written in the form

$$\dot{x} = Ax + BHz + Pw + \phi(x, z, w)$$
$$\dot{z} = GCx + Fz + GQw + \chi(x, z, w)$$
$$\dot{w} = Sw + \psi(w)$$

where $\phi(x, z, w)$, $\chi(x, z, w)$ and $\psi(w)$ vanish at the origin with their first order derivatives, and C, Q, H, F, G are matrices defined by

$$C = \left[\frac{\partial h}{\partial x}\right]_{x=0} \qquad H = \left[\frac{\partial \theta}{\partial z}\right]_{z=0} \qquad Q = \left[\frac{\partial q}{\partial w}\right]_{w=0}$$

$$F = \left[\frac{\partial \eta}{\partial z}\right]_{z=0, e=0} \qquad G = \left[\frac{\partial \eta}{\partial e}\right]_{z=0, e=0}.$$

Using this notation, it is immediately realized that the requirement (i) corresponds to the fact that the Jacobian matrix of (2.8) at $x = 0$,

$$J = A + BK$$

has all eigenvalues with negative real part, and the requirement (iii) corresponds to the fact that the Jacobian matrix of (2.9) at $(x, z) = (0, 0)$,

$$J = \begin{bmatrix} A & BH \\ GC & F \end{bmatrix}$$

has all eigenvalues with negative real part.

From the theory of linear systems, it is well-known that the latter can be achieved if and only if the pair of matrices (A, B) is *stabilizable* (i.e. there exists K such that all the eigenvalues of $(A+BK)$ have negative real part) and the pair of matrices (C, A) is *detectable* (i.e. there exists G such that all the eigenvalues of $(A + GC)$ have negative real part). Thus, it is clear that these two properties of the linear approximation of the plant (2.1) at $x = 0$ are indeed necessary conditions for the solvability of the Error Feedback Regulator Problem.

The subsequent developments are based on a certain set of standing hypotheses. In view of the previous discussion it is assumed, of course, that the pair (A, B) is stabilizable (*Hypothesis* (H1)). However, for reasons that will become more clear later in the proof of Theorem 2.9, the assumption that the pair (C, A) is detectable is replaced by the *stronger* condition that the pair

$$[C \quad Q], \qquad \begin{bmatrix} A & P \\ 0 & S \end{bmatrix}$$

is detectable (*Hypothesis* (H2)). In other words, detectability of the linear approximation of the *full* system (2.3) is assumed.

In addition, an hypothesis about the asymptotic properties of the exosystem (2.2b) is introduced, whose role will be discussed in more detail in the Remarks 2.2 and 2.3. The hypothesis in question is based on the notion of Poisson stability. Given a differential equation

$$\dot{x} = f(x)$$

a point x° is said to be *Poisson stable* if $\Phi_t^f(x^\circ)$ is defined for all $t \in \mathbf{R}$ (recall that $\Phi_t^f(x)$ denotes the flow of the vector field $f(x)$) and, for each neighborhood U° of x° and for each real number $T > 0$, there exists a time $t_1 > T$ such that $\Phi_{t_1}^f(x^\circ) \in U^\circ$, and a time $t_2 < -T$ such that $\Phi_{t_2}^f(x^\circ) \in U^\circ$. In other words, a point x° is Poisson stable if the trajectory $x(t)$ which originates in x° passes arbitrarily close to x° for arbitrarily large times, in forward and backward direction.

In what follows, we assume the following (*Hypothesis* (H3)): the point $w = 0$ is a stable equilibrium (in the ordinary sense) of the exosystem, and there is an open neighborhood of the point $w = 0$ in which every point is Poisson stable.

Remark 2.2. The hypothesis (H3) implies that the matrix S—which describes the linear approximation of the exosystem at $w = 0$—*has all its eigenvalues on the imaginary axis*. In fact, no eigenvalue of S can have positive real part, because otherwise the equilibrium $w = 0$ would be unstable. Moreover, the assumed Poisson stability of each point in a neighborhood of $w = 0$ implies that no trajectory of the exosystem can converge to $w = 0$ as time tends to infinity, and this, in turn, implies the absence of eigenvalues of S with negative real part. In fact, if S had eigenvalues with negative real part, the exosystem would admit a stable invariant manifold near the equilibrium, and the trajectories originating on this manifold would converge to $w = 0$ as time tends to infinity. Having limited in this way the class of the exosystems is helpful, as it will be shown in a moment, in rendering necessary certain sufficient conditions for the existence of solutions to the regulator problems. On the other hand, the class of exosystems satifying the hypothesis (H3) is sufficiently adequate in the present setup, for it includes—for instance—the systems in which every solution of (2.2b) is a periodic solution (and, accordingly, the reference output $y_R(t)$ is a periodic function of t). Note also, in particular, that assuming that no trajectory of the exosystem is convergent to 0 as time tends to infinity does not involve loss of generality, because none of such trajectories would affect the asymptotic behavior of the error $e(t)$. •

Remark 2.3. Stability of the exosystem has been clearly imposed in order to avoid that the plant is driven by unbounded inputs. Note that the closed loop systems (2.5) and (2.7) have a triangular structure, as the one of the equations (B.2.6). Thus, using the result indicated in Section B.2, it is easy to conclude that, if condition (i) (respectively (iii)) is satisfied, the stability of the exosystem, required by hypothesis (H3), implies the stability of the equilibrium (x, w) of the closed loop (2.5) (respectively, of the equilibrium $(x, z, w) = (0, 0, 0)$ of the closed loop system (2.7)). •

We discuss now conditions for the solvability of both regulator problems. We begin with the State Feedback Regulator Problem.

Lemma 2.4. Suppose (H3) holds and assume that, for some $\alpha(x, w)$, the condition (i) is satisfied. Then, the condition (ii) is also satisfied if and only if there exists a C^k $(k \geq 2)$ mapping $x = \pi(w)$, with $\pi(0) = 0$, defined in a neighborhood $W^\circ \subset W$ of 0, satisfying the conditions

$$\frac{\partial \pi}{\partial w} s(w) = f(\pi(w)) + g(\pi(w))\alpha(\pi(w), w) + p(\pi(w))w \quad (2.10a)$$

$$h(\pi(w)) + q(w) = 0. \quad (2.10b)$$

Proof. By assumption, the eigenvalues of the matrix $(A + BK)$ have negative real part, and those of the matrix S are on the imaginary axis. Thus, there exists a matrix T which solves the linear equation

$$TS = (A + BK)T + (P + BL).$$

A linear change of variable $\tilde{x} = x - Tw$, modifies the closed loop system (2.5) into a system of the form

$$\dot{\tilde{x}} = (A + BK)\tilde{x} + \tilde{\phi}(\tilde{x}, w)$$
$$\dot{w} = Sw + \psi(w)$$

in which $\tilde{\phi}(\tilde{x}, w)$ and $\psi(w)$ vanish at the origin with their first order derivatives. Using the results of Section **B**.1, we deduce the existence, for the system (2.5), of a local center manifold at $(0,0)$. This manifold can be expressed as the graph of a C^k mapping

$$\tilde{x} = \tilde{\pi}(w)$$

with $\tilde{\pi}(w)$ satisfying an equation of the form (**B**.1.3b). A straightforward calculation shows that, in the original coordinates, this center manifold is the graph of the mapping

$$\pi(w) = Tw + \tilde{\pi}(w)$$

with $\pi(w)$ such that condition (2.10a) is satisfied.

Choose a real number $R > 0$, and let w° be a point of W°, with $\| w^\circ \| < R$. Since, by the hypothesis (H3), the equilibrium $w = 0$ of the exosystem is stable, it is possible to choose R so that the solution $w(t)$ of (2.5b) satisfying $w(0) = w^\circ$ remains in W° for all $t \geq 0$. If $x(0) = x^\circ = \pi(w^\circ)$, the corresponding solution $x(t)$ of (2.5a) will be such that $x(t) = \pi(w(t))$ for all $t \geq 0$ because the manifold $x = \pi(w)$ is by definition invariant under the flow of (2.5). Note that the mapping

$$\mu : W^\circ \to U \times W^\circ$$
$$w \to (\pi(w), w)$$

(whose rank is equal to the dimension s of W° at each point of W°), defines a diffeomorphism of a neighborhood of W° onto its image. Thus, the restriction of the flow of (2.5) to its center manifold is a diffeomorphic copy of the flow of the exosystem, and any point on the center manifold sufficiently close to the origin is Poisson stable by hypothesis (H3). We will show that this and the fulfillment of requirement (ii) imply (2.10b).

For suppose (2.10b) is not true at some $(\pi(w^\circ), w^\circ)$ sufficiently close to $(0,0)$. Then,

$$M = \| h(\pi(w^\circ)) + q(w^\circ) \| > 0$$

and there exists a neighborhood V of $(\pi(w^\circ), w^\circ)$ such that

$$\| h(\pi(w)) + q(w) \| > M/2$$

at each $(\pi(w), w) \in V$. If (ii) holds for a trajectory starting at $(\pi(w^\circ), w^\circ)$, there exists T such that

$$\| h(\pi(w(t))) + q(w(t)) \| < M/2$$

for all $t > T$. But if $(\pi(w^\circ), w^\circ)$ is Poisson stable, then for some $t' > T$, $(\pi(w(t')), w(t')) \in V$ and this contradicts the previous inequality. As a consequence (2.10b) must be true.

In order to prove the sufficiency, observe that, if (2.10a) is satisfied, the graph of the mapping $y = \pi(x)$ is by construction a center manifold for (2.5). Moreover, by (2.10b), the error satisfies

$$e(t) = h(x(t)) - h(\pi(w(t))).$$

Observe that, by assumption, the point $(x, w) = (0, 0)$ is a stable equilibrium of (2.5). Then, for sufficiently small $(x(0), w(0))$, the solution $(x(t), w(t))$ of (2.5) remains in any arbitrarily small neighborhood of $(0, 0)$ for all $t \geq 0$. Using a property of center manifolds illustrated in Section **B**.1, it is deduced that there exist real numbers $M > 0$ and $a > 0$ such that

$$\| x(t) - \pi(w(t)) \| \leq M e^{-at} \| x(0) - \pi(w(0)) \|$$

for all $t \geq 0$. By continuity of $h(x)$, $\lim_{t \to \infty} e(t) = 0$, i. e. the condition (ii) is satisfied. •

Using this result, it is possible to prove a necessary and sufficient condition for the solution of the State Feedback Regulator Problem.

Theorem 2.5. Under the hypotheses (H1) and (H3), the State Feedback Regulator Problem is solvable if and only if there exist C^k $(k \geq 2)$ mappings $x = \pi(w)$, with $\pi(0) = 0$, and $u = c(w)$, with $c(0) = 0$, both defined in a neighborhood $W^\circ \subset W$ of 0, satisfying the conditions

$$\frac{\partial \pi}{\partial w} s(w) = f(\pi(w)) + g(\pi(w))c(w) + p(\pi(w))w \qquad (2.11a)$$

$$h(\pi(w)) + q(w) = 0. \qquad (2.11b)$$

Proof. The necessity has already been proven in Lemma 2.4. About the sufficiency, observe that, by hypothesis (H1), there exists a matrix K such that $(A + BK)$ has eigenvalues with negative real part. Suppose the conditions (2.11) of the Theorem are satisfied for some $\pi(w)$ and $c(w)$, and define a state feedback in the following way

$$\alpha(x, w) = c(w) + K(x - \pi(w)).$$

It is immediate to check that this is a solution of the State Feedback Regulator Problem. In fact, this choice clearly satisfies the requirement (i), because

$\alpha(x, 0) = Kx$. Moreover, by construction

$$\alpha(\pi(w), w) = c(w)$$

and, therefore, the condition (2.11a) becomes identical to the condition to (2.10a). On the other hand, condition (2.11b) is already exactly equal to the condition (2.10b). Thus, again using Lemma 2.4, we conclude that also the requirement (ii) is satisfied. •

Remark 2.6. The first one of the two conditions (2.11) expresses the fact that there is a submanifold in the state space of the plant, namely the graph of the mapping $x = \pi(w)$, which is rendered locally *invariant* by means of a suitable feedback law, namely $u = c(w)$. The second condition expresses the fact that the error map, i.e. the output of the composite system (2.3), is zero at each point of this manifold. Altogether, conditions (i) and (ii) express the property that the graph of the mapping $x = \pi(w)$ is an *output zeroing submanifold* of the system (2.3). •

Remark 2.7. If the system (2.3) is a linear system, the conditions (2.11) reduce to linear matrix equations. In this case the system in question can be written in the form

$$\dot{x} = Ax + Bu + Pw \qquad (2.12a)$$
$$\dot{w} = Sw \qquad (2.12b)$$
$$e = Cx + Qw \qquad (2.12c)$$

and, if the mappings $x = \pi(w)$, $u = c(w)$, are expanded as

$$\pi(w) = \Pi w + \tilde{\pi}(w)$$
$$c(w) = \Gamma w + \tilde{c}(w)$$

with

$$\Pi = \left[\frac{\partial \pi}{\partial w} \right]_{w=0} \qquad \Gamma = \left[\frac{\partial c}{\partial w} \right]_{w=0}.$$

the equations (2.11) have a solution if and only if the linear matrix equations

$$\Pi S = A\Pi + B\Gamma + P \qquad (2.13a)$$
$$C\Pi + Q = 0 \qquad (2.13b)$$

are solved by Π and Γ. Note that, if this is the case, the mappings $\pi(w)$ and $c(w)$ which solve (2.11) are linear mappings. •

The proof of the sufficiency, in Theorem 2.5, shows in particular that, once a solution $\pi(w)$, $c(w)$ of the equations (2.11) is known, a State Feedback which solves the Regulator Problem is provided by

$$\alpha(x, w) = c(w) + K(x - \pi(w))$$

where K is any matrix which places the eigenvalues of $(A + BK)$ in the open left-half complex plane. A block-diagram interpretation of the feedback law thus found is described in Fig. 7.2 where, for simplicity, it is assumed that $\pi(w) = 0$. From the theory developed so far, it is concluded that, if the gain matrix K asymptotically stabilizes the closed loop, from any (sufficiently small) initial condition, the output response $y(t)$ converges to a *steady state response* which exactly coincides with $q(w(t))$.

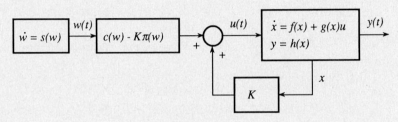

Fig. 7.2

Remark 2.8. It might be instructive to compare the results obtained here with those illustrated in Section 4.5. In that case, convergence to zero of the error was implied by the fact that $e(t)$ was a solution of a certain linear differential equation. Here, in the proof of Theorem 2.5 (and Lemma 2.4) the error $e(t)$ has been shown to converge to zero as a consequence of a general property of center manifolds. The approach followed here, which is much less demanding, shows that there is no need to impose that $e(t)$ obeys a linear differential equation. In particular, $e(t)$ may be 0 at some t and nonzero for larger values of t. •

The solution of the Error Feedback Regulator Problem can be derived along the same lines.

Lemma 2.9. Suppose (H3) holds and assume that, for some $\theta(z)$ and $\eta(z, e)$ the condition (iii) is fulfilled. Then, the condition (iv) is also fulfilled if and only if there exist C^k $(k \geq 2)$ mappings $x = \pi(w)$, with $\pi(0) = 0$, and $z = \sigma(w)$, with $\sigma(0) = 0$, both defined in a neighborhood $W^\circ \subset W$ of 0, satisfying the conditions

$$\frac{\partial \pi}{\partial w} s(w) = f(\pi(w)) + g(\pi(w))\theta(\sigma(w)) + p(\pi(w))w \qquad (2.14a)$$

$$\frac{\partial \sigma}{\partial w} s(w) = \eta(\sigma(w), 0) \qquad (2.14b)$$

$$h(\pi(w)) + q(w) = 0. \qquad (2.14c)$$

Proof. By assumption, all the eigenvalues of the matrix

$$\begin{bmatrix} A & BH \\ GC & F \end{bmatrix}$$

have negative real part and those of the matrix S are on the imaginary axis. Thus, the closed loop system (2.7) has a center manifold at $(0,0,0)$, the graph of a mapping

$$x = \pi(w)$$
$$z = \sigma(w).$$

with $\pi(w)$ and $\sigma(w)$ satisfying (2.14a) and

$$\frac{\partial \sigma}{\partial w} s(w) = \eta(\sigma(w), h(\pi(w)) + q(w)) \tag{2.14d}$$

As in the proof of Lemma 2.4, the hypothesis (H3) and the fulfillment of (iv) imply that the mapping $x = \pi(w)$ must satisfy (2.14c) and this, together with (2.14d), implies (2.14b). The sufficiency can be proved exactly as in Lemma 2.4. •

Theorem 2.10. Under the hypotheses (H1), (H2) and (H3), the Error Feedback Regulator Problem is solvable if and only if there exists C^k ($k \geq 2$) mappings $x = \pi(w)$, with $\pi(0) = 0$, and $u = c(w)$, with $c(0) = 0$, both defined in a neighborhood $W^\circ \subset W$ of 0, satisfying the conditions (2.11).

Proof. The necessity was already expressed by Lemma 2.9. The proof of sufficiency is constructive, as the proof of Theorem 2.5. By hypothesis (H1) and (H2), it is possible to find matrices H, G_1, G_2 such that

$$A + BH \qquad \text{and} \qquad \begin{bmatrix} A - G_1 C & P - G_1 Q \\ -G_2 C & S - G_2 Q \end{bmatrix}$$

have all eigenvalues with negative real part.

Suppose conditions (2.11) are satisfied by some $\pi(w)$ and $c(w)$, and define an error feedback in the following way

$$u = \theta(z)$$
$$\dot{z} = \eta(z, e)$$

with z, $\theta(z)$ and $\eta(z,e)$ given by

$$z = \mathrm{col}(z_1, z_2)$$
$$\theta(z) = c(z_2) + H(z_1 - \pi(z_2))$$
$$\eta(z_1, z_2, e) = \mathrm{col}(\eta_1(z_1, z_2, e), \eta_2(z_1, z_2, e))$$
$$\eta_1(z_1, z_2, e) = f(z_1) + p(z_1)z_2 + g(z_1)(c(z_2)+$$
$$H z_1 - H\pi(z_2)) - G_1(h(z_1) + q(z_2) - e)$$
$$\eta_2(z_1, z_2, e) = s(z_2) - G_2(h(z_1) + q(z_2) - e).$$

An immediate calculation shows that the jacobian matrix of (2.9) has the form

$$\begin{bmatrix} A & BH & BK \\ G_1C & A + BH - G_1C & P + BK - G_1Q \\ G_2C & -G_2C & S - G_2Q \end{bmatrix}$$

with

$$K = \left[\frac{\partial c}{\partial w}\right]_{w=0} - H\left[\frac{\partial \pi}{\partial w}\right]_{w=0}$$

The matrix in question is similar to

$$\begin{bmatrix} A + BH & \star \\ 0 & \begin{bmatrix} A - G_1C & P - G_1Q \\ -G_2C & S - G_2Q \end{bmatrix} \end{bmatrix}$$

and therefore all its eigenvalues have negative real part. This shows that the condition (iii) is satisfied. The fulfillment of (iv) is a consequence of Lemma 2.9, because, as an easy calculation shows, the conditions (2.14) are satisfied by $\pi(w)$ and $\sigma(w) = \mathrm{col}(\pi(w), w)$. •

It is interesting to observe that the solvability of both the State Feeedback and Error Feedback Regulator Problems depends on the same set of conditions, namely the conditions (2.11). This is fact is stressed, for completeness, in the following statement.

Corollary 2.11. Under the hypotheses (H1), (H2) and (H3), the Error Feedback Regulator Problem is solvable if and only the State Feedback Regulator Problem is solvable.

We reserve the last part of this Section to the illustration of how the existence conditions (2.11) can be tested in the particular case in which the system (2.1) is single-input single-output, has relative degree r at $x = 0$, and the vector $p(x)$ is zero, i.e. the only requirement is the tracking of a prescribed reference output.

Suppose the system (2.1) is put in normal form and observe that, accordingly, the composite system (2.3) becomes

$$\dot{z}_1 = z_2$$
$$\cdots$$
$$\dot{z}_{r-1} = z_r$$
$$\dot{z}_r = b(\xi, \eta) + a(\xi, \eta)u$$
$$\dot{\eta} = q(\xi, \eta)$$
$$\dot{w} = s(w)$$
$$e = z_1 + \hat{q}(w)$$

(the output map of the exosystem here is denoted by $\hat{q}(w)$ in order to avoid confusion with the notation $q(\xi, \eta)$ used in the normal form).

In order to check whether or not the equations (2.11) can be solved, it is convenient to set

$$\pi(w) = \text{col}(k(w), \lambda(w))$$

with

$$k(w) = \text{col}(k_1(w), \ldots, k_r(w)).$$

In this case, the equations in question reduce to

$$\frac{\partial k_1(x)}{\partial w} s(w) = k_2(w)$$
$$\cdots$$
$$\frac{\partial k_{r-1}(x)}{\partial w} s(w) = k_r(w)$$
$$\frac{\partial k_r(x)}{\partial w} s(w) = b(k(w), \lambda(w)) + a(k(w), \lambda(w))c(w)$$
$$\frac{\partial \lambda(x)}{\partial w} s(w) = q(k(w), \lambda(w))$$
$$k_1(w) + \hat{q}(w) = 0.$$

The last one of these, together with the first $r - 1$, yields immediately

$$k_i(w) = -L_s^{i-1} \hat{q}(w) \tag{2.15}$$

for all $1 \leq i \leq r$. The r-th one can be solved by

$$c(w) = \frac{L_s k_r(w) - b(k(w), \lambda(w))}{a(k(w), \lambda(w))} \tag{2.16}$$

and, therefore, we can conclude that the solvability of equations (2.11) is in this case equivalent to the solvability of

$$\frac{\partial \lambda}{\partial w} s(w) = q(k(w), \lambda(w)) \tag{2.17}$$

for some C^k ($k \geq 2$) mapping $\lambda(w)$.

We formalize this in the following statement.

Corollary 2.12. Suppose the hypotheses (H1) and (H3) hold. Suppose also that the system (2.1) is single-input single-output, with relative degree r at $x = 0$, and $p(w) = 0$. Define $k_i(w)$, $1 \leq i \leq r$, as in (2.15). Then, the State Feedback Regulator Problem is solvable if and only if the equation (2.17) can be solved by some C^k ($k \geq 2$) mapping $\lambda(w)$. Under the additional hypothesis (H2), the Error Feedback Regulator Problem is solvable if and only if the State Feedback Regulator Problem is solvable.

Recall that the linear approximation at $w = 0$ of the exosystem has by assumption *all* eigenvalues on the imaginary axis. Thus, if the linear approximation of

$$\dot{\eta} = q(0, \eta)$$

at $\eta = 0$ has *no* eigenvalue on the imaginary axis, the equation (2.17) is exactly the equation which must be satisfied by any center manifold (see Section **B**.1) for the system

$$\dot{\eta} = q(k(w), \eta)$$
$$\dot{w} = s(w).$$

Thus, we have

Corollary 2.13. Suppose hypotheses (H1) and (H3) hold. Suppose also that the system (2.1) is single-input single-output, with relative degree r at $x = 0$, and $p(w) = 0$. If the linear approximation at $x = 0$ of the zero dynamics of (2.1) has no eigenvalue on the imaginary axis, the State Feedback Regulator Problem is solvable. Under the additional hypothesis (H2), the Error Feedback Regulator Problem is also solvable.

As illustrated in Section **B**.1, the mapping $\eta = \lambda(w)$ can be approximated up to an arbitrary degree of accuracy by a polynomial in w. Once $\lambda(w)$ has been determined, the mapping $c(w)$ can be constructed without extra effort using (2.16). From the mapping thus found, one immediately obtains a solution of either the State Feedback or Error Feedback Regulator Problem, as shown in the proofs of Theorems 2.5 and 2.10.

We conclude the Section with a simple example of application.

Example 2.1. Consider the system already in normal form

$$\dot{z}_1 = z_2$$
$$\dot{z}_2 = u$$
$$\dot{\eta} = \eta + z_1 + z_2^2$$
$$y = z_1$$

and suppose it is desired to track asymptotically any reference output of the form

$$y_R(t) = M \sin(at + \phi)$$

where a is a fixed (positive) number, and M, ϕ arbitrary parameters.

Note that the zero dynamics of this system are unstable and, therefore, the approach described in Section 4.5 cannot be pursued. Note also that the system is not exactly linearizable via feedback, because the distribution $\text{span}\{g, ad_f g\}$ is not involutive, as a simple calculation shows. Thus, it is not possible to solve the problem by reduction of the plant to a linear system.

In this case, any desired reference output can be imagined as the output of an exosystem defined by

$$\begin{bmatrix} \dot{w}_1 \\ \dot{w}_2 \end{bmatrix} = \begin{bmatrix} aw_2 \\ -aw_1 \end{bmatrix} = s(w)$$
$$\hat{q} = -w_1$$

and therefore we could try to solve the problem via the theory developed in this Section, e.g. posing a State Feedback Regulator Problem.

The plant satisfies the assumption (H1), the exosystem satisfies the assumption (H3) and, moreover, the (single) eigenvalue of the linear approximation of the zero dynamics of the plant is not on the imaginary axis. Therefore (see Corollary 2.13), the problem in question is solvable.

Following the procedure illustrated above, one has to set

$$k_1(w) = -\hat{q}(w) = w_1$$
$$k_2(w) = L_s k_1(w) = aw_2$$

and then search for a solution $\lambda(w_1, w_2)$ of the partial differential equation (2.17), i.e.

$$\frac{\partial \lambda}{\partial w_1} aw_2 - \frac{\partial \lambda}{\partial w_2} aw_1 = \lambda(w_1, w_2) + w_1 + (aw_2)^2$$

A tedious, but elementary, calculation shows that this equation can be solved by a complete polynomial of second degree, i.e.

$$\lambda(w_1, w_2) = c_1 w_1 + c_2 w_2 + d_1 w_1^2 + d_2 w_2^2 + d_{12} w_1 w_2$$

Once $\lambda(w_1, w_2)$ has been calculated, from the previous theory it follows that the mapping

$$\pi(w) = \begin{bmatrix} k_1(w) \\ k_2(w) \\ \lambda(w_1, w_2) \end{bmatrix} = \begin{bmatrix} w_1 \\ aw_2 \\ \lambda(w_1, w_2) \end{bmatrix}$$

and the function

$$c(w) = L_s k_2(w) = -a^2 w_1$$

are solutions of the equations (2.11). In particular, a solution of the regulator problem is provided by

$$\alpha(x, w) = c(w) + K(x - \pi(w))$$

in which K is any matrix which places the eigenvalues of

$$\begin{bmatrix} 0 & 1 & 0 \\ 0 & 0 & 0 \\ 1 & 0 & 1 \end{bmatrix} + \begin{bmatrix} 0 \\ 1 \\ 0 \end{bmatrix} K$$

in the left-half complex plane.

As expected, the difference

$$\begin{bmatrix} x_1 \\ x_2 \\ x_3 \end{bmatrix} = \begin{bmatrix} z_1 \\ z_2 \\ \eta \end{bmatrix} - \pi(w) = \begin{bmatrix} z_1 - w_1 \\ z_2 - aw_2 \\ \eta - \lambda(w_1, w_2) \end{bmatrix}$$

is asymptotically decaying to zero, and so is the error $e(t)$, which in this case is exactly equal to x_1. In fact, the variables x_1, x_2, x_3 satisfy

$$\begin{bmatrix} \dot{x}_1 \\ \dot{x}_2 \\ \dot{x}_3 \end{bmatrix} = \begin{bmatrix} 0 & 1 & 0 \\ k_1 & k_2 & k_3 \\ 1 & 0 & 1 \end{bmatrix} \begin{bmatrix} x_1 \\ x_2 \\ x_3 \end{bmatrix} + \begin{bmatrix} 0 \\ 0 \\ x_2^2 + 2ax_2 w_2(t) \end{bmatrix} . \bullet$$

7.3 Disturbance Decoupling

A major outcome of theory of controlled invariant distributions, developed in Section 6.3, is the synthesis of feedback control laws which render the output of a system independent of certain disturbances (for each value of the time, not only asymptotically, as achieved in the previous Section).

Given, as in Sections 4.6 and 6.3, a system of the form

$$\dot{x} = f(x) + \sum_{i=1}^{m} g_i(x)u_i + p(x)w \tag{3.1a}$$

$$y = h(x) \tag{3.1b}$$

the matter is to solve following problem.

Disturbance Decoupling Problem. Consider a system of the form (3.1) and a point x°. Find, if possible, a regular feedback of the form $u = \alpha(x) + \beta(x)v$, defined in a neighborhood U of x°, which renders the output y independent of the disturbance w.

The discussions at the beginning of Section 6.3, together with the properties established in Lemma 6.3.4, already provide the desired answer which, for convenience, is summarized in the following statement.

Proposition 3.1. Suppose $\Delta^\star$ is finitely computable and G, $\Delta^\star$, $\Delta^\star + G$ are nonsingular in a neighborhood of x°. Then, the disturbance decoupling problem is solvable if and only if

$$p \in \Delta^\star \tag{3.2}$$

in a neighborhood of x°.

Note that, under the slightly stronger assumption that the point x° is a regular point of the controlled invariant distribution algorithm one can easily construct, by means of the procedure described in the Proposition 6.3.7, a state feedback which solves this problem. Note also that the solvability of the problem does not require at all that the system has some relative degree at the point x°. Of course, if the system *has* a relative degree at x°, then the distribution $\Delta^\star$ is rendered invariant by the standard noninteracting feedback (see Corollary 6.3.14), and if the condition (3.2) is satisfied this feedback provides also a solution to the disturbance decoupling problem. We recover in this way the preliminary results already established in Chapters 4 and 5.

Example 3.1. Consider again the system described in the Example 6.3.2 and note that the system in question does not have a relative degree at x°. The disturbance decoupling problem will be solvable if and only if the vector field $p(x)$ has the form

$$p(x) = \mathrm{col}(0, 0, 0, p_4(x), p_5(x)).$$

Suppose this is the case. In order to solve the problem, one has to find a feedback which renders $\Delta^\star$ invariant. To this end, note that, performing the controlled invariant distribution algorithm, we already obtained

$$\Delta^{\star\perp}(x) = \Omega_1(x) = \mathrm{span}\{d\lambda_1, d\lambda_2, d\lambda_3\}$$

with

$$\lambda_1 = x_1, \quad \lambda_2 = x_3, \quad \lambda_3 = x_2$$

and

$$\text{span}\{d\lambda_1, d\lambda_2\} \cap G^\perp = 0.$$

Thus, according to the results illustrated in the last part of Proposition 6.3.7, a feedback which renders $\Delta^\star$ invariant is a solution of

$$\begin{bmatrix} d\lambda_1 \\ d\lambda_2 \end{bmatrix} (f(x) + g(x)\alpha(x)) = \begin{bmatrix} 0 \\ 0 \end{bmatrix}$$

$$\begin{bmatrix} d\lambda_1 \\ d\lambda_2 \end{bmatrix} g(x)\beta(x) = \begin{bmatrix} 1 & 0 \\ 0 & 1 \end{bmatrix}$$

i.e.

$$\alpha(x) = \begin{bmatrix} -x_2 \\ -x_1 x_4 \end{bmatrix}$$

and $\beta(x)$ the identity matrix. The corresponding closed loop system will then be

$$\dot{x}_1 = v_1$$
$$\dot{x}_2 = -x_2 x_3 + x_3 v_1$$
$$\dot{x}_3 = v_2$$
$$\dot{x}_4 = x_3^2 - x_2 x_5 - x_1^2 x_4 + x_5 v_1 + x_1 v_2 + p_4(x)w$$
$$\dot{x}_5 = x_1 - x_2 - x_1 x_2 x_3 x_4 + v_1 + x_2 x_3 v_2 + p_5(x)w$$

and its decomposed structure shows that the output depends on a set of state variables (x_1, x_2, x_3) which are independent of the others (x_4, x_5) and not affected by the disturbance. $\bullet$

7.4 Noninteracting Control with Stability via Static Feedback

In Section 5.3, we have addressed the problem of finding a feedback law which renders the input-output behavior of a nonlinear system with m inputs and m outputs equivalent to that of an aggregate of m independent single-input single-output subsystems. In particular, we have seen that this problem is solvable (locally around a point x° in the state space) by means of static feedback, i.e. by means of a feedback of the form

$$u = \alpha(x) + \beta(x)v \tag{4.1}$$

if and only if the matrix (5.1.2) is invertible at x°, i.e. if and only if the system has some vector relative degree $\{r_1, \ldots, r_m\}$ at this point. A solution of this problem is provided by the standard noninteractive feedback

$$u = A^{-1}(x)(-b(x) + v) \tag{4.2}$$

in which $A(x)$ and $b(x)$ are given by (5.1.2) and (5.1.8).

At the end of the same Section, we have also pointed out that, if the zero dynamics of the nonlinear system are asymptotically stable, it is easy to find a feedback which, simultaneously, renders the system noninteractive from the input-output point of view, and also *asymptotically stable*. In fact, it suffices to add to the standard noninteractive feedback law a control of the form (see (5.3.3))

$$v = \mathrm{col}(v_1, \ldots, v_m)$$

with

$$v_i = -c_0^i h_i(x) - c_1^i L_f h_i(x) - \ldots - c_{r_i-1}^i L_f^{r_i-1} h_i(x) + \bar{v}_i.$$

However, as stressed in the Remark 5.3.4, the hypothesis that the zero dynamics are asymptotically stable may not be necessary in order to obtain noninteracting control with stability and, in fact, there may be cases of systems having unstable zero dynamics, in which the simultaneous achievement of these two goals is still possible.

We wish now to discuss this problem in more detail. For convenience, we start with a formal definition. As in Section 7.2, we suppose that the point x° at which the problem is to be solved is an equilibrium point of the vector field $f(x)$, and that the feedback (4.1) preserves this equilibrium, i.e. is such that $\alpha(x^\circ) = 0$.

Problem of Noninteracting Control with Stability (via Static Feedback). Consider a nonlinear system of the form (1.1), an initial point x° and suppose $f(x^\circ) = 0$. Find a regular feedback of the form (4.1), defined in a neighborhood U of x°, with $\alpha(x^\circ) = 0$, such that
  (i) the equilibrium $x = x^\circ$ of

$$\dot{x} = f(x) + g(x)\alpha(x)$$

is asymptotically stable in the first approximation,
 (ii) in the closed loop system

$$\dot{x} = f(x) + g(x)\alpha(x) + g(x)\beta(x)v$$
$$y = h(x)$$

each output channel y_i, $1 \leq i \leq m$, is affected only by the corresponding input channel v_i and not by v_j, if $j \neq i$.

Remark 4.1. Note that, in view of the results illustrated in Section B.2

(and already utilized in a similar way in Remark 5.3.5), the fulfillment of (i) guarantees that for each ε there exist δ and K such that

$$\| x^\circ \| < \delta, \quad |v_i(t)| \leq K \qquad \text{for all } t \geq 0, 1 \leq i \leq m$$

implies $\| x(t) \| < \varepsilon$, for all $t \geq 0$, in the corresponding noninteractive closed loop. •

We begin by identifying a necessary condition for the solution of such a problem. The idea is to establish some common features of all feedback laws which solve the noninteracting control problem, i.e. satisfy requirement (ii), and then to check whether or not the fulfillment of requirement (i) is compatible with the features thus found.

First of all, it will be shown that there exists a distribution, denoted $P^\star$, which is rendered invariant by any feedback which solves the noninteracting control problem.

Lemma 4.2. Suppose the system (1.1) has relative degree $\{r_1, \ldots, r_m\}$ at x°. For each $1 \leq i \leq m$, define a distribution $\Delta_i^\star$ as follows

$$\Delta_i^\star(x) = \bigcap_{k=1}^{r_i} \ker(dL_f^{k-1} h_i(x)). \tag{4.3}$$

Let $\mathcal{S}(\Delta_i^\star)$ be the distribution associated to $\Delta_i^\star$ by means of the controllability distribution algorithm (see (6.4.2)). Suppose $\mathcal{S}(\Delta_i^\star)$ is finitely computable and x° is a regular point of $\mathcal{S}(\Delta_i^\star)$. Then, in a neighborhood of x°, $\mathcal{S}(\Delta_i^\star)$ is the largest local controllability distribution contained in $\ker(dh_i)$.

Let $u = \alpha(x) + \beta(x)v$ be *any* regular feedback which solves the noninteracting control problem at x°. Set

$$\tilde{f}(x) = f(x) + g(x)\alpha(x)$$

$$\tilde{g}_i(x) = (g(x)\beta(x))_i \qquad 1 \leq i \leq m.$$

Then, in a neighborhood of x°,

$$\mathcal{S}(\Delta_i^\star) = \langle \tilde{f}, \tilde{g}_1, \ldots, \tilde{g}_m | \text{span}\{\tilde{g}_j : j \neq i\} \rangle. \tag{4.4}$$

In particular, the distribution

$$P^\star = \bigcap_{i=1}^{m} \mathcal{S}(\Delta_i^\star)$$

is contained in $\Delta^\star$, and is invariant under $\tilde{f}, \tilde{g}_1, \ldots, \tilde{g}_m$, i.e. satisfies

$$[\tilde{f}(x), P^\star] \subset P^\star$$
$$[\tilde{g}_i(x), P^\star] \subset P^\star \qquad 1 \le i \le m.$$

Proof. As observed in the Remark **6.3.15**, the distribution (4.3) is the largest locally controlled invariant distribution contained in $\ker(dh_i)$. Note that, around x°, the distribution G has constant dimension m (because the matrix (5.1.2) is nonsingular), the distribution (4.3) has constant dimension $n - r_i$ (see Lemma **5.1.2**), and the distribution $\Delta_i^\star \cap G$ has constant dimension $m - 1$ (in fact, the latter is spanned by vectors of the form $g(x)\gamma$, with γ such that $\langle dL_f^{k-1}h_i, g(x)\rangle\gamma = 0$ for all $1 \le k \le r_i$, and the set of all γ's which satisfy this condition is an $(m - 1)$-dimensional subspace of $\mathbf{R}^m$). By Lemma **6.4.5**, $\mathcal{S}(\Delta_i^\star)$ is the largest local controllability distribution contained in $\Delta_i^\star$, and then in $\ker(dh_i)$.

To prove (4.4), observe first of all that, by definition of relative degree (see proof of Lemma **5.2.1**)

$$L_{\tilde{f}}^k h_i = L_f^k h_i \qquad \text{for all } 0 \le k \le r_i - 1.$$

If (α, β) is any regular feedback which solves the noninteracting control problem at x°, then (see Theorem **3.3.2**, condition (iii))

$$\langle \tilde{f}, \tilde{g}_1, \ldots, \tilde{g}_m | \text{span}\{\tilde{g}_j : j \ne i\}\rangle \subset (\text{span}\{dL_f^k h_i : 0 \le k \le r_i - 1\})^\perp = \Delta_i^\star.$$

Consider now the sequence of distributions Δ_k generated by means of the following algorithm

$$\Delta_0 = \text{span}\{\tilde{g}_j : j \ne i\}$$
$$\Delta_k = \Delta_{k-1} + \sum_{s=0}^{m}[\tilde{g}_s, \Delta_{k-1}]$$

(here $\tilde{g}_0 = \tilde{f}$) and note that (recall Lemma **1.8.2**)

$$\Delta_k \subset \langle \tilde{f}, \tilde{g}_1, \ldots, \tilde{g}_m | \text{span}\{\tilde{g}_j : j \ne i\}\rangle \subset \Delta_i^\star.$$

We show now that the distributions S_k generated by means of the controllability distribution algorithm starting from $\Delta_i^\star$ satisfy

$$S_k = \Delta_k.$$

This is certainly true for $k = 0$ because

$$S_0 = \Delta_i^\star \cap G = \text{span}\{\tilde{g}_j : j \ne i\}.$$

Suppose is true for some k and note that $[\tilde{g}_s, \Delta_k] \subset \Delta_i^\star$, because $\langle \tilde{f}, \tilde{g}_1, \ldots, \tilde{g}_m | \text{span}\{\tilde{g}_j : j \neq i\}\rangle$ is invariant under $\tilde{g}_s$. Then

$$S_{k+1} = \Delta_i^\star \cap \left(\sum_{s=0}^{m}[\tilde{g}_s, \Delta_k] + \Delta_k + G\right) = \Delta_{k+1} + \Delta_i^\star \cap G = \Delta_{k+1}$$

We obtain in this way

$$\mathcal{S}(\Delta_i^\star) \subset \langle \tilde{f}, \tilde{g}_1, \ldots, \tilde{g}_m | \text{span}\{\tilde{g}_j : j \neq i\}\rangle \subset \Delta_i^\star.$$

However, since $\langle \tilde{f}, \tilde{g}_1, \ldots, \tilde{g}_m | \text{span}\{\tilde{g}_j : j \neq i\}\rangle$ is by construction a controllability distribution contained in $\Delta_i^\star$ and $\mathcal{S}(\Delta_i^\star)$ is the largest local controllability distribution contained in $\Delta_i^\star$, necessarily

$$\mathcal{S}(\Delta_i^\star) = \langle \tilde{f}, \tilde{g}_1, \ldots, \tilde{g}_m | \text{span}\{\tilde{g}_j : j \neq i\}\rangle$$

i.e. (4.4) holds.

The rest of the statement is a straightforward consequence. •

Combining this result with a uniqueness property of all feedback functions which render invariant a given distribution (Lemma 6.2.4), it is easy to arrive at the important conclusion described in the following statement. For notational convenience, we shall henceforth denote the distributions $\mathcal{S}(\Delta_i^\star)$ by a simpler symbol, setting

$$P_i^\star = \mathcal{S}(\Delta_i^\star).$$

Lemma 4.3. Suppose the system (1.1) has relative degree $\{r_1, \ldots, r_m\}$ at x°, the distributions $P_i^\star$, $1 \leq i \leq m$, are finitely computable and x° is a regular point of $P_1^\star, \ldots, P_m^\star, P^\star$. In a neighborhood U° of the point x°, $P^\star$ is involutive, and U° can be partitioned into maximal integral submanifolds of $P^\star$. Let x° be an equilibrium of the vector field $f(x)$ and let $S^\star$ denote the integral submanifold of $P^\star$ which contains the point x°. Let $u = \alpha(x) + \beta(x)v$ be any regular feedback which solves the noninteracting control problem at x°, with $\alpha(x^\circ) = 0$. The submanifold $S^\star$ is locally invariant under the vector field $f(x) + g(x)\alpha(x)$, and the restriction of $f(x) + g(x)\alpha(x)$ to $S^\star$ does not depend upon the particular choice of α.

Proof. Since each $P_i^\star$ is involutive, so is $P^\star$ By Lemma 4.2, $P^\star$ is invariant under the vector field $\tilde{f}(x)$, and therefore, according to the interpretation of invariance given in Section 1.6, the flow of $\tilde{f}(x)$ locally carries the integral submanifold $S^\star$ into another integral submanifold of $P^\star$. But the point x° is fixed under the flow of $\tilde{f}(x)$ and therefore it is concluded that the flow of $\tilde{f}(x)$ carries $S^\star$ into itself; in other words, $S^\star$ is locally invariant under $\tilde{f}(x)$. The last part of the statement follows from Lemma 6.2.4 because the nonsingularity of the matrix (5.1.2) implies

$$\dim(G) = m$$
$$P^\star \cap G \subset \Delta^\star \cap G = \{0\}. \quad \bullet$$

The property thus found immediately implies a necessary condition for the existence of solutions to the problem of noninteracting control with stability. In fact, this property essentially establishes that for each system in which the Noninteracting Control Problem is solvable it is possible to identify a submanifold $S^\star$, the integral submanifold of $P^\star$ through x°, and a well-defined vector field $s^\star$ of $S^\star$, with the property that, in any closed loop system which has been rendered noninteractive via static state feedback, the vector field $f + g\alpha$ satisfies

$$f(x) + g(x)\alpha(x) = s^\star(x) \qquad \text{for all } x \in S^\star.$$

As a consequence, independently of what $\alpha(x)$ has been chosen in order to achieve noninteracting control, the trajectories of $\dot{x} = f(x) + g(x)\alpha(x)$ starting at a point $x(0) \in S^\star$ will coincide with the solutions of

$$\dot{x} = s^\star(x) \qquad x(0) \in S^\star \tag{4.5}$$

and the condition (i) can be achieved only if the equilibrium $x = x^\circ$ of (4.5) is asymptotically stable in the first approximation.

In other words, we have proved the following result.

Theorem 4.4. Suppose the system (1.1) has relative degree $\{r_1, \ldots, r_m\}$ at x°, the distributions $P_i^\star$, $1 \leq i \leq m$, are finitely computable and x° is a regular point of $P_1^\star, \ldots, P_m^\star, P^\star$. Then the Problem of Noninteracting Control with Stability via Static Feedback is solvable at x° only if the equilibrium $x = x^\circ$ of (4.5) is asymptotically stable in the first approximation.

Remark 4.5. In the statement of Lemma 4.2, we have observed that the distribution $P^\star$ is contained into the distribution $\Delta^\star$, the largest locally controlled invariant distribution contained in $\ker(dh)$. If both these distributions are nonsingular and the inclusion $\Delta^\star \supset P^\star$ is proper, i.e. the dimension of $\Delta^\star$ exceeds that of $P^\star$, the integral submanifolds of $P^\star$ are proper submanifolds of the integral submanifolds of $\Delta^\star$. More precisely, each of the integral submanifolds of $\Delta^\star$ is *partitioned* into integral submanifolds of $P^\star$. Since, in the case of systems having a relative degree at the point x°, the integral submanifold of $\Delta^\star$ through x° locally coincides with the zero dynamics submanifold $Z^\star$ (see Corollary 6.3.14), we conclude that $S^\star$ is a proper submanifold of $Z^\star$. Moreover, it is also known that the standard noninteractive feedback (4.2) renders the submanifold $Z^\star$ invariant under the vector field $f + g\alpha$, and therefore the restriction of $f + g\alpha$ to $Z^\star$ coincides with the zero dynamics vector field (see again Corollary 6.3.14) of the system in question. Thus, the vector field $s^\star$ is nothing else than the vector field which describes *the restriction of the zero dynamics* of the system (1.1) *to its invariant manifold* $S^\star$. In other words, the dynamical system (4.5) is a subsystem of the system

$$\dot{x} = f^\star(x) \qquad x(0) \in Z^\star \tag{4.6}$$

which describes the zero dynamics of (1.1).

Of course, if the zero dynamics of (1.1) are asymptotically stable in the first approximation, so are those of any subsystem of (4.6) and the necessary condition established in Theorem 4.5 will be automatically satisfied. •

In order to test whether or not the condition expressed by the previous Theorem is satisfied, it is convenient to introduce suitable local coordinates. To this end, consider, in addition to the distributions $P_1^\star, \ldots, P_m^\star, P^\star$ defined before also the distribution

$$P = \langle f, g_1, \ldots, g_m | \mathrm{span}\{g_j(x) : 1 \le j \le m\}\rangle$$

introduced in Section 1.8. All these distributions, if nonsingular, are also involutive, by Lemma 1.8.7. Therefore they are completely integrable by Frobenius' Theorem, and it is possible to find suitable sets of real-valued functions whose differentials span $(P_1^\star)^\perp, \ldots, (P_m^\star)^\perp, (P^\star)^\perp, P^\perp$. The following Lemma illustrates that using these functions it is possible to construct a local coordinates transformation which induces special forms for the vectors of $P_1^\star, \ldots, P_m^\star, P^\star, P$.

Lemma 4.6. Suppose the distributions $P_i^\star, P_i^\star + (\bigcap_{j \ne i} P_j^\star)$, for all $1 \le i \le m$, $P^\star$ and P are nonsingular in a neighborhood of the point x°. Then there exist a neighborhood U° of x° and a coordinates transformation defined on U°,

$$z = \mathrm{col}(z^1, \ldots, z^m, z^{m+1}, z^{m+2}) = \Phi(x)$$
$$= \mathrm{col}(z^1(x), \ldots, z^m(x), z^{m+1}(x), z^{m+2}(x))$$

such that

$$P^\perp = \mathrm{span}\{dz^{m+2}\} \tag{4.7a}$$
$$(P_i^\star)^\perp = \mathrm{span}\{dz^i, dz^{m+2}\} \tag{4.7b}$$
$$(P^\star)^\perp = \mathrm{span}\{dz^1, \ldots, dz^m, dz^{m+2}\}. \tag{4.7c}$$

In particular, it is possible to choose for each $1 \le i \le m$, the coordinate functions

$z^i(x)$ in the form

$$z^i(x) = \mathrm{col}(\xi^i(x), \phi^i(x)) \tag{4.8}$$

with

$$\xi^i(x) = \mathrm{col}(h_i(x), L_f h_i(x), \dots, L_f^{r_i-1} h_i(x))$$

as in the local normal form (5.1.6).

Proof. Recall that the distribution P does not change if the vector fields f and g_i, $1 \le i \le m$, are modified by means of a regular feedback, i.e. that

$$P = \langle \tilde{f}, \tilde{g}_1, \dots, \tilde{g}_m | \mathrm{span}\{\tilde{g}_j : 1 \le j \le m\}\rangle.$$

Note that the distribution $P_i^\star + (\bigcap_{j \ne i} P_j^\star)$, contains $\mathrm{span}\{\tilde{g}_j : 1 \le j \le m\}$, is invariant under $\tilde{f}, \tilde{g}_1, \dots, \tilde{g}_m$, and is contained in the distribution P. Therefore, it must coincide with P, i.e.

$$P_i^\star + \bigcap_{j \ne i} P_j^\star = P$$

and also

$$\sum_{i=1}^{m} P_i^\star = P$$

i.e., by duality,

$$(P_i^\star)^\perp \cap \left(\sum_{j \ne i} (P_j^\star)^\perp\right) = P^\perp \tag{4.9}$$

$$\bigcap_{i=1}^{m} (P_i^\star)^\perp = P^\perp.$$

Let $z^{m+2}(x)$ be a collection of functions whose differentials span $P^\perp$. For each $1 \le i \le m$, since $P_i^\star$ and P are simultaneously integrable, it is possible (see Corollary 1.4.2) to find a collection of functions $z^i(x)$ such that dz^i and dz^{m+2} span $(P_i^\star)^\perp$, i.e. satisfy (4.7b) (and therefore also (4.7c)). The property (4.9) guarantees that the differentials of all the functions thus defined are linearly independent at x°, so that they can be considered as a partial set of local coordinates in a neighborhood of x°. For suppose they were linearly dependent at x°. Then there would exist row vectors $c_1, \dots, c_m, c_{m+2}$, with $c_i \ne 0$ for some

$1 \leq i \leq m$, such that

$$c_i dz^i + c_{m+2} dz^{m+2} = \sum_{\substack{j=1 \\ j \neq i}}^{m} c_j dz^j$$

The vector on the left-hand side belongs to $(P_i^\star)^\perp$, by construction, and that on the right-hand side belongs to

$$\sum_{j \neq i} (P_j^\star)^\perp.$$

Thus, by (4.9), the vector on the left-hand side is a vector in $P^\perp$, and c_i is necessarily 0, i.e. a contradiction.

If the number of functions thus constructed is not exactly equal to n, one can find an additional collection $z^{m+1}(x)$ of functions which completes the coordinates transformation near x^o.

The last part of the statement is an immediate consequence of the fact that $P_i^\star \subset \Delta_i^\star$. Since by definition $(\Delta_i^\star)^\perp$ is spanned by the differentials of the elements of $\xi^i(x)$, it is indeed possible (using again Corollary 1.4.2) to find $\phi^i(x)$ in order to have (4.8) satisfied. •

Consider now a system which satisfies the assumptions of Theorem 4.4, and suppose that a feedback which solves the noninteracting control problem has been implemented. Using the coordinates $z^1, \ldots, z^m, z^{m+1}, z^{m+2}$ introduced in the previous Lemma, it is possible to represent the equations describing the corresponding closed loop system in a particularly interesting form.

Proposition 4.7. Suppose the system (1.1) has relative degree $\{r_1, \ldots, r_m\}$ at x^o, the distributions $P_i^\star$, $1 \leq i \leq m$, are finitely computable and x^o is a regular point of $P_i^\star$, $P_i^\star + (\bigcap_{j \neq i} P_j^\star)$, for all $1 \leq i \leq m$, $P^\star$ and P. Let $u = \alpha(x) + \beta(x)v$ be any regular feedback which solves the noninteracting control problem around x^o. In the coordinates $z = \Phi(x)$ defined by Lemma 4.6, the closed loop system

$$\dot{x} = f(x) + g(x)\alpha(x) + g(x)\beta(x)v$$
$$y = h(x)$$

is represented by equations of the form

$$\dot{z}^1 = f_1(z^1, z^{m+2}) + g_{11}(z^1, z^{m+2})v_1$$

$$\cdots \tag{4.10}$$

$$\dot{z}^m = f_m(z^m, z^{m+2}) + g_{mm}(z^m, z^{m+2})v_m$$
$$\dot{z}^{m+1} = f_{m+1}(z) + g_{m+1,1}(z)v_1 + \ldots + g_{m+1,m}(z)v_m$$
$$\dot{z}^{m+2} = f_{m+2}(z^{m+2})$$

$$y_1 = h_1(z^1, z^{m+2})$$

$$\cdots \tag{4.10}$$

$$y_m = h_m(z^m, z^{m+2}).$$

In these coordinates, the submanifold $S^\star$ is the set

$$S^\star = \{x \in U^\circ : z^1(x) = 0, \ldots, z^m(x) = 0, z^{m+2}(x) = 0\}$$

and the system (4.5) is represented by the differential equation

$$\dot{z}^{m+1} = f_{m+1}(0, \ldots, 0, z^{m+1}, 0). \tag{4.11}$$

Proof. The proof is based on arguments essentially identical to those used in the Remark 1.6.8. Let $f(z) = \mathrm{col}(f_1(z), \ldots, f_{m+2}(z))$ denote the representation of the vector field $f(x) + g(x)\alpha(x)$ in the new coordinates, and note that

$$df_i(z) = L_f dz^i(z).$$

Since $(P_i^\star)^\perp$ is invariant under $f(z)$ by construction, for $1 \leq i \leq m$, then

$$L_f dz^i(z) \subset \mathrm{span}\{dz^i, dz^{m+2}\}$$

and, thus, $f_i(z)$ depends only on z^i and z^{m+2}. For the same reason, the invariance of $P^\perp$ under $f(z)$ proves that $f_{m+2}(z)$ depends only on z^{m+2}.

$(P_i^\star)^\perp$, $1 \leq i \leq m$, and $P^\perp$ are also invariant under all vector fields $(g(x)\beta(x))_j$, and therefore the representation $g_j(z) = \mathrm{col}(g_{1j}(z), \ldots, g_{m+2,j}(z))$ of the latter in the new coordinates has similar properties, i.e. $g_{ij}(z)$ depends only on z^i and z^{m+2}. Moreover, since $g_j(z)$ in contained in $P_i^\star$, for all $1 \leq i \leq m$ with $i \neq j$, and also in P, we have

$$\langle dz^i, g_j(z) \rangle = 0$$

for all $1 \leq i \leq m$ with $i \neq j$, and for $i = m + 2$. This proves that $g_j(z)$ has nonzero entries only on the j-th and $(m + 1)$-th block.

Finally, dh_i belongs to $(P_i^\star)^\perp$ by construction. Thus, $h_i(z)$ depends only on z^i and z^{m+2}. The last part of the statement is an immediate consequence of the choice of the new coordinates. •

Remark 4.8. It may be interesting to compare the form of the equations described in the previous Proposition with the local normal form introduced in Section 5.1. To this end, observe that the equations (5.1.6) and (5.1.7) are simply a local description of the system (1.1) in suitable coordinates, while the ones introduced above describe a closed loop system which has been rendered noninteractive by means of state feedback. Thus, in order to compare the two sets of equations, it is necessary to impose on (5.1.6) and (5.1.7) a feedback (any one can be used to this purpose) which solves the noninteracting control problem. Suppose the standard noninteracting control feedback is imposed on (5.1.6) and (5.1.7). Then, one obtains a system of equations of the form

$$\dot{\xi}^1 = b_{11} v_1$$
$$\cdots$$
$$\dot{\xi}^m = b_{mm} v_m$$
$$\dot{\eta} = q(\xi, \eta) + p_{m1}(\xi, \eta) v_1 + \ldots + p_{mm}(\xi, \eta) v_m$$
$$y_1 = c_1 \xi^1$$
$$\cdots$$
$$y_m = c_m \xi^m$$

in which, for all $1 \leq i \leq m$,

$$b_{ii} = \mathrm{col}(0, \ldots, 0, 1)$$
$$c_i = [\, 1 \quad 0 \quad \cdots \quad 0 \,].$$

On the other hand, if the functions $z^i(x)$, $1 \leq i \leq m$, are chosen in the way specified by (4.8), i.e. with the first r_i components exactly equal to those utilized to derive the normal form (5.1.6), it is immediate to realize that each of the first m sets of equations introduced in the Proposition 4.7 can be decomposed as

$$\dot{\xi}^i = b_{ii} v_i$$
$$\dot{\phi}^i = \bar{f}_i(\xi^i, \phi^i, z^{m+2}) + \bar{g}_{ii}(\xi^i, \phi^i, z^{m+2}) v_i$$

In other words, the functions $f_i(z^i, z^{m+2})$, $g_{ii}(z^i, z^{m+2})$, $h_i(z^i, z^{m+2})$ can be expressed in the form

$$f_i(z^i, z^{m+2}) = \begin{bmatrix} 0 \\ \bar{f}_i(\xi^i, \phi^i, z^{m+2}) \end{bmatrix}$$

$$g_{ii}(z^i, z^{m+2}) = \begin{bmatrix} b_{ii} \\ \bar{g}_{ii}(\xi^i, \phi^i, z^{m+2}) \end{bmatrix}$$

$$h_i(z^i, z^{m+2}) = c_i \xi^i.$$

We deduce from the comparison of the two forms thus obtained that the set of equations

$$\dot{\phi}^1 = \bar{f}_1(\xi^1, \phi^1, z^{m+2}) + \bar{g}_{11}(\xi^1, \phi^1, z^{m+2})v_1$$

$$\ldots$$

$$\dot{\phi}^m = \bar{f}_m(\xi^m, \phi^m, z^{m+2}) + \bar{g}_{mm}(\xi^m, \phi^m, z^{m+2})v_m$$
$$\dot{z}^{m+1} = f_{m+1}(z) + g_{m+1,1}(z)v_1 + \ldots + g_{m+1,m}(z)v_m$$
$$\dot{z}^{m+2} = f_{m+2}(z^{m+2})$$

is nothing else than a decomposition of the equation

$$\dot{\eta} = q(\xi, \eta) + p_{m1}(\xi, \eta)v_1 + \ldots + p_{mm}(\xi, \eta)v_m.$$

In particular, setting $v_i = 0$ and $\xi^i = 0$, for all $1 \leq i \leq m$, one finds a decomposed description of the zero dynamics of the system in the form

$$\dot{\phi}^1 = \bar{f}_1(0, \phi^1, z^{m+2})$$

$$\ldots$$

$$\dot{\phi}^m = \bar{f}_m(0, \phi^m, z^{m+2})$$
$$\dot{z}^{m+1} = f_{m+1}(0, \phi^1, \ldots, 0, \phi^m, z^{m+1}, z^{m+2})$$
$$\dot{z}^{m+2} = f_{m+2}(z^{m+2})$$

The set $S^\star$ corresponds to the subset of points having $\phi^i = 0$ and $z^{m+2} = 0$. This is clearly an invariant set of the zero dynamics manifold $Z^\star$, and the restriction of the zero dynamics vector field $f^\star$ to this set coincides with the vector field $s^\star$, as already shown—in a coordinate-free manner—in the Remark 4.5. •

We show now that the necessary condition indicated in Theorem 4.4 is essentially also sufficient for the solvability of the problem under consideration. This fact is based on the following property.

Lemma 4.9. Suppose the assumptions of Proposition 4.7 are satisfied. Suppose the linear approximation of (1.1) at $x = x^\circ$ is stabilizable. Then, for each $1 \leq i \leq m$, the linear approximation of the subsystem

$$\dot{z}^i = f_i(z^i, 0) + g_{ii}(z^i, 0)v_i \tag{4.12}$$

of (4.10) is stabilizable at $z^i(x^\circ)$.

Proof. Clearly, if the linear approximation of (1.1) at $x = x^\circ$ is stabilizable, so is that of the system (4.10), which has been obtained from (1.1) via regular feedback and coordinates tranformations, at the point $\Phi(x^\circ)$. Without loss of generality, we may suppose $\Phi(x^\circ) = 0$. The linear approximation of (4.10) at

$z = 0$ has a form

$$
\dot{z} =
\begin{bmatrix}
A_1 & \cdots & 0 & 0 & A_{1,m+2} \\
\cdots & \cdots & \cdots & \cdots & \cdots \\
0 & \cdots & A_m & 0 & A_{m,m+2} \\
\star & \cdots & \star & A_{m+1,m+1} & A_{m+1,m+2} \\
0 & \cdots & 0 & 0 & A_{m+2,m+2}
\end{bmatrix}
z +
\begin{bmatrix}
b_1 & \cdots & 0 \\
\cdots & \cdots & \cdots \\
0 & \cdots & b_m \\
\star & \cdots & \star \\
0 & \cdots & 0
\end{bmatrix}
v
$$

If the latter is stabilizable, then the matrix $A_{m+2,m+2}$ has all eigenvalues with negative real part and the pairs (A_i, b_i) are stabilizable. Since the latter define the linear approximation at $z_i = 0$ of (4.12), the result follows. •

From this Lemma, we deduce that, if the linear approximation of (1.1) at $x = x^\circ$ is stabilizable, it is possible to find matrices $K_1, \ldots, K_m$ such that the linear feedback

$$v_i = K_i z^i + \bar{v}_i \tag{4.13}$$

stabilizes in the first approximation the subsystem (4.12). This feedback preserves the noninteractive structure of (4.10) (because v_i depends only on z^i and $\bar{v}_i$ and y_i is affected only by z^i).

The linear approximation of

$$\dot{z}^{m+2} = f_{m+2}(z^{m+2})$$

at $z^{m+2}(x^\circ)$ is already asymptotically stable in the first approximation, as a consequence of the stabilizability of the linear approximation of (1.1) at x° (see proof of Lemma 4.9). Thus, because of the special structure of the equations (4.10), we can conclude that, if also the system (4.11) is asymptotically stable in the first approximation at $z^{m+1}(x^\circ)$, imposing the feedback (4.13) on (4.10) yields a closed loop system which satisfies both the requirements (i) and (ii) of the Problem of Noninteracting Control with Stability. In other words, the composition of the standard noninteractive control feedback, which induces the structure described by (4.10), with the additional feedback (4.13) solves the problem under consideration.

We formalize this result in the following statement.

Theorem 4.10. Suppose the system (1.1) has relative degree $\{r_1, \ldots, r_m\}$ at x°, the distributions $P_i^\star$, $1 \leq i \leq m$, are finitely computable and x° is a regular point of $P_i^\star$, $P_i^\star + (\bigcap_{j \neq i} P_j^\star)$, for all $1 \leq i \leq m$, $P^\star$ and P. Then, the Problem of Noninteracting Control with Stability via Static Feedback is solvable if and only if

(i) the linear approximation of (1.1) at $x = x^\circ$ is stabilizable,

(ii) the linear approximation of (4.5) at $x = x^\circ$ is asymptotically stable.

Proof. The necessity of (i) follows immediately by the requirement of achieving asymptotic stability in the first approximation for $f(x) + g(x)\alpha(x)$. Necessity of (ii) and sufficiency of both (i) and (ii) have already been proved. •

Remark 4.11. Note that the standard noninteractive feedback renders the distribution $\Delta^\star$ invariant under the vector fields of the corresponding closed loop system. However, the composition of this feedback with the law (4.13)— which, as we have shown, solves the Problem Noninteracting Control with Stability via Static Feedback—does not anymore leave $\Delta^\star$ invariant. •

We conclude the Section with a simple example of application of the ideas illustrated so far.

Example 4.1. Consider the system

$$\dot{x} = \begin{bmatrix} x_1 + x_1 x_4 \\ x_2 e^{x_3} \\ x_2 + x_3^2 \\ x_1 + x_2 - x_4 + x_1 x_4 \end{bmatrix} + \begin{bmatrix} x_3 \\ 1 \\ 0 \\ 1 + x_3 \end{bmatrix} u_1 + \begin{bmatrix} 1 \\ 0 \\ 0 \\ 1 \end{bmatrix} u_2$$

$$y_1 = x_1$$
$$y_2 = x_2.$$

A simple calculation shows that this system has relative degree $\{1, 1\}$ at $x^\circ = 0$. In fact,

$$L_g h(x) = A(x) = \begin{bmatrix} x_3 & 1 \\ 1 & 0 \end{bmatrix}.$$

The zero dynamics of the system are defined on the submanifold

$$Z^\star = \{x \in \mathbf{R}^4 : x_1 = x_2 = 0\}$$

and the zero dynamics vector field $f^\star(x)$ is given by the restriction of the vector field

$$f(x) + g(x)(-A^{-1}(x)b(x))$$

to $Z^\star$. Since

$$b(x) = \begin{bmatrix} x_1 + x_1 x_4 \\ x_2 e^{x_3} \end{bmatrix}$$

the representation of the zero dynamics, in the (x_3, x_4) coordinates of $Z^\star$, is following one

$$\dot{x}_3 = x_3^2$$
$$\dot{x}_4 = -x_4.$$

Note that the point $x = 0$ is an unstable equilibrium of these equations, and therefore the approach to noninteracting control used in Section 5.3 would yield an unstable closed loop.

In order to check whether or not the Problem of Noninteracting Control with Stability is solvable, we have to calculate the vector field $s^\star(x)$. This requires first the calculation of the distributions $P_1^\star$, $P_2^\star$ and $P^\star$. We have

$$P_1^\star = \langle \tilde{f}, \tilde{g}_1, \tilde{g}_2 | \text{span}\{\tilde{g}_2\}\rangle$$
$$P_2^\star = \langle \tilde{f}, \tilde{g}_1, \tilde{g}_2 | \text{span}\{\tilde{g}_1\}\rangle$$

where $\tilde{f}(x) = f(x) + g(x)\alpha(x)$, $\tilde{g}_1(x) = (g(x)b(x))_1$, $\tilde{g}_2(x) = (g(x)b(x))_2$, and $\alpha(x), b(x)$ is any feedback solving the Noninteracting Control Problem. Choosing the standard noninteracting feedback, one obtains

$$\tilde{f}(x) = \begin{bmatrix} 0 \\ 0 \\ x_2 + x_3^2 \\ x_2 - x_2 e^{x_3} - x_4 \end{bmatrix} \qquad \tilde{g}_1(x) = \begin{bmatrix} 1 \\ 0 \\ 0 \\ 1 \end{bmatrix} \qquad \tilde{g}_2(x) = \begin{bmatrix} 0 \\ 1 \\ 0 \\ 1 \end{bmatrix}.$$

The calculation of $P_1^\star$ and $P_2^\star$ can be carried out by means of the algorithm (1.8.1). In order to obtain $P_1^\star$, we set

$$\Delta_0 = \text{span}\{\tilde{g}_2(x)\}$$

and then we iterate, using

$$\Delta_k = \Delta_{k-1} + [\tilde{f}, \Delta_{k-1}] + [\tilde{g}_1, \Delta_{k-1}] + [\tilde{g}_2, \Delta_{k-1}].$$

Standard calculations show that

$$\Delta_2(x) = \text{span}\{ \begin{bmatrix} 0 \\ 1 \\ 0 \\ 1 \end{bmatrix}, \begin{bmatrix} 0 \\ 0 \\ -1 \\ e^{x_3} \end{bmatrix}, \begin{bmatrix} 0 \\ 0 \\ 2x_3 \\ (x_3^2 + 1)e^{x_3} \end{bmatrix} \}.$$

This distribution, which is nonsingular in a neighborhood of $x = 0$ and invariant under the vector fields $\tilde{f}(x), \tilde{g}_1(x), \tilde{g}_2(x)$, is the required distribution $P_1^\star$. Note that it is possible to simplify the expression of the vectors which span this distribution and obtain for instance

$$P_1^\star = \mathrm{span}\left\{ \begin{bmatrix} 0 \\ 1 \\ 0 \\ 0 \end{bmatrix}, \begin{bmatrix} 0 \\ 0 \\ 1 \\ 0 \end{bmatrix}, \begin{bmatrix} 0 \\ 0 \\ 0 \\ 1 \end{bmatrix} \right\}.$$

Proceeding in a similar way, one obtains

$$P_2^\star = \mathrm{span}\left\{ \begin{bmatrix} 1 \\ 0 \\ 0 \\ 0 \end{bmatrix}, \begin{bmatrix} 0 \\ 0 \\ 0 \\ 1 \end{bmatrix} \right\}$$

and concludes that

$$P^\star = P_1^\star \cap P_2^\star = \mathrm{span}\{[0 \quad 0 \quad 0 \quad 1]^T\}.$$

The integral submanifold of $P^\star$ which contains the point $x = 0$ is clearly the set

$$S^\star = \{x \in \mathbf{R}^4 : x_1 = x_2 = x_3 = 0\}.$$

This is an invariant manifold for $\tilde{f}(x) = f(x) + g(x)\alpha(x)$, and the restriction of this vector field to $S^\star$ is by the definition the vector field $s^\star(x)$ whose properties determine the solvability of the Problem of Noninteracting Control with Stability. Note that $S^\star$ is also an invariant manifold of the zero dynamics vector field (see Remark 4.5), and therefore we can immediately obtain a representation of $s^\star(x)$ by setting $x_3 = 0$ in the representation of the vector field $f^\star(x)$. This yields

$$\dot{x}_4 = -x_4$$

This system has an asymptotically stable equilibrium at the origin and therefore, by Theorem 4.10, the problem in question *is* solvable.

In order to find a solution, it is convenient to put the closed loop system

$$\dot{x} = \tilde{f}(x) + \tilde{g}_1(x)v_1 + \tilde{g}_2(x)v_2$$

(obtained by means of the standard noninteractive feedback) in the form (4.10). To this end, note that

$$(P_1^\star)^\perp = \mathrm{span}\{dx_1\} = \mathrm{span}\{dh_1\}$$
$$(P_2^\star)^\perp = \mathrm{span}\{dx_2, dx_3\} = \mathrm{span}\{dh_2\} + \mathrm{span}\{dx_3\}.$$

Thus, one can set

$$z^1 = x_1$$
$$z^2 = \mathrm{col}(x_2, x_3)$$
$$z^3 = x_4$$

(and no variable z^4 exists, because $(P)^\perp = (P_1^\star)^\perp \cap (P_2^\star)^\perp = 0$). Accordingly, one obtains a system in the form (4.10)

$$\dot{x}_1 = v_1$$
$$\dot{x}_2 = v_2$$
$$\dot{x}_3 = x_2 + x_3^2$$
$$\dot{x}_4 = x_2 - x_2 e^{x_3} - x_4 + v_1 + v_2$$

(see Fig. 7.3).

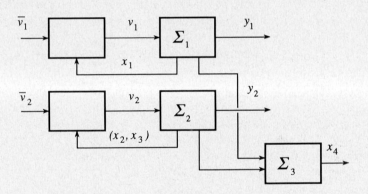

Fig. 7.3

At this point, it suffices to stabilize—by means of linear feedback—the two subsystems with state variables z^1 and z^2. One can set, for instance,

$$v_1 = -x_1 + \bar{v}_1$$
$$v_2 = -x_2 - x_3 + \bar{v}_2.$$

This additional feedback preserves the noninteractive structure and stabilizes the system. In summary, a feedback law which solves the Problem of Noninteracting Control with Stability, obtained by composition of the feedback just determined with the standard noninteractive feedback, has the form

$$u_1 = -x_2 e^{x_3} - x_2 - x_3 + \bar{v}_2$$
$$u_2 = x_1 - x_1 x_4 + x_2 x_3 e^{x_3} - x_1 + \bar{v}_1 + x_3 x_2 + x_3^2 - x_3 \bar{v}_2.$$

7.5 Achieving Relative Degree via Dynamic Extension

The analysis developed in Chapter 5 has shown that a nonlinear system of the form (5.1.1) which has a (vector) relative degree at the point x° lends itself to the implementation of some relevant control strategies. For instance, this system can be rendered noninteractive (from an input-output point of view) via state feedback. If, in addition, the equality $r_1 + \ldots + r_m = n$ is satisfied, this system can be changed into a fully linear and controllable one by means of feedback and coordinates transformation. Note that the latter condition, in view of a property illustrated in Corollary (6.3.14), is exactly the condition under which the system (by assumption with some relative degree at the point x°) has a 0-dimensional zero dynamics manifold (in this case, in fact, $Z^\star = \{x^\circ\}$), i.e., more briefly has *no zero dynamics*.

The purpose of this Section is to show that, under certain assumptions, it is possible to modify—by means of control laws which are more general than those considered so far—a system which does not have a vector relative degree in such a way as to obtain a new system which does have a relative degree. Of course, this cannot be achieved by means of static state feedback of the form (5.2.3) because, as shown for instance in the proof of Lemma 5.2.1, the property—for a system—of having relative degree is invariant under this type of feedback. We will use rather a feedback structure which incorporates an additional set of state variables, namely a *dynamic* state feedback. As anticipated in Section 4.5 (see in particular (4.5.10a)), this type of feedback is modeled by equations of the form

$$u = \alpha(\zeta, x) + \beta(\zeta, x)v \tag{5.1a}$$

$$\dot\zeta = \gamma(\zeta, x) + \delta(\zeta, x)v. \tag{5.1b}$$

The reason why the addition of auxiliary state variables may be helpful in achieving relative degree can be easily motivated with the aid of a simple example.

Example 5.1. Consider again the system of the Example 6.1.1. This system does not have a relative degree at any point because the matrix (5.1.2), which in this case is

$$L_g h(x) = \begin{bmatrix} 1 & 0 \\ x_3 & 0 \end{bmatrix}$$

has rank 1 for all x.

The reason why this system has no relative degree is that the lowest derivatives of y_1 and y_2 which are affected by the input (in this case $y_1^{(1)}$ and $y_2^{(1)}$), are affected both by u_1 and none by u_2. Thus, in order to obtain a relative degree, one could try to render $y_1^{(1)}$ and $y_2^{(1)}$ independent of u_1, that is to "delay" the appearance of u_1 to higher order derivatives of y_1 and y_2 , and hope that when this happens also u_2 shows up. In order to render $y_1^{(1)}$ and $y_2^{(1)}$ independent of the input, in particular of its first component u_1, it suffices to set u_1 equal to the output of another (auxiliary) dynamical system, with some internal state ζ, and driven by a new reference input v_1. The simplest way in which this result can be achieved is to set u_1 equal to the output of an "integrator" driven by v_1, i.e. to set

$$u_1 = \zeta$$
$$\dot{\zeta} = v_1$$

(see Fig. 7.4).

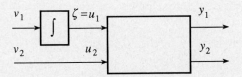

Fig. 7.4

For consistency of notation it is also set, for the second input channel which has been left unchanged,

$$u_2 = v_2.$$

The composed system thus obtained is described by equations of the form

$$\dot{\tilde{x}} = \tilde{f}(\tilde{x}) + \tilde{g}_1(\tilde{x})v_1 + \tilde{g}_2(\tilde{x})v_2$$
$$y = h(x)$$

with $\tilde{x} = (x, \zeta)$ and

$$\tilde{f}(x,\zeta) = \begin{bmatrix} \zeta \\ x_4 + x_3\zeta \\ \lambda x_3 + x_4 \\ 0 \\ 0 \end{bmatrix} \quad \tilde{g}_1(x,\zeta) = \begin{bmatrix} 0 \\ 0 \\ 0 \\ 0 \\ 1 \end{bmatrix} \quad \tilde{g}_2(x,\zeta) = \begin{bmatrix} 0 \\ 0 \\ 0 \\ 1 \\ 0 \end{bmatrix}.$$

Straightforward calculations show that now

$$L_{\tilde{g}}h(x,\zeta) = 0$$
$$L_{\tilde{g}}L_{\tilde{f}}h(x,\zeta) = \begin{bmatrix} 1 & 0 \\ x_3 & 1 \end{bmatrix}$$

i.e. that the system in question *has* (vector) relative degree $\{2, 2\}$. •

Having explained why the addition of auxiliary state variables, in particular the addition of *integrations* on certain input channels, is helpful in obtaining a relative degree, we describe now a recursive procedure which essentially identifies the channels on which the integrations must be added and the number of integrators needed in order to achieve the desired goal, that is some (vector) relative degree. As we shall see, the procedure in question incorporates also a feedback-type modification of the original system and thus, the entire control structure that will be determined is that of a dynamic feedback of the form (5.1).

In what follows, we consider, as done throughout most of these notes, a multivariable system with the same number m of inputs and outputs channels. The symbol r_i is still used to denote the largest integer such that

$$L_{g_j} L_f^k h_i(x) = 0$$

for all $k < r_i - 1$, all $1 \leq j \leq m$, all x near x° but, of course, it is not necessarily assumed that the system has relative degree $\{r_1, \ldots, r_m\}$ (i.e. that the matrix (5.1.2) is nonsingular). Moreover, it is supposed that the vector fields $f(x), g_1(x), \ldots, g_m(x)$ and the mapping $h(x)$ are *analytic* in their domain U of definition. This will make it possible identify certain open and dense subsets of U on which the ranks of certain matrices are constant.

Dynamic extension algorithm. Consider the matrix $A(x)$ defined by (5.1.2). The rank of this matrix is constant on an open and dense subset of U. Let q denote this rank and suppose $q < m$ (otherwise, the system already has vector relative degree $\{r_1, \ldots, r_m\}$ at each point x° of a dense subset of U). Without loss of generality, suppose that the first q rows of $A(x)$ are linearly independent at each x in an open and dense subset U° of U (this can always be achieved by changing the order of the output channels) and let $\alpha(x), \beta(x)$ be a pair of analytic functions (with nonsingular $\beta(x)$) which solve the equations

$$\langle dL_f^{r_i-1} h_i(x), f(x) + g(x)\alpha(x) \rangle = 0 \qquad \text{for all } 1 \leq i \leq q \tag{5.2a}$$

$$\langle dL_f^{r_i-1} h_i(x), g(x)\beta(x) \rangle = \delta_{ij} \quad \text{for all } 1 \leq i \leq q,\ 1 \leq j \leq m \tag{5.2b}$$

for all $x \in U^\circ$. Set

$$\zeta = \text{col}(\zeta_1, \ldots, \zeta_q)$$
$$v' = \text{col}(v_1, \ldots, v_q)$$
$$v'' = \text{col}(v_{q+1}, \ldots, v_m)$$

and let $\beta'(x)$ (respectively $\beta''(x)$) denote the matrix formed by the first q columns (respectively last $(q - m)$ columns) of $\beta(x)$.

Define the dynamic feedback

$$u = \alpha(x) + \beta'(x)\zeta + \beta''(x)v'' \tag{5.3a}$$

$$\dot{\zeta} = v' \tag{5.3b}$$

and the corresponding closed-loop system

$$\dot{x} = f(x) + g(x)\alpha(x) + g(x)\beta'(x)\zeta + g(x)\beta''(x)v''$$
$$\dot{\zeta} = v'$$
$$y = h(x). \quad \bullet$$

The purpose of the Dynamic Extension Algorithm is to construct, starting with a system in which the rank of the matrix (5.1.2) has a certain value q, an extended (and feedback-modified) system in which the rank of the corresponding matrix (5.1.2) is possibly larger than q. However, this may not be immediately the case, for the rank of the extended system may continue to be equal to q. If the rank of the matrix (5.1.2) remains equal to q, one can iterate once more (or few more times, if needed) the construction, until the rank actually increases. As a matter of fact, it is not difficult to show that the number of iterations needed to have the rank increasing, whenever this is possible, is bounded by a number that depends on the dimension n of the state space of the system on which the first iteration was performed.

Proposition 5.1. The following two events are mutually exclusive
(i) after at most $n - (r_1 + \ldots + r_q + \min\{r_j : q + 1 \leq j \leq m\})$ iterations, the system obtained by means of the Dynamic Extension Algorithm has a matrix (5.1.2) whose rank is larger than or equal to $q + 1$,
(ii) after any arbitrary number of iterations, the system obtained by means of the Dynamic Extension Algorithm has a matrix (5.1.2) whose rank is always equal to q.

Proof. A simple calculation shows that, by construction, the closed loop system obtained at the end of the first iteration of the algorithm is such that

$$y_1^{(r_1)} = \zeta_1$$
$$\ldots$$
$$y_q^{(r_q)} = \zeta_q$$

and

$$y_i^{(r_i)} = \phi_i^{r_i}(x) + \psi_{i1}^{r_i}(x)\zeta_1 + \ldots + \psi_{iq}^{r_i}(x)\zeta_q$$

for all $q + 1 \leq i \leq m$ (and at least one of the $\psi_{ij}^{r_i}(x)$'s is nonzero by definition of

r_i). If the outputs are differentiated once more, one obtains

$$y_1^{(r_1+1)} = v_1$$

$$\dots$$

$$y_q^{(r_q+1)} = v_q$$

$$y_i^{(r_i+1)} = \psi_{i1}^{r_i}(x)v_1 + \dots + \psi_{iq}^{r_i}(x)v_q + \frac{\partial\left(\phi_i^{r_i}(x) + \sum\limits_{j=1}^{q} \psi_{ij}^{r_i}(x)\zeta_j\right)}{\partial x}\dot{x}.$$

Thus, the addition of the dynamic feedback (5.3) has the effect of increasing by exactly one unit all the r_i's, $1 \leq i \leq m$. The rank of the corresponding matrix (5.1.2) will increase if and only if, for some value of i between $q+1$ and m, one of the inputs $v_{q+1}, \dots, v_m$ appears explicitly in the expression of the $(r_i + 1)$-th derivative of y_i.

If this is not the case, one iterates again the construction. The "new" equations (5.2) can be solved immediately by $\alpha(x) = 0$ and $\beta(x) = I$, and the associated dynamic feedback simply corresponds to the addition of another integrator on each of the inputs $v_1, \dots, v_q$.

Suppose that $k+1$ iterations are needed in order to obtain a system, denoted for convenience

$$\dot{\tilde{x}} = \tilde{f}(\tilde{x}) + \tilde{g}(\tilde{x})\tilde{u}$$
$$y = \tilde{h}(\tilde{x})$$

in which the $(r_i + k + 1)$-th derivative of y_i, for some value of i such that $q+1 \leq i \leq m$, depends explicitly on one of the inputs $v_{q+1}, \dots, v_m$. In this system the matrix

$$\begin{bmatrix} dL_{\tilde{f}}^{r_1+k}h_1(\tilde{x}) \\ \vdots \\ dL_{\tilde{f}}^{r_q+k}h_q(\tilde{x}) \\ dL_{\tilde{f}}^{r_i+k}h_i(\tilde{x}) \end{bmatrix}\tilde{g}(\tilde{x})$$

has rank $q+1$ for all $\tilde{x}$ in an open and dense subset U° of U. As a consequence (see Lemma 5.1.2 and Remark 5.1.5) at each point of U°, the differentials of the functions

$$\{L_{\tilde{f}}^s h_j(\tilde{x}) : 0 \leq s \leq r_j + k, 1 \leq j \leq q, j = i\}$$

are linearly independent. The dimension of the state variable $\tilde{x}$ is equal to $n + (k+1)q$, whereas the number of these functions is equal to $(r_1 + \dots + r_q + r_i) + (k+1)(q+1)$. The linear independence of the differentials of these functions implies

$$(r_1 + \dots + r_q + r_i) + (k+1)(q+1) \leq n + (k+1)q$$

i.e.

$$k < n - (r_1 + \ldots + r_q + r_i)$$

This completes the proof. •

The previous discussion clearly shows that, after a certain number of iterations, the Dynamic Extension Algorithm may produce a feedback-modified system in which the corresponding matrix (5.1.2) has full rank m at all points in an open and dense subset of the state space. This system will have thus a well-defined (vector) relative degree at each of these point. Moreover, the bound given by Proposition 5.1 specifies also how many times it is necessary to try in order to decide whether or not the procedure should be further pursued or abandoned. Note that the upper bound given by Proposition 5.1 refers to the number of iterations needed in order to increase the rank of the matrix (5.1.2) by at least one unit and depends on the dimension of the state space of the system on which the iterations were started. If, as a result of dynamic extension, the rank of the matrix (5.1.2) has increased, the upper bound on the number of iterations needed to achieve a further increase must be recalculated, on the basis of the (larger) dimension of the state space of the extended system.

Of course, one might pose the question of whether or not there are other procedures yielding extended systems with some well-defined relative degree, perhaps with better chances of success. As a matter of fact, this is not the case, in the sense that either the procedure just described ends up with a system having a (vector) relative degree at any point of an open and dense subset in the state space, or there is no other way to achieve this result under a dynamic feedback of the form (5.1).

Proposition 5.2. Suppose there exists a dynamic feedback of the form (5.1) with the property that the closed loop system

$$\dot{x} = f(x) + g(x)\alpha(\zeta, x) + g(x)\beta(\zeta, x)v \qquad (5.4a)$$

$$\dot{\zeta} = \gamma(\zeta, x) + \delta(\zeta, x)v \qquad (5.4b)$$

$$y = h(x) \qquad (5.4c)$$

has some vector relative degree at each point (x°, ζ°) of an open and dense set in the state space. Then, after a finite number of iterations, the Dynamic Extension Algorithm produces a feedback-modified system in which the matrix (5.1.2) has full rank m at all points in an open and dense subset of the state space.

Proof. By contradiction, suppose the Dynamic Extension Algorithm does not succeed in producing a system with a vector relative degree. This means that, at a certain stage, one has obtained a system on which all further iterations were unsuccessful. For simplicity (and without loss of generality), let this system still be denoted by

$$\dot{x} = f(x) + g(x)u$$
$$y = h(x)$$

and suppose the corresponding matrix (5.1.2) has rank $q < m$. For this system we have, as already observed in the proof of Proposition 5.1,

$$y_1^{(r_1)} = u_1$$
$$\cdots$$
$$y_q^{(r_q)} = u_q$$

and

$$y_i^{(r_i)} = \phi_i^{r_i}(x) + \psi_{i1}^{r_i}(x)u_1 + \ldots + \psi_{iq}^{r_i}(x)u_q$$

for all $q + 1 \leq i \leq m$. If all the subsequent iterations of the Dynamic Extension Algorithm are unsuccessful, then for every $k \geq 0$, the $(r_i + k)$-th derivative of the output y_i, for all $q + 1 \leq i \leq m$, will never depend explicitly on $u_{q+1}, \ldots, u_m$, but only on $x(t)$ and on the values of $u_1, \ldots, u_q$, and their derivatives up to order k, at time t. Since

$$y_j^{(r_j+s)} = u_j^{(s)} \qquad \text{for all } 1 \leq j \leq q, \text{ all } s > 0$$

it is concluded, for instance, that for every $k \geq 0$ and for all $q + 1 \leq i \leq m$, $y_i^{(r_i+k)}(0)$ is completely determined by $x(0)$ and $y_j^{(r_j+s)}(0)$, $1 \leq j \leq q$ and $s \leq k$.

This fact contradicts the existence of a dynamic feedback yielding a closed loop system having a vector relative degree. If this were the case, in fact, the closed loop system could have been rendered noninteractive by means of an additional state feedback, and in the system thus obtained the outputs would satisfy relations of the type

$$y_i^{(s_i)} = v_i \qquad \text{for all } 1 \leq i \leq m$$

where the v_i's are independent controls and the s_i's suitable integers. This shows that the $y_j^{(s_j+s)}(0)$'s, $1 \leq i \leq m$, $s > 0$, are independent $\bullet$

Of course, if there exists a feedback of the form (5.1) which changes the original system (5.1.1) into a new system (of the form (5.4)) having some vector relative degree at a point (x°, ζ°) of the extended state space, then an additional static feedback (determined on the basis of the results illustrated in Section 5.3, e.g. a standard noninteractive feedback) of the form

$$v = \bar{\alpha}(\zeta, x) + \bar{\beta}(\zeta, x)\bar{v}$$

can make each output y_i depending only on the i-th component of the new reference input $\bar{v}$ and not on the other ones. In other words, the original system (5.1.1) can be rendered *noninteractive* by means of *dynamic* state feedback.

Another property of the systems having a (vector) relative degree is that, if

$$r_1 + \ldots + r_m = n \qquad (5.5)$$

where n is the dimension of the state space, there exist a feedback and a coordinates transformation that can change the system into a fully linear and controllable one. Thus, if relative degree can be achieved via dynamic feedback and the condition (5.5) is satisfied in the extended system, then the original system can be changed into a fully linear and controllable one via *dynamic* feedback and coordinates transformations.

If the relative degree has been achieved via dynamic feedback, then the data included in the previous condition, namely the integers n and the r_i's, are not known until the dynamic feedback has been constructed, i.e. for instance until all the iterations of the Dynamic Extension Algorithm have been successfully completed. However, it is possible to prove, under rather mild assumptions, that the *fulfillment* of the condition (5.5) depends directly on a simple property of the original system, namely the absence of zero dynamics.

In order to see this, all we have to show is that the zero dynamics are left unchanged under dynamic extension. Of course, the zero dynamics are left unchanged by any regular static state feedback, simply because the mapping from v to u defined by

$$u = \alpha(x) + \beta(x)v$$

is invertible. Thus, the point is to show that the zero dynamics are left unchanged by addition of integrators, and it is sufficient to check that this happens in the simple case in which only one integrator is added on one input channel, say the first one.

Suppose that the system

$$\dot{x} = f(x) + g(x)u$$
$$y = h(x)$$

satisfies the hypotheses of Propositions 6.1.1 and 6.1.2 . Then, if the initial state x° is sufficiently close to the origin, $y(t) = 0$ for all sufficiently small values of t if and only if $x^\circ \in Z^\star$ and the input to the system is

$$u(t) = u^\star(x(t))$$

where $u^\star(x)$ is a uniquely defined function, and $x(t)$ satisfies the equation

$$\dot{x} = f^\star(x) \qquad x^\circ \in Z^\star.$$

Consider now the extended system

$$\dot{x} = f(x) + g_1(x)\zeta + g_2(x)v_2 + \ldots + g_m(x)v_m$$
$$\dot{\zeta} = v_1$$
$$y = h(x)$$

On this system, the constraint $y(t) = 0$ implies (and is implied by) $x^\circ \in Z^\star$ and

$$v_i(t) = u_i^\star(x(t)) \qquad \text{for all } 2 \leq i \leq m$$
$$\zeta(t) = u_1^\star(x(t))$$

with $x(t)$ defined as before, and thus

$$v_1(t) = \frac{\partial\zeta}{\partial t} = \frac{\partial u_1^\star}{\partial x}\dot{x} = \frac{\partial u_1^\star}{\partial x}f^\star(x(t)) = L_{f^\star}u_1^\star(x(t)).$$

We deduce from this argument that in the extended system, the manifold

$$\bar{Z}^\star = \{(x,\zeta) \in U \times \mathbf{R} : x \in Z^\star, \zeta = u_1^\star(x)\}$$

is a locally maximal output zeroing submanifold, and that the unique input $v^\star$ which renders the vector field

$$\bar{f}^\star(x) = \begin{bmatrix} f(x) + g_1(x)\zeta \\ 0 \end{bmatrix} + \begin{bmatrix} 0 & g_2(x) & \cdots & g_m(x) \\ 1 & 0 & \cdots & 0 \end{bmatrix} v^\star(x)$$

tangent to $\bar{Z}^\star$ is the input

$$v_1^\star(x) = L_{f^\star}u_1^\star(x)$$
$$v_i^\star(x) = u_i^\star(x) \qquad \text{for all } 2 \leq i \leq m.$$

Therefore, the zero dynamics are left unchanged by addition of integrators on input channels.

Suppose now that, in particular, $Z^\star = \{x^\circ\}$, i.e. the original system has no zero dynamics. From the previous argument it follows that $\bar{Z}^\star = \{(x^\circ, 0)\}$ (recall that $u^\star(x^\circ) = 0$). This property is preserved under any further dynamic extension and therefore it is possible to conclude that after any iteration of the Dynamic Extension Algorithm a system is obtained which has no zero dynamics as well. If the final extended system has a vector relative degree at the point $(x, \zeta) = (x^\circ, 0)$, using Corollary 5.3.14 it is deduced that its zero dynamics

manifold is given by

$$\{(x,\zeta) : L_f^{k-1}h_i(x,\zeta) = 0, 1 \le k \le r_i, 1 \le 1 \le m\}$$

and, this, together with the fact that the zero dynamics manifold must degenerate to the single point $\{(x^\circ, 0)\}$, implies the fulfillment of (5.5).

We shall see applications of these ideas in the next Section.

7.6 Examples

We conclude the Chapter by showing the application of some of the results developed in the last Section to the control of a general aviation aircraft and to the control of a two-link robot arm with nonnegligible joint elasticity.

The dynamical model of an aircraft can be described by means of three sets of first order differential equations, involving the following sets of state variables:

- the angles (ψ, ϑ, ϕ) which characterize the attitude of the aircraft with respect to the so-called *wind axes* (these three angles are respectively called *yaw angle, pitch angle* and *roll angle*),

- the components, denoted (p, q, r) of the angular velocity vector ω with respect to a reference frame fixed with the aircraft (these three quantities are respectively called *roll rate, pitch rate* and *yaw rate*),

- the amplitude V of the *velocity* along the flying path, and two angles α and β which identify the direction of the tangent vector to the flying path with respect to the main symmetry axis of the aircraft (which are respectively called *angle of attack* and *sideslip angle*): α is the angle between the tangent to the flying path and the longitudinal axis in the pitch direction (Fig. 7.5), and β is the angle between the tangent to the flying path and the longitudinal axis in the yaw direction).

Fig. 7.5

The derivatives with respect to time of the angles (ψ, ϑ, ϕ) can be expressed in the form

$$\mathrm{col}(\dot\psi, \dot\vartheta, \dot\phi) = M(\psi, \vartheta, \phi)\omega^\star \qquad (6.1a)$$

where $\omega^\star = \mathrm{col}(p^\star, q^\star, r^\star)$ is the angular velocity vector expressed with respect to the wind axes and $M(\psi, \vartheta, \phi)$ is the matrix already introduced in Section 5.5.

The derivative with respect to time of the angular velocity vector $\omega = \mathrm{col}(p, q, r)$ can be expressed in the form

$$\dot{\omega} = J^{-1}S(\omega)J\omega + J^{-1}T \tag{6.1b}$$

in which $S(\omega)$ is the matrix already introduced in Section 1.5, J is the inertia matrix, which in this case has the form

$$J = \begin{bmatrix} I_x & 0 & -I_{xz} \\ 0 & I_y & 0 \\ -I_{xz} & 0 & I_z \end{bmatrix}$$

and T represents the vector of external torques.

Finally, the derivatives of V, α, β with respect to time have the form

$$\begin{aligned} \dot{V} &= -(D/m) - g\sin\vartheta \\ \dot{\alpha} &= q - q^*\sec\beta - (p\cos\alpha + r\sin\alpha)\tan\beta \\ \dot{\beta} &= r^* + p\sin\alpha - r\cos\alpha \end{aligned} \tag{6.1c}$$

in which D is a scalar quantity called the *drag* force, m is the mass of the aircraft, and is g the gravity acceleration.

In order to complete the model it is necessary to specify how the three rates (p^*, q^*, r^*), which appear in the first and third set of equations, are related to the other state variables. The relations in question have the form

$$\begin{aligned} p^* &= p\cos\alpha\cos\beta + (q - \dot{\alpha})\sin\beta + r\sin\alpha\cos\beta \\ q^* &= \frac{1}{mV}(L - mg\cos\vartheta\cos\phi) \\ r^* &= \frac{1}{mV}(-S + mg\cos\vartheta\sin\phi) \end{aligned} \tag{6.1d}$$

in which S and L are two scalar quantities called the *side* and *lift* forces. Replacing (p^*, q^*, r^*) in the previous equations and solving for $\dot{\alpha}$ (which appears in both (6.1c) and (6.1d)), one obtains a system of nine first order differential equations in the state variables ψ, ϑ, ϕ, p, q, r, V, α, β, which describes the dynamics of the aircraft.

The vector T of the external torques and the vector $\mathrm{col}(D, L, S)$ of the external forces contain the input variables. The first one of these two vectors can be given an approximate expression of the following form

$$
T = \begin{bmatrix} a_{11}V^2 \sin\beta + a_{12}rV + a_{13}pV \\ a_{21}V^2 + a_{22}V^2 \sin\alpha + a_{23}qV \\ a_{31}V^2 \sin\beta + a_{32}rV + a_{33}pV \end{bmatrix} + V^2 \begin{bmatrix} b_{11}\cos\beta & 0 & b_{13}\cos\beta \\ 0 & b_{22}\cos\alpha & 0 \\ 0 & 0 & b_{33}\cos\beta \end{bmatrix} \begin{bmatrix} \delta_a \\ \delta_e \\ \delta_r \end{bmatrix}
$$

in which the a_{ij}'s and the b_{ij}'s are fixed aerodynamic parameters (dependent on the geometry of the aircraft, the air density, etc.) and δ_a, δ_e, δ_r, denote the deflections of the *aileron*, of the *elevator* and of the *rudder*. The vector $\mathrm{col}(D, L, S)$ can be given an approximate expression of the form

$$
\begin{bmatrix} D \\ L \\ S \end{bmatrix} = \begin{bmatrix} c_{11}V^2 + c_{12}V^2 \cos\alpha \\ c_{21}V^2 + c_{22}V^2 \sin 2\alpha \\ c_{31}V^2 \sin 2\beta \end{bmatrix} + P \begin{bmatrix} -\cos\alpha \cos\beta \\ \sin\alpha \\ \cos\alpha \cos\beta \end{bmatrix} \delta_P
$$

in which the c_{ij}'s are again fixed parameters, P indicates the maximal thrust and δ_P the setting of the *throttle*. Note that, in the previous description, the effect of the thrust on the vector T of external torques is neglected, and so are the effects of the deflections $(\delta_a, \delta_e, \delta_r)$ on the vector $\mathrm{col}(D, L, S)$ of the external forces.

The equations thus illustrated describe a system whose state is defined in a certain open neighborhood U of $\mathbf{R}^9$, subject to the action of the 4-dimensional input vector

$$
u = \mathrm{col}(\delta_P, \delta_a, \delta_e, \delta_r).
$$

Our purpose is to show that this system can be locally modified, via dynamic feedback and coordinates transformation, into a fully linear and controllable system. To this end, we first observe that, if the nine state variables are rearranged into the following three subsets

$$
x_1 = (V, \vartheta, \psi)
$$
$$
x_2 = (\phi, \alpha, \beta)
$$
$$
x_3 = (p, q, r)
$$

and the input variables into the following two subsets

$$
u_1 = \delta_P
$$
$$
u_2 = (\delta_a, \delta_e, \delta_r)
$$

then the previous equations exhibit a structure of the form

$$\dot{x}_1 = F_1(x_1, x_2) + G_1(x_1, x_2)u_1$$
$$\dot{x}_2 = F_2(x_1, x_2, x_3) + G_2(x_1, x_2)u_1 \qquad\qquad (6.2)$$
$$\dot{x}_3 = F_3(x_1, x_2, x_3) + G_3(x_1, x_2, x_3)u_2$$

in which the F_i's are 3×1 vectors, G_1, G_2 are 3×1 vectors and G_3 is a 3×3 matrix (for reasons of space, the explicit expressions of the functions involved in these relations—whose determination involves no difficulty—are omitted).

It will be shown that, for a suitable choice of output functions, the system in question is such that the design procedure illustrated in the previous Section can be successfully applied. To begin with, consider the output

$$y = x_1 = \operatorname{col}(V, \vartheta, \psi)$$

(note that this output is 3-dimensional, whereas the input to the system is 4-dimensional) and observe that, by definition

$$y^{(1)} = \operatorname{col}(y_1^{(1)}, y_2^{(1)}, y_3^{(1)}) = F_1(x_1, x_2) + G_1(x_1, x_2)u_1.$$

Since none of the three entries of the 3×1 vector $G_1(x_1, x_2)$ is identically zero, it is concluded that $r_1 = r_2 = r_3 = 1$, but the system cannot have relative degree $\{1, 1, 1\}$ because the matrix (5.1.2) in this case has the form

$$A(x) = \begin{bmatrix} (G_1)_1 & 0 & 0 & 0 \\ (G_1)_2 & 0 & 0 & 0 \\ (G_1)_3 & 0 & 0 & 0 \end{bmatrix}$$

where $(G_1)_i$ denotes the i-th entry of G_1. We apply once the Dynamic Extension Algorithm. In the present case, $q = 1$, and since $(G_1)_1$ is nonzero, the first row of $A(x)$ is nonzero. Thus equations (5.2) reduce to

$$(F_1)_1(x_1, x_2) + (G_1)_1(x_1, x_2)\alpha_1(x) = 0$$
$$(G_1)_1(x_1, x_2)\beta_{1j}(x) = \delta_{1j} \qquad 1 \le j \le 4.$$

Clearly one can choose

$$\alpha(x) = -\frac{(F_1)_1(x_1, x_2)}{(G_1)_1(x_1, x_2)}$$

$\alpha_i(x) = 0$ for $2 \le i \le 4$,

$$\beta_{11}(x) = \frac{1}{(G_1)_1(x_1, x_2)}$$

and $\beta_{ij}(x) = \delta_{ij}$ for all other values of $1 \le i, j \le 4$. Note that the feedback thus defined depends only on (x_1, x_2) and, therefore, the composition of the latter with the system (6.2) yields a set of equations having the same structure. Note also that choosing the matrices $\beta(x)$ and $\alpha(x)$ as

$$\beta(x) = \begin{bmatrix} \dfrac{1}{(G_1)_1(x_1, x_2)} & 0 \\ 0 & G_3^{-1}(x_1, x_2, x_3) \end{bmatrix} \tag{6.3a}$$

$$\alpha(x) = -\beta(x) \begin{bmatrix} (F_1)_1(x_1, x_2) \\ F_3(x_1, x_2, x_3) \end{bmatrix} \tag{6.3b}$$

(which is indeed possible because the matrix $G_3(x_1, x_2, x_3)$ is invertible at each point of an open and dense subset of U), one may obtain a simpler set of equations. The composition of (6.2) with the feedback $u = \alpha(x) + \beta(x)v$ thus defined is a system of equations which, using again—in order to avoid an overload of notation—the same symbols used in (6.2), has the form

$$\begin{aligned} \dot{x}_1 &= F_1(x_1, x_2) + G_1(x_1, x_2)v_1 \\ \dot{x}_2 &= F_2(x_1, x_2, x_3) + G_2(x_1, x_2)v_1 \\ \dot{x}_3 &= v_2 \end{aligned} \tag{6.4}$$

(in which v_1 is a scalar, v_2 is 3-dimensional vector, and the first equation of the first set is simply $\dot{x}_{11} = v_1$). The dynamic extension, which corresponds to the addition of one integrator on the input channel v_1, changes (6.4) into

$$\begin{aligned} \dot{x}_1 &= F_1(x_1, x_2) + G_1(x_1, x_2)\zeta \\ \dot{\zeta} &= v_1 \\ \dot{x}_2 &= F_2(x_1, x_2, x_3) + G_2(x_1, x_2)\zeta \\ \dot{x}_3 &= v_2. \end{aligned}$$

We now recalculate the r_i's, in order to check whether or not the extended system has some relative degree. By construction,

$$y^{(1)} = F_1(x_1, x_2) + G_1(x_1, x_2)\zeta$$

does not depend on the input. In order to obtain a shortened expression for $y^{(2)}$ we set

$$F_1(x_1, x_2) + G_1(x_1, x_2)\zeta = B_1(x_1, x_2, \zeta)$$

thus obtaining

$$y^{(2)} = \frac{\partial B_1}{\partial x_1}(F_1 + G_1\zeta) + \frac{\partial B_1}{\partial x_2}(F_2 + G_2\zeta) + G_1 v_1.$$

Again, we see that all the entries of $y^{(2)}$ depend on v_1 but not on v_2. Thus $r_1 = r_2 = r_3 = 2$, but the system cannot have relative degree $\{2, 2, 2\}$ because the corresponding matrix (5.1.2) is identical to the one calculated before the dynamic extension. We proceed with another cycle of dynamic extension (note that the bound specified by Proposition 5.1 is in the present case equal to 6). Of course there is no need of an additional feedback, so we proceed by adding a second integrator on the input channel v_1, and this yields an extended system

$$\dot{x}_1 = F_1(x_1, x_2) + G_1(x_1, x_2)\zeta$$
$$\dot{\zeta} = \xi$$
$$\dot{\xi} = v_1$$
$$\dot{x}_2 = F_2(x_1, x_2, x_3) + G_2(x_1, x_2)\zeta$$
$$\dot{x}_3 = v_2.$$

In this new system, $y^{(1)}$ and $y^{(2)}$ do not the depend on the input, by construction. In order to obtain a shortened expression for $y^{(3)}$ we set

$$\frac{\partial B_1}{\partial x_1}(F_1 + G_1\zeta) + \frac{\partial B_1}{\partial x_2}(F_2 + G_2\zeta) + G_1\xi = B_2(x_1, x_2, x_3, \zeta, \xi)$$

thus obtaining

$$y^{(3)} = \frac{\partial B_2}{\partial x_1}(F_1 + G_1\zeta) + \frac{\partial B_2}{\partial x_2}(F_2 + G_2\zeta) + \frac{\partial B_2}{\partial x_3}v_2 + \frac{\partial B_2}{\partial \zeta}\xi + G_1 v_1.$$

Now $r_1 = r_2 = r_3 = 3$, and the input v_2 appears explicitly. The matrix (5.1.2) has the form

$$A(x, \zeta) = \begin{bmatrix} G_1 \\ \dfrac{\partial B_2}{\partial x_3} \end{bmatrix} = \begin{bmatrix} G_1 \\ \left(\dfrac{\partial F_1}{\partial x_2} + \dfrac{\partial G_1}{\partial x_2}\zeta\right)\left(\dfrac{\partial F_2}{\partial x_3}\right) \end{bmatrix}$$

A simple (but tedious) calculation shows that this matrix has rank 3 at any point (in the extended state space) characterized by $\zeta = \alpha = \beta = \psi = \vartheta = \phi = 0$ and $V \neq 0$. Thus the extended system has relative degree $\{3, 3, 3\}$ at any point of an open and dense subset of the state space.

We introduce now a fourth output function

$$y_4 = \phi$$

(which is the first component of the vector x_2) and we note that, in the extended system, $y_4^{(1)}$ does not depend on the input, whereas $y_4^{(2)}$ does. More specifically, setting

$$(F_2)_1 + (G_2)_1\zeta = D_2(x_1, x_2, x_3, \zeta)$$

we find that

$$y_4^{(2)} = \frac{\partial D_2}{\partial x_1}(F_1 + G_1\zeta) + \frac{\partial D_2}{\partial x_2}(F_2 + G_2\zeta) + \frac{\partial D_2}{\partial x_3}v_2 + (G_2)_1\xi.$$

The extended system with the four outputs

$$y_1 = V, \qquad y_2 = \vartheta, \qquad y_3 = \psi, \qquad y_4 = \phi \qquad\qquad (6.5)$$

has $r_1 = r_2 = r_3 = 3$ and $r_4 = 2$. The associated matrix (5.1.2) has the form

$$A(x,\zeta) = \begin{bmatrix} G_1 & \dfrac{\partial B_2}{\partial x_3} \\ 0 & \dfrac{\partial(F_2)_1}{\partial x_3} \end{bmatrix}$$

and, as a simple calculation shows, is nonsingular at any point (in the extended state space) characterized by $\zeta = \alpha = \beta = \psi = \vartheta = \phi = 0$ and $V \neq 0$. The system thus defined has relative degree $\{3,3,3,2\}$ at any point of an open and dense subset of the state space.

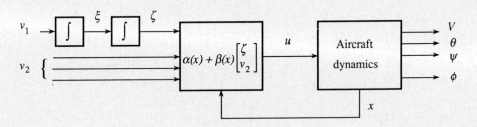

Fig. 7.6

In summary, we can conclude the following. The system composed (see Fig. 7.6) by the original equations (6.1) describing the dynamics of the aircraft with a dynamic state feedback of the form

$$\begin{bmatrix} \delta_P \\ \delta_a \\ \delta_e \\ \delta_r \end{bmatrix} = \alpha(x) + \beta(x) \begin{bmatrix} \zeta \\ v_{21} \\ v_{22} \\ v_{23} \end{bmatrix}$$

$$\dot{\zeta} = \xi$$

$$\dot{\xi} = v_1$$

with $\alpha(x)$ and $\beta(x)$ defined by (6.3), has relative degree $\{r_1, r_2, r_3, r_4\} = \{3,3,3,2\}$ with respect to the choice of outputs (6.5). Since the extended system thus defined has dimension $n = 11$, the condition

$$r_1 + r_2 + r_3 + r_4 = n$$

is fulfilled and, therefore, by means of an additional static state feedback (see Section 5.2) the system in question can be transformed (Fig. 7.7) into a system which, in suitable coordinates, is linear and controllable (Fig. 7.8).

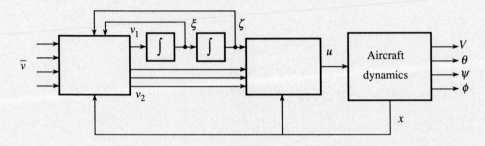

Fig. 7.7

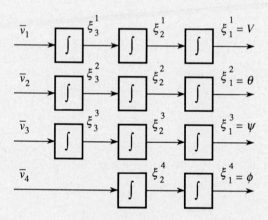

Fig. 7.8

The second example of a system which can be rendered both noninteractive (from the input-output point of view) and linear (in suitable cooordinates) by means of dynamic state feedback is that of a multi-link robot arm with nonnegligible elasticity between actuators and links. We have already briefly illustrated the phenomenon of elastic coupling between actuators and links of a robot arm in Section 4.10, where we showed that the elementary model of a single-link arm can be exactly linearized via static feedback and change of coordinates. This is not anymore the case—in general—when the arm consists of two or more links. As we shall see in a moment on a specific case, even in very simple configurations the features of the models are such that not only the exact linearization problem but even the less demanding noninteracting control problem is not solvable via *static* state feedback. However, by means of the design methodologies described in the previous Section, these two goals can still be achieved via *dynamic* feedback.

The simplest model on which these features can be illustrated is the one of an arm consisting of two links moving on an horizontal plane. The first link is rotating (about a fixed point) of the base frame, and the second link is rotating about the end point of the first link. For simplicity it is assumed that only the coupling between the first and second link (i.e. the second joint) exhibits significant elasticity.

The description of this system requires three angular coordinates, that can be chosen in the following way: the rotation q_1 of the first link with respect to the base frame, the rotation q_2 of the axis of the actuator which moves the second link (with respect to its own base fixed to the first link), the rotation q_3 of the second link with respect to the first link. The equations describing the motion of the arm, whose derivation is not within the scope of these notes and can be found in the appropriate literature, have the form

$$B(q)\ddot{q} + C(q, \dot{q}) + r(q) = T$$

in which $B(q)$ (the so-called *inertia matrix*) is a 3×3 symmetric and positive definite matrix of the form

$$B(q) = \begin{bmatrix} A_1 + 2A_3 \cos q_3 & A_4 & A_2 + A_3 \cos q_3 \\ A_4 & A_4 & 0 \\ A_2 + A_3 \cos q_3 & 0 & A_2 \end{bmatrix}$$

$C(q, \dot{q})$ (the *Coriolis* and *centrifugal forces*) is a 3×1 vector of the form

$$C(q, \dot{q}) = \begin{bmatrix} -A_3 \sin q_3 (2\dot{q}_1 \dot{q}_3 + \dot{q}_3^2) \\ 0 \\ A_3 \sin q_3 \dot{q}_1^2 \end{bmatrix}$$

and, finally, $r(q)$ is a 3×1 vector of the form

$$r(q) = \begin{bmatrix} 0 \\ -\dfrac{K}{N}[q_3 - \dfrac{q_2}{N}] \\ K[q_3 - \dfrac{q_2}{N}] \end{bmatrix}.$$

The coefficients A_i , $1 \leq i \leq 4$, which appear in these expressions are parameters related to the mass distribution in the arm. K is an elasticity constant and N represents the gear ratio of the coupling between the second actuator and the second link. The 3×1 vector T on the right-hand side includes the two control forces u_1 and u_2 imposed by the two actuators, and has the form

$$T = \text{col}(u_1, u_2, 0).$$

Note that the third entry of this vector is 0 because there is no independent input available for the coordinate q_3.

Choosing the state variables x_i , $1 \leq i \leq 6$ as

$$\begin{cases} x_i = q_i & \text{for } 1 \leq i \leq 3 \\ x_i = \dot{q}_{i-3} & \text{for } 4 \leq i \leq 6 \end{cases}$$

the system of equations can be put in the customary form

$$\dot{x} = f(x) + g_1(x)u_1 + g_2(x)u_2.$$

More precisely, it is easy to check that

$$\dot{x} = \begin{bmatrix} x_4 \\ x_5 \\ x_6 \\ f_4(x_2, x_3, x_4, x_6) \\ f_5(x_2, x_3, x_4, x_6) \\ f_6(x_2, x_3, x_4, x_6) \end{bmatrix} + \begin{bmatrix} 0 & 0 \\ 0 & 0 \\ 0 & 0 \\ g_{41}(x_3) & -g_{41}(x_3) \\ g_{51}(x_3) & g_{52}(x_3) \\ g_{61}(x_3) & -g_{61}(x_3) \end{bmatrix} \begin{bmatrix} u_1 \\ u_2 \end{bmatrix}.$$

A very natural choice of outputs in this system (as in any robot arm) is the set of the angular coordinates which define the relative positions of the links, namely

$$y_1 = q_1 = x_1$$
$$y_2 = q_3 = x_3.$$

With respect to these outputs the system does not have a relative degree because, as immediate calculations show,

$$L_g h(x) = 0$$

and

$$A(x) = L_g L_f h(x) = \begin{bmatrix} g_{41}(x_3) & -g_{41}(x_3) \\ g_{61}(x_3) & -g_{61}(x_3) \end{bmatrix}$$

has rank 1 for all x.

However, relative degree can be achieved after two iterations of the Dynamic Extension Algorithm. Standard calculations, that are left as an exercise to the reader, show that cascading the system in question with a compensator of the form

$$u_1 = \frac{1}{g_{41}(x)}(-f_4(x) + \zeta) + v_2$$

$$u_2 = v_2$$

$$\dot{\zeta} = \xi$$

$$\dot{\xi} = v_1$$

a system having relative degree $\{r_1, r_2\} = \{4, 4\}$ is obtained. The composite system has dimension $n = 8$ and therefore the condition $r_1 + r_2 = n$ is also fulfilled. It is concluded that by means of an additional static state feedback (see Section 5.2) the system in question can be transformed into a system which, in suitable coordinates, is linear and controllable.

Appendix A

A.1 Some Facts from Advanced Calculus

Let A be an open subset of $\mathbf{R}^n$ and $f : A \to \mathbf{R}$ a function. The value of f at $x = (x_1, \ldots, x_n)$ is denoted $f(x) = f(x_1, \ldots, x_n)$. The function f is said to be a function of class C^∞ (or, simply, C^∞ or, also, a *smooth* function) if its partial derivatives of any order with respect to $x_1, \ldots, x_n$ exist and are continuous. A function f is said to be *analytic* (sometimes denoted as C^ω) if it is C^∞ and for each point $x^\circ \in A$ there exists a neighborhood U of x°, such that the Taylor series expansion of f at x° converges to $f(x)$ for all $x \in U$.

Example. A typical example of a function which is C^∞ but not analytic is the function $f : \mathbf{R} \to \mathbf{R}$ defined by

$$
\begin{cases}
f(x) = 0 & \text{if } x \leq 0 \\
f(x) = \exp(-\dfrac{1}{x}) & \text{if } x > 0.
\end{cases}
$$

A mapping $F : A \to \mathbf{R}^m$ is a collection $(f_1, \ldots, f_m)$ of functions $f_i : A \to \mathbf{R}$. The mapping F is C^∞ if all f_i's are C^∞. •

Let $U \in \mathbf{R}^n$ and $V \in \mathbf{R}^n$ be open sets. A mapping $F : U \to V$ is a diffeomorphism if it is bijective (i.e. one-to-one and onto) and both F and F^{-1} are of class C^∞. The *jacobian matrix* of F at a point x is the matrix

$$
\frac{\partial F}{\partial x} = \begin{bmatrix} \dfrac{\partial f_1}{\partial x_1} & \cdots & \dfrac{\partial f_1}{\partial x_n} \\ \cdots & \cdots & \cdots \\ \dfrac{\partial f_n}{\partial x_1} & \cdots & \dfrac{\partial f_n}{\partial x_n} \end{bmatrix}
$$

The value of $\frac{\partial F}{\partial x}$ at a point $x = x^\circ$ is sometimes denoted $\left[\frac{\partial F}{\partial x}\right]_{x^\circ}$. •

Theorem (inverse function theorem). Let A be an open set of $\mathbf{R}^n$ and $F : A \to \mathbf{R}^n$ a C^∞ mapping. If $\left[\frac{\partial F}{\partial x}\right]_{x^\circ}$ is nonsingular at some $x^\circ \in A$, then there exists an open neighborhood U of x° in A such that $V = F(U)$ is open in $\mathbf{R}^n$ and the restriction of F to U is a diffeomorphism onto V.

Theorem (rank theorem). Let $A \subset \mathbf{R}^n$ and $B \subset \mathbf{R}^m$ be open sets, $F : A \to B$ a C^∞ mapping. Suppose $\left[\frac{\partial F}{\partial x}\right]_x$ has rank k for all $x \in A$. For each point $x^\circ \in A$ there exist a neighborhood A_0 of x° in A and a neighborhood B_0

of $F(x^\circ)$ in B, two open sets $U \subset \mathbf{R}^n$ and $V \subset \mathbf{R}^m$, and two diffeomorphisms $G : U \to A_0$ and $H : B_0 \to V$ such that $H \circ F \circ G(U) \subset V$ and such that for all $(x_1, \ldots, x_n) \in U$

$$(H \circ F \circ G)(x_1, \ldots, x_n) = (x_1, \ldots, x_k, 0, \ldots, 0).$$

Remark. Let P_k denote the mapping $P_k : \mathbf{R}^n \to \mathbf{R}^m$ defined by

$$P_k(x_1, \ldots, x_n) = (x_1, \ldots, x_k, 0, \ldots, 0).$$

Then, since H and G are invertible, one may restate the previous expression as

$$F = H^{-1} \circ P_k \circ G^{-1}$$

which holds at all points of A_0. •

Theorem (implicit function theorem). Let $A \subset \mathbf{R}^m$ and $B \subset \mathbf{R}^n$ be open sets. Let $F : A \times B \to \mathbf{R}^n$ be a C^∞ mapping. Let $(x, y) = (x_1, \ldots, x_m, y_1, \ldots, y_n)$ denote a point of $A \times B$. Suppose that for some $(x^\circ, y^\circ) \in A \times B$

$$F(x^\circ, y^\circ) = 0$$

and the matrix

$$\frac{\partial F}{\partial y} = \begin{bmatrix} \dfrac{\partial f_1}{\partial y_1} & \cdots & \dfrac{\partial f_1}{\partial y_n} \\ \cdots & \cdots & \cdots \\ \dfrac{\partial f_n}{\partial y_1} & \cdots & \dfrac{\partial f_n}{\partial y_n} \end{bmatrix}$$

is nonsingular at (x°, y°). Then, there exist open neighborhoods A_0 of x° in A and B_0 of y° in B and a unique C^∞ mapping $G : A_0 \to B_0$ such that

$$F(x, G(x)) = 0$$

for all $x \in A_0$.

Remark. As an application of the implicit function theorem, consider the following corollary. Let A be an open set in $\mathbf{R}^n$, let M be a $k \times n$ matrix whose entries are real-valued C^∞ functions defined on A and b a k-vector whose entries are also real-valued C^∞ functions defined on A. Suppose that for some $x^\circ \in A$

$$\mathrm{rank} M(x^\circ) = k.$$

Then, there exist an open neighborhood U of x° and a C^∞ mapping $G : U \to \mathbf{R}^n$ such that

$$M(x)G(x) = b(x)$$

for all $x \in U$.

In other words, the equation

$$M(x)y = b(x)$$

has at least a solution which is a C^∞ function of x in a neighborhood of x°. If $k = n$ this solution is unique. •

A.2 Some Elementary Notions of Topology

This Section is a review of the most elementary topological concepts that will be encountered later on.

Let S be a set. A *topological structure*, or a *topology*, on S is a collection of subsets of S, called *open* sets, satisfying the axioms

(i) the union of any number of open sets is open

(ii) the intersection of any finite number of open sets is open

(iii) the set S and the empty set $\emptyset$ are open

A set S with a topology is called a *topological space*.

A *basis* for a topology is a collection of open sets, called *basic open sets*, with the following properties

(i) S is the union of basic open sets

(ii) a nonempty intersection of two basic open sets is an union of basic open sets

A *neighborhood* of a point p of a topological space is any open set which contains p.

Let S_1 and S_2 be topological spaces and F a mapping $F : S_1 \to S_2$. The mapping F is *continuous* if the inverse image of every open set of S_2 is an open set of S_1. The mapping F is *open* if the image of an open set of S_1 is an open set of S_2. The mapping F is an *homeomorphism* if it is a bijection and both continuous and open.

If F is an homeomorphism, the inverse mapping F^{-1} is also an homeomorphism.

Two topological spaces S_1, S_2 such that there is an homeomorphism $F : S_1 \to S_2$ are said to be *homeomorphic*.

A subset U of a topological space is said to be *closed* if its complement $\bar{U}$ in S is open. It is easy to see that the intersection of any number of closed sets is closed , the union of any finite number of closed sets is closed, and both S and $\emptyset$ are closed.

If S_0 is a subset of a topological space S, there is a unique open set, noted $\text{int}(S_0)$ and called the *interior* of S_0, which is contained in S_0 and contains any other open set contained in S_0. As a matter of fact, $\text{int}(S_0)$ is the union of all open sets contained in S_0. Likewise, there is a unique closed set, noted $\text{cl}(S_0)$

and called the *closure* of S_o, which contains S_o and is contained in any other closed set which contains S_o. Actually, $\mathrm{cl}(S_o)$ is the intersection of all closed sets which contain S_o.

A subset of S is said to be *dense* in S if its closure coincides with S.

If S_1 and S_2 are topological spaces, then the cartesian product $S_1 \times S_2$ can be given a topology taking as a basis the collection of all subsets of the form $U_1 \times U_2$, with U_1 a basic open set of S_1 and U_2 a basic open set of S_2. This topology on $S_1 \times S_2$ is sometimes called the *product topology*.

If S is a topological space and S_1 a subset of S, then S_1 can be given a topology taking as open sets the subsets of the form $S_1 \cap U$ with U any open set in S. This topology on S_1 is sometimes called the *subset topology*.

Let $F : S_1 \to S_2$ be a continuous mapping of topological spaces, and let $F(S_1)$ denote the image of F. Clearly, $F(S_1)$ with the subset topology is a topological space. Since F is continuous, the inverse image of any open set of $F(S_1)$ is an open set of S_1. However, not all open sets of S_1 are taken onto open sets of $F(S_1)$. In other words, the mapping $F' : S_1 \to F(S_1)$ defined by $F'(p) = F(p)$ is continuous but not necessarily open. The set $F(S_1)$ can be given another topology, taking as open sets in $F(S_1)$ the images of open sets in S_1. It is easily seen that this new topology, sometimes called the *induced topology*, contains the subset topology (i.e. any set which is open in the subset topology is open also in the induced topology), and that the mapping F' is now open. If F is an injection, then S_1 and $F(S_1)$ endowed with the induced topology are homeomorphic.

A topological space S is said to satisfy the *Hausdorff separation axiom* (or, briefly, to be an Hausdorff space) if any two different points p_1 and p_2 have disjoint neighborhoods.

A.3 Smooth Manifolds

Definition. A *locally Euclidean space* X of dimension n is a topological space such that, for each $p \in X$, there exists a homeomorphism ϕ mapping some open neighborhood of p onto an open set in $\mathbf{R}^n$.

Definition. A *manifold* N of dimension n is a topological space which is locally Euclidean of dimension n, is Hausdorff and has a countable basis.

It is not possible that an open subset U of $\mathbf{R}^n$ be homeomorphic to an open subset V of $\mathbf{R}^m$, if $n \neq m$ (Brouwer's theorem on invariance of domain). Therefore, the *dimension* of a locally Euclidean space is a well-defined object.

A *coordinate chart* on a manifold N is a pair (U, ϕ), where U is an open set of N and ϕ a homeomorphism of U onto an open set of $\mathbf{R}^n$. Sometimes ϕ is represented as a set $(\phi_1, \ldots, \phi_n)$ and $\phi_i : U \to \mathbf{R}$ is called the *i*-th *coordinate function*. If $p \in U$, the n-tuple of real numbers $(\phi_1(p), \ldots, \phi_n(p))$ is called the set of *local coordinates* of p in the coordinate chart (U, ϕ). A coordinate chart

(U, ϕ) is called a *cubic* coordinate chart if $\phi(U)$ is an open cube about the origin in $\mathbf{R}^n$. If $p \in U$ and $\phi(p) = 0$, then the coordinate chart is said to be *centered* at p.

Let (U, ϕ) and (V, ψ) be two coordinate charts on a manifold N, with $U \cap V \neq 0$. Let $(\psi_1, \dots, \psi_n)$ be the set of coordinate functions associated with the mapping ψ. The homeomorphism

$$\psi \circ \phi^{-1} : \phi(U \cap V) \to \psi(U \cap V)$$

taking, for each $p \in U \cap V$, the set of local coordinates $(\phi_1(p), \dots, \phi_n(p))$ into the set of local coordinates $(\psi_1(p), \dots, \psi_n(p))$, is called a *coordinates transformation* on $U \cap V$. Clearly, $\phi \circ \psi^{-1}$ gives the inverse mapping, which expresses $(\phi_1(p), \dots, \phi_n(p))$ in terms of $(\psi_1(p), \dots, \psi_n(p))$.

Frequently, the set $(\phi_1(p), \dots, \phi_n(p))$ is represented as an n-vector $x = \text{col}(x_1, \dots, x_n)$, and the set $(\psi_1(p), \dots, \psi_n(p))$ as an n-vector $y = \text{col}(y_1, \dots, y_n)$. Consistently, the coordinates transformation $\psi \circ \phi^{-1}$ can be represented in the form

$$y = \begin{bmatrix} y_1 \\ \vdots \\ y_n \end{bmatrix} = \begin{bmatrix} y_1(x_1, \dots, x_n) \\ \vdots \\ y_n(x_1, \dots, x_n) \end{bmatrix} = y(x)$$

and the inverse transformation $\phi \circ \psi^{-1}$ in the form

$$x = x(y)$$

Two coordinate charts (U, ϕ) and (V, ψ) are C^∞-*compatible* if, whenever $U \cap V \neq 0$, the coordinates transformation $\psi \circ \phi^{-1}$ is a diffeomorphism, i.e. if $y(x)$ and $x(y)$ are both C^∞ maps (see Fig. A.1).

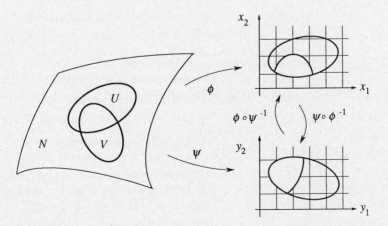

Fig. A.1

A C^∞ *atlas* on a manifold N is a collection $\mathcal{A} = \{(V_i, \phi_i) : i \in I\}$ of pairwise C^∞-compatible coordinate charts, with the property that $\bigcup_{i \in I} U_i = N$. An atlas is *complete* if not properly contained in any other atlas.

Definition. A smooth or C^∞ *manifold* is a manifold equipped with a complete C^∞ atlas.

Remark. if $\mathcal{A}$ is any C^∞ atlas on a manifold N, there exists a *unique* complete C^∞ atlas $\mathcal{A}^*$ containing $\mathcal{A}$. The latter is defined as the set of all coordinate charts (U, ϕ) which are compatible with every coordinate chart (U_i, ϕ_i) of $\mathcal{A}$. This set contains $\mathcal{A}$, is a C^∞ atlas, and is complete by construction. •

Some elementary examples of smooth manifolds are the ones described below.

Example. Any open set U of $\mathbf{R}^n$ is a smooth manifold, of dimension n. For, consider the atlas $\mathcal{A}$ consisting of the (single) coordinate chart $(U, \text{identity map on } U)$ and let $\mathcal{A}^*$ denote the unique complete atlas containing $\mathcal{A}$. In particular, $\mathbf{R}^n$ is a smooth manifold. •

Remark. One may define different complete C^∞ atlases on the same manifold, as the following example shows. Let $N = \mathbf{R}$, and consider the coordinate charts $(\mathbf{R}, \phi)$ and $(\mathbf{R}, \psi)$, with

$$\phi(x) = x$$
$$\psi(x) = x^3.$$

Since $\phi^{-1}(x) = x$ and $\psi^{-1}(x) = x^{1/3}$,

$$\phi \circ \psi^{-1}(x) = x^{1/3}$$

and the two charts are not compatible. Therefore the unique complete atlas $\mathcal{A}_\phi^*$ which includes $(\mathbf{R}, \phi)$ and the unique complete atlas $\mathcal{A}_\psi^*$ which includes $(\mathbf{R}, \psi)$ are different. This means that the same manifold N may be considered as a substrate of two different objects (two *smooth* manifolds), one arising with the atlas $\mathcal{A}_\phi^*$ and the other with the atlas $\mathcal{A}_\psi^*$. •

Example. Let U be an open set of $\mathbf{R}^m$ and let $\lambda_1, \ldots, \lambda_{m-n}$ be real-valued C^∞ functions defined on U. Let N denote the (closed) subset of U on which all functions $\lambda_1, \ldots, \lambda_{m-n}$ vanish, i.e. let

$$N = \{x \in U : \lambda_i(x) = 0, \ 1 \le i \le m - n\}.$$

Suppose the rank of the jacobian matrix

$$\begin{bmatrix} \dfrac{\partial \lambda_1}{\partial x_1} & \cdots & \dfrac{\partial \lambda_1}{\partial x_m} \\ \cdots & \cdots & \cdots \\ \dfrac{\partial \lambda_{m-n}}{\partial x_1} & \cdots & \dfrac{\partial \lambda_{m-n}}{\partial x_m} \end{bmatrix}$$

is $m - n$ at all $x \in N$. Then N is a smooth manifold of dimension n.

The proof of this essentially depends on the Implicit Function Theorem, and uses the following arguments. Let $x^\circ = (x_1^\circ, \ldots, x_n^\circ, x_{n+1}^\circ, \ldots, x^\circ{}_m)$ be a point of N and assume, without loss of generality, that the matrix

$$\begin{bmatrix} \dfrac{\partial \lambda_1}{\partial x_{n+1}} & \cdots & \dfrac{\partial \lambda_1}{\partial x_m} \\ \cdots & \cdots & \cdots \\ \dfrac{\partial \lambda_{m-n}}{\partial x_{n+1}} & \cdots & \dfrac{\partial \lambda_{m-n}}{\partial x_m} \end{bmatrix}$$

is nonsingular at x°. Then, there exist neighborhoods $A_\circ$ of $(x_1^\circ, \ldots, x_n^\circ)$ in $\mathbf{R}^n$ and $B_\circ$ of $(x_{n+1}^\circ, \ldots, x_m^\circ)$ in $\mathbf{R}^{m-n}$ and a C^∞ mapping $G : A_\circ \to B_\circ$ such that

$$\lambda_i(x_1, \ldots, x_n, g_1(x_1, \ldots, x_n), \ldots, g_{m-n}(x_1, \ldots, x_n)) = 0$$

for all $1 \leq i \leq m - n$. This makes it possible to describe points of N around x° as m-tuples $(x_1, \ldots, x_m)$ such that $x_{n+i} = g_i(x_1, \ldots, x_n)$ for $1 \leq i \leq m - n$. In this way one can construct a coordinate chart around each point x° of N and the coordinate charts thus defined form a C^∞ atlas.

A manifold of this type is sometimes called a smooth *hypersurface* in $\mathbf{R}^m$. An important example of hypersurface is the *sphere* S^{m-1}, defined by taking $n = m - 1$ and

$$\lambda_1 = x_1^2 + x_2^2 + \ldots + x_m^2 - 1.$$

The set of points of $\mathbf{R}^m$ on which $\lambda_1(x) = 0$ consists of all the points on a sphere of radius 1 centered at the origin. Since

$$\left[\dfrac{\partial \lambda_1}{\partial x_1} \quad \cdots \quad \dfrac{\partial \lambda_1}{\partial x_m} \right]$$

never vanishes on this set, the required conditions are satisfied and the set is a smooth manifold, of dimension $m - 1$. •

Example. An open subset N' of a smooth manifold N is itself a smooth manifold. The topology of N' is the subset topology. If (U, ϕ) is a coordinate chart of a complete C^∞ atlas of N, such that $U \cap N' \neq \emptyset$, then the pair (U', ϕ') defined as

$$U' = U \cap N'$$

$$\phi' = \text{restriction of } \phi \text{ to } U'$$

is a coordinate chart of N'. In this way, one may define a complete C^∞ atlas of N'. The dimension of N' is the same as that of N. •

Example. Let M and N be smooth manifolds, of dimension m and n. Then the cartesian product $M \times N$ is a smooth manifold. The topology of $M \times N$ is the product topology. If (U, ϕ) and (V, ψ) are coordinate charts of M and N, the pair $(U \times V, (\phi, \psi))$ is coordinate chart of $M \times N$. The dimension of $M \times N$ is clearly $m + n$.

An important example of this type of manifold is the *torus* $T^2 = S^1 \times S^1$, the cartesian product of two circles. •

Let λ be a real-valued function defined on a manifold N. If (U, ϕ) is a coordinate chart on N, the composed function

$$\hat{\lambda} = \lambda \circ \phi^{-1} : \phi(U) \to \mathbf{R}$$

taking, for each $p \in U$, the set of local coordinates $(x_1, \ldots, x_n)$ of p into the real number $\lambda(p)$, is called an *expression of λ in local coordinates.*

In practice, whenever no confusion arises, one often uses the same symbol λ to denote $\lambda \circ \phi^{-1}$, and write $\lambda(x_1, \ldots, x_n)$ to denote the value of λ at a point p of local coordinates $(x_1, \ldots, x_n)$.

If N and M are manifolds, of dimension n and m, $F : N \to M$ is a mapping, (U, ϕ) a coordinate chart on N and (V, ψ) a coordinate chart on M, the composed mapping

$$\hat{F} = \psi \circ F \circ \phi^{-1}$$

is called an expression of F in local coordinates. Note that this definition makes sense only if $F(U) \cap V \neq \emptyset$. If this is the case, then $\hat{F}$ is well defined for all n-tuples $(x_1, \ldots, x_n)$ whose image under $F \circ \phi^{-1}$ is a point in V.

Here again, one often uses F to denote $\psi \circ F \circ \phi^{-1}$, writes $y_i = f_i(x_1, \ldots, x_n)$ to denote the value of the i-th coordinate of $F(p)$, p being a point of local coordinates $(x_1, \ldots, x_n)$, and also

$$y = \begin{bmatrix} y_1 \\ \vdots \\ y_m \end{bmatrix} = \begin{bmatrix} f_1(x_1, \ldots, x_n) \\ \vdots \\ f_m(x_1, \ldots, x_n) \end{bmatrix} = F(x).$$

Definition. Let N and M be smooth manifolds. A mapping $F : N \to M$ is a *smooth* mapping if for each $p \in N$ there exist coordinate charts (U, ϕ) of N and (V, ψ) of M, with $p \in U$ and $F(p) \in V$, such that the expression of F in local coordinates is C^∞.

Remark. Note that the property of being smooth is independent of the choice of the coordinate charts on N and M. Different coordinate charts (U', ϕ') and

(V', ψ') are by definition C^∞ compatible with the former and

$$
\begin{aligned}
\hat{F}' &= \psi' \circ F \circ \phi'^{-1} \\
&= \psi' \circ \psi^{-1} \circ \psi \circ F \circ \phi^{-1} \circ \phi \circ \phi'^{-1} \\
&= (\psi' \circ \psi^{-1}) \circ \hat{F} \circ (\phi' \circ \phi^{-1})^{-1}
\end{aligned}
$$

being a composition of C^∞ functions is still C^∞. •

Definition. Let N and M be smooth manifolds, both of dimension n. A mapping $F : N \to M$ is a *diffeomorphism* if F is bijective and both F and F^{-1} are smooth mappings. Two manifolds N and M are *diffeomorphic* if there exists a diffeomorphism $F : N \to M$.

The *rank* of a mapping $F : N \to M$ at a point $p \in N$ is the rank of the jacobian matrix

$$
\begin{bmatrix}
\dfrac{\partial f_1}{\partial x_1} & \cdots & \dfrac{\partial f_1}{\partial x_n} \\
\cdots & \cdots & \cdots \\
\dfrac{\partial f_m}{\partial x_1} & \cdots & \dfrac{\partial f_m}{\partial x_n}
\end{bmatrix}
$$

at $x = \phi(p)$. It must be stressed that, although apparently dependent on the choice of local coordinates, the notion of rank thus defined is actually coordinate-independent. The reader may easily verify that the ranks of the jacobian matrices of two different expressions of F in local coordinates are equal.

Theorem. Let N and M be smooth manifolds both of dimension n. A mapping $F : N \to M$ is a diffeomorphism if and only if F is bijective, F is smooth and $\mathrm{rank}(F) = n$ at all points of N.

Remark. In some cases, the assumption that functions, mappings, etc. are C^∞, may be replaced by the stronger assumption that functions, mappings, etc. are analytic. In this way one may define the notion of analytic manifold, analytic mappings of manifolds, and so on. We shall make this assumption explicitly whenever needed. •

A.4 Submanifolds

Definitions. Let $F : N \to M$ be a smooth mapping of manifolds.
(i) F is an *immersion* if $\mathrm{rank}(F) = \dim(N)$ for all $p \in N$
(ii) F is an *univalent immersion* if F is an immersion and is injective
(iii) F is an *embedding* if F is an univalent immersion and the topology induced on $F(N)$ by the one of N coincides with the topology of $F(N)$ as a subset of M.

Remark. The mapping F, being smooth, is in particular a continuous mapping of topological spaces. Therefore (see Section **A**.2) the topology induced on $F(N)$ by the one of N may properly contain the topology of $F(N)$ as a subset of M. This motivates the definition (iii). ●

The difference between (i), (ii) and (iii) is clarified by the following examples.

Examples. Let $N = \mathbf{R}$ and $M = \mathbf{R}^2$. Let t denote a point in N and (x_1, x_2) a point in M. The mapping F is defined by (Fig. A.2)

$$x_1(t) = at - \sin t$$
$$x_2(t) = \cos t$$

and, then,

$$\text{rank}(F) = \text{rank} \begin{bmatrix} a - \cos t \\ - \sin t \end{bmatrix}.$$

If $a = 1$ this mapping is *not* an immersion because $\text{rank}(F) = 0$ at $t = 2k\pi$ (for any integer k).

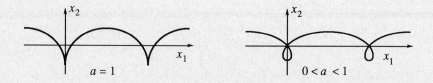

Fig. A.2

If $0 < a < 1$ the mapping is an immersion, because $\text{rank}(F) = 1$ for all t, but *not* an univalent immersion, because $F(t_1) = F(t_2)$ for all t_1, t_2 such that $t_1 = 2k\pi - \tau$, $t_2 = 2k\pi + \tau$ and $\sin \tau = a\tau$.

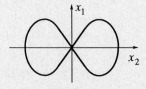

Fig. A.3

As a second example we consider the so-called "figure-eight" (Fig. A.3). Let N be the open interval $(0, 2\pi)$ of the real line and $M = \mathbf{R}^2$. Let t denote a point in N and (x_1, x_2) a point in M. The mapping F is defined by

$$x_1(t) = \sin 2t$$
$$x_2(t) = \sin t.$$

This mapping is an immersion because

$$\text{rank}(F) = \text{rank} \begin{bmatrix} \dfrac{dx_1}{dt} \\ \dfrac{dx_2}{dt} \end{bmatrix} = \text{rank} \begin{bmatrix} 2\cos 2t \\ \cos t \end{bmatrix} = 1$$

for all $0 < t < 2\pi$. It is also univalent because

$$F(t_1) = F(t_2) \Rightarrow t_1 = t_2$$

However, the mapping is *not* an embedding. For, consider the image of F. The mapping F takes the open set $(\pi - \varepsilon, \pi + \varepsilon)$ of N onto a subset U' of $F(N)$ which is open *by definition* in the topology induced by the one of N, but is not an open set in the topology of $F(N)$ as a subset of M. This is because U' cannot be seen as the intersection of $F(N)$ with an open set of $\mathbf{R}^2$.

As a third example one may consider the mapping $F : \mathbf{R} \to \mathbf{R}^3$ given by

$$x_1(t) = \cos 2\pi t$$
$$x_2(t) = \sin 2\pi t$$
$$x_3(t) = t$$

whose image is an "helix" winding on an infinite cylinder whose axis is the x_3 axis. The reader may easily check that this is an embedding. •

The following theorem shows that every immersion *locally* is an embedding.

Theorem. Let $F : N \to M$ be an immersion. For each $p \in N$ there exists a neighborhood U of p with the property that the restriction of F to U is an embedding.

Example. Consider again the "figure-eight" discussed above. If U is any interval of the type $(\delta, 2\pi - \delta)$, then the critical situation we had before disappears and the image U' of $(\pi - \varepsilon, \pi + \varepsilon)$ is now open also in the topology of $F(N)$ as a subset of $\mathbf{R}^2$. •

The notions of univalent immersion and of embedding are used in the following way.

Definition. The image $F(N)$ of a univalent immersion is called an *immersed submanifold* of M. The image $F(N)$ of an embedding is called an *embedded submanifold* of M.

Remark. Conversely, one may say that a subset M' of M is an immersed (respectively, embedded) submanifold of M if there is another manifold N and a univalent immersion (respectively, embedding) $F : N \to M$ such that $F(N) = M'$. ●

The use of the word "submanifold" in the above definition clearly indicates the possibility of giving $F(N)$ the structure of a smooth manifold, and this may actually be done in the following way. Let $M' = F(N)$ and $F' : N \to M'$ denote the mapping defined by

$$F'(p) = F(p)$$

for all $p \in N$. Clearly, F' is a bijection. If the topology of M' is the one induced by the one of N (i.e. open sets of M' are the images under F' of open sets of N), F' is a homeomorphism. Consequently, any coordinate chart (U, ϕ) of N induces a coordinate chart (V, ψ) of M', defined as

$$V = F'(U), \quad \psi = \phi \circ (F')^{-1}$$

C^∞-compatible charts of N induce C^∞-compatible charts of M' and so complete C^∞-atlases induce complete C^∞-atlases. This gives M' the structure of a smooth manifold.

The smooth manifold M' thus defined is *diffeomorphic* to the smooth manifold N. A diffeomorphism between M' and N is indeed F' itself, which is bijective, smooth and has rank equal to the dimension of N at each $p \in N$.

Embedded submanifolds can also be characterized in a different way, based on the following considerations.

Let M be a smooth manifold of dimension m and (U, ϕ) a cubic coordinate chart. Let n be an integer, $0 < n < m$, and p a point of U. The subset of U

$$S_p = \{q \in U : x_i(q) = x_i(p), i = n+1, \dots, m\}$$

is called an n-dimensional *slice* of U passing through p. In other words, a slice of U is the locus of all points of U for which some coordinates (e.g. the last $m - n$) are constant.

Theorem. Let M be a smooth manifold of dimension m. A subset M' of M is an embedded submanifold of dimension $n < m$ if and only if for each $p \in M'$ there exists a cubic coordinate chart (U, ϕ) of M, with $p \in U$, such that $U \cap M'$ coincides with an n-dimensional slice of U passing through p.

This theorem provides a more "intrinsic" characterization of the notion of an embedded submanifold (of a manifold M), directly related to the existence of special coordinate charts (of M). Note that, if (U, ϕ) is a coordinate chart of M

such that $U \cap M'$ is an n-dimensional slice of U, the pair (U', ϕ') defined as

$$U' = U \cap M'$$
$$\phi'(p) = (x_1(p), \ldots, x_n(p))$$

is a coordinate chart of M'. This is illustrated in Fig. A.4 (where $M = \mathbf{R}^3$ and $n = 2$).

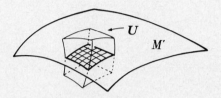

Fig. A.4

Remark. Note that an open subset M' of M is indeed an embedded submanifold of M, of the same dimension m. Thus, a submanifold M' of M may be a *proper* subset of M, although being a manifold of the same dimension. •

Remark. It can be proved that any smooth hypersurface in $\mathbf{R}^m$ is an embedded submanifold of $\mathbf{R}^m$. Moreover, it has also been shown that if N is an n-dimensional smooth manifold, there exist an integer $m \geq n$ and a mapping $F : N \to \mathbf{R}^m$ which is an embedding (Whitney's embedding theorem). In other words, any manifold is diffeomorphic to an embedded submanifold of $\mathbf{R}^m$, for a suitably large m. •

Remark. Let V be a n-dimensional subspace of $\mathbf{R}^m$. Any subset of $\mathbf{R}^m$ of the form

$$x^\circ + V = \{x \in \mathbf{R}^m : x = x' + x^\circ; x' \in V\}$$

where x° is some fixed point of $\mathbf{R}^m$, is indeed a smooth hypersurface and so an embedded submanifold of $\mathbf{R}^m$, of dimension n. This is sometimes called a *flat* submanifold of $\mathbf{R}^m$. •

A.5 Tangent Vectors

Let N be a smooth manifold of dimension n. A real-valued function λ is said to be *smooth in a neighborhood* of p, if the domain of λ includes an open set U of N containing p and the restriction of λ to U is a smooth function. The set of all smooth functions in a neighborhood of p is denoted $C^\infty(p)$. Note that $C^\infty(p)$

forms a vector space over the field $\mathbf{R}$. For, if λ, γ are functions in $C^\infty(p)$ and a, b are real numbers, the function $a\lambda + b\gamma$ defined as

$$(a\lambda + b\gamma)(q) = a\lambda(q) + b\gamma(q)$$

for all q in a neighborhood of p, is again a function in $C^\infty(p)$. Note also that two functions $\lambda, \gamma \in C^\infty(p)$ may be multiplied to give another element of $C^\infty(p)$, written $\lambda\gamma$ and defined as

$$(\lambda\gamma)(q) = \lambda(q) \cdot \gamma(q)$$

for all q in a neighborhood of p.

Definition. A *tangent vector* v at p is a map $v : C^\infty(p) \to \mathbf{R}$ with the following properties:

(i) (linearity): $v(a\lambda + b\gamma) = av(\lambda) + bv(\gamma)$ for all $\lambda, \gamma \in C^\infty(p)$ and $a, b \in \mathbf{R}$

(ii) (Leibnitz rule): $v(\lambda\gamma) = \gamma(p)v(\lambda) + \lambda(p)v(\gamma)$ for all $\lambda, \gamma \in C^\infty(p)$

Definition. Let N be a smooth manifold. The *tangent space* to N at p, written T_pN, is the set of all tangent vectors at p.

Remark. A map which satisfies the properties (i) and (ii) is also called a *derivation.* •

Remark. The set T_pN forms a vector space over the field $\mathbf{R}$ under the rules of scalar multiplication and addition defined in the following way. If v_1, v_2 are tangent vectors and c_1, c_2 real numbers, $c_1v_1 + c_2v_2$ is a new tangent vector which takes the function $\lambda \in C^\infty(p)$ into the real number

$$(c_1v_1 + c_2v_2)(\lambda) = c_1v_1(\lambda) + c_2v_2(\lambda).$$

Remark. We shall see later on that, if the manifold N is a smooth hypersurface in $\mathbf{R}^m$, the object previously defined may be naturally identified with the intuitive notion of "tangent hyperplane" at a point. •

Let (U, ϕ) be a (fixed) coordinate chart around p. With this coordinate chart one may associate n tangent vectors at p, denoted

$$\left(\frac{\partial}{\partial\phi_1}\right)_p, \ldots, \left(\frac{\partial}{\partial\phi_n}\right)_p$$

defined in the following way

$$\left(\frac{\partial}{\partial\phi_i}\right)_p(\lambda) = \left[\frac{\partial(\lambda \circ \phi^{-1})}{\partial x_i}\right]_{x=\phi(p)}$$

for $1 \leq i \leq n$. The right-hand side is the value taken at $x = (x_1, \ldots, x_n) = \phi(p)$ of the partial derivative of the function $\lambda \circ \phi^{-1}(x_1, \ldots, x_n)$ with respect to x_i (recall that the function $\lambda \circ \phi^{-1}$ is an expression of λ in local coordinates).

Theorem. Let N be a smooth manifold of dimension n. Let p be any point of N. The tangent space $T_p N$ to N at p is an n-dimensional vector space over the field $\mathbf{R}$. If (U, ϕ) is a coordinate chart around p, then the tangent vectors $\left(\frac{\partial}{\partial \phi_1}\right)_p, \ldots, \left(\frac{\partial}{\partial \phi_n}\right)_p$ form a basis of $T_p N$.

The basis $\left\{\left(\frac{\partial}{\partial \phi_1}\right)_p, \ldots, \left(\frac{\partial}{\partial \phi_n}\right)_p\right\}$ of $T_p N$ is sometimes called the *natural basis* induced by the coordinate chart (U, ϕ).

Let v be a tangent vector at p. From the above theorem it is seen that

$$v = \sum_{i=1}^{n} v_i \left(\frac{\partial}{\partial \phi_i}\right)_p$$

where $v_1, \ldots, v_n$ are real numbers. One may compute the v_i's explicitly in the following way. Let ϕ_i be the i-th coordinate function. Clearly $\phi_i \in C^{\infty}(p)$, and then

$$v(\phi_i) = \sum_{j=1}^{n} v_j \left(\frac{\partial}{\partial \phi_j}\right)_p (\phi_i)$$

$$= \sum_{j=1}^{n} v_j \left[\frac{\partial(\phi_i \circ \phi^{-1})}{\partial x_j}\right]_{x=\phi(p)} = v_i$$

because $\phi_i \circ \phi^{-1}(x_1, \ldots, x_n) = x_i$. Thus the real number v_i coincides with the value of v at ϕ_i, the i-th coordinate function.

A change of coordinates around p clearly induces a *change of basis* in $T_p N$. The computations involved are the following ones. Let (U, ϕ) and (V, ψ) be coordinate charts around p. Let $\left\{\left(\frac{\partial}{\partial \psi_1}\right)_p, \ldots, \left(\frac{\partial}{\partial \psi_n}\right)_p\right\}$ denote the natural basis of $T_p N$ induced by the coordinate chart (V, ψ). Then

$$\left(\frac{\partial}{\partial \psi_i}\right)_p (\lambda) = \left[\frac{\partial(\lambda \circ \psi^{-1})}{\partial y_i}\right]_{y=\psi(p)} = \left[\frac{\partial(\lambda \circ \phi^{-1} \circ \phi \circ \psi^{-1})}{\partial y_i}\right]_{y=\psi(p)}$$

$$= \sum_{j=1}^{n} \left[\frac{\partial(\lambda \circ \phi^{-1})}{\partial x_j}\right]_{x=\phi(p)} \cdot \left[\frac{\partial(\phi_j \circ \psi^{-1})}{\partial y_i}\right]_{y=\psi(p)}$$

$$= \sum_{j=1}^{n} \left[\left(\frac{\partial}{\partial \phi_j}\right)_p (\lambda)\right] \left[\frac{\partial(\phi_j \circ \psi^{-1})}{\partial y_i}\right]_{y=\psi(p)}.$$

In other words

$$\left(\frac{\partial}{\partial \psi_i}\right)_p = \sum_{j=1}^{n} \left[\frac{\partial(\phi_j \circ \psi^{-1})}{\partial y_i}\right]_{y=\psi(p)} \left(\frac{\partial}{\partial \phi_j}\right)_p.$$

Note that the quantity

$$\frac{\partial(\phi_j \circ \psi^{-1})}{\partial y_i}$$

is the element on the j-th row and i-th column of the *jacobian matrix* of the coordinates transformation

$$x = x(y).$$

So the elements of the columns of the jacobian matrix of $x = x(y)$ are the coefficients which express the vectors of the "new" basis as linear combinations of the vectors of the "old" basis.

If v is a tangent vector, and $(v_1, \ldots, v_n)$, $(w_1, \ldots, w_n)$ the n-tuples of real numbers which express v in the form

$$v = \sum_{i=1}^{n} v_i \left(\frac{\partial}{\partial \phi_i} \right)_p = \sum_{i=1}^{n} w_i \left(\frac{\partial}{\partial \psi_i} \right)_p$$

then

$$\begin{bmatrix} v_1 \\ \vdots \\ v_n \end{bmatrix} = \begin{bmatrix} \dfrac{\partial x_1}{\partial y_1} & \cdots & \dfrac{\partial x_1}{\partial y_n} \\ \cdots & \cdots & \cdots \\ \dfrac{\partial x_n}{\partial y_1} & \cdots & \dfrac{\partial x_n}{\partial y_n} \end{bmatrix} \begin{bmatrix} w_1 \\ \vdots \\ w_n \end{bmatrix}.$$

Definition. Let N and M be smooth manifolds. Let $F : N \to M$ be a smooth mapping. The *differential* of F at $p \in N$ is the map

$$F_\star : T_p N \to T_{F(p)} M$$

defined as follows. For $v \in T_p N$ and $\lambda \in C^\infty(F(p))$,

$$(F_\star(v))(\lambda) = v(\lambda \circ F)$$

Remark. $F_\star$ is a map of the tangent space of N at a point p into the tangent space of M at the point $F(p)$. If $v \in T_p N$, the value $F_\star(v)$ of $F_\star$ at v is a tangent vector in $T_{F(p)} M$. So one has to express the way in which $F_\star(v)$ maps the set $C^\infty(F(p))$, of all functions which are smooth in a neighborhood of $F(p)$, into $\mathbf{R}$. This is actually what the definition specifies. Note that there is one of such maps for *each* point p of N (see Fig. A.5). •

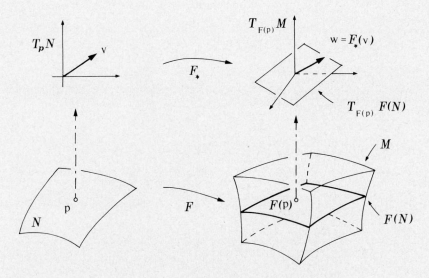

Fig. A.5

Theorem. The differential $F_\star$ is a linear map.

Since $F_\star$ is a linear map, given a basis for T_pN and a basis for $T_{F(p)}M$ one may wish to find its matrix representation. Let (U, ϕ) be a coordinate chart around p, (V, ψ) a coordinate chart around $q = F(p)$, $\left\{ \left(\frac{\partial}{\partial \phi_1}\right)_p, \ldots, \left(\frac{\partial}{\partial \phi_n}\right)_p \right\}$ the natural basis of T_pN and $\left\{ \left(\frac{\partial}{\partial \psi_1}\right)_q, \ldots, \left(\frac{\partial}{\partial \psi_m}\right)_q \right\}$ the natural basis of T_qM. In order to find a matrix representation of $F_\star$ one has simply to see how $F_\star$ maps $\left(\frac{\partial}{\partial \phi_i}\right)_p$ for each $1 \leq i \leq n$.

$$\left[F_\star \left(\frac{\partial}{\partial \phi_i} \right)_p \right] (\lambda) = \left(\frac{\partial}{\partial \phi_i} \right)_p (\lambda \circ F) = \left[\frac{\partial (\lambda \circ F \circ \phi^{-1})}{\partial x_i} \right]_{x = \phi(p)}$$

$$= \left[\frac{\partial (\lambda \circ \psi^{-1} \circ \psi \circ F \circ \phi^{-1})}{\partial x_i} \right]_{x = \phi(p)}$$

$$= \sum_{j=1}^{m} \left[\frac{\partial (\lambda \circ \psi^{-1})}{\partial y_j} \right]_{y = \psi(q)} \left[\frac{\psi_j \circ F \circ \phi^{-1})}{\partial x_i} \right]_{x = \phi(p)}$$

$$= \sum_{j=1}^{m} \left(\left(\frac{\partial}{\partial \psi_j} \right)_q (\lambda) \right) \left[\frac{\partial (\psi_j \circ F \circ \phi^{-1})}{\partial x_i} \right]_{x = \phi(p)}$$

in other words

$$F_\star \left(\frac{\partial}{\partial \phi_i} \right)_p = \sum_{j=1}^{m} \left[\frac{\partial (\psi_j \circ F \circ \phi^{-1})}{\partial x_i} \right]_{x = \phi(p)} \left(\frac{\partial}{\partial \psi_j} \right)_q .$$

Now, recall that $\psi \circ F \circ \phi^{-1}$ is an expression of F in local coordinates. Then, the quantity

$$\frac{\partial(\psi_j \circ F \circ \phi^{-1})}{\partial x_i}$$

is the element on the j-th row and i-th column of the *jacobian matrix* of the mapping expressing F in local coordinates. Using again

$$F(x) = F(x_1, \ldots, x_n) = \begin{bmatrix} F_1(x_1, \ldots, x_n) \\ \vdots \\ F_m(x_1, \ldots, x_n) \end{bmatrix}$$

to denote $\psi \circ F \circ \phi^{-1}$, one has simply

$$F_\star \left(\frac{\partial}{\partial \phi_i} \right)_p = \sum_{j=1}^m \left[\frac{\partial F_j}{\partial x_i} \right] \left(\frac{\partial}{\partial \psi_j} \right)_q.$$

If $v \in T_p N$ and $w = F_\star(v) \in T_{F(p)} M$ are expressed as

$$v = \sum_{i=1}^n v_i \left(\frac{\partial}{\partial \phi_i} \right)_p, \qquad w = \sum_{i=1}^m w_i \left(\frac{\partial}{\partial \psi_i} \right)_q$$

then

$$\begin{bmatrix} w_1 \\ \vdots \\ w_m \end{bmatrix} = \begin{bmatrix} \dfrac{\partial F_1}{\partial x_1} & \cdots & \dfrac{\partial F_1}{\partial x_n} \\ \cdots & \cdots & \cdots \\ \dfrac{\partial F_m}{\partial x_1} & \cdots & \dfrac{\partial F_m}{\partial x_n} \end{bmatrix} \begin{bmatrix} v_1 \\ \vdots \\ v_n \end{bmatrix}.$$

Remark. The matrix representation of $F_\star$ is exactly the jacobian of its expression in local coordinates. From this, it is seen that the rank of a mapping coincides with the rank of the corresponding differential. •

Remark (chain rule). It is easily seen that, if F and G are smooth mappings, then

$$(G \circ F)_\star = G_\star F_\star. \quad •$$

The following examples may clarify the notion of tangent space and the one of differential.

Example. The tangent vectors on $\mathbf{R}^n$. Let $\mathbf{R}^n$ be equipped with the "natural" complete atlas already considered in previous examples (i.e. the one including the chart $(\mathbf{R}^n$, identity map on $\mathbf{R}^n))$. Then, if v is a tangent vector at a point

x and λ a smooth function

$$v(\lambda) = \sum_{i=1}^{n} v_i \left(\frac{\partial}{\partial x_i} \right)_x (\lambda) = \sum_{i=1}^{n} \left[\frac{\partial \lambda}{\partial x_i} \right]_x v_i.$$

So, $v(\lambda)$ is just the value of the *derivative* of λ *along the direction* of the vector

$$\mathrm{col}(v_1, \ldots, v_n)$$

at the point x. •

Remark. Let $F : N \to M$ be a univalent immersion. Let $n = \dim(N)$ and $m = \dim(M)$. By definition, $F_\star$ has rank n at each point. Therefore the image $F_\star(T_p N)$ of $F_\star$, at each point p, is a subspace of $T_{F(p)} M$ isomorphic to $T_p N$. The subspace $F_\star(T_p N)$ can actually be *identified* with the tangent space at $F(p)$ to the submanifold $M' = F(N)$. In order to understand this point, let F' denote the function $F' : N \to M'$ defined as

$$F'(p) = F(p)$$

for all $p \in N$. F' is a diffeomorphism and so $F'_\star$ is an isomorphism. Therefore the image $F'_\star(T_p N)$ is exactly the tangent space at $F'(p)$ to M'. Any tangent vector in $T_{F(p)} M'$ is the image $F'_\star(v)$ of a (unique) vector $v \in T_p N$ and can be identified with the (unique) vector $F_\star(v)$ of $F_\star(T_p N)$.

In other words, the tangent space at p to a *submanifold* M' of M can be identified with a *subspace* of the tangent space at p to M.

The same considerations can be repeated in local coordinates. It is known that an immersion is locally an embedding. Therefore, around every point $p \in M'$ it is possible to find a coordinate chart (U, ϕ) of M, with the property that the pair (U', ϕ') defined by

$$U' = \{ q \in U : \phi_i(q) = \phi_i(p), i = n+1, \ldots, m \}$$
$$\phi' = (\phi_1, \ldots, \phi_n)$$

is a coordinate chart of M'. According to this choice, the tangent space to M' at p is identified with the n-dimensional subspace of $T_p M$ spanned by the tangent vectors $\left\{ \left(\frac{\partial}{\partial \phi_1} \right)_p, \ldots, \left(\frac{\partial}{\partial \phi_n} \right)_p \right\}$. •

Example. The tangent vector to a smooth curve in $\mathbf{R}^n$. We define first the notion of a smooth curve in $\mathbf{R}^n$. Let $N = (t_1, t_2)$ be an open interval on the real line. A *smooth curve* in $\mathbf{R}^n$ is the image of a univalent immersion $\sigma : N \to \mathbf{R}^n$. Thus, a smooth curve is an immersed submanifold of $\mathbf{R}^n$. In N and $\mathbf{R}^n$ one may choose natural local coordinates as usual and, letting t denote an element of N, express σ by means of an n-tuple of real-valued functions $\sigma_1, \ldots, \sigma_n$ of t.

A smooth curve is a 1-dimensional immersed submanifold of $\mathbf{R}^n$. At a point $\sigma(t_o)$, the tangent space to the curve is a 1-dimensional vector space which, as we have seen, may be identified with a subspace of the tangent space to $\mathbf{R}^n$ at this point. A basis of the tangent space to the curve at $\sigma(t_o)$ is given by the image under $\sigma_\star$ of $\left(\frac{d}{dt}\right)_{t_o}$, a tangent vector at t_o to N. This image is computed as follows

$$\sigma_\star(\frac{d}{dt})_{t_o} = \sum_{i=1}^{n} (\frac{d\sigma_i}{dt})_{t_o} \left(\frac{\partial}{\partial x_i}\right)_{\sigma(t_o)}.$$

Thinking of $t \in N$ as time and $\sigma(t)$ as a point moving in $\mathbf{R}^n$, we may interpret the vector

$$\text{col}((\frac{d\sigma_1}{dt})_{t_o}, \ldots, (\frac{d\sigma_n}{dt})_{t_o})$$

as the velocity along the curve, evaluated at the point $\sigma(t_o)$. So, we have that the velocity vector at a point of the curve spans the tangent space to the curve at this point. From this point of view, we see that the notion of tangent space to a 1-dimensional manifold may be identified with the geometric notion of tangent line to a curve in a Euclidean space (Fig. A.6). •

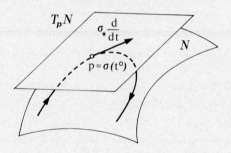

Fig. A.6

Example. Let h be a smooth function $h : \mathbf{R}^2 \rightarrow \mathbf{R}$ and $F : \mathbf{R}^2 \rightarrow \mathbf{R}^3$ a mapping defined by

$$F(x_1, x_2) = (x_1, x_2, h(x_1, x_2)).$$

This mapping is an embedding and therefore $F(\mathbf{R}^2)$, a surface in $\mathbf{R}^2$, is an embedded submanifold of $\mathbf{R}^3$. At each point $F(x)$ of this surface, the tangent space, identified as a subspace of the tangent space to $\mathbf{R}^3$ at this point, may be computed as

$$\text{span}\{F_\star\left(\frac{\partial}{\partial x_1}\right)_x, F_\star\left(\frac{\partial}{\partial x_2}\right)_x\}.$$

Now,

$$F_\star\left(\frac{\partial}{\partial x_1}\right)_x = \sum_{i=1}^{3}\left[\frac{\partial F_i}{\partial x_1}\right]\left(\frac{\partial}{\partial x_i}\right)_{F(x)} = \left(\frac{\partial}{\partial x_1}\right)_{F(x)} + \left[\frac{\partial h}{\partial x_1}\right]\left(\frac{\partial}{\partial x_3}\right)_{F(x)}$$

$$F_\star\left(\frac{\partial}{\partial x_2}\right)_x = \left(\frac{\partial}{\partial x_2}\right)_{F(x)} + \left[\frac{\partial h}{\partial x_2}\right]\left(\frac{\partial}{\partial x_3}\right)_{F(x)}$$

The tangent space to $F(\mathbf{R}^2)$ at some point $(x_1^o, x_2^o, h(x_1^o, x_2^o))$ is the set of tangent vectors whose expressions in local coordinates are from the form

$$v = \begin{bmatrix} \alpha \\ \beta \\ (\frac{\partial h}{\partial x_1})\alpha + (\frac{\partial h}{\partial x_2})\beta \end{bmatrix}$$

α, β being real numbers and $\frac{\partial h}{\partial x_1}$, $\frac{\partial h}{\partial x_2}$ being evaluated at $x_1 = x_1^o$ and $x_2 = x_2^o$. From this point of view, we see that the notion of tangent space to a 2-dimensional manifold may be identified with the geometric notion of tangent plane to a surface in a Euclidean space. ●

One may define objects dual to the ones considered so far.

Definition. Let N be a smooth manifold. The *cotangent space* to N at p, written $T_p^\star N$, is the dual space of $T_p N$. Elements of the cotangent space are called *tangent covectors*.

Remark. Recall that a dual space $V^\star$ of a vector space V is the space of all linear functions from V to $\mathbf{R}$. If $v^\star \in V^\star$, then $v^\star : V \to \mathbf{R}$ and the value of $v^\star$ at $v \in V$ is written as $\langle v^\star, v \rangle$. $V^\star$ forms a vector space over the field $\mathbf{R}$, with rules of scalar multiplication and addition which define $c_1 v_1^\star + c_2 v_2^\star$ in the following terms

$$\langle c_1 v_1^\star + c_2 v_2^\star, v \rangle = c_1 \langle v_1^\star, v \rangle + c_2 \langle v_2^\star, v \rangle.$$

If $e_1, \ldots, e_n$ is a basis of V, the unique basis $e_1^\star, \ldots, e_n^\star$ of $V^\star$ which satisfies

$$\langle e_i^\star, e_j \rangle = \delta_{ij}$$

is called a *dual basis*.

If V and W are vector spaces, $F : V \to W$ a linear mapping, $v \in V$ and $w^\star \in W^\star$, the mapping $F^\star : W^\star \to V^\star$ defined by

$$\langle F^\star(w^\star), v \rangle = \langle w^\star, F(v) \rangle$$

is called the *dual* mapping (of F). ●

Let λ be a smooth function $\lambda : N \to \mathbf{R}$. There is a natural way of identifying the differential λ_* of λ at p with an element of $T_p^* N$. For, observe that λ^* is a linear mapping

$$\lambda_* : T_p N \to T_{\lambda(p)} \mathbf{R}$$

and that $T_{\lambda(p)} \mathbf{R}$ is isomorphic to $\mathbf{R}$. The natural isomorphism between $\mathbf{R}$ and $T_{\lambda(p)} \mathbf{R}$ is the one in which the element c of $\mathbf{R}$ corresponds to the tangent vector $c \left(\frac{d}{dt} \right)_t$. If $c \left(\frac{d}{dt} \right)_t$ is the value at v of the differential λ_* at p, then c must depend linearly on v, i.e. there must exist a covector, denoted $(d\lambda)_p$, such that

$$\lambda_*(v) = \langle (d\lambda)_p, v \rangle \left(\frac{d}{dt} \right)_t.$$

Given a basis of $T_p N$, the covector $(d\lambda)_p$ (like any other covector), may be represented in matrix form. Let $\{ \left(\frac{\partial}{\partial \phi_1} \right)_p, \ldots, \left(\frac{\partial}{\partial \phi_n} \right)_p \}$ be the natural basis of $T_p N$ induced by the coordinate chart (U, ϕ). The image under λ_* of a vector

$$v = \sum_{i=1}^{n} v_i \left(\frac{\partial}{\partial \phi_i} \right)_p$$

is the vector

$$\lambda_* = \left(\sum_{i=1}^{n} \frac{\partial \lambda}{\partial x_i} v_i \right) \left(\frac{d}{dt} \right)_t$$

and this shows that

$$\langle (d\lambda)_p, v \rangle = \begin{bmatrix} \dfrac{\partial \lambda}{\partial x_1} & \cdots & \dfrac{\partial \lambda}{\partial x_n} \end{bmatrix} \begin{bmatrix} v_1 \\ \vdots \\ v_n \end{bmatrix}.$$

Remark. Note also that the value at λ of a tangent vector v is equal to the value at v of the tangent covector $(d\lambda)_p$, i.e.

$$v(\lambda) = \langle (d\lambda)_p, v \rangle. \; \bullet$$

The dual basis of $\{ \left(\frac{\partial}{\partial \phi_1} \right)_p, \ldots, \left(\frac{\partial}{\partial \phi_n} \right)_p \}$ is computed as follows. From the equality $v(\lambda) = \langle (d\lambda)_p, v \rangle$ we deduce that

$$\langle (d\phi_i)_p, \left(\frac{\partial}{\partial \phi_j} \right)_p \rangle = \left(\frac{\partial}{\partial \phi_j} \right)_p (\phi_i) = \frac{\partial (\phi_i \circ \phi^{-1})}{\partial x_j} = \frac{\partial x_i}{\partial x_j} = \delta_{ij}$$

so that the desired dual basis is exactly provided by the set of tangent covectors $\{ (d\phi_1)_p, \ldots, (d\phi_n)_p \}$.

If $v^\star$ is any tangent covector, expressed as

$$v_\star = \sum_{i=1}^{n} v_i^\star (d\phi_i)_p$$

the real numbers $v_1^\star, \ldots, v_n^\star$ are such that

$$v_i^\star = \langle v^\star, \left(\frac{\partial}{\partial \phi_i}\right)_p \rangle.$$

Note also that, if v is any tangent vector expressed as

$$v = \sum_{i=1}^{n} v_i \left(\frac{\partial}{\partial \phi_i}\right)_p$$

the real numbers $v_1, \ldots, v_n$ are such that

$$v_i = \langle (d\phi_i)_p, v \rangle$$

A.6 Vector Fields

Definition. Let N be a smooth manifold, of dimension n. A *vector field* f on N is a mapping assigning to each point $p \in N$ a tangent vector $f(p)$ in $T_p N$. A vector field f is *smooth* if for each $p \in N$ there exists a coordinate chart (U, ϕ) about p and n real-valued smooth functions $f_1, \ldots, f_n$ defined on U such that, for all $q \in U$

$$f(q) = \sum_{i=1}^{n} f_i(q) \left(\frac{\partial}{\partial \phi_i}\right)_q.$$

Remark. Because of C^∞-compatibility of coordinate charts, given any coordinate chart (V, ψ) about p other than (U, ϕ), one may find a neighborhood $V' \subset V$ of p and n real-valued smooth functions $f_1', \ldots, f_n'$ defined on V', such that, for all $q \in V'$

$$f(q) = \sum_{i=1}^{n} f_i'(q) \left(\frac{\partial}{\partial \psi_i}\right)_q.$$

Thus, the notion of smooth vector field is independent of the coordinates used. •

Remark. If (U, ϕ) is a coordinate chart of N, on the *submanifold* U of N one may define a special set of smooth vector fields, denoted $\left(\frac{\partial}{\partial \phi_1}\right), \ldots, \left(\frac{\partial}{\partial \phi_n}\right)$ in the following way

$$\left(\frac{\partial}{\partial \phi_i}\right) : p \mapsto \left(\frac{\partial}{\partial \phi_i}\right)_p.$$

It must be stressed, however, that such a set of vector fields is an object defined only in U. •

For any fixed coordinate chart (U, ϕ), the set of tangent vectors $\{(\frac{\partial}{\partial \phi_1})_q, \ldots, (\frac{\partial}{\partial \phi_n})_q\}$ is a basis of $T_q N$ at each $q \in U$ and, therefore, there is a *unique* set of smooth functions $\{f_1, \ldots, f_n\}$ that makes it possible to express the value of a vector field f at q in the form

$$f(q) = \sum_{i=1}^{n} f_i(q) \left(\frac{\partial}{\partial \phi_i}\right)_q.$$

Expressing each f_i in local coordinates, as

$$\hat{f}_i = f_i \circ \phi^{-1}$$

provides an expression in local coordinates of the vector field f itself. So, if p is a point of coordinates $(x_1, \ldots, x_n)$ in the chart (U, ϕ), $f(p)$ is a tangent vector of coefficients $(\hat{f}_1(x_1, \ldots, x_n), \ldots, \hat{f}_n(x_1, \ldots, x_n))$ in the natural basis $\{(\frac{\partial}{\partial \phi_1})_p, \ldots, (\frac{\partial}{\partial \phi_n})_p\}$ of $T_p N$ induced by (U, ϕ). Most of the times, whenever possible, the symbol f_i replaces $f_i \circ \phi^{-1}$ and the expression of f in local coordinates is given a form of an n-vector $f = \mathrm{col}(f_1, \ldots, f_n)$.

Remark. Let f be a smooth vector field, (U, ϕ) and (V, ψ) two coordinate charts about p and $f(x) = f(x_1, \ldots, x_n)$, $f'(y) = f'(y_1, \ldots, y_n)$ the corresponding expressions of f in local coordinates. Then

$$f'(y) = \left[\frac{\partial y}{\partial x} f(x)\right]_{x=x(y)}. \quad •$$

The notion of vector field makes it possible to introduce the concept of *differential equation on a manifold* N. For, let f be a smooth vector field. A smooth curve $\sigma : (t_1, t_2) \to N$ is an *integral curve* of f if

$$\sigma_\star \left(\frac{d}{dt}\right)_t = f(\sigma(t))$$

for all $t \in (t_1, t_2)$. The left-hand side is a tangent vector to the submanifold $\sigma((t_1, t_2))$ at the point $\sigma(t)$; the right-hand side is a tangent vector to N at $\sigma(t)$. As usual, we identify the tangent space to a submanifold of N at a point with a subspace of the tangent space to N at this point.

In local coordinates, $\sigma(t)$ is expressed as an n-tuple $(\sigma_1(t), \ldots, \sigma_n(t))$, and $f(\sigma(t))$ as

$$f(\sigma(t)) = \sum_{i=1}^{n} f_i(\sigma_1(t), \ldots, \sigma_n(t)) \left(\frac{\partial}{\partial \phi_i}\right)_{\sigma(t)}.$$

Moreover

$$\sigma_\star\Big(\frac{d}{dt}\Big)_t = \sum_{i=1}^n \frac{d\sigma_i}{dt}\Big(\frac{\partial}{\partial\phi_i}\Big)_{\sigma(t)}.$$

Therefore, the expression of σ in local coordinates is such that

$$\frac{d\sigma_i}{dt} = f_i(\sigma_1(t),\ldots,\sigma_n(t))$$

for all $1 \le i \le n$. This shows that the notion of integral curve of a vector field corresponds to the notion of solution of a set of n ordinary differential equations of the first order.

For this reason one often uses the notation

$$\dot\sigma(t) = \sigma_\star\Big(\frac{d}{dt}\Big)_t$$

to indicate the image of $\big(\frac{d}{dt}\big)_t$ under the differential $\sigma_\star$ at t.

The following theorem contains all relevant informations about the properties of integral curves of vector fields.

Theorem. Let f be a smooth vector field on a manifold N. For each $p \in N$ there exists an open interval—depending on p and written I_p— of $\mathbf{R}$ such that $0 \in I_p$ and a smooth mapping

$$\Phi : W \to N$$

defined on the subset W of $\mathbf{R} \times N$

$$W = \{(t,p) \in \mathbf{R} \times N : t \in I_p\}$$

with the following properties
 (i) $\Phi(0,p) = p$,
(ii) for each p the mapping $\sigma_p : I_p \to N$ defined by

$$\sigma_p(t) = \Phi(t,p)$$

is an integral curve of f,
(iii) if $\mu : (t_1,t_2) \to N$ is another integral curve of f satisfying the condition $\mu(0) = p$, then $(t_1,t_2) \subset I_p$ and the restriction of σ_p to (t_1,t_2) coincides with μ,
(iv) $\Phi(s,(t,p)) = \Phi(s+t,p)$ whenever both sides are defined,
 (v) whenever $\Phi(t,p)$ is defined, there exists an open neighborhood U of p such that the mapping $\Phi_t : U \to N$ defined by

$$\Phi_t(q) = \Phi(t,q)$$

is a diffeomorphism onto its image, and

$$\Phi_t^{-1} = \Phi_{-t}.$$

Remark. Properties (i) and (ii) say that σ_p is an integral curve of f passing through p at $t = 0$. Property (iii) says that this curve is unique and that the domain I_p on which σ_p is defined is maximal. Properties (iv) and (v) say that the family of mappings $\{\Phi_t\}$ is a one-parameter (namely, the parameter t) group of local diffeomorphisms, under the operation of composition. •

Example. Let $N = \mathbf{R}$ and use x to denote a point in $\mathbf{R}$. Consider the vector field

$$f(x) = (x^2 + 1)\left(\frac{\partial}{\partial x}\right)_x$$

so

$$\frac{d\sigma}{dt} = \sigma^2 + 1$$

A solution of this equation has the form

$$\sigma(t) = \tan(t + \arctan(x^\circ))$$

with x° being indeed the value of σ at $t = 0$. Clearly, for each x° the solution is defined for

$$-\frac{\pi}{2} < t + \arctan(x^\circ) < \frac{\pi}{2}.$$

Thus W is the set

$$W = \{(t, x^\circ) : t \in (-\frac{\pi}{2} - \arctan(x^\circ), \frac{\pi}{2} - \arctan(x^\circ))\}$$

which has the form indicated in Fig. A.7. •

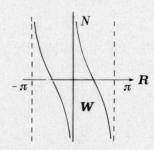

Fig. A.7

The mapping Φ is called the *flow* of f. Often, for practical purposes, the notation Φ_t replaces Φ, with the understanding that t is a variable. To stress the dependence on f, sometimes Φ_t is written as Φ_t^f.

Definition. A vector field f is *complete* if, for all $p \in N$, the interval I_p coincides with $\mathbf{R}$, i.e.—in other words—if the flow Φ of f is defined on the whole cartesian product $\mathbf{R} \times N$.

The integral curves of a complete vector field are thus defined, whatever the initial point p is, for all $t \in \mathbf{R}$.

Definition. Let f be a smooth vector field on N and λ a smooth real-valued function on N. The *derivative of λ along f* is a function $N \to \mathbf{R}$, written $L_f\lambda$ and defined as

$$(L_f\lambda)(p) = (f(p))(\lambda)$$

(i.e. $(L_f\lambda)(p)$ is the value at λ of the tangent vector $f(p)$ at p).

The function $L_f\lambda$ is a smooth function. In local coordinates, $L_f\lambda$ is represented by

$$(L_f\lambda)(x_1,\ldots,x_n) = \left[\frac{\partial\lambda}{\partial x_1} \quad \cdots \quad \frac{\partial\lambda}{\partial x_n}\right] \begin{bmatrix} f_1 \\ \vdots \\ f_n \end{bmatrix}.$$

If f_1, f_2 are vector fields and λ a real-valued function, we denote

$$L_{f_1}L_{f_2}\lambda = L_{f_1}(L_{f_2}\lambda).$$

The set of all smooth vector fields on a manifold N is denoted by the symbol $V(N)$. This set is a *vector space* over $\mathbf{R}$ since if f, g are vector fields and a, b are real numbers, their linear combination $af + bg$ is a vector field defined by

$$(af + bg)(p) = af(p) + bg(p).$$

If a, b are smooth real-valued functions on N, one may still define a linear combination $af + bg$ by

$$(af + bg)(p) = a(p)f(p) + b(p)g(p)$$

and this gives $V(N)$ the structure of a *module* over the ring, denoted $C^\infty(N)$, of all smooth real-valued functions defined on N. The set $V(N)$ can be given, however, a more interesting algebraic structure in this way.

Definition. A vector space V over $\mathbf{R}$ is a *Lie algebra* if in addition to its vector space structure it is possible to define a binary operation $V \times V \to V$, called a product and written $[\cdot, \cdot]$, which has the following properties

(i) it is skew commutative, i.e.

$$[v, w] = -[w, v]$$

(ii) it is bilinear over **R**, i.e.

$$[\alpha_1 v_1 + \alpha_2 v_2, w] = \alpha_1 [v_1, w] + \alpha_2 [v_2, w]$$

(where α_1, α_2 are real numbers)
(iii) it satisfies the so called *Jacobi identity*, i.e.

$$[v, [w, z]] + [w, [z, v]] + [z, [v, w]] = 0.$$

The set $V(N)$ forms a Lie algebra with the vector space structure already discussed and a product $[\cdot, \cdot]$ defined in the following way. If f and g are vector fields, $[f, g]$ is a new vector field whose value at p, a tangent vector in $T_p N$, maps $C^\infty(p)$ into **R** according to the rule

$$([f, g](p))(\lambda) = (L_f L_g \lambda)(p) - (L_g L_f \lambda)(p).$$

In other words, $[f, g](p)$ takes λ into the real number $(L_f L_g \lambda)(p) - (L_g L_f \lambda)(p)$. Note that one may write more simply

$$L_{[f,g]} \lambda = L_f L_g \lambda - L_g L_f \lambda.$$

Theorem. $V(M)$ with the product $[f, g]$ thus defined is a Lie algebra.

The product $[f, g]$ is called the *Lie bracket* of the two vector fields f and g.
The reader may easily find that the expression of $[f, g]$ in local coordinates is given by the n-vector

$$\begin{bmatrix} \dfrac{\partial g_1}{\partial x_1} & \cdots & \dfrac{\partial g_1}{\partial x_n} \\ \cdots & \cdots & \cdots \\ \dfrac{\partial g_n}{\partial x_1} & \cdots & \dfrac{\partial g_n}{\partial x_n} \end{bmatrix} \begin{bmatrix} f_1 \\ \vdots \\ f_n \end{bmatrix} - \begin{bmatrix} \dfrac{\partial f_1}{\partial x_1} & \cdots & \dfrac{\partial f_1}{\partial x_n} \\ \cdots & \cdots & \cdots \\ \dfrac{\partial f_n}{\partial x_1} & \cdots & \dfrac{\partial f_n}{\partial x_n} \end{bmatrix} \begin{bmatrix} g_1 \\ \vdots \\ g_n \end{bmatrix} = \dfrac{\partial g}{\partial x} f - \dfrac{\partial f}{\partial x} g.$$

If, in particular, $N = \mathbf{R}^n$ and

$$f(x) = Ax \qquad g(x) = Bx$$

then

$$[f, g](x) = (BA - AB)x.$$

The matrix $[A, B] = (BA - AB)$ is called the *commutator* of A, B.

The importance of the notion of Lie bracket of vector fields is very much related to its applications in the study of nonlinear control systems. For the moment, we give hereafter two interesting properties.

Theorem. Let N' be an embedded submanifold of N. Let U' be an open set of N' and f, g two smooth vector fields of N such that for all $p \in U'$

$$f(p) \in T_p N' \qquad \text{and} \qquad g(p) \in T_p N'$$

Then also

$$[f, g](p) \in T_p N'$$

for all $p \in U'$.

In other words, the Lie bracket of two vector fields "tangent" to a fixed submanifold is still tangent to that submanifold.

Theorem. Let f, g be two smooth vector fields on N. Let Φ_t^f denote the flow of f. For each $p \in N$

$$\lim_{t \to 0} \frac{1}{t} \left[(\Phi_{-t}^f)_\star g(\Phi_t^f(p)) - g(p) \right] = [f, g](p).$$

Remark. The first term of the expression under bracket is a tangent vector at p, obtained in the following way. With p, the mapping Φ_t^f (always defined for sufficiently small t) associates a point $q = \Phi_t^f(p)$. The vector field g is evaluated at q and the value $g(q) \in T_q N$ is taken back to $T_p N$ via the differential $(\Phi_{-t}^f)_\star$ (which maps the tangent space at q onto the tangent space at $p = \Phi_{-t}^f(q)$). •

Let f be a smooth vector field on N, g a smooth vector field on M and $F : N \to M$ a smooth function. The vector fields f, g are said to be F-related if

$$F_\star f = g \circ F.$$

Note that the vector field $(\Phi_{-t}^f)_\star g(\Phi_t^f(p))$ considered in the above Remark is Φ_{-t}^f-related to g.

Remark. If $\bar{f}$ is F-related to f and $\bar{g}$ is F-related to g, then $[\bar{f}, \bar{g}]$ is F-related to $[f, g]$. •

Remark. The Lie bracket of g and f may be interpreted as the value at $t = 0$ of the derivative with respect to t of a function defined as

$$W(t) = (\Phi_{-t}^f)_\star g(\Phi_t^f(p)) - g(p).$$

Moreover, it is easily seen that for any $k \geq 0$

$$\left(\frac{d^k W(t)}{dt^k}\right)_{t=0} = ad_f^k g(p).$$

If $W(t)$ is analytic in a neighborhood of $t = 0$, then $W(t)$ can be expanded in the form

$$W(t) = \sum_{k=0}^{\infty} ad_f^k g(p) \frac{t^k}{k!}$$

known as the *Campbell-Baker-Hausdorff formula*. •

One may define an object which dualizes the notion of a vector field.

Definition. Let N be a smooth manifold of dimension n. A *covector field* (also called *one-form*) ω on N is a mapping assigning to each point $p \in N$ a tangent covector $\omega(p)$ in $T_p^* N$. A covector field f is *smooth* if for each $p \in N$ there exists a coordinate chart (U, ϕ) about p and n real-valued smooth functions $\omega_1, \ldots, \omega_n$ defined on U, such that, for all $q \in U$

$$\omega(q) = \sum_{i=1}^{n} \omega_i(q)(d\phi_i)_q.$$

The notion of smooth covector field is clearly independent of the coordinates used. The expression of a covector field in local coordinates is often given the form of a row vector $\omega = \text{row}(\omega_1, \ldots, \omega_n)$ in which the ω_i's are real-valued functions of $x_1, \ldots, x_n$.

If ω is a covector field and f is a vector field, $\langle \omega, f \rangle$ denotes the smooth real-valued function defined by

$$\langle \omega, f \rangle(p) = \langle \omega(p), f(p) \rangle.$$

With any smooth function $\lambda : N \to \mathbf{R}$ one may associate a covector field by taking at each p the cotangent vector $(d\lambda)_p$. The covector field thus defined is usually still represented by the symbol $d\lambda$. However, the converse is *not always true*.

Definition. A covector field ω is *exact* if there exists a smooth real-valued function $\lambda : N \to \mathbf{R}$ such that

$$\omega = d\lambda.$$

The set of all smooth covector fields on a manifold N is denoted by the symbol $V^*(N)$.

One may also define the notion of derivative of a covector field ω along a vector field f. In order to do this, one has to introduce first the notion of a covector field ω. Let p be a point of the domain of Φ_t^f. Recall that $(\Phi_t^f)_* : T_p N \to T_{\Phi_t^f(p)} N$

is a linear mapping and let $(\Phi_t^f)^\star : T_{\Phi_t^f(p)}^\star \to T_p^\star N$ denote the dual mapping. With ω and Φ_t^f we associate a new covector field whose value at a point p in the domain of Φ_t^f is defined by

$$(\Phi_t^f)^\star \omega(\Phi_t^f(p)).$$

The covector field thus defined is said to be Φ_t^f-related to ω.

Theorem. Let f be a smooth vector field and ω a smooth covector field on N. For each $p \in N$ the limit

$$\lim_{t \to 0} \frac{1}{t} \big[(\Phi_t^f)^\star \omega(\Phi_t^f(p)) - \omega(p) \big]$$

exists.

Definition. The *derivative of ω along f* is a covector field on N, written $L_f\omega$, whose value at p is set equal to the value of the limit

$$\lim_{t \to 0} \frac{1}{t} \big[(\Phi_t^f)^\star \omega(\Phi_t^f(p)) - \omega(p) \big].$$

The expression of $L_f\omega$ in local coordinates is given by the (row) n-vector

$$[f_1 \quad \cdots \quad f_n] \begin{bmatrix} \dfrac{\partial \omega_1}{\partial x_1} & \cdots & \dfrac{\partial \omega_n}{\partial x_1} \\ \cdots & \cdots & \cdots \\ \dfrac{\partial \omega_1}{\partial x_n} & \cdots & \dfrac{\partial \omega_n}{\partial x_n} \end{bmatrix} + [\omega_1 \quad \cdots \quad \omega_n] \begin{bmatrix} \dfrac{\partial f_1}{\partial x_1} & \cdots & \dfrac{\partial f_1}{\partial x_n} \\ \cdots & \cdots & \cdots \\ \dfrac{\partial f_n}{\partial x_1} & \cdots & \dfrac{\partial f_n}{\partial x_n} \end{bmatrix} =$$

$$= \left[\frac{\partial \omega^T}{\partial x} f \right]^T + \omega \frac{\partial f}{\partial x}$$

where the superscript "T" denotes "transpose".

Appendix B

B.1 Center Manifold Theory

Consider a nonlinear system

$$\dot{x} = f(x) \tag{1.1}$$

where f is a C^r vector field ($r \geq 2$) defined on an open subset U of $\mathbf{R}^n$, and let $x^\circ \in U$ be a point of *equilibrium* for f, i.e. a point such that $f(x^\circ) = 0$. Without loss of generality we may assume $x^\circ = 0$. It is well known that the (local) asymptotic stability of this point can be determined, to some extent, by the behavior of the linear approximation of f at $x = 0$. For, let

$$F = \left[\frac{\partial f}{\partial x} \right]_{x=0}$$

denote the jacobian matrix of f at $x = 0$. Then

(i) if all the eigenvalues of F are in the (open) left-half complex plane, then $x = 0$ is an asymptotically stable equilibrium of (1.1),

(ii) if one or more eigenvalues of F are in the right-half complex plane, then $x = 0$ is an unstable equilibrium of (1.1).

This important result is commonly known as the *Principle of Stability in the First Approximation*. It is also well understood that this principle does not completely cover the analysis of the local stability of the equilibrium $x = 0$, because nothing can be inferred—in general—about the asymptotic properties of (1.1) when some eigenvalue of F has zero real part. The case of a system whose matrix F has some eigenvalue with zero real part is commonly referred to as a *critical case* of the asymptotic analysis.

In this Section we describe an interesting set of results—known as Center Manifold Theory—that in many instances is of great help in analyzing critical cases. We begin with some definitions.

Definition. A C^r submanifold S of U is said to be *locally invariant* for (1.1), if for each $x^\circ \in S$, there exist $t_1 < 0 < t_2$ such that the integral curve $x(t)$ of (1.1) satisfying $x(0) = x^\circ$ is such that $x(t) \in S$ for all $t \in (t_1, t_2)$.

Suppose the matrix F has n° eigenvalues with zero real part, n^- eigenvalues with negative real part and n^+ eigenvalues with positive real part. Then, it is well known from linear algebra that the domain of the linear mapping F can be decomposed into the direct sum of three invariant subspaces, noted E°,

E^-, E^+, of dimension n^o, n^-, n^+ respectively. If the linear mapping F is viewed as a representation of the differential (at $x = 0$) of the nonlinear mapping $f : x \in U \rightarrow f(x) \in \mathbf{R}^n$, its domain is the tangent space $T_o U$ to U at $x = 0$, and the three subspaces in question can be viewed as subspaces of $T_o U$ satisfying

$$T_o U = E^o \oplus E^- \oplus E^+.$$

Definition. Let $x = 0$ be an equilibrium of (1.1). A manifold S, passing through $x = 0$, is said to be a *center manifold* for (1.1) at $x = 0$, if it is locally invariant and the tangent space to S at 0 is exactly E^o.

In what follows, we will consider only cases in which the matrix F has all eigenvalues with nonpositive real part, because these are the only cases in which $x = 0$ can be a stable equilibrium. In any of these cases, one can always choose coordinates in U such that the system (1.1) is represented in the form

$$\dot{y} = Ay + g(y, z) \tag{1.2a}$$
$$\dot{z} = Bz + f(y, z) \tag{1.2b}$$

where A is an $(n^- \times n^-)$ matrix having all eigenvalues with negative real part, B is a $(n^o \times n^o)$ matrix having all eigenvalues with zero real part, and the functions g and f are C^r functions vanishing at $(y, z) = (0, 0)$ together with all their first order derivatives. In fact, it suffices to expand the rigth-hand side of (1.1) in the form

$$f(x) = Fx + \tilde{f}(x)$$

where $\tilde{f}(x)$ vanishes at $x = 0$ together with all its first order derivatives, and then to reduce F to a block diagonal form

$$TFT^{-1} = \begin{bmatrix} A & 0 \\ 0 & B \end{bmatrix}$$

by means of a linear change of coordinates

$$\begin{bmatrix} y \\ z \end{bmatrix} = Tx.$$

We shall henceforth consider only systems in the form (1.2). Existence of center manifolds for (1.2) is illustrated in the following statement.

Theorem. There exist a neighborhood $V \subset \mathbf{R}^{n^o}$ of $z = 0$ and a C^r mapping $\pi : V \rightarrow \mathbf{R}^{n^-}$ such that

$$S = \{(y, z) \in (\mathbf{R}^{n^-}) \times V : y = \pi(z)\}$$

is a center manifold for (1.2).

By definition, a center manifold for the system (1.2) passes through $(0,0)$ and is tangent to the subset of points whose y coordinate is 0. Thus, the mapping π satisfies

$$\pi(0) = 0 \qquad \frac{\partial \pi}{\partial z}(0) = 0. \qquad (1.3a)$$

Moreover, this manifold is locally invariant for (1.2), and this imposes on the mapping π a constraint that can be easily deduced in the following way. Let $(y(t), z(t))$ be a solution curve of (1.2) and suppose this curve belongs to the manifold S, i.e. is such that $y(t) = \pi(z(t))$. Differentiating this with respect to time we obtain the relation

$$\frac{dy}{dt} = A\pi(z(t)) + g(\pi(z(t)), z(t)) = \frac{\partial \pi}{\partial z}\frac{dz}{dt} = \frac{\partial \pi}{\partial z}(Bz(t) + f(\pi(z(t)), z(t))).$$

Since a relation of this type must be satisfied for any solution curve of (1.2) contained in S, we conclude that the mapping π satisfies the partial differential equation

$$\frac{\partial \pi}{\partial z}(Bz + f(\pi(z), z)) = A\pi(z) + g(\pi(z), z). \qquad (1.3b)$$

The previous statement describes the existence, but not the uniqueness of a center manifold for (1.2). As a matter of fact, a system may have many center manifolds, as the following example shows.

Example. Consider the system

$$\dot{y} = -y$$
$$\dot{z} = -z^3.$$

The C^∞ function $y = \pi(z)$ defined as

$$\begin{cases} \pi(z) = c\exp(-\tfrac{1}{2}z^{-2}) & \text{if } z \neq 0 \\ \pi(z) = 0 & \text{if } z = 0 \end{cases}$$

is a center manifold for every value of $c \in \mathbf{R}$. •

Note also that if g and f are C^∞ functions, the system (1.2) has a C^k center manifold for any $k \geq 2$, but not necessarily a C^∞ center manifold.

Lemma. Suppose $y = \pi(z)$ is a center manifold for (1.2) at $(0,0)$. Let $(y(t), z(t))$ be a solution of (1.2). There exist a neighborhood U° of $(0,0)$ and real numbers $M > 0$, $K > 0$ such that, if $(y(0), z(0)) \in U^\circ$, then

$$\| y(t) - \pi(z(t)) \| \leq Me^{-Kt} \| y(0) - \pi(z(0)) \|$$

for all $t \geq 0$, as long as $(y(t), z(t)) \in U^\circ$.

This Lemma shows that any trajectory of the system (1.2) starting at a point sufficiently close to $(0,0)$, i.e. close to the point at which the center manifold has been defined, converges to the center manifold as t tends to ∞, with exponential decay (Fig. B.1). In particular, this shows that if (y°, z°) is an equilibrium point of (1.2) sufficiently close to $(0,0)$, then this point must belong to any center manifold for (1.2) passing through $(0,0)$. In fact, in this case the solution curve of (1.2) satisfying $(y(0), z(0)) = (y^\circ, z^\circ)$ is such that

$$y(t) = y^\circ \qquad z(t) = z^\circ \qquad \text{for all } t \geq 0$$

and this is compatible with the estimate given by the Lemma only if $y^\circ = \pi(z^\circ)$. For the same reasons, if Γ is a periodic orbit of (1.2) all contained in a sufficiently small neighborhood of $(0,0)$, then Γ must lie on any center manifold for (1.2) at $(0,0)$. Thus, despite of the non uniqueness of center manifolds, there are points that must always belong to any center manifold.

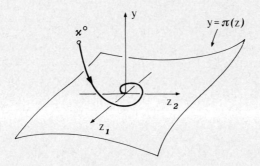

Fig. B.1

The following theorem provides a more detailed picture about the role of the center manifold in the analysis of the asymptotic properties of the system (1.2) near $(0,0)$. Recall that, by definition, if $(y(0), z(0))$ is any initial condition on the center manifold $y = \pi(z)$, then necessarily $y(t) = \pi(z(t))$ for all t in a neighborhood of $t = 0$. As a consequence, any trajectory of (1.2) starting at a point $y^\circ = \pi(z^\circ)$ of this center manifold can be described in the form

$$y(t) = \pi(\zeta(t)) \qquad z(t) = \zeta(t)$$

where $\zeta(t)$ is the solution of the differential equation

$$\dot{\zeta} = B\zeta + f(\pi(\zeta), \zeta) \tag{1.4}$$

satifying the initial condition $\zeta(0) = z^\circ$. The essence of the following results is that the asymptotic behavior of (1.2)—for small initial conditions—is completely determined by its behavior for initial conditions *on the center manifold*, i.e. by the asymptotic behavior of (1.4).

Theorem (reduction principle). Suppose $\zeta = 0$ is a stable (resp. asymptotically stable, unstable) equilibrium of (1.4). Then $(y, z) = (0, 0)$ is a stable (resp. asymptotically stable, unstable) equilibrium of (1.2).

Example. As an immediate application of the reduction principle to the analysis of critical cases, consider a system of the form (1.2), with g such that

$$g(0, z) = 0.$$

In this case, the center manifold equation (1.3b) is trivially solved by $\pi(z) = 0$, and the reduction principle establishes that the stability properties of (1.2) at $(0, 0)$ can be completely determined from those of the reduced system

$$\dot{\zeta} = B\zeta + f(0, \zeta). \bullet$$

This Theorem is rather important, for it reduces the stability analysis of an n-dimensional system to that of a lower dimensional (namely, n°-dimensional) one, but its practical application requires solving the center manifold equation, and this—in cases other than the one illustrated in the previous example—is in general quite difficult. It is however always possible to approximate the solution $y = \pi(z)$ of the equation (1.3b) to any required degree of accuracy and, then, to use the approximate solution thus found in the reduced equation (1.4). In this way, one may still be able to determine the asymptotic properties of the equilibrium $\zeta = 0$ of (1.4)

Theorem. Let $y = \pi_k(z)$ be a polynomial of degree k, $1 < k < r$, satisfying

$$\pi_k(0) = 0 \qquad \frac{\partial \pi_k}{\partial z}(0) = 0$$

and suppose

$$\frac{\partial \pi_k}{\partial z}(Bz + f(\pi_k(z), z)) - A\pi_k(z) - g(\pi_k(z), z) = R_k(z)$$

where $R_k(z)$ is some (possibly unknown) function vanishing at 0 together with all partial derivatives of order less than or equal to k. Then, any solution $\pi(z)$ of the center manifold equation (1.3b) is such that the difference

$$D_k(z) = \pi(z) - \pi_k(z)$$

vanishes at 0 together with all partial derivatives of order less than or equal to k.

The practical application of this result is illustrated in the following examples. In all of them, the reduced equation (1.4) is 1-dimensional, and its stability can be easily determined on the basis of the following property.

Proposition. Consider the one-dimensional system

$$\dot{x} = ax^m + Q_m(x)$$

with $m \geq 2$, and $Q_m(x)$ a function vanishing at 0 together with all partial derivatives of order less than or equal to m. The point of equilibrium $x = 0$ is asymptotically stable if m is odd and $a < 0$. The equilibrium is unstable if m is odd and $a > 0$, or if m is even.

Example. Consider the system

$$\dot{y} = -y + z^2$$
$$\dot{z} = azy.$$

The center manifold equation (1.3b) is in this case

$$\frac{\partial \pi}{\partial z}(az\pi(z)) = -\pi + z^2.$$

The simplest approximation we may try for $\pi(z)$ is a polynomial of the second order, namely $\pi_2(z) = \alpha z^2$, where α must be such as to satisfy the center manifold equation at least up to terms of order 2. This yields

$$\frac{\partial \pi_2}{\partial z}(az\pi_2(z)) - (-\pi_2(z) + z^2) = (\alpha - 1)z^2 + a\alpha^2 z^4.$$

Setting $\alpha = 1$ we obtain, on the right-hand side of this expression, a remainder $R_3(z)$ that vanishes at 0 together with all derivatives of order less than or equal to 3. We may thus set

$$\pi(z) = z^2 + D_3(z)$$

where $D_3(z)$ is some unspecified function of z (vanishing at 0 together with all derivatives of order less than or equal to 3). Replacing $\pi(z)$ in (1.4), we obtain

$$\dot{\zeta} = a\zeta^3 + Q_4(\zeta)$$

where $Q_4(\zeta)$ is an unknown remainder, vanishing at 0 together with all the derivatives of order less than or equal to 4. On the basis of the previous Proposition we deduce that the equilibrium $\zeta = 0$ of the reduced equation (1.4) is asymptotically stable if and only if $a < 0$. At this point, on the basis of the Reduction Principle, we can conclude that the full system is asymptotically stable at the equilibrium $(y, z) = (0, 0)$ if and only if $a < 0$. •

Example. Consider the system

$$\dot{y} = -y + y^2 - z^3$$
$$\dot{z} = az^3 + z^m y$$

where m is any positive integer. The center manifold equation (1.3b) is in this case

$$\frac{\partial \pi}{\partial z}(az^3 + z^m \pi(z)) = -\pi(z) + \pi^2(z) - z^3.$$

Again, we start trying for $\pi(z)$ an approximation of the second order, namely $\pi_2(z) = \alpha z^2$, with α such as to satisfy the center manifold equation at least up to terms of order 2. However, since,

$$\frac{\partial \pi_2}{\partial z}(az^3 + z^m \pi_2(z)) - (-\pi_2(z) + \pi_2^2(z) - z^3) = \alpha z^2 + R_2(z)$$

we deduce that necessarily $\alpha = 0$. Thus an approximation of the second order is meaningless, and we have to try with a polynomial of the third order . We set $\pi_3(z) = \beta z^3$ because we already know that the coefficient of z^2 must be zero. In this case we have

$$\frac{\partial \pi_3}{\partial z}(az^3 + z^m \pi_3(z)) - (-\pi_3(z) + \pi_3^2(z) - z^3) = (\beta + 1)z^3 + R_3(z)$$

and the center manifold equation (1.3b) will be satisfied up to terms of order 3 if $\beta = -1$. Thus we may set

$$\pi(z) = -z^3 + D_3(z).$$

Replacing $\pi(\zeta)$ in (1.4), we obtain

$$\dot{\zeta} = a\zeta^3 - \zeta^{m+3} + Q_{m+3}(\zeta)$$

where $Q_{m+3}(\zeta)$ is an unknown remainder, vanishing at 0 together with all the derivatives of order less than or equal to $m + 3$. Since $m \geq 1$, we can invoke the previous Proposition and conclude that, for any m, the equation (1.4) is asymptotically stable at $\zeta = 0$ if and only if $a < 0$. As a consequence of the Reduction Principle, this is true also for the equilibrium $(y, z) = (0, 0)$ of the full system. •

Example. Consider the system

$$\dot{y} = -y + ayz + bz^2$$
$$\dot{z} = cyz - z^3.$$

Again, we try first an approximation for $\pi(z)$ of the form $\pi_2(z) = \alpha z^2$. In this case we find

$$\frac{\partial \pi_2}{\partial z}(cz\pi_2(z) - z^3) - (-\pi_2(z) + az\pi_2(z) + bz^2) = (\alpha - b)z^2 + R_2(z)$$

and therefore $\alpha = b$. Replacing

$$\pi(\zeta) = b\zeta^2 + D_2(\zeta)$$

in the equation (1.4), we obtain

$$\dot{\zeta} = (cb - 1)\zeta^3 + Q_3(\zeta)$$

Again, on the basis of the previous Proposition, we can conclude that the reduced equation—and so the full system—is asymptotically stable if $(cb - 1) < 0$, and unstable if $(cb - 1) > 0$. If $cb = 1$, the right-hand side of this equation is totally unspecified, and thus we have to find a better approximation for the center manifold. Choosing $\pi_3(z) = bz^2 + \beta z^3$, we find now

$$\frac{\partial \pi_3}{\partial z}(cz\pi_3(z) - z^3) - (-\pi_3(z) + az\pi_3(z) + bz^2) = (ab - \beta)z^3 + R_3(z)$$

and so $\beta = ab$. Replacing

$$\pi(\zeta) = b\zeta^2 + ab\zeta^3 + D_3(\zeta)$$

in (1.4) we obtain (assuming $cb = 1$)

$$\dot{\zeta} = a\zeta^4 + Q_4(\zeta)$$

and we can conclude that the system is unstable if $a \neq 0$. If $a = 0$ we don't known yet, because the right-hand side of this equation is unspecified Thus, the only case left is the one in which $cb = 1$ and $a = 0$. In this particular situation, however, the center manifold equation (1.3b) is satisfied *exactly* by the function $\pi(z) = bz^2$ and the reduced system is then

$$\dot{\zeta} = 0.$$

Its equilibrium $\zeta = 0$ is stable (not asymptotically) and so is the equilibrium of the full system. ●

Exercise. Consider the system

$$\dot{y} = az + u(y)$$
$$\dot{z} = -z^3 + byz^m$$

where $m \geq 0$, and $u(y)$ represents a feedback, depending on the state variable y only. Choose

$$u(y) = -Ky$$

and show that the equilibrium $(y, z) = (0,0)$ of the full system
- if $m = 0$ and $ab < 0$, is asymptotically stable for all values of $K > 0$,
- if $m = 1$, is always unstable,
- if $m = 2$, is asymptotically stable for all values of $K > \max(0, ab)$,
- if $m \geq 3$, is asymptotically stable for all values of $K > 0$.

Show that these conclusions remain unchanged if

$$u(y) = Ky + f(y)$$

where $f(y)$ is a function of y vanishing at 0 together with its first derivative. •

B.2 Some Useful Lemmas

We present in this Section some interesting results about the asymptotic properties of certain nonlinear systems, that are used several times throughout the text.

Lemma. Consider a system

$$\dot{z} = f(z, y) \tag{2.1a}$$
$$\dot{y} = Ay + p(z, y) \tag{2.1b}$$

and suppose that $p(z, 0) = 0$ for all z near 0 and

$$\frac{\partial p}{\partial y}(0, 0) = 0.$$

If $\dot{z} = f(z, 0)$ has an asymptotically stable equilibrium at $z = 0$ and the eigenvalues of A all have negative real part, then the system (2.1) has an asymptotically stable equilibrium at $(z, y) = (0, 0)$.

Proof. Expand $f(z, y)$ as

$$f(z, y) = Fz + Gy + g(z, y).$$

Using a linear change of coordinates $(z_1, z_2) = Tz + Ky$ it is possible to rewrite the system (2.1) in the form

$$\dot{z}_1 = F_1 z_1 + g_1(z_1, z_2, y)$$
$$\dot{z}_2 = F_2 z_2 + G_2 y + g_2(z_1, z_2, y)$$
$$\dot{y} = Ay + p(z_1, z_2, y)$$

with F_2 having all the eigenvalues with negative real part and F_1 having all the eigenvalues with zero real part. Moreover, the functions g_1, g_2 vanish at $(0,0,0)$ together with their first-order partial derivatives.

By assumption, the equilibrium $(0,0)$ of

$$\dot{z}_1 = F_1 z_1 + g_1(z_1, z_2, 0) \tag{2.2a}$$
$$\dot{z}_2 = F_2 z_2 + g_2(z_1, z_2, 0) \tag{2.2b}$$

is asymptotically stable. Let $z_2 = \pi_2(z_1)$ be a center manifold for (2.2) at $(0,0)$. By assumption, π_2 satisfies

$$\left[\frac{\partial \pi_2}{\partial z_1}\right](F_1 z_1 + g_1(z_1, \pi_2(z_1), 0)) = F_2 \pi_2(z_1) + g_2(z_1, \pi_2(z_1), 0)$$

and then—by the reduction principle—the reduced dynamics

$$\dot{x} = F_1 x + g_1(x, \pi_2(x), 0)$$

has necessarily an asymptotically stable equilibrium at $x = 0$. Consider now the full system (2.1). A center manifold for this system is a pair

$$z_2 = k_2(z_1), \qquad y = k_1(z_1)$$

such that

$$\left[\frac{\partial k_2}{\partial z_1}\right](F_1 z_1 + g_1(z_1, k_2(z_1), k_1(z_1))) = F_2 k_2(z_1) + G_2 k_1(z_1) +$$
$$+ g_2(z_1, k_2(z_1), k_1(z_1))$$
$$\left[\frac{\partial k_1}{\partial z_1}\right](F_1 z_1 + g_1(z_1, k_2(z_1), k_1(z_1))) = A k_1(z_1) + p(z_1, k_2(z_1), k_1(z_1)).$$

A trivial calculation shows that these equations are solved by

$$k_2(z_1) = \pi_2(z_1), \qquad k_1(z_1) = 0$$

As a consequence, using again the reduction principle, we see that the dynamics (2.1) has an asymptotically stable equilibrium at $(0,0)$ if the reduced dynamics

$$\dot{x} = F_1 x + g_1(x, \pi_2(x), 0)$$

has. But this reduced dynamics is exactly the reduced dynamics of (2.2) and the claim follows. ●

Remark. We stress that the result of this Lemma requires, for the dynamics of

$$\dot{z} = f(z, 0)$$

just asymptotic stability, and not necessarily asymptotic stability *in the first approximation*, i.e. a jacobian matrix

$$\left[\frac{\partial f(z, 0)}{\partial z}\right]_{z=0}$$

having all the eigenvalues in the open left-half plane. •

In the next Lemma the asymptotic properties of a *time-varying* system are illustrated. To this end, recall that the equilibrium $x = 0$ of a time-varying system

$$\dot{x} = f(x, t) \tag{2.3}$$

is said to be *uniformly stable* if, for all $\varepsilon > 0$, there exist a $\delta > 0$ (possibly dependent on ε but independent of t°) such that

$$\| x^\circ \| < \delta \Rightarrow \| x(t, t^\circ, x^\circ) \| < \varepsilon \text{ for all } t \geq t^\circ \geq 0$$

where $x(t, t^\circ, x^\circ)$ denotes the solution of (2.3) satisfying $x(t^\circ, t^\circ, x^\circ) = x^\circ$. The equilibrium $x = 0$ of (2.3) is said to be *uniformly asymptotically stable*, if it is uniformly stable and, in addition, there exist $\gamma > 0$ and, for all $M > 0$, a $T > 0$ (possibly dependent on M but independent of x° and t°) such that

$$\| x^\circ \| < \gamma \Rightarrow \| x(t, t^\circ, x^\circ) \| < M \text{ for all } t \geq t^\circ + T, \ t^\circ \geq 0.$$

Lemma. Consider the system

$$\dot{x} = f(x, t) + p(x, t). \tag{2.4}$$

Suppose the equilibrium $x = 0$ of $\dot{x} = f(x, t)$ is uniformly asymptotically stable. Suppose $f(x, t)$ is locally Lipschitzian in x, uniformly with respect to t, i.e. there exists L (independent of t) such that

$$\| f(x', t) - f(x'', t) \| < L(\| x' - x'' \|)$$

for all x', x'' in a neighborhood of $x = 0$ and all $t \geq 0$. Then, for all $\varepsilon > 0$, there exist $\delta_1 > 0$ and $\delta_2 > 0$ (both δ_1 and δ_2 possibly depend on ε but are independent of t°) such that, if $\| x^\circ \| < \delta_1$ and $\| p(x, t) \| < \delta_2$ for all (x, t) such that $\| x \| < \varepsilon$ and $t \geq t^\circ$, the solution $x(t, t^\circ, x^\circ)$ of (2.4) satisfies

$$\| x(t, t^\circ, x^\circ) \| < \varepsilon \text{ for all } t \geq t^\circ \geq 0.$$

The property expressed by this statement is sometimes referred to as *total stability*, or *stability under persistent disturbances*. Note that the function $p(x, t)$ need not to be zero for $x = 0$. From this Lemma it is easy to deduce some applications of interest for systems in triangular form.

Corollary. Consider the system

$$\dot{z} = q(z, y, t) \tag{2.5a}$$
$$\dot{y} = g(y). \tag{2.5b}$$

Suppose
(i) $(z, y) = (0, 0)$ is an equilibrium of (2.5), and the function $q(z, y, t)$ is locally lipschitzian in (z, y), uniformly with respect to t, i.e. there exists L (independent of t) such that

$$\| q(z', y', t) - q(z'', y'', t) \| < L(\| z' - z'' \| + \| y' - y'' \|)$$

for all z', z'' in a neighborhood of $z = 0$, all y', y'' in a neighborhood of $y = 0$, and all $t \geq 0$,
(ii) the equilibrium $z = 0$ of $\dot{z} = q(z, 0, t)$ is uniformly asymptotically stable,
(iii) the equilibrium $y = 0$ of (2.5b) is stable.
Then the equilibrium $(z, y) = (0, 0)$ of (2.5) is uniformly stable.

Proof. Is a simple consequence of the previous Lemma. For, set

$$f(z, t) = q(z, 0, t)$$
$$p(z, t) = q(z, y(t), t) - q(z, 0, t)$$

where $y(t)$ is the solution of (2.5b) satisfying $y(t^\circ) = y^\circ$. Thus (2.5a) has the form (2.4). Note that if $\| y(t) \| < \varepsilon_y$ for all $t \geq t^\circ$, then, by assumption (i), $p(z, t)$ satisfies

$$\| p(z, t) \| = \| q(z, y(t), t) - q(z, 0, t) \| < L\varepsilon_y$$

for all z in a neighborhood of $z = 0$ and all $t \geq t^\circ$. By assumption (ii) and the previuos Lemma, for all $\varepsilon_z > 0$, there exists $\delta_1 > 0$ and $\delta_2 > 0$, such that $\| z^\circ \| < \delta_1$ and $\| p(z, t) \| < \delta_2$, for all (z, t) such that $\| z \| < \varepsilon_z$ and $t \geq t^\circ$, imply

$$\| z(t, t^\circ, z^\circ) \| < \varepsilon_z \text{ for all } t \geq t^\circ \geq 0.$$

By assumption (iii) one can find δ_y such that $\| y^\circ \| < \delta_y$ implies $\| y(t) \| < \delta_2/L$ for all $t \geq t^\circ$, and this completes the proof. •

Remark. This result has an obvious counterpart in the study of the *stability of* (nonequilibrium) *solutions* of a differential equation. To this end, recall that a solution $x^*(t)$ (defined for all $t \geq 0$) of a differential equation of the form (2.3) is said to be *uniformly stable* if, for all $\varepsilon > 0$, there exists a $\delta > 0$ (possibly dependent on ε but independent of t°) such that

$$\| x^\circ - x^*(t^\circ) \| < \delta \Rightarrow \| x(t, t^\circ, x^\circ) - x^*(t) \| < \varepsilon \text{ for all } t \geq t^\circ \geq 0$$

where $x(t, t^\circ, x^\circ)$ denotes the solution of (2.3) satisfying $x(t^\circ, t^\circ, x^\circ) = x^\circ$. The solution $x^*(t)$ of (2.3) is said to be *uniformly asymptotically stable*, if it is uniformly stable and, in addition, there exist $\gamma > 0$ and, for all $M > 0$, a $T > 0$ (possibly dependent on M but independent of x° and t°) such that

$$\| x^\circ - x^*(t^\circ) \| < \gamma \Rightarrow \| x(t, t^\circ, x^\circ) - x^*(t) \| < M \text{ for all } t \geq t^\circ + T, t^\circ \geq 0. \quad \bullet$$

The study of the stability of a solution $x^*(t)$ of (2.3) can be reduced to the study of the stability of the equilibrium of a suitable differential equation. For, it suffices to set

$$w = x - x^*$$

and discuss the stability of the equilibrium $w = 0$ of

$$\dot{w} = f(w + x^*(t), t) - f(x^*(t), t)$$

Thus, the previous Corollary is helpful also in determining the uniform stability of some nonequilibrium solution of equations of the form (2.5). For instance, suppose $\dot{z} = q(z, 0, t)$ has a uniformly asymptotically stable solution $z^*(t)$, defined for all t. Set

$$F(w, y, t) = q(w + z^*(t), y, t) - q(z^*(t), 0, t)$$

Then $w = 0$ is a uniformly asymptotically stable equilibrium of $\dot{w} = F(w, 0, t)$. If $q(z, y, t)$ is locally lipschitzian in (z, y), uniformly in t, so is $F(w, y, t)$, provided that $z^*(t)$ is sufficiently small for all $t \geq 0$. Assumptions (i), (ii) and (iii) are satisfied, and it is possible to conclude that the solution $(z^*(t), 0)$ of (2.5) is uniformly stable.

If the system (2.5) is time invariant, then the result of the previous Corollary can be expressed in a simpler form.

Corollary. Consider the system

$$\dot{z} = q(z, y) \tag{2.6a}$$

$$\dot{y} = g(y). \tag{2.6b}$$

Suppose $(z, y) = (0, 0)$ is an equilibrium of (2.6), the equilibrium $z = 0$ of $\dot{z} = q(z, 0)$ is asymptotically stable, the equilibrium $y = 0$ of (2.6b) is stable.

Then the equilibrium $(z, y) = (0, 0)$ of (2.6) is stable.

Another interesting application of the previous Lemma is the one described in the following statement.

Corollary. Consider the system

$$\dot{x} = f(x) + g(x)u(t). \tag{2.7}$$

Suppose $x = 0$ is an asymptotically stable equilibrium of $\dot{x} = f(x)$. Then, for all $\varepsilon > 0$ there exist $\delta_1 > 0$ and $K > 0$ such that, if $\| x^\circ \| < \delta_1$ and $|u(t)| < K$ for all $t \geq t^\circ$, the solution $x(t, t^\circ, x^\circ)$ of (2.7) satisfies

$$\| x(t, t^\circ, x^\circ) \| < \varepsilon \text{ for all } t \geq t^\circ \geq 0$$

Proof. Since $g(x)$ is smooth, there exists a real number $M > 0$ such that $\| g(x) \| < M$ for all x such that $\| x \| < \varepsilon$. Choosing $K = \delta_2/M$ yields $\| g(x)u(t) \| < \delta_2$ and the result follows from the Lemma.

B.3 Local Geometric Theory of Singular Perturbations

Consider a system of differential equations of the form

$$\varepsilon \dot{y} = g(y, z, \varepsilon) \tag{3.1a}$$
$$\dot{z} = f(y, z, \varepsilon) \tag{3.1b}$$

with (y, z) defined on an open subset of $\mathbf{R}^\nu \times \mathbf{R}^\mu$, and ε a small positive real parameter. A system of this type is called a *singularly perturbed system*. In fact, at $\varepsilon = 0$, this system degenerates to a set of only μ differential equations

$$\dot{z} = f(y, z, 0) \tag{3.2a}$$

subject to a constraint of the form

$$0 = g(y, z, 0). \tag{3.2b}$$

Let K denote the set of solution points of the equation (3.2b), and suppose

$$\mathrm{rank}\left(\frac{\partial g}{\partial y}\right) = n$$

at some point (y°, z°) of K. By the implicit function Theorem, there exist neighborhoods A° of z° and B° of y° and a unique smooth mapping $h : A^\circ \to B^\circ$, such that $g(h(z), z, 0) = 0$ for all $z \in A^\circ$. Therefore, locally around (y°, z°)

the degenerate system (3.2) is equivalent to a μ-dimensional differential system defined on the graph of the mapping h, i.e. on the set

$$S = \{(y, z) \in B^\circ \times A^\circ : y = h(z)\}$$

and thereby represented by the equation

$$\dot{z} = f(h(z), z, 0). \tag{3.3a}$$

This system is called the *reduced system*.

Note that, after a change of variables

$$w = y - h(z)$$

the set S can be identified as the set of pairs (w, z) such that $w = 0$. In the new variables (3.1) is represented by the system

$$\varepsilon \dot{w} = g(w + h(z), z, \varepsilon) - \varepsilon \frac{\partial h}{\partial z} f(w + h(z), z, \varepsilon) = g_0(w, z, \varepsilon)$$
$$\dot{z} = f(w + h(z), z, \varepsilon) = f_0(w, z, \varepsilon).$$

Since $g_0(0, z, 0) = 0$ by construction, the reduced system is now described by

$$\dot{z} = f_0(0, z, 0).$$

The form of the singularity of (3.1) suggests also a change of variable in the time axis, namely the replacement of t by a "rescaled" time variable τ defined as

$$\tau = t/\varepsilon.$$

Since ε is a small number, the variables t and τ are usually referred to as the "slow" time and "fast" time. Moreover, to indicate differentiation with respect to t, the superscript "'" is used.

The substitution of t by τ, together with that of y by w, yield a system of the form

$$w' = g_0(w, z, \varepsilon) \tag{3.4a}$$
$$z' = \varepsilon f_0(w, z, \varepsilon) \tag{3.4b}$$

in which, since $g_0(0, z, 0) = 0$, any point $(0, z)$ (i.e. each point of the set S) is an *equilibrium point* at $\varepsilon = 0$. Note that the behavior of the system (3.4) at $\varepsilon = 0$ is characterized by the family of ν-dimensional differential equations

$$w' = g_0(w, z, 0) \tag{3.3b}$$

(in which z can be regarded as a constant parameter).

The two equations (3.3a) and (3.3b) (which, it must be stressed, are defined on two different time axis) represent in some sense two kinds of "extreme" behaviors associated with the original system (3.1). The purpose of the singular perturbation theory is the study of the behavior of a singularly perturbed system for small (nonzero) values of ε and, if possible, to infer its asymptotic properties from the knowledge of the asymptotic behavior of the two "limit" systems (3.3a) and (3.3b).

Before proceeding further, it is convenient to observe that a system of the form (3.1) is a particular case of a more general class of systems that can be characterized in a coordinate-free manner, without explicitly asking for a separation of the variables into groups z and y. For, consider a system

$$x' = F(x, \varepsilon) \tag{3.5}$$

with x defined on an open set U of $\mathbf{R}^n$, ε a "parameter" ranging on an interval $(-\varepsilon_0, +\varepsilon_0)$ of $\mathbf{R}$ and $F : U \times (-\varepsilon_0, +\varepsilon_0) \to \mathbf{R}^n$ a C^r mapping. Suppose also there exists a μ-dimensional (with $\mu < n$) submanifold E of U consisting entirely of equilibrium points of $x' = F(x, 0)$, i.e. such that

$$F(x, 0) = 0 \text{ for all } x \in E.$$

The class of systems thus defined contains as a special case the system (3.1), with time rescaled; in fact, the set S is exactly a μ-dimensional submanifold of $\mathbf{R}^\nu \times \mathbf{R}^\mu$ consisting of equilibrium points of the rescaled system at $\varepsilon = 0$. In view of this fact, we shall henceforth proceed with the study of the more general class of systems (3.5).

Let

$$J_x = \left[\frac{\partial F(x, 0)}{\partial x} \right]$$

denote the Jacobian matrix of $F(x, 0)$ at a point x of E. It is easy to verify that the tangent space $T_x E$ to E at x is contained in the kernel of this matrix. For, let $\sigma : \mathbf{R} \to E$ be a smooth curve such that $\sigma(0) = x$ and note that, since every point of E annihilates $F(x, 0)$, by definition $F(\sigma(t), 0) = 0$ for all t. Differentiating this with respect to time yields

$$\left[\frac{\partial F(x, 0)}{\partial x} \right]_{x = \sigma(t)} \dot{\sigma}(t) = 0.$$

At $t = 0$, we have $J_x \dot{\sigma}(0) = 0$, and this, in view of the arbitrariness of σ, proves that $T_x E \subset \ker(J_x)$.

From this property, we deduce that 0 is an eigenvalue of J_x with multiplicity at least μ. The μ eigenvalues of J_x associated with the eigenvectors which span the subspace $T_x E$ are called the *trivial eigenvalues* of J_x, whereas the remaining $n - \mu$ are called the *nontrivial eigenvalues*.

From now on, we assume that all the nontrivial eigenvalues of J_x have negative real part. As a consequence, the two sets of trivial and nontrivial eigenvalues are disjoint sets and, from linear algebra, it follows that there exists a unique subspace of $T_x U$, noted V_x, which is invariant under J_x and complementary to $T_x E$, i.e. such that

$$T_x U = T_x E \oplus V_x.$$

As a matter of fact, V_x is exactly the subspace of $T_x U$ spanned by the eigenvectors associated with the nontrivial eigenvalues of J_x. Let P_x denote the projection of $T_x U$ onto $T_x E$ along V_x, i.e. the unique linear mapping satisfying

$$\ker(P_x) = V_x \qquad \text{and} \qquad \text{Im}(P_x) = T_x E$$

We use P_x to define a vector field on E. Namely, we set

$$f_R : x \in E \to f_R(x) = P_x \left[\frac{\partial F(x, \varepsilon)}{\partial \varepsilon} \right]_{\varepsilon = 0}.$$

This vector field is called the *reduced vector field* of the system (3.5). Note that this definition agrees with the one given at the beginning. As a matter of fact, if the system is in the special form (3.4), with $g_0(0, z, 0) = 0$, the jacobian matrix J_x has the form

$$J_x = \begin{bmatrix} G & 0 \\ 0 & 0 \end{bmatrix} \qquad \text{with} \quad G = \frac{\partial g_0}{\partial w}$$

and its nontrivial eigenvalues are those of G. The subspace $T_x E$ is the set of all vectors whose first ν coordinates are zero, the subspace V_x is the set of all vectors whose last μ coordinates are zero, and P_x is described, in matrix form, as

$$P_x = \begin{bmatrix} 0 & I \end{bmatrix}.$$

Then, it is clear that

$$P_x \left[\frac{\partial F(x, \varepsilon)}{\partial \varepsilon} \right]_{\substack{\varepsilon = 0 \\ x \in E}} = \left[\frac{\partial \varepsilon f_0(w, z, \varepsilon)}{\partial \varepsilon} \right]_{\substack{\varepsilon = 0 \\ w = 0}} = f_0(0, z, 0).$$

The following statement describes conditions under which the local asymptotic behavior of the system (3.5) for small nonzero ε, in particular its asymptotic stability at some equilibrium point, can be described in terms of properties of the two "limit" systems

$$x' = F(x, 0) \tag{3.6a}$$

$$\dot{x} = f_R(x) \qquad x \in E. \tag{3.6b}$$

Theorem. Let E° be a subset of E such that, for all $x \in E^\circ$, the nontrivial eigenvalues of J_x have negative real part. Suppose $x^\circ \in E^\circ$ is an equilibrium point of the reduced system (3.6b) and suppose that all the eigenvalues of the jacobian matrix

$$A_R = \left[\frac{\partial f_R(x)}{\partial x}\right]_{x=x^\circ}$$

have negative real part. Then there exists $\varepsilon_0 > 0$ such that, for each $\varepsilon \in (0, \varepsilon_0)$ the system (3.5) has an equilibrium point x_ε near x°, with the following properties

  (i) x_ε is the unique equilibrium of $x' = F(x, \varepsilon)$ contained in a suitable neighborhood of the point x°,

  (ii) x_ε is an asymptotically stable equilibrium of $x' = F(x, \varepsilon)$.

Deferring the proof for a moment, we show first that by means of suitable (local) changes of coordinates, a system of the form (3.5), satisfying the assumptions of the previous Theorem, can be put into a form that closely resembles the one considered at the beginning, namely the form (3.4). To this end, begin by choosing local coordinates (ξ, η) on U in such a way that E is represented, locally around x°, in the form $E = \{(\xi, \eta) : \xi = 0\}$, and $x^\circ = (0,0)$. Accordingly, the system (3.4) is represented in the form

$$\xi' = g(\xi, \eta, \varepsilon)$$
$$\eta' = f(\xi, \eta, \varepsilon).$$

By construction, since E consists of points of equilibrium for $F(x, 0)$,

$$g(0, \eta, 0) = 0$$
$$f(0, \eta, 0) = 0$$

for all η. Thus, it is possible to expand f and g near $(0, 0, 0)$ in the form

$$g(\xi, \eta, \varepsilon) = G\xi + g_0\varepsilon + g_2(\xi, \eta, \varepsilon)$$
$$f(\xi, \eta, \varepsilon) = F\xi + f_0\varepsilon + f_2(\xi, \eta, \varepsilon)$$

where f_2 and g_2 vanish at $(0, 0, 0)$ together with their first order derivatives, and $f_2(0, \eta, 0) = 0$, $g_2(0, \eta, 0) = 0$ for all η. By construction, the eigenvalues of G are the nontrivial eigenvalues of J_x at $x = x^\circ$.

The equations of this system can be simplified by means of the Center Manifold Theory. For, consider the "extended" system

$$\xi' = G\xi + g_0\varepsilon + g_2(\xi, \eta, \varepsilon)$$
$$\eta' = F\xi + f_0\varepsilon + f_2(\xi, \eta, \varepsilon)$$
$$\varepsilon' = 0$$

and note that after a linear change of variables

$$y = \xi + \lambda\varepsilon$$
$$z = \eta + K\xi$$

(with $K = -FG^{-1}$ and $\lambda = G^{-1}g_0$) the system in question can be rewritten in the form

$$y' = Gy + q(y, z, \varepsilon)$$

$$\begin{bmatrix} z' \\ \varepsilon' \end{bmatrix} = \begin{bmatrix} 0 & k \\ 0 & 0 \end{bmatrix} \begin{bmatrix} z \\ \varepsilon \end{bmatrix} + \begin{bmatrix} p(y, z, \varepsilon) \\ 0 \end{bmatrix}.$$

By construction, q and p vanish at $(0, 0, 0)$ together with their first derivatives, and $q(0, z, 0) = 0$, $p(0, z, 0) = 0$. Note that, in the new coordinates, points of the set E correspond to points having $y = 0$.

Choose now a center manifold $y = \pi(z, \varepsilon)$ for this system at $(0, 0, 0)$, and note that points of the form $(0, z, 0)$—being equilibrium points of the extended system—if z is sufficiently small belong to any center manifold through $(0, 0, 0)$. Therefore

$$0 = \pi(z, 0)$$

for small z. After a new change of variables

$$w = y - \pi(z, \varepsilon)$$

the extended system becomes

$$w' = a(w, z, \varepsilon) \qquad (3.7a)$$
$$z' = b(w, z, \varepsilon) \qquad (3.7b)$$
$$\varepsilon' = 0 \qquad (3.7c)$$

and by construction

$$a(0, z, \varepsilon) = G\pi(z, \varepsilon) + q(\pi(z, \varepsilon), z, \varepsilon) - \frac{\partial\pi}{\partial z}(k\varepsilon + p(\pi(z, \varepsilon), z, \varepsilon)) = 0$$
$$b(0, z, 0) = p(0, z, 0) = 0$$

for small (z, ε). Note that, in the new coordinates, the center manifold is the set of points having $w = 0$, and that $b(w, z, \varepsilon)$ can be represented in the following way

$$b(w, z, \varepsilon) = \int_0^1 \frac{\partial}{\partial\alpha} b(\alpha w, z, \alpha\varepsilon)\, d\alpha$$

$$= \varepsilon \int_0^1 b_\varepsilon(\alpha w, z, \alpha\varepsilon)\, d\alpha + \int_0^1 b_w(\alpha w, z, \alpha\varepsilon)\, d\alpha \cdot w$$

$$= \varepsilon f_0(w, z, \varepsilon) + F_1(w, z, \varepsilon)w$$

We can therefore conclude that, choosing suitable local coordinates, the system (3.5) can be put in the form

$$w' = a(w, z, \varepsilon)$$
$$z' = \varepsilon f_0(w, z, \varepsilon) + F_1(w, z, \varepsilon)w$$

with $a(0, z, \varepsilon) = 0$.

Note that we don't have used yet the assumption on the jacobian matrix A_R.

Proof of the Theorem. Suppose that $f_0(0, 0, 0) = 0$ and that the jacobian matrix

$$\left[\frac{\partial f_0(0, z, 0)}{\partial z} \right]_{z=0} \tag{3.8}$$

has all the eigenvalues in the left-half plane (we will show at the end that this is implied by the corresponding assumption on the reduced vector field f_R). Then, if ε is sufficiently small, the equation $f_0(0, z, \varepsilon) = 0$ has a root z_ε for each ε near 0 (with $z_0 = 0$). The point $(0, z_\varepsilon)$ is an equilibrium of the system (3.7a)-(3.7b)

$$w' = a(w, z, \varepsilon)$$
$$z' = \varepsilon f_0(w, z, \varepsilon) + F_1(w, z, \varepsilon)w.$$

Observe that $(0, z_\varepsilon)$ is the point x_ε whose original coordinates are

$$\xi_\varepsilon = \pi(z_\varepsilon, \varepsilon) - \lambda \varepsilon$$
$$\eta_\varepsilon = z_\varepsilon - K\pi(z_\varepsilon, \varepsilon) + K\lambda \varepsilon.$$

Clearly, x_ε is the unique possible equilibrium point of $F(x, \varepsilon)$—for small fixed ε—in a neighborhood of the point x^0 (i.e. of the equilibrium point of the reduced system). In fact, suppose x_1 is another equilibrium point of the system (3.5) close to x^0. Then the point (x_1, ε) (an equilibrium of the extended system) must belong to any center manifold (for the extended system) passing through $(x^0, 0)$. Since in the coordinates (w, z, ε) the set of points having $w = 0$ describes exactly a center manifold of this type, we deduce that in these coordinates the point (x_1, ε) must be represented in the form $(0, z_1, \varepsilon)$. Being two equilibria, the two points $(0, z_1, \varepsilon)$ and $(0, z_\varepsilon, \varepsilon)$ must satisfy

$$b(0, z_1, \varepsilon) = b(0, z_\varepsilon, \varepsilon)$$

i.e.

$$f_0(0, z_1, \varepsilon) = f_0(0, z_\varepsilon, \varepsilon)$$

but this, if ε is sufficiently small, implies $z_1 = z_\varepsilon$, in view of the nonsingularity of the matrix (3.8). This, in turn, implies $x_1 = x_\varepsilon$.

Since all the eigenvalues of the matrix (3.8) have negative real part, also those of

$$\left[\frac{\partial f_o(w, z, \varepsilon)}{\partial z}\right]_{\substack{w=0 \\ z=z_\varepsilon}}$$

all have negative real part for small ε. Moreover, since

$$\left[\frac{\partial a(w, z, 0)}{\partial w}\right]_{\substack{w=0 \\ z=0}} = G$$

also the matrix

$$\left[\frac{\partial a(w, z, \varepsilon)}{\partial w}\right]_{\substack{w=0 \\ z=z_\varepsilon}}$$

has all eigenvalues with negative real part for small ε.

If ε is positive, the equilibrium $(0, z_\varepsilon)$ of (3.7) is asymptotically stable. In fact, the jacobian matrix of the right-hand side, evaluated at this point, has the form

$$\left[\begin{array}{cc} \left[\dfrac{\partial a(w, z, \varepsilon)}{\partial w}\right]_{\substack{w=0 \\ z=z_\varepsilon}} & 0 \\ \star & \varepsilon\left[\dfrac{\partial f_o(w, z, \varepsilon)}{\partial w}\right]_{\substack{w=0 \\ z=z_\varepsilon}} \end{array}\right]$$

and all the eigenvalues in the left-half plane.

In order to complete the proof we have to show that the matrix (3.8) has all the eigenvalues in the left half plane. To this end, recall that in the (w, z, ε) coordinates, points of the set E correspond to points having $w = 0$ and $\varepsilon = 0$. Note that

$$\left[F(x, \varepsilon)\frac{\partial}{\partial x}, \frac{\partial}{\partial \varepsilon}\right]_{\substack{\varepsilon=0 \\ x \in E}} = -\left[\frac{\partial F(x, \varepsilon)}{\partial \varepsilon}\right]_{\substack{\varepsilon=0 \\ x \in E}}$$

and that the right-hand side of this expression, in the (w, z, ε) coordinates, becomes

$$\left[b(w, z, \varepsilon)\frac{\partial}{\partial z}, \frac{\partial}{\partial \varepsilon}\right]_{\substack{\varepsilon=0 \\ x \in E}} = -f_o(0, z, 0)\frac{\partial}{\partial z}.$$

Thus, it is easily deduced that the tangent vector

$$f_o(0, z, 0)\frac{\partial}{\partial z}$$

represents exactly the vector field f_R at the point $(0, z, 0)$ of E. From this, the conclusion follows immediately. ●

Remark. Note that, for a system given in the simplified form (3.1), the previous Theorem establishes that if a point $(h(z^o), z^o)$ satisfies

$$g(h(z^o), z^o, 0) = 0$$
$$f(h(z^o), z^o, 0) = 0,$$

if the system

$$w' = g(w + h(z^o), z^o, 0)$$

is asymptotically stable in the first approximation at $w = 0$, and the reduced system

$$\dot{z} = f(h(z), z, 0)$$

is asymptotically stable in the first approximation at $z = z^o$, then for each sufficiently small $\varepsilon > 0$, there exists an equilibrium of (3.1) near $(h(z^o), z^o)$ which is asymptotically stable in the first approximation. •

Remark. Observe that, in a sufficiently small neighborhood of $(x^o, 0)$, the equilibrium points of the extended system

$$x' = F(x, \varepsilon) \tag{3.9a}$$
$$\varepsilon' = 0 \tag{3.9b}$$

are only those of the set E, and those of the graph of the function $\psi : \varepsilon \to x_\varepsilon$. •

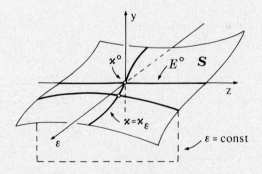

Fig. B.2

Fig B.2 illustrates some of the ingredients introduced in the previous discussion, in the particular case of a system having $n = 2$, $\mu = 1$. In the (y, z, ε) coordinates, it shows the set E^o, the center manifold S of the extended system, and the location of the equilibrium points $(x_\varepsilon, \varepsilon)$. The trajectories of the extended system are contained in planes parellel to the (y, z) plane and, on each of these planes, obviously coincide with those of the system (3.5) for a specific value of ε. Note that, since the center manifold S is by definition an invariant manifold, the intersections of S with planes parellel to the (y, z) plane are invariant manifolds of (3.5) for the corresponding value of $\varepsilon : E^\circ$ is an invariant manifold consisting of equilibrium points, the other ones contain only one equilibrium, which is asymptotically stable. Fig. B.3 shows a possible behavior of these trajectories for some $\varepsilon > 0$. Clearly, for $\varepsilon = 0$ different trajectories converge to different equilibrium points on E°, whereas, for $\varepsilon > 0$ all trajectories locally converge to the equilibrium x_ε.

Note that if the reduced system is asymptotically stable *but not in the first approximation* (at the point x°), i.e. A_R has not all the eigenvalues in the left half plane, the results illustrated above are not anymore true. This is illustrated for instance in the following simple example.

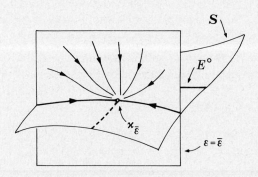

Fig. B.3

Example. Consider the system

$$\varepsilon \dot{y} = -y + \varepsilon z^2$$
$$\dot{z} = (y^2 - z^5).$$

In this case the set S is the set of points with $y = 0$, and the reduced system, given by

$$\dot{z} = -z^5$$

is asymptotically stable at $z = 0$. In order to study the behavior of the entire system, we rescale the time

$$y' = -y + \varepsilon z^2$$
$$z' = \varepsilon(y^2 - z^5)$$

and we note that the rescaled system has two equilibria, one at $(y, z) = (0, 0)$ and the other at $(y, z) = (\varepsilon^5, \varepsilon^2)$. The first one is a critical point, and its stability may be analyzed by means of the Center Manifold Theory. A center manifold at $(0,0)$ is a function $y = \pi(z)$ satisfying

$$\frac{\partial \pi}{\partial z}\varepsilon(\pi^2(z) - z^5) = -\pi(z) + \varepsilon z^2$$

and it is easily seen that

$$\pi(z) = \varepsilon z^2 + R_5(z)$$

where $R_5(z)$ is a remainder of order 5. The flow on the center manifold is given by

$$z' = \varepsilon(\varepsilon^2 z^4 - z^5) + R_7(z)$$

(with $R_7(z)$ a remainder of order 7) and is unstable at $z = 0$ for any ε. Thus, from Center Manifold Theory, we conclude that the point $(0,0)$ is an unstable equilibrium of the system. The analysis of the stability at $(\varepsilon^5, \varepsilon^2)$ is simpler because, as the reader can easily verify, the linear approximation of the system at this point is asymptotically stable. Thus the point in question is an asymptotically stable equilibrium of the system.

We conclude that, for any arbitrarily small value of $\varepsilon > 0$, there exist always two equilibria of the system near the equilibrium of the reduced system, one being unstable and the other asymptotically stable. •

We conclude the Section by stating another interesting result, that provides an additional "geometric" insight to the previuos analysis.

Theorem. Suppose the assumptions of the previous Theorem are satisfied. Then, in a neighborhood of the point $(x^o, 0)$ in $U \times (-\varepsilon_o, +\varepsilon_o)$, there exists a smooth integrable distribution Δ with the following properties
(i) $\dim(\Delta) = n - \mu$
(ii) if S is a center manifold for the extended system (3.9) at $(x^o, 0)$, at each point x of S

$$T_x S \cap \Delta(x) = 0$$

(iii) Δ is invariant under the vector field

$$f(x, \varepsilon) = \begin{bmatrix} F(x, \varepsilon) \\ 0 \end{bmatrix}.$$

In other words, this Theorem says that, in a neighborhood of the point $(x^o, 0)$, there is a partition of $U \times (-\varepsilon_o, +\varepsilon_o)$ into submanifolds of dimension $n - \mu$ (the integral submanifolds of Δ) and that each of these manifolds intersects S transversally, exactly in one point. Moreover, by property (iii), each of these submanifolds is contained in a subset of the form $U \times \{\varepsilon\}$ and the flow of $F(x, \varepsilon)$ carries submanifolds into submanifolds (the *partition* being *invariant* under the flow of $f(x, \varepsilon)$). In particular, every submanifold belonging to the set $U \times \{0\}$ is a *locally invariant submanifold* of $F(x, 0)$.

Bibliographical Notes

Chapter 1. The definition of distribution used here is taken from Sussmann (1973); in most of the references quoted in the Appendix A, the term "distribution" without any further specification is used to indicate what we mean here for "nonsingular distribution". Different proofs of Frobenius' Theorem are available. The one used here is adapted from Lobry (1970) and Sussmann (1973). Additional results on simultaneous integrability of distributions can be found in Respondek (1982).

The importance in control theory of the notion of invariance of a distribution under a vector field was pointed out independently by Hirschorn (1981) and by Isidori et al. (1981a). A more general notion of invariance, under a group of local diffeomorphisms, was given earlier by Sussmann (1973). The local decompositions described in Section 1.7 are consequences of ideas of Krener (1977).

Theorems 1.8.13 and 1.8.14 were first proved by Sussmann-Jurdjevic (1972). The proof described here is due to Krener (1974). An earlier version of Theorem 1.8.13, dealing with trajectories traversed in either time direction, was given by Chow (1939). Additional and more complete results on local controllability can be found in the work of Sussmann (1983), (1987). Controllability of systems evolving on Lie groups was studied by Brockett (1972a). Controllability of polynomial systems was studied by Ballieul (1981) and Jurdjevic-Kupca (1985). Theorem 1.9.8, although in a slightly different version, is due to Hermann-Krener (1977). Additional results on observability, dealing with the problem of identifying the initial state from the response under a fixed input function, can be found in Sussmann (1979b).

Chapter 2. The proof of Theorems 2.1.2 and 2.1.4 may be found in Sussmann (1973). An independent proof of Theorem 2.1.7. was given earlier by Hermann (1962) and an independent proof of Corollary 2.1.9 by Nagano (1970). The relevance of the control Lie algebra in the analysis of global reachability derives from the work of Chow (1939) and was subsequently elucidated by Lobry (1970), Haynes-Hermes (1970), Elliott (1971) and Sussmann-Jurdjevic (1972). The properties of the observation space were studied by Hermann-Krener (1977), and, in the case of discrete-time systems, by Sontag (1979). Reachability, observability and decompositions of bilinear systems were studied by Brockett (1972b), Goka et al. (1973) and d'Alessandro et al. (1974). The application to the study of attitude control of spacecraft is adapted from Crouch (1984).

Chapter 3. The functional expansions illustrated in Section 3.1 were introduced in a series of works by Fliess; a comprehensive exposition of the subject, together with several additional results, can be found in Fliess (1981). The expressions of the kernels of the Volterra series expansion were discovered by Lesjak-Krener (1978); the expansions 3.2.9 are due to Fliess et al. (1983). The structure of the Volterra kernels was earlier analyzed by Brockett (1976), who proved that any individual kernel can always be interpreted as a kernel of a suitable bilinear system, and related results may also be found in Gilbert (1977). The expressions of the kernels of bilinear systems were first calculated by Bruni et al. (1970). Multivariable Laplace transforms of Volterra kernels and their properties are estensively studied by Rugh (1981). Functional expansions for nonlinear discrete-time systems have been studied by Sontag (1979) and Monaco-Normand Cyrot (1986).

The conditions under which the output of a system is unaffected by some specific input channel were studied by Isidori et al. (1981a) and Claude (1982); the former contains, in particular, a different proof of Theorem **3**.3.5.

Definitions and properties of generalized Hankel matrices were developed by Fliess (1974). Theorem **3**.4.4 was proved independently by Isidori (1973) and Fliess. The notion of Lie rank and Theorem **3**.4.5 are due to Fliess (1983). Equivalence of minimal realizations was extensively studied by Sussmann (1977); the version given here of the uniqueness Theorem essentially develops an idea of Hermann-Krener (1977); related results may also be found in Fliess (1983). An independent approach to realization theory was followed by Jakubczyk (1980), (1986), who also provided an alternative proof of Theorem **3**.4.5. Additional results on this subject can be found in Celle-Gauthier (1987). Realization of finite Volterra series was studied by Crouch (1981). Constructive realization methods from the Laplace transform of a Volterra kernel may be found in the work of Rugh (1983). Realization theory of discrete-time response maps was extensively studied by Sontag (1979).

Chapter 4. The convenience of describing a system in the special local coordinates considered in Sections **4**.1 and **5**.1 was first explicitly suggested in the work of Isidori et al. (1981a). Additional material on this and similar subjects can be found in the work of Zeitz (1983), Bestle-Zeitz (1983) and of Krener (1987). The exact state-space linearization problem was proposed and solved, for single-input systems, by Brockett (1978). A complete solution for multi-input systems was found by Jakubczyk-Respondek (1980). Independent work of Su (1982) and Hunt-Su-Meyer (1983a) led to a slightly different formulation, together with a procedure for the construction of the linearizing transformation. The possibility of using noninteracting control techniques for the solution of such a problem was pointed out in Isidori et al. (1981a). Additional results on this subject can be found in the work of Sommer (1980) and Marino et al. (1985). The existence of globally defined transformations was investigated by Dayawansa et al. (1985). Exact linearization of discrete-time systems was studied by Lee et al. (1986) and by Jakubczyk (1987).

The notion of zero dynamics was introduced by Byrnes-Isidori (1984). Its application to the solution of critical problems of asymptotic stabilization was described in Byrnes-Isidori (1988a). Additional material on this subject can be found in the work of Aeyels (1985), where for the first time the usefulness of center manifold theory for the solvability of critical problems of asymptotic stabilization was pointed out, and Marino (1988). The concept of zero dynamics for discrete-time systems and its properties are developed in the works of Glad (1987) and Monaco-Normand Cyrot (1988).

The subject of asymptotic stabilization via state feedback is only marginally touched in these notes, and there are several important issues that have not been covered here for reasons of space. These include, for instance, the problem of equivalence between stabilizability and controllability (see Jurdjevic-Quinn (1979) and Brockett (1983)), the smoothness properties of a stabilizing feedback (see Sussmann (1979a) and Sontag-Sussmann (1980)), the input-output approach to stability of feedback systems (see Hammer (1986), (1987) and Sontag (1989)).

The singular perturbation analysis of high-gain feedback systems was independently studied by Byrnes-Isidori (1984) and by Marino (1985). Global results can be found in the recent work of Knobloch (1988b). A link with the so-called variable structure control theory developed by Utkin (1977) can be found in the work of Marino (1985). An application of singular perturbation theory to the design of adaptive control can be found in the work of Khalil-Saberi (1987); an application to the so-called almost disturbance decoupling problem can be found in the work of Marino et al. (1989).

The problem of finding the largest linearizable subsystem in a single-input nonlinear system was addressed and solved by Krener et al. (1983). The solution of the corresponding problem for a multi-input system was found by Marino (1986). The problem of exact linearization of a nonlinear system with outputs is extensively discussed by Cheng et al. (1988). The use of output injection in order to obtain observers with linear error dynamics was independently suggested by Krener-Isidori (1983) and Bestle-Zeitz (1983) for single-output systems. A complete analysis of the corresponding problem for multi-output systems can be found in the work of Krener-Respondek (1985). Hammouri-Gauthier (1988) have suggested the use of output injection in order to obtain bilinear error dynamics. Additional results on the design of nonlinear observers can be found in the work of Zeitz (1985).

Chapter 5. The solution of the problem of noninteracting control is due to Porter (1970). Additional results can be found in Singh-Rugh (1972) and Freund (1975). The work of Isidori et al. (1981a) showed how the solution of the noninteracting control problem can be analyzed from the differential-geometric viewpoint. Related results can be found in Knobloch (1988a). In the case of discrete-time nonlinear systems, the problem was studied by Grizzle (1985b) and Monaco-Normand Cyrot (1986).

The exact linearization of the input-output response was studied by Isidori-Ruberti (1984), who proposed an approach (exposed in Section 5.4) inspired by the works of Silverman (1969) on the inversion of linear systems and Van Dooren et al. (1979) on the calculation of the so-called zero structure at infinity. For discrete-time systems, the corresponding problem was investigated by Monaco-Normand Cyrot (1983) and Lee-Markus (1987). An interesting approach, alternative to exact linearization, is the one based on approximate linearization around an operating point, considered as a smoothly varying parameter. This approach, which is not included here for reasons of space, was pursued by Baumann-Rugh (1986), Réboulet et al. (1986), Wang-Rugh (1987), Sontag (1987a), (1987b). The matching of the input-output behavior of a prescribed system was studied by Isidori (1985), Di Benedetto-Isidori (1986) and Di Benedetto (1988).

Chapter 6. Controlled invariant submanifold and controlled invariant distribution are nonlinear versions of the notion of controlled invariant subspace, introduced independently by Basile-Marro (1969) and by Wonham-Morse (1970). As illustrated in Section 6.3, the two notions are not equivalent in a nonlinear setting: the former lends itself to the definition of the nonlinear analogue of the notion of transmission zero, while the latter is particularly suited to the study of decoupling and noninteracting control problems.

The properties of controlled invariant distribution were studied earlier. The notion of controlled invariant distribution was introduced by Isidori et al. (1981a) and independently (although in a less general form) by Hirschorn (1981). The proof of Lemma 6.2.1 is adapted form the ones proposed by Hirschorn (1981), Isidori et al. (1981b) and Nijmeijer (1981). The calculation of the largest controllability distribution contained in ker(dh) by means of the controlled invariant distribution algorithm was suggested by Isidori et al. (1981a). The simpler procedure described in

Proposition **6**.3.7 is due to Krener (1985). Lemma **6**.3.13 is due to Claude (1981). The notion of controllability distribution, the nonlinear version of the one of controllability subspace, and the corresponding properties were studied by Isidori-Krener (1982) and by Nijmeijer (1982). The theory of globally controlled invariant distributions can be found in the work of Dayawansa et al. (1988).

The calculation of the largest output zeroing submanifold by means of the zero dynamics algorithm was suggested by Isidori-Moog (1988). The works of Byrnes-Isidori (1988a), (1988b) have shown how this algorithm is useful in order to derive the normal forms illustrated in Proposition **6**.1.9.

Controlled invariance for general nonlinear systems (i.e. systems in which the control does not enter linearly) has been studied by Nijmeijer-Van der Schaft (1983). Controlled invariance for discrete-time nonlinear systems has been studied by Grizzle (1985a) and Monaco-Normand Cyrot (1985).

Chapter 7. The results described in Section **7**.1, which are the multivariable version of the results illustrated in Section **4**.4, have been adapted from Byrnes-Isidori (1988a). The nonlinear regulator theory illustrated in Section **7**.2 is taken from a very recent work of Byrnes-Isidori (1989). The special case of constant reference signals was treated earlier by Jie-Rugh (1988). Necessary conditions for the existence of error feedback nonlinear regulators were investigated by Hepburn and Wonham (1984).

The usefulness of the differential geometric approach in the solution of the nonlinear disturbance decoupling problem was pointed out by Hirschorn (1981) and Isidori et al. (1981a). The solution of the problem of noninteracting control with stability via static state feedback is due to Isidori-Grizzle (1988). Lemma **7**.4.2 incorporates some earlier results by Nijmeijer-Schumacher (1986) and Ha-Gilbert (1986).

The possibility of using dynamic feedback in order to achieve relative degree was first shown by Singh (1980), and by Descusse-Moog (1985), (1987). The exposition in Section **7**.5 follows closely the approach suggested by Nijmeijer-Respondek (1986), (1988). The notions of left and right invertibility proposed by Fliess (1986a), (1986b), together with the introduction of differential-algebraic methods in the analysis of control systems, provide a precise conceptual framework in which the equivalence between (right) invertibility and the possibility of achieving noninteracting control via dynamic feedback can be estabilshed. Additional results on the use of differential algebra in control theory can be found in the work of Pommaret (1988). Additional results on the subject of system invertibility can be found in the works of Hirschorn (1979a), (1979b), Singh (1981), Isidori-Moog (1988), Moog (1988) and Di Benedetto et al. (1989). The latter, in particular, clarifies the relationships between different algorithms proposed in the literature.

An important property of linear systems is that the possibility of achieving noninteracting control via dynamic feedback implies the possibility of achieving noninteracting control together with asymptotic stability (see Wonham (1979)). This property is not anymore true in a nonlinear setting. In other words, there are nonlinear systems for which, as shown in Grizzle-Isidori (1988), it is possible to obtain noninteracting control but no (either static or dynamic) feedback exists which yields a stable noninteractive closed loop. The obstruction to the achievement of noninteracting control with stability via dynamic state feedback, which depends on certain Lie brackets of the vector fields which characterize the noninteractive system, has been studied by Wagner (1989).

As shown at the end Section **7**.5, the absence of zero dynamics, together with the possibility of achieving relative degree via dynamic feedback, are properties which imply the existence of a feedback and coordinates change which transform the system into a fully linear and controllable one. This property, that was recognized by Isidori et al. (1986), finds a very natural application to the control of the nonlinear dynamics of an aircraft as well as to the control of a robot arm with joints elasticity. The first application was pursued by Meyer-Cicolani (1980) and, more recently, by Lane-Stengel (1988) and Hauser et al. (1988). The second application was developed by De Luca et al. (1985). Other relevant applications of the theory discussed in this Chapter to process control are those pursued by Hoo-Kantor (1986) and by Lévine-Rouchon (1989).

Appendix A. A comprehensive exposition of all the subjects summarized in this Appendix can be found in the books of Boothby (1975), Brickell-Clark (1970), Singer-Thorpe (1967), Warner (1979).

Appendix B. For a comprehensive introduction to the stability theory, the reader is referred e.g. to the books of Hahn (1967) and Vidyasagar (1978). The purpose of this Appendix is to cover some specific subjects, that are frequently used throughout the text, which are not usually treated in standard reference books on stability of control systems. The exposition of center manifold theory follows closely the one of Carr (1981). The concepts of stability under persistent disturbances and a proof of the second Lemma of Section **B**.2 can be found in Hahn (1967), pages 275-276 (see also Vidyasagar (1980)). Section **B**.3 is essentially a synthesis of some results taken from the work of Fenichel (1979). Additional material on this subject can be found in the works of Knobloch-Aulbach (1984) and Marino-Kokotovic (1988). A comprehensive exposition of theory and applications of singular perturbation methods in control can be found in the book of Kokotovic-Khalil-O'Reilly (1986).

References

D. Aeyels
(1985) Stabilization of a class of nonlinear systems by a smooth feedback control, *Syst. Contr. Lett.* **5**, pp. 289-294.

A. Andreini, A. Bacciotti, G. Stefani
(1988) Global stabilizability of homogeneous vector fields of odd degree, *Syst. Contr. Lett.* **10**, pp. 251-256.

J. Ballieul
(1981) Controllability and observability of polynomial systems, *Nonlin. Anal.* **5**, pp. 543-552.

G. Basile, G. Marro
(1969) Controlled and conditioned invariant subspaces in linear systems theory, *J. Optimiz. Th. Appl.* **3**, pp. 306-315.

W.T. Baumann, W.J. Rugh
(1986) Feedback control of nonlinear systems by extended linearization, *IEEE Trans. Aut. Contr.* **AC-31** pp. 40-46.

D. Bestle, M. Zeitz
(1983) Canonical form observer design for nonlinear time-variable systems, *Int. J. Contr.* **38**, pp. 419-431.

W.A. Boothby
(1975) *An Introduction to Differentiable Manifolds and Riemaniann Geometry*, Academic Press.

F.R. Brickell, R.S. Clark
(1970) *Differentiable Manifolds*, Van Nostrand.

R.W. Brockett
(1972a) System theory on group manifolds and coset spaces, *SIAM J. Contr.* **10**, pp. 265-284.
(1972b) On the algebraic structure of bilinear systems, *Theory and Applications of Variable Structure Systems*, R. Mohler and A. Ruberti eds., Academic Press, pp. 153-168.
(1976) Volterra series and geometric control theory, *Automatica* **12**, pp. 167-176.
(1978) Feedback invariants for non-linear systems, *IFAC Congress* **6**, pp. 1115-1120.
(1983) Asymptotic stability and feedback stabilization, *Differential Geometric Control Theory*, R.W. Brockett, R.S. Millman and H. Sussmann eds., Birkhauser, pp. 181-191.

C. Bruni, G. Di Pillo, G. Koch
(1971) On the mathematical models of bilinear systems, *Ricerche di Automatica* **2**, pp. 11-26.

P. Brunovsky
(1970) A classification of linear controllable systems, *Kybernetika* **6**, pp. 173-188.

C.I. Byrnes, A. Isidori
(1984) A frequency domain philosophy for nonlinear systems, *IEEE Conf. Dec. Contr.* **23**, pp. 1569-1573.
(1988a) Local stabilization of minimum-phase nonlinear systems, *Syst. Contr. Lett.* **11**, pp. 9-17.
(1988b) Feedback design from the zero dynamics point of view, *Bozeman Work. Comp. Contr.*
(1989a) Analysis and design of nonlinear feedback systems, *IEEE Trans. Aut. Contr.*, to appear.
(1989b) Output regulation of nonlinear systems, preprint.

J. Carr
(1981) *Applications of Centre Manifold Theory*, Springer Verlag.

F. Celle, J.P. Gauthier
(1987) Realizations of nonlinear analytic input-output maps, *Math. Syst. Theory* **19**, pp. 227-237.

R. Charlet, J. Lévine, R. Marino
(1989) On dynamic feedback linearization, *Syst. Contr. Lett.*, to appear.

D. Cheng, A. Isidori, W. Respondek, T.J. Tarn
(1988) On the linearization of nonlinear systems with outputs, *Math. Syst. Theor.* **21**, pp. 63-83.

D. Cheng, T.J. Tarn, A. Isidori
(1985) Global external linearization of nonlinear systems via feedback, *IEEE Trans. Aut. Contr.* **AC-30**, pp. 808-811.

W.L. Chow
(1938) Uber systeme von linearen partiellen differentialgleichungen ester ordnung, *Math. Ann.* **117**, pp. 98-105.

D. Claude
(1982) Decoupling of nonlinear systems, *Syst. Contr. Lett.* **1**, pp. 242-248.

D. Claude, M. Fliess, A. Isidori
(1983) Immersion, directe et par bouclage, d'un système nonlinéaire dans un linéaire, *C. R. Acad. Sci. Paris* **296**, pp. 237-240.

P. Crouch
(1981) Dynamical realizations of finite Volterra Series, *SIAM J. Contr. Optimiz.* **19**, pp. 177-202.
(1984) Spacecraft attitude control and stabilization, *IEEE Trans. Aut. Contr.* **AC-29**, pp. 321-331.

P. D'Alessandro, A. Isidori, A. Ruberti
(1974) Realization and structure theory of bilinear dynamical systems, *SIAM J. Contr.* **12**, pp. 517-535.

W.P. Dayawansa, W.M. Boothby, D. Elliott
(1985) Global state and feedback equivalence of nonlinear systems, *Syst. Contr. Lett.* **6**, pp. 229-234.

W.P. Dayawansa, D. Cheng, T.J. Tarn, W.M. Boothby
(1988) Global (f, g)-invariance of nonlinear systems, *SIAM J. Contr. Optimiz.* **26**, pp. 1119-1132.

A. De Luca, A. Isidori, F. Nicol
(1985) Control of robot arms with elastic joints via nonlinear dynamic feedback, *IEEE Conf. Decis. Contr.* **24**, pp. 1671-1679.

J. Descusse, C.H. Moog
(1985) Decoupling with dynamic compensation for strong invertible affine nonlinear systems, *Int. J. Contr.* **43**, pp. 1387-1398.
(1987) Dynamic decoupling for right-invertible nonlinear systems, *Syst. Contr. Lett.* **8**, pp. 345-349.

M.D. Di Benedetto
(1988) A condition for the solvability of the model matching problem, preprint.

M.D. Di Benedetto, J.W. Grizzle, C.H. Moog
(1989) Rank invariants of nonlinear systems, *SIAM J. Contr. Optimiz.*, to appear.

M.D. Di Benedetto, A. Isidori
(1986) The matching of nonlinear models via dynamic state feedback, *SIAM J. Contr. Optimiz.* **24**, pp. 1063-1075.

D.L. Elliott
(1970) A consequence of controllability, *J. Diff. Eqs.* **10**, pp. 364-370.

N. Fenichel
(1979) Geometric singular perturbation theory for ordinary differential equations, *J. Diff. Eqs.* **31**, pp. 53-93.

M. Fliess
(1974) Matrices de Hankel, *J. Math. Pures Appl.* **53**, pp. 197-224.
(1981) Fonctionnelles causales non linéaires et indéterminées non commutatives, *Bull. Soc. Math. France* **109**, pp. 3-40.
(1983) Réalisation locale des systèmes non linéaires, algèbres de Lie filtrées transitives et séries génératrices non commutatives, *Invent. Math.* **71**, pp. 521-537.
(1986a) A note on the invertibility of nonlinear input-output differential systems, *Syst. Contr. Lett.* **8**, pp. 147-151.
(1986b) Some remarks on nonlinear invertibility and dynamic state feedback, *Theory and Applications of Nonlinear Control Systems*, C.I.Byrnes and A.Lindquist eds., North Holland, pp. 115-122.

M. Fliess, M. Lamnabhi, F. Lamnabhi-Lagarrigue
(1983) An algebraic approach to nonlinear functional expansions, *IEEE Trans. Circ. Syst.* **CAS-30**, pp. 554-570.

E. Freund
(1975) The structure of decoupled nonlinear systems, *Int. J. Contr.* **21**, pp. 443-450.

E.G. Gilbert
(1977) Functional expansion for the response of nonlinear differential systems, *IEEE Trans. Aut. Contr.* **AC-22**, pp. 909-921.

T. Glad
(1987) Output dead-beat control for nonlinear systems with one zero at infinity, *Syst. Contr. Lett.* **9**, pp. 249-255.

T. Goka, T.J. Tarn, J. Zaborszky
(1973) On the controllability of a class of discrete bilinear systems, *Automatica* **9**, pp. 615-622.

J.W. Grizzle
(1985a) Controlled invariance for discrete time nonlinear systems with an application to the disturbance decoupling problem, *IEEE Trans. Aut. Contr.* **AC-30**, pp. 868-874.
(1985b) Local input output decoupling of discrete time nonlinear systems, *Int. J. Contr.* **43**, pp. 1517-1530.

J.W. Grizzle, A. Isidori
(1989) Block noninteracting control with stability via static state feedback, *Math. Contr. Sign. Syst.*, to appear.

I.J. Ha
(1989) The standard decomposed system and noninteracting feedback control of nonlinear systems, *SIAM J. Contr. Optimiz.*, to appear.

I.J. Ha, E.G. Gilbert
(1986) A complete characterization of decoupling control laws for a general class of nonlinear systems, *IEEE Trans. Aut. Contr.* **AC-31**, pp. 823-830.

W. Hahn
(1967) *Stability of motion*, Springer Verlag.

J. Hammer
(1986) Stabilization of nonlinear systems, *Int. J. Contr.* **44**, pp. 1349-1381.
(1987) Fraction representations of nonlinear systems: a simplified approach, *Int. J. Contr.* **46**, pp. 455-472.

H. Hammouri, J.P. Gauthier
(1988) Bilinearization up to output injection, *Syst. Contr. Lett.* **11**, pp. 139-150.

J. Hauser, S.S. Sastry, G. Meyer
(1988) Nonlinear controller design for flight control systems, *UCB/ERL Memo* **M88/76**.

G.W. Haynes, H. Hermes
(1970) Nonlinear controllability via Lie theory, *SIAM J. Contr.* **8**, pp. 450-460.

J.S.A. Hepburn, W.M. Wonham
(1984) Error feedback and internal models on differentiable manifolds, *IEEE Trans. Aut. Contr.* **AC-29**, pp. 397-403.

R. Hermann
(1962) The differential geometry of foliations, *J. Math. Mech.* **11**, pp. 302-316.

R. Hermann, A.J. Krener
(1977) Nonlinear controllability and observability, *IEEE Trans. Aut. Contr.* **AC-22**, pp. 728-740.

H. Hermes
(1980) On a stabilizing feedback attitude control, *SIAM J. Contr. Optimiz.* **18**, pp. 352-361.

R.M. Hirschorn
(1979a) Invertibility of nonlinear control systems, *SIAM J. Contr. Optimiz.* **17**, pp. 289-297.
(1979b) Invertibility for multivariable nonlinear control systems, *IEEE Trans. Aut. Contr.* **AC-24**, pp. 855-865.
(1981) (A,B)-invariant distributions and disturbance decoupling of nonlinear systems, *SIAM J. Contr. Optimiz.* **19**, pp. 1-19.

K.H. Hoo, J.C. Kantor
(1986) Global linearization and control of a mixed-culture bioreactor with competition and external inhibition, *Math. Bioscie.* **82**, pp. 43-62.

L.R. Hunt, R. Su, G. Meyer
(1983a) Design for multi-input nonlinear systems, *Differential Geometric Control Theory*, R.W. Brockett, R.S. Millman and H. Sussmann eds., Birkhauser, pp. 268-298.
(1983b) Global transformations of nonlinear systems, *IEEE Trans. Aut. Contr.* **AC-28**, pp. 24-31.

A. Isidori
(1973) Direct construction of minimal bilinear realizations from nonlinear input-output maps, *IEEE Trans. Aut. Contr.* **AC-18**, pp. 626-631.
(1985) The matching of a prescribed linear input-output behavior in a nonlinear system, *IEEE Trans. Aut. Contr.* **AC-30**, pp. 258-265.
(1985) *Nonlinear Control Systems: an Introduction*, Springer Verlag: Lec. notes Contr. Info. Scie. **72**.
(1987) *Lectures on Nonlinear Control*, Carl Cranz Gesellschaft.

A. Isidori, J.W. Grizzle
(1988) Fixed modes and nonlinear noninteracting control with stability, *IEEE Trans. Aut. Contr.* **AC-33**, pp. 907-914.

A. Isidori, A.J. Krener, C. Gori Giorgi, S. Monaco
(1981a) Nonlinear decoupling via feedback: a differential geometric approach, *IEEE Trans. Aut. Contr.* **AC-26**, pp. 331-345.
(1981b) Locally (f,g)-invariant distributions, *Syst. Contr. Lett.* **1**, pp. 12-15.

A. Isidori, C. Moog
(1988) On the nonlinear equivalent of the notion of transmission zeros, *Modelling and Adaptive Control*, C.I.Byrnes and A.Kurzhanski eds., Springer Verlag: Lec. notes Contr. Info. Scie. **105**, pp. 445-471.

A. Isidori, C.H. Moog, A. De Luca
(1986) A sufficient condition for full linearization via dynamic state feedback, *IEEE Conf. Dec. Contr.* **25**, pp. 203-208.

A. Isidori, A. Ruberti
(1984) On the synthesis of linear input-output responses for nonlinear systems, *Syst. Contr. Lett.* **4**, pp. 17-22.

B. Jakubczyk
(1980) Existence and uniqueness of realizations of nonlinear systems, *SIAM J. Contr. Opimiz.* **18**, pp. 445-471.
(1986) Local realization of nonlinear causal operators, *SIAM J. Contr. Optimiz.* **24**, pp. 230-242.
(1987) Feedback linearization of discrete-time systems, *Syst. Contr. Lett.* **9**, pp. 441-446.

B. Jakubczyk, W. Respondek
(1980) On linearization of control systems, *Bull. Acad. Polonaise Sci. Ser. Sci. Math.* **28**, pp. 517-522.

B. Jakubczyk, E. Sontag
(1988) Controllability of nonlinear discrete time systems: a Lie-algebraic approach, preprint.

H. Jie, W.J. Rugh
(1988) On a nonlinear multivariable servomechanism problem, *Tech. Rep.* JHU/ECE 88/04.2.

V. Jurdjevic, I. Kupca
(1985) Polynomial control systems, *Math. Ann.* **272**, pp. 361-368.

V. Jurdjevic, J.P. Quinn
(1979) Controllability and stability, *J. Diff. Eqs.* **28**, pp. 381-389.

R.E. Kalman
(1972) Kronecker invariants and feedback, *Ordinary differential equations*, C. Weiss ed., Academic Press, pp. 459-471.

H. Khalil, A. Saberi
(1987) Adaptive stabilization of a class of nonlinear systems using high-gain feedback, *IEEE Trans. Aut. Contr.* **AC-32**, pp. 1031-1035.

H.W. Knobloch
(1988a) On the dependence of solutions upon the right hand side of an ordinary differential equation, *Aequ. Math.* **35**, pp. 140-163.
(1988b) Stabilization of nonlinear control systems by means of "high-gain" feedback, preprint

H.W. Knobloch, B. Aulbach
(1984) Singular perturbations and integral manifolds, *J. Math. Phys. Sci.* **18**, pp. 415-424.

P. Kokotovic
(1984) Applications of singular perturbations techniques to control problems, *SIAM Rev.* **26**, pp. 501-550.
(1985) Control theory in the 80's: trends in feedback design, *Automatica* **21**, pp. 225-236.

P. Kokotovic, H.K. Khalil, J. O'Reilly
(1986) *Singular perturbation methods in control: analysis and design*, Academic Press.

S.R. Kou, D.L. Elliot, T.J. Tarn
(1973) Observability of nonlinear systems, *Inform. Contr.* **22**, pp. 89-99.

A.J. Krener
(1974) A generalization of Chow's theorem and the bang-bang theorem to nonlinear control systems, *SIAM J. Contr.* **12**, pp. 43-52.
(1977) A decomposition theory for differentiable systems, *SIAM J. Contr. Optimiz.* **15**, pp. 289-297.
(1985) (Adf, g), (adf, g) and locally (adf, g)-invariant and controllability distributions, *SIAM J. Contr. Optimiz.* **23**, pp. 523-549.
(1987) Normal forms for linear and nonlinear systems, *Contempor. Math.* **68**, pp. 157-189.

A.J. Krener, A. Isidori
(1982) (Adf, G) invariant and controllability distributions, *Feedback Control of Linear and Nonlinear Systems*, D. Hinrichsen and A. Isidori eds., Springer Verlag: Lec. notes Contr. Info. Scie **39**, pp. 157-164
(1983) Linearization by output injection and nonlinear observers, *Syst. Contr. Lett.* **3**, pp. 47-52.

470   References

A.J. Krener, A. Isidori, W. Respondek
(1983) Partial and robust linearization by feedback, *IEEE Conf. Dec. Contr.* **22**, pp. 126-130.

A.J. Krener, W. Respondek
(1985) Nonlinear observers with linearizable error dynamics, *SIAM J. Contr. Optimiz.* **23**, pp. 197-216.

S.H. Lane, R.F. Stengel
(1988) Flight control design using nonlinear inverse dynamics, *Automatica* **24**, pp. 471-483.

H.G. Lee, A. Arapostathis, S.I. Marcus
(1986) Linearization of discrete-time systems, *Int. J. Contr.* **45**, pp. 1803-1822.
C. Lesjak, A.J. Krener

H.G. Lee, S.I. Marcus
(1987) On input-output linearization of discrete-time nonlinear systems, *Syst. Contr. Lett.* **8**, pp. 249-259.

C. Lesjak, A.J. Krener
(1978) The existence and uniqueness of Volterra series for nonlinear systems, *IEEE Trans. Aut. Contr.* **AC-23**, pp. 1091-1095.

J. Lévine, P. Rouchon
(1989) Quality control of binary distillation columns based on nonlinear aggregation models, preprint

C. Lobry
(1979) Contrôlabilité des systèmes non linéaires, *SIAM J. Contr.* **8**, pp. 573-605.

R. Marino
(1985) High-gain feedback in nonlinear control systems, *Int. J. Contr.* **42**, pp. 1369-1385.
(1986) On the largest feedback linearizable subsystem, *Syst. Contr. Lett.* **7**, pp. 345-351.
(1988) Feedback stabilization of single-input nonlinear systems, *Syst. Contr. Lett.* **10**, pp. 201-206.

R. Marino, W.M. Boothby, D.L. Elliott
(1985) Geometric properties of linearizable control systems, *Math. Syst. Theory* **18**, pp. 97-123.

R. Marino, P.V. Kokotovic
(1988) A geometric approach to nonlinear singularly perturbed control systems, *Automatica* **24**, pp. 31-41.

R. Marino, W. Respondek, A.J. Van der Schaft
(1989) Almost disturbance decoupling for single-input single-output nonlinear systems, *IEEE Trans. Aut. Contr.*, to appear.

G. Meyer, L. Cicolani
(1980) Application of nonlinear system inverses to automatic flight control design, *Theory and Application of Optimal Control in Aerospace Systems*, P. Kant ed., NATO AGARD - AG251, pp 10.1-10.29.

S.H. Mikhail, W.M. Wonham
(1978) Local decomposability and the disturbance decoupling problem in nonlinear autonomous systems, *Allerton Conf. Comm. Contr. Comp.* **16**, pp. 664-669.

S. Monaco, D. Normand-Cyrot
(1983) The immersion under feedback of a multidimensional discrete-time nonlinear system into a linear system, *Int. J. Contr.* **38**, pp. 245-261.
(1985) Invariant distributions for discrete-time nonlinear system, *Syst. Contr. Lett.* **5**, pp. 191-196.
(1986) Nonlinear systems in discrete-time, *Algebraic and Geometric Methods in Nonlinear Control Theory*, M.Fliess and M.Hazewinkel eds., Reidel, pp. 411-430.
(1988) Zero dynamics of sampled nonlinear systems, *Syst. Contr. Lett.* **11**, pp. 229-234.

C.H. Moog
(1988) Nonlinear decoupling and structure at infinity, *Math. Contr. Sign. Syst.* **1**, pp. 257-268.

T. Nagano
(1966) Linear differential systems with singularities and applications to transitive Lie algebras, *J. Math. Soc. Japan* **18**, pp. 398-404.

H. Nijmeijer
(1981) Controlled invariance for affine control systems, *Intr. J. Contr.* **34**, pp. 824-833.
(1982) Controllability distributions for nonlinear systems, *Syst. Contr. Lett.* **2**, pp. 122-129.

H. Nijmeijer, W. Respondek
(1986) Decoupling via dynamic compensation for nonlinear control systems, *IEEE Conf. Dec. Contr.* **25**, pp. 192-197.
(1988) Dynamic input-output decoupling of nonlinear control systems, *IEEE Trans. Aut. Contr.* **AC-33**, pp. 1065-1070.

H. Nijmeijer, J.M. Schumacher
(1985) Zeros at infinity for affine nonlinear systems, *IEEE Trans Aut. Contr.* **AC-30**, pp. 566-573.
(1986) The regular local noninteracting control problem for nonlinear systems, *SIAM J. Contr. Optimiz.* **24**, pp. 1232-1245.

H. Nijmeijer, A.J. Van der Schaft
(1982) Controlled invariance for nonlinear systems, *IEEE Trans Aut. Contr.* **AC-27**, pp. 904-914.

J.F. Pommaret
(1986) Géometrie différentielle algébrique et théorie du contrôle, *C. R. Acad. Sci. Paris* **302**, pp. 547-550.

W.M. Porter
(1970) Diagonalization and inverses for nonlinear systems, *Int. J. Control* **10**, pp. 252-264.

C. Réboulet, P. Mouyon, C. Champetier
(1986) About the local linearization of nonlinear systems, *Algebraic and Geometric Methods in Nonlinear Control Theory*, M.Fliess and M.Hazewinkel eds., Reidel, pp. 311-322.

W. Respondek
(1982) On decomposition of nonlinear control systems, *Syst. Contr. Lett.* **1**, pp. 301-308.

W.J. Rugh

(1981) *Nonlinear System Theory: the Volterra-Wiener Approach*, Johns Hopkins Press.

(1983) A method for constructing minimal linear-analitic realizations for polynomial systems, *IEEE Trans. Aut. Contr.* **AC-28**, pp. 1036-1043.

S.S. Sastry, A. Isidori

(1989) Adaptive control of linearizable systems, *IEEE Trans. Aut. Contr.*, to appear.

L.M. Silverman

(1969) Inversion of multivariable linear systems, *IEEE Trans. Aut. Contr.* **AC-14**, pp. 270-276.

L.M. Singer, J.A. Thorpe

(1967) *Lecture Notes on Elementary Topology and Geometry*, Scott, Foresman, Hill.

S.N. Singh

(1980) Decoupling of invertible nonlinear systems with state feedback and precompensation, *IEEE Trans. Aut. Contr.* **AC-25**, pp. 1237-1239.

(1981) A modified algorithm for invertibility in nonlinear systems, *IEEE Trans. Aut. Contr.* **AC-26**, pp. 595-598.

S.N. Singh, W.J. Rugh

(1972) Decoupling in a class of nonlinear systems by state variable feedback, *Trans. ASME J. Dyn. Syst. Meas. Contr.* **94**, pp. 323-329.

R. Sommer

(1980) Control design for multivariable nonlinear time-varying systems, *Int. J. Contr.* **31**, pp. 883-891.

E. Sontag

(1979) *Polynomial Response Maps*, Springer Verlag.

(1987a) Controllability and linearized regulation, *IEEE Trans. Aut. Contr.* **AC-32**, pp. 877-888.

(1987b) Nonlinear control via equilinearization, *IEEE Conf. Dec. Contr.* **26**, pp. 1363-1367.

(1989) Smooth stabilization implies coprime factorization, *IEEE Trans. Aut. Contr.*, to appear.

E. Sontag, H. Sussmann

(1980) Remarks on continuous feedback, *IEEE Conf. Dec. Contr.* **19**, pp. 916-921.

R. Su

(1982) On the linear equivalents of nonlinear systems, *Syst. Contr. Lett.* **2**, pp. 48-52.

H. Sussmann

(1973) Orbits of families of vector fields and integrability of distributions, *Trans. Am. Math. Soc.* **180**, pp. 171-188.

(1977) Existence and uniqueness of minimal realizations of nonlinear system, *Math. Syst. Theory* **10**, pp. 263-284.

(1979a) Subanalytic sets and feedback control, *J. Diff. Eqs.* **31**, pp. 31-52.

(1979b) Single input observability of continuous time systems, *Math. Syst. Theory* **12**, pp. 371-393.

(1983) Lie brackets and local controllability, *SIAM J. Contr. Optimiz.* **21**, pp. 686-713.

(1987) On a general theorem on local controllability, *SIAM J. Contr. Optimiz.* **25**, pp. 158-194.

H. Sussmann, V. Jurdjevic

(1972) Controllability of nonlinear systems, *J. Diff. Eqs.* **12**, pp. 95-116.

V.I. Utkin

(1977) Variable structure systems with sliding modes; a survey, *IEEE Trans. Aut. Contr.* **AC-22**, pp. 212-222.

A. Van der Schaft

(1982) Observability and controllability for smooth nonlinear systems, *SIAM J. Contr. Optimiz.* **20**, pp. 338-354.

(1988) On clamped dynamics of nonlinear systems, *Analysis and Control of Nonlinear Systems*, C.I. Byrnes, C.F.Martin and R.E.Saeks eds., North Holland, pp. 499-506.

(1989) Representing a nonlinear state space system as a set of higher order differential equations in the inputs and outputs, *Syst. Contr. Lett.* **12**, pp. 151-160.

P.M. Van Dooren, P. Dewilde, J. Wandewalle

(1979) On the determination of the Smith-MacMillan from of a rational matrix from its Laurent expansion, *IEEE Trans. Circ. Syst.* **CT-26**, pp. 180-189.

M. Vidyasagar

(1978) *Nonlinear Systems Analysis*, Prentice-Hall.

(1980) Decomposition techniques for large-scale systems with nonadditive interactions: stability and stabilizability, *IEEE Trans. Aut. Contr.* **AC-25**, pp. 773-779.

J. Wang, W.J. Rugh

(1987) Feedback linearization families for nonlinear systems, *IEEE Trans. Aut. Contr.* **AC-32**, pp. 935-940.

K.G. Wagner

(1989) Nonlinear noninteraction with stability by dynamic state feedback, preprint.

F.W. Warner

(1979) *Foundations of differentiable manifolds and Lie groups*, Scott, Foresman, Hill.

W.M. Wonham

(1979) *Linear Multivariable Control: a Geometric Approach*, Springer Verlag.

W.M. Wonham, A.S. Morse

(1970) Decoupling and pole assignment in linear multivariable systems: a geometric approach, *SIAM J. Contr.* **8**, pp. 1-18.

M. Zeitz

(1983) Controllability canonical (phase-variable) forms for nonlinear time-variable systems, *Int. J. Contr.* **37**, pp. 1449-1457.

(1987) The extended Luenberger observer for nonlinear systems, *Syst. Contr. Lett.* **9**, pp. 149-156.

Subject Index

Aileron, 393
Analytic
- function, 403
- mapping, 403
- system, 6
Angle of attack, 391
Annihilator,
- of a codistribution, 21
- of a distribution, 21
Asymptotic
- Model Matching (Pbm of), 194
- Output Tracking (Pbm of), 190, 347
- stability of interconnected systems, 442
- Stabilization (Pbm of), 183, 344
Atlas, 408

Basic open sets, 405
Basis, of a topology, 405
Bilinear system, 101
Body frame, 36
Brower's Theorem, 406
Brunowsky canonical form, 247

Campbell-Baker-Hausdorff formula, 123, 432
Center
- manifold, 435
- - theory, 434
Centrifugal forces, 399
Change of basis, in the tangent space, 417
C^∞-compatible coordinate charts, 407
Closed set, 405
Closure, of a set, 406
Codistribution, 21
Complete
- atlas, 408
- submodule, 41

- vector field, 429
Completely integrable distribution, 25
Commutator, 430
Control Lie Algebra, 90
Continuous mapping, 405
Controllability
- distribution, 338
- - algorithm, 338
- rank condition, 91
Controllable, weakly, 92
Controlled invariant
- distribution, 311
- - algorithm, 322
- submanifold, 289
Convolution integral, 120
Coordinate
- chart, 406
- function, 406
Coordinates
-, change of, 11
- transformation, 11, 407
Coriolis forces, 399
Cotangent space, 423
Covector field, 7, 432
Covectors 7, 423
Critical
- case, of asymptotic analysis, 434
- problem, of asymptotic stabilization, 185
Cubic coordinate neighborhood, 48, 407

DC Motor, 266
Decompositions
- of bilinear systems, 101
- of linear systems, 1, 97
- of nonlinear systems, 53, 54
Decoupling matrix, 263
Dense subset, 406

Derivative
- of a covector field, 10, 433
- of a real-valued function, 8, 421
Detectable, 352
Diffeomorphism
- of manifolds, 411
- global, 12
- local, 12
Differential
- of a mapping, 418
- of a real-valued function, 8
- equation, on a manifold, 426
Distribution, 14, 41
Disturbance Decoupling
- (Pbm of), 196, 364
- for MIMO systems, 282
- via disturbance measurements, 200
- with stability, 200
Drag force, 392
Dual
- basis, 423
- space, 7, 423
Dynamic
- equation, 38
- extension algorithm, 384
- feedback, *see* Feedback

Elevator, 393
Embedded submanifold, 83, 413
Embedding, 411
Equilibrium point, 434
Exact
- covector field, 432
- differential, 8
- linearization
- - of multi-input systems, 245
- - of single-input systems, 161
- - of the input-output response, 171, 212, 265, 268
Exosystem, 348
Expression in local coordinates,
- of a function, 410
- of a mapping, 410

Fast time, 448
Feedback
-, dynamic
- -, output, 207

- -, state, 156, 196, 382
-, static
- -, output, 202
- -, state, 156, 244
Finitely computable, 324, 339
Fliess functional expansion, 119
Flow, 429
Formal power series , 113, 270
Frobenius Theorem, 25
Fundamental formula, 119

Gradient, 8
Growth condition, 114, 120

Hankel
- matrix, 133
- rank, 132
Hausdorff separation axiom, 406
High gain, 202
Homeomorphism, 405
Hypersurface, 409

Image of a matrix, 2
Immersed submanifold, 83, 413
Immersion, 411
Implicit function Theorem, 404
Indistinguishability, 4, 56
Induced topology, 406
Inertia matrix, 38, 399
Inner product, 7
Input-output Exact Linearization (Pbm of), 268
Integral
- curve, 426
- submanifold, 82
Integrator, 384
Interior, of a set, 405
Invariance, of the output, 124
Invariant
- codistribution, 51
- distribution, 44
- subspace, 1
Inverse
- function Theorem, 403
- of a MIMO system, 243
- of a SISO system, 183
Invertibility condition, 293

Involutive closure, 20, 208
Involutive distribution, 18
Iterated integral, 113

Jacobi identity, 10, 430
Jacobian matrix, 9, 418
Joint elasticity, 230, 398

Kernel
- of a matrix, 4
- of a Volterra series, 121
Kinematic equation, 38

Lie
- algebra, 429
- - of vector fields, 39, 430
- bracket 10, 430
- rank, 133
- subalgebra, 89
Lift force, 392
Linear
- approximation, 168, 434
- system, 1
Linearizing
- coordinates, 165, 246
- feedback, 165, 246
Lipschitzian, 444
Local coordinates, 406
Locally
- controlled invariant, *see* Controlled
  invariant
- euclidean space, 406
- finitely generated distribution, 86
- invariant manifold, 434
- lipschitzian, *see* Lipschitzian

Manifold, 406
Maximal integral manifold property, 82
Maximal linear subsystem, 208
Model matching
-, for a MIMO system, 282
-, for a SISO system, 194
Module, 39, 429
Multiindex, 112

Natural basis, of the tangent space, 417

Neighborhood, 405
Nested sequence of distributions, 34
Noninteracting Control (Problem of),
  259
- with Stability, 366
Noninteractive feedback, standard, 262
Nonsingular distribution, 16
Nontrivial eigenvalues, 450
Normal form
- of a general nonlinear system, 308
- of a MIMO system, 240
- of a SISO system, 152

Observability, 4, 74
- rank condition, 97
Observable
-, locally, 97
- pair, 4
Observation space, 92
Observer
- Linearization (Pbm of), 218
- with linear error dynamics, 218
One-form, 432
Open
- mapping, 405
- set, 405
Orthogonal
- group, 39
- matrix, 37
Output
- feedback, *see* Feedback
- tracking, 190
- zeroing submanifold, 290

Partition of state space
- global
- -, into integral submanifolds, 82
- -, into parallel planes, 3, 5
- local, into slices of a coordinates
  neighborhood, 43, 48
Pitch, 284, 391
Plant, 347
Poisson stable, 352
Product topology, 406

Rank Theorem, 403
Reachability, 2, 55, 56

Reachable pair, 3
Realizability conditions,
-, via bilinear systems, 136
-, via nonlinear systems, 138
Realization, 130
-, diffeomorphism of, 141
-, minimality of, 141
-, uniqueness of, 141
Reduced
- system, 448
- vector field, 450
Reduction principle, 440
Reference
- frame, 36
- model, 194
- output, 182
Regular
- feedback, 244
- point
- - of a distribution, 16
- - of the controlled invariant distribution
    algorithm, 327
- - of the zero dynamics algorithm, 304
Regulation, 348
Regulator Problem
- via Error Feedback, 350
- via State Feedback, 350
Related
- covector fields, 433
- vector fields, 431
Relative degree
- of a MIMO system, 235
- of a SISO system, 145
Reproducing a Reference Output (Pbm
    of)
- for a MIMO system, 241
- for a SISO system, 182
Rescaled time variable, 448
Restriction of a system to a submanifold,
    91
Rigid body, 36, 106
Robot arm, 230
Roll, 284, 391
Rotation matrix, 37
Row reduction, 271
Rudder, 393

Side force, 392

Sideslip angle, 391
Singular pertubations theory, 202, 447
Singularly perturbed system, 447
Skew symmetric matrix, 110
Slice of a neighborhood, 46, 48
Slow time, 448
Small time constant, 207
Smooth
- curve, 421
- distribution, see Distribution
- function, 403
- manofold, 408
- mapping, 403
- system, 6
Smoothing of a distribution, 16
Sphere, 409
Stability in the First Approximation
    (Principle of), 184, 434
Stabilizable, 352
State feedback, see Feedback
State Space Exact Linearization (Pbm
    of)
-, for multi-input systems, 245
-, for single-input systems, 161
Static feedback, see Feedback
Steady state response, 357
Structure Algorithm, 272
Submanifolds, 411
Submodule, 40
Subset topology, 406
Sussmann Theorem, 84
System
- defined on a manifold, 36
- matrix, 294

Tangent
- space
- - to a manifold, 416
- - to SO(3), 109
- vector, 416
Time-varying system, 444
Throttle, 393
Toeplitz matrix, 271
Topological
- space, 405
- structure, 405
Topology, 405
Total stability, 445

Torus, 410
Transmission
- polynomials, 294
- zeros, 294
Triangular decompositions, 47
Trivial eigenvalues, 450

Uncontrollable modes, 184
Uniformly
- asymptotically stable, 444
- stable, 444
Univalent immersion, 411

Vector
- field 7, 425
- relative degree, 235
Volterra series, 121

Whitney's theorem, 415

Yaw, 284, 391

Zero dynamics
- algorithm, 290
- of a general nonlinear system, 292
- of a MIMO system, 242
- of a SISO system, 174
- submanifold, 292
- vector field, 292
Zeroing the Output (Problem of)
-, for a general nonlinear system, 290
-, for a MIMO system, 241
-, for a SISO system, 173
Zeros of a transfer function, 176